Organic Waste: Recycling and Treatment

Organic Waste: Recycling and Treatment

Editor: Ollie Hall

STATES
ACADEMIC PRESS
www.statesacademicpress.com

States Academic Press,
109 South 5th Street,
Brooklyn, NY 11249, USA

Visit us on the World Wide Web at:
www.statesacademicpress.com

ISBN: 978-1-63989-400-0 (Hardback)

Cataloging-in-Publication Data

Organic waste : recycling and treatment / edited by Ollie Hall.
p. cm.
Includes bibliographical references and index.
ISBN 978-1-63989-400-0
1. Organic wastes--Recycling. 2. Compost. 3. Factory and trade waste. 4. Recycling (Waste, etc.).
5. Refuse and refuse disposal. I. Hall, Ollie.

TD794.5 .O74 2022

363.728--dc23

Table of Contents

Preface

Organic waste or biodegradable waste refers to all those waste materials which originate from living sources such as plants, animals and microorganisms. They can be broken down into simple molecules like water, methane and carbon dioxide. They can be classified into cattle wastes, food wastes, and municipal solid wastes. Organic waste recycling refers to all those processes which aid in managing organic waste, and converting it into useful matter using various treatments and techniques. Some of the commonly used methods are composting, anaerobic digestion, rendering, immobilized enzyme reaction, rapid thermophilic digestion, and using organic waste as animal fodder. Some of the challenges associated with this discipline are generation of odor and air pollution, accumulation of heavy metals, and formation of bioaerosols. This book unravels the recent studies in the area of organic waste recycling and treatment. It is a compilation of chapters that discuss the most vital concepts, and methods of recycling and treatment of organic waste. This book aims to serve as a resource guide for students and experts alike and contribute to the growth of the discipline.

This book is a comprehensive compilation of works of different researchers from varied parts of the world. It includes valuable experiences of the researchers with the sole objective of providing the readers (learners) with a proper knowledge of the concerned field. This book will be beneficial in evoking inspiration and enhancing the knowledge of the interested readers.

In the end, I would like to extend my heartiest thanks to the authors who worked with great determination on their chapters. I also appreciate the publisher's support in the course of the book. I would also like to deeply acknowledge my family who stood by me as a source of inspiration during the project.

Editor

1

Effect of abattoir waste water on metabolic and antioxidant profiles of cowpea seedlings grown in crude oil contaminated soil

F. I. Achuba[1] · M. O. Ja-anni[2]

Abstract

Purpose The pollution of agricultural land by crude oil is a wide spread phenomenon in petroleum producing countries of the world. Therefore, there is the need to find effective and eco-friendly means of mitigating the toxic effect of crude oil on crop plants.

Methods Green house experiment was conducted with cowpea seeds planted in polybags containing crude oil contaminated soil and similar concentrations treated with known amount of abattoir wastewater. The effect of abattoir wastewater (AWW) treatment of the crude oil contaminated soil on the metabolic and antioxidant properties of the leaves of cowpea seedlings was investigated. The concentrations of total sugar, total protein, total amino acid and β-carotene, chlorophyll, starch metabolizing enzymes as well as oxidative stress markers, lipid peroxidation, catalase, xanthine oxidase, aldehyde oxidase and superoxide dismutase were determined using standard methods.

Results Treatment of the crude oil contaminated soil with AWW resulted in decreases in total sugar, total protein and total amino acid as well as increase in the activities of both α-amylase and starch phosphorylase and total chlorophyll and β-carotene levels compared with those in crude oil alone treated groups. Treatment of the crude oil contaminated soil with AWW also significantly decreased lipid peroxidation and Cu/ZnSOD, but increased MnSOD, catalase, xanthine oxidase and aldehyde oxidase activities.

Conclusion These observations suggest that abattoir wastewater has the ability to remediate the adverse effects of crude oil polluted soil on cowpea seedlings. Therefore, abattoir wastewater should be harnessed and used in treating crude oil polluted farmland so as to reduce its toxicity on crop plants.

Keywords Abattoir · Cowpea · Crude oil · Metabolism · Antioxidants

Introduction

Soil pollution by petroleum hydrocarbon is common in crude oil producing areas (Sivaraj et al. 2014), which results in changes in soil properties such as water and oxygen shortage, soil enzymes activities as well as deficiency in nitrogen and phosphorous (Wyszkowski et al. 2004; Wyszkowski and Wyszkowska 2005; Achuba 2006; Wyszkowski and Ziolkowska 2008; Achuba 2014; Achuba and Okoh 2014; Otitoju et al. 2017).

Plants growing in soils contaminated with crude oil and its refined products have been reported to exhibit several anomalies in their metabolism and growth indices. The notable symptoms observed in plants growing on soil treated with petroleum hydrocarbons include decrease in chlorophyll content (Achuba 2006; Peretiemo-Clarke and Achuba 2007; Al-Hawas et al. 2012), decrease in the activities of starch metabolizing enzymes (Achuba 2006), increase in total carbohydrates, total proteins and amino acids (Peretiemo- Clarke and Achuba 2007; Al-Hawas et al. 2012) and in the activities of peroxidase, ascorbate peroxidase and superoxide dismutase, and inhibition of catalase activity (Al-Hawas et al. 2012).

Bioremediation is one of the best methods used to restore contaminated soil (Romantschuk et al. 2000). Crude oil

✉ F. I. Achuba
achubabch@yaoo.com

M. O. Ja-anni
jaannimary@gmail.com

1 Department of Biochemistry, Delta State University, PMB 1, Abraka, Nigeria

2 Department of Biochemistry, Delta State University, Abraka, Nigeria

contaminated soils are modifiable by bioremediation since microorganisms are competent in degrading petroleum hydrocarbons (Sangeetha and Thangadurai 2014). Bioremediation of crude oil contaminated soil can be enhanced by stimulation of the indigenous microorganism, provision of nutrients and oxygen into the soil (biostimulation) (Seklemova et al. 2001) and through inoculation of an enhanced microbial consortium into the soil (bioaugmentation) (Richard and Vogel 1999). Abattoir wastewater contains high levels of microorganisms, such as fungi and bacteria that are capable of degrading petroleum hydrocarbons (Ogbonna et al. 2012; Umanu and Owoseni 2013). The aim of this study was to determine the effect of abattoir wastewater on cowpea seedlings grown in crude oil contaminated soil, in a bid to establish the possible use of abattoir wastewater as a bio-resource in crude oil contaminated farm lands.

Materials and methods

Materials

The reagents used were high-quality analytical grades. The *Vigna unguiculata* (L) (IT845-2246-4 accession) seeds, experimental soil and crude oil were obtained as reported in Achuba (2006). Abattoir wastewater was obtained from local slaughterhouses in Obiaruku, Delta State, Nigeria. It was collected from four different sites and pooled together. The wastewater was filtered to remove solid materials as well as suspended animal fats and oil. The temperature was determined with mercury thermometer while the pH was measured with pH meter (Model, Jenway 3020). Conductivity meter (C0150) was used to determine total dissolved solid and conductivity. Portable turbidity meter (Lamotte 2020) was used to determine turbidity. Dissolved oxygen meter (model Acorn DO) was used to determine biochemical oxygen demand (BOD).Closed reflux method was used to determine chemical oxygen demand (COD). Total suspended solid (TSS) was determined as earlier described (Anon 1992).The physicochemical properties of the wastewater are shown in Table 1:

Study location

The study was carried out in the research laboratory of the Department of Biochemistry, Delta State University, Abraka, Nigeria. It is located between latitude 5°45′ and 5°5 0′N and longitude 6° and 6°15′E. It is in the rainforest region with annual rainfall of 3000 mm and temperature of between 28 ℃ and 33 ℃ (UNDP (2006). The topography is lowland plain about above 45 m sea level. The soil type is made up of ferrosal red and brown soil with abundant free iron oxide (Osakwe 2012).

Preliminary assessment of abattoir wastewater

A preliminary study was undertaken to determine the exact volume of abattoir wastewater that would be suitable for the amelioration of the toxic effects of crude oil treatments. Plant growth parameters (data not shown) and metabolic status of cowpea seedlings exposed to 40 ml of abattoir wastewater treatment of soil was found to perform better compared to other levels of treatments (10, 20, 30, 50, 60, 80 and 100 ml) (Achuba and Ekute 2017).

Experimental design

One hundred planting bags (volume and height 1178.3 cm^3 15 cm) filled soil (1.6 kg) were used for this study. The experiment was divided into two series. In series A, each bag (Groups 1–6) was treated with 0.10, 0.25, 0.50, 1.00, 1.50, 2.00 and 0.00%(v/w) of whole crude oil. Each treatment was replicated five times. Series B had same groups 1–6 treatments of crude oil in addition to treatment with abattoir wastewater (40 ml each) and each treatment replicated five times as well. Five seeds of the *V. unguiculata* were planted in each test bag. During the experiment, the setup was watered (80 cm^3) daily to keep the soil moist. Record of germination was taken every 4 days until the 12th day and seeds which did not germinate were regarded as not germinable.

Determination of total sugar content, amino acid and total protein

The homogenate used for the determination of total sugar, amino acid and protein was prepared as reported by Achuba (2006) The sugar was determined as described by Tietz (1982), while total free amino acid was determined by ninhydrin after deproteination with 5% trichloroacetic acid (TCA). The amount of total protein was determined according to Lowry et al. (1951).

Table 1 Physicochemical properties of abattoir wastewater used

Appearance	Temp (℃)	pH	Conductivity (dS m^{-1})	Turbidity (NTU)	TDS (mg L^{-1})	TSS (mg L^{-1})	BOD (mg L^{-1})	COD (mg L^{1})
Turbid	29 ± 2	5.6 ± 0.3	35.0 ± 1	8.3 ± 0.1	740 ± 11.7	884 ± 3.0	48 ± 2.3	150 ± 10.1

Temp temperature, *TDS* total dissolved solid, *TSS* total suspended solid, *BOD* biological oxygen demand, *COD* chemical oxygen demand

Determination of photosynthetic pigments (chlorophyll and total carotene)

The leaves (1.0 g) were homogenized using a mortar and pestle in 96% methanol and the homogenate obtained was centrifuged at 2500 rpm for 10 min. The supernatant obtained diluted. The concentration of total chlorophyll was determined as described by Lichtenthaler (1987). To estimate the concentrations of carotenoid and chlorophyll, the absorbance of the supernatant was read at 662, 646 and 470 nm. The concentrations of these molecules were estimated by the formula of Duxbury and Yentsch (1956).

Assay for alpha-amylase activity and determination of starch phophorylase activity in the cotyledons of cowpea seedlings

The alpha-amylase assay was carried out by the method of Gupta et al. (2003) using the supernatant obtained from the homogenization and centrifugation of the pooled cotyledons (0.5 g, kept in ice). The activity was calculated using the formula proposed by Xiao et al. (2006). The method of Singh and Steinnes (1976) was adopted for the determination of starch phosphorylase activity.

Preparation of homogenate for the determination of oxidative stress makers

The cowpea leaves (0.05 g), 0.05 M phosphate buffer pH 7.5 and few drops of butylated hydroxyl toluene (BHT), were homogenized in mortar (that was previously chilled at 4 °C) with pestle. The mixture was filtered with cheese cloth and the filtrate centrifuged at 5000 g for 10 min.

Determination of oxidative stress markers

Lipid peroxidation was determined according to the method of Gutteridge and Wilkins (1982).The activity of total superoxide dismutase was determined as described by Misra and Fredorich (1972), while MnSOD and Cu/ZnSOD were determined as reported by Crapo et al. (1978). The activities of catalase and xanthine oxidase were determined as reported by Rani et al. (2004) and were expressed as units/g wet tissue.

Statistical analysis

The results were analyzed using one way analysis of variance (ANOVA) and post Hoc Fisher's test for multiple comparison with the aid of statistical package for social science (SPSS), version 20. Level of significance was set at p values < 0.05.

Results

The effect of adding abattoir wastewater (AWW) to crude oil treated soil on the concentrations of sugar, protein, amino acid and β-carotene in the leaves of cowpea seedlings planted in crude oil polluted soil is shown in Table 2. The increase in the concentration of total sugar of the seedlings was significant ($p < 0.05$) at all the levels of contaminations

Table 2 Effect of abattoir wastewater treatment on the level of sugar, protein, amino acid and β-carotene in cowpea seedlings exposed to crude oil treated soil

Concentration of crude in soil (% v/w ml g^{-1})	Total sugar (mg g^{-1})		Total protein (mg g^{-1})		Total amino acid (mg g^{-1})		β-carotene (mg g^{-1})	
	Crude	Crude + AWW	Crude	Crude + AWW	Crude	Crude + AWW	Crude	Crude + AWW
0.00	33.27 ± 0.86	32.40 ± 0.27	19.60 ± 0.20	20.73 ± 0.78	3.77 ± 0.15	3.40 ± 0.20	22.57 ± 0.87	25.00 ± 1.31
0.10	41.20 ± 0.78[a]	38.53 ± 2.29[a]	18.77 ± 0.15	19.00 ± 0.36	3.67 ± 0.06	3.70 ± 0.01[a]	18.87 ± 1.43	21.27 ± 0.91[d]
0.25	43.57 ± 0.75[a]	42.23 ± 5.13[a]	19.87 ± 0.25	19.23 ± 0.15	3.83 ± 0.06	3.75 ± 0.06[a]	16.37 ± 0.67	20.23 ± 0.75[d]
0.50	47.63 ± 0.67[a]	45.60 ± 0.20[a]	20.70 ± 0.10	19.83 ± 0.25	3.93 ± 0.06	3.80 ± 0.07[a]	17.48 ± 1.18	20.90 ± 0.96[d]
1.00	56.40 ± 0.44[a]	46.07 ± 0.15[ab]	21.57 ± 0.45[a]	23.40 ± 0.66[a]	4.13 ± 0.06[a]	4.13 ± 0.02[a]	13.33 ± 0.75	18.10 ± 0.82[d]
1.50	57.57 ± 0.76[a]	46.93 ± 0.71[ab]	24.50 ± 0.47[a]	22.70 ± 0.20[ab]	4.50 ± 0.10[a]	4.10 ± 0.05[a]	12.43 ± 0.49	17.53 ± 1.16[d]
2.00	59.30 ± 0.76[a]	49.43 ± 0.71[ab]	26.43 ± 0.47[a]	23.60 ± 0.20[ab]	4.80 ± 0.10[a]	4.25 ± 0.05[ab]	12.43 ± 0.49	17.53 ± 1.16[d]

Results are expressed as mean ± SD of three determinations

[a]Significantly higher than control

[b]Significantly lower than crude oil alone treatment

[c]Significantly lower than the control

[d]Significantly higher than crude oil alone treatment

of soil with crude oil. However, the increase became significant at 1.00% concentration of crude oil treatment of soil for both total protein and total amino acid. On the other hand, a significant decrease ($p < 0.05$) in concentration of β-carotene in the leaves of cowpea seedlings planted in crude oil treated soil relative to the control values was noted. Treatment of the crude oil contaminated soil with AWW resulted in decrease in total sugar, total protein and total amino acid compared with those in crude oil alone groups. The decrease was significant at higher concentrations of crude oil in soil (1.00, 1.50 and 2.00%). Moreover, treatment of the crude oil contaminated soil with AWW significantly increased the β-carotene content of the cowpea seedlings except at 0.10% concentration of crude oil in soil.

Crude oil treatment of soil decreased the amount of chlorophyll in the leaves of cowpea seedlings compared to the control (Table 3). The decrease was significant ($p < 0.05$) for chlorophyll except at lower concentrations of crude oil in soil. However, seedlings exposed to AWW had significantly higher chlorophyll content at higher levels of soil contamination. Similarly, crude oil contaminated soil decreased the activities of α-amylase and starch phosphorylase in the cotyledon of cowpea seedlings. Treatment of crude oil contaminated soil with AWW resulted in increase in the activities of starch metabolizing enzymes studied. However, the increase was not significant for starch phosphorylase and was only significant for α-amylase at 2.00% concentration of crude oil.

Lipid peroxidation significantly increased in the leaves of cowpea seedlings as the concentration of crude oil in soil increased (Table 4). The addition of AWW to the crude oil treated soil significantly decreased lipid peroxidation in the leaves of the cowpea seedlings as compared to the corresponding crude oil alone treatment. The activity of xanthine oxidase increased significantly at low levels of soil treatment with crude oil (0.10 and 0.25%). However, the activity of the enzyme decreased at higher level of crude oil in soil (0.50, 1.00 and 2.00%). Treatment with AWW caused significant increase in the activity of xanthine oxidase as compared with the corresponding crude oil alone treatment. Table 3 also indicates that planting cowpea seedlings in crude oil contaminated soil decreased the activities of catalase and aldehyde oxidase and the decrease was significant at all contamination levels for catalase while it was only significant at 2.00% level soil contamination for aldehyde oxidase. Treatment with AWW increased catalase activity significantly ($p < 0.05$) compared to the corresponding crude oil alone treatment while it was only significant at 1.50 and 2.00% level of contamination for aldehyde oxidase.

Table 5 shows that contamination of soil with crude oil significantly ($P < 0.05$) increased total superoxide dismutase (SOD) at low levels of soil contamination with crude oil (0.10, 0.25 and 0.50%) but significantly decreased total SOD at higher levels of crude oil treatment of soil (1.00, 1.50 and 2.00%). While contamination with crude oil significantly increased CuZnSOD activity at all levels of soil contamination, but decreased MnSOD activity, although not significant relative to control value. Treatment of contaminated soil with AWW significantly decreased total SOD at lower concentrations of crude oil in soil (0.10, 0.25 and 0.50%) contamination levels but significantly increased total SOD at higher level of soil contamination (1.00, 1.50 and 2.00%). Moreover, treatment with AWW decreased CuZnSOD activity significantly at higher levels of soil contamination while MnSOD activity was not significantly increased.

Table 3 Effect of abattoir wastewater on the content of chlorophyll, α-amylase and starch phosphorylase in cowpea seedlings exposed to crude oil polluted soil

Concentration of crude oil in soil (% v/w ml g^{-1})	Total chlorophyll (mg g^{-1})		α-amylase activity (Unit ml^{-1})		Starch phosphorylase activity (mg min^{-1} gfw)	
	Crude	Crude + AWW	Crude	Crude + AWW	Crude	Crude + AWW
0.00	288.33 ± 2.08	303.00 ± 1.00	20.87 ± 1.44	24.83 ± 1.43	1.07 ± 0.05	1.17 ± 0.04
0.10	280.33 ± 3.06	290.33 ± 2.52^{d}	18.63 ± 0.45	21.07 ± 1.60	1.02 ± 0.09	1.03 ± 0.03
0.25	277.33 ± 4.73	291.67 ± 1.53^{d}	18.13 ± 0.15	19.50 ± 0.66	0.93 ± 0.07	1.04 ± 0.06^{d}
0.50	238.67 ± 7.09^{a}	289.00 ± 1.00^{bd}	19.17 ± 1.37	20.00 ± 1.59	1.18 ± 0.28	1.26 ± 0.09^{d}
1.00	215.67 ± 0.57^{a}	258.00 ± 7.00^{abd}	17.67 ± 1.12	19.13 ± 0.8^{d}	0.83 ± 0.09	0.99 ± 0.05^{d}
1.50	211.67 ± 2.08^{a}	236.00 ± 9.54^{abd}	15.40 ± 0.96^{a}	18.50 ± 0.75^{d}	0.68 ± 0.05^{a}	0.93 ± 0.09^{d}
2.00	208.33 ± 2.08^{a}	217.00 ± 9.54^{d}	13.23 ± 1.04^{a}	17.20 ± 0.90^{abd}	0.52 ± 0.04^{a}	0.79 ± 0.06^{d}

Results are expressed as mean ± SD of three determinations

[a]Significantly higher than control

[b]Significantly lower than crude oil alone treatment

[c]Significantly lower than the control

[d]Significantly higher than crude oil alone treatment

Table 4 Effect of abattoir wastewater treatment on lipid peroxidation and activities of xanthine oxidase, catalase and aldehyde oxidase in cowpea seedlings exposed to crude oil treated soil

Concentration of crude oil in soil (% v/w ml g^{-1})	Lipid peroxidation (nMol cm^{-3})		Xanthine oxidase activity(Unit cm^{-3})		Catalase activity (nmol min^{-1} gfw)		Aldehyde oxidase activity (Unit cm^{-3})	
	Crude	Crude + AWW	Crude	Crude + AWW	Crude	Crude + AWW	Crude	Crude + AWW
0.00	0.98 ± 0.12	0.92 ± 0.07	3.78 ± 0.57	4.65 ± 0.15	1.08 ± 0.13	1.15 ± 0.04	1.29 ± 0.17	1.66 ± 0.065
0.10	1.40 ± 0.1[a]	1.08 ± 0.13[b]	4.47 ± 0.44[a]	4.71 ± 0.26[a]	0.83 ± 0.11[c]	1.25 ± 0.08[d]	1.23 ± 0.08	1.53 ± 0.08[d]
0.25	1.74 ± 0.0[a]	1.32 ± 0.11[ab]	4.74 ± 0.09[a]	4.74 ± 0.09[a]	0.66 ± 0.05[c]	1.16 ± 0.01[d]	1.23 ± 0.07	1.48 ± 0.06[d]
0.50	1.93 ± 0.0[a]	1.36 ± 0.04[ab]	3.98 ± 0.09	4.74 ± 0.09[ad]	0.94 ± 0.06	1.14 ± 0.01[d]	1.34 ± 0.02	1.58 ± 0.03[d]
1.00	2.09 ± 0.1[a]	1.41 ± 0.05[ab]	3.18 ± 0.04	4.45 ± 0.05[ad]	0.71 ± 0.04[c]	1.01 ± 0.09[d]	1.40 ± 0.40	1.39 ± 0.02
1.50	2.23 ± 0.2[a]	1.44 ± 0.01[ab]	3.11 ± 0.09	4.26 ± 0.08[d]	0.61 ± 0.02[c]	0.91 ± 0.04[d]	0.95 ± 0.05	1.18 ± 0.07[d]
2.00	2.54 ± 0.0[a]	1.46 ± 0.02[ab]	3.10 ± 0.12[a]	4.15 ± 0.16[d]	0.53 ± 0.05[c]	0.85 ± 0.11[d]	0.72 ± 0.053	1.33 ± 0.41[d]

Results are expressed as mean ± SD of three determinations

[a]Significantly higher than control

[b]Significantly lower than crude oil alone treatment

[c]Significantly lower than the control

[d]Significantly higher than crude oil alone treatment

Table 5 Effect of abattoir wastewater on superoxide dismutase activities cowpea seedlings planted in crude oil treated soil

Concentration of crude oil in soil (% v/w ml g^{-1})	Total SOD (Unit g^{-1}fw)		CuZnSOD (Unit g^{-1}fw)		MnSOD (Unit g^{-1}fw)	
	Crude	Crude + AWW	Crude	Crude + AWW	Crude	Crude + AWW
0.00	2.28 ± 0.15	2.21 ± 0.17	2.10 ± 0.09	2.06 ± 0.02	0.97 ± 0.13	0.84 ± 0.11
0.10	2.73 ± 0.13[a]	2.37 ± 0.06	2.23 ± 0.07	2.09 ± 0.08	0.44 ± 0.51	0.37 ± 0.43
0.25	3.01 ± 0.09[a]	2.54 ± 0.37[c]	2.37 ± 0.04[a]	2.25 ± 0.06	0.74 ± 0.48	0.84 ± 0.57
0.50	3.18 ± 0.17[a]	2.64 ± 0.05[c]	2.45 ± 0.07[a]	2.13 ± 0.12[c]	0.48 ± 0.21	0.81 ± 0.11[d]
1.00	1.96 ± 0.07	2.69 ± 0.09[ad]	2.44 ± 0.07[a]	2.28 ± 0.07	0.73 ± 0.08	0.82 ± 0.01[d]
1.50	1.73 ± 0.12[b]	2.46 ± 0.07[d]	2.34 ± 0.08[a]	2.07 ± 0.03[c]	0.63 ± 0.06	0.75 ± 0.11[d]
2.00	1.45 ± 0.10[b]	2.62 ± 0.29[d]	2.49 ± 0.11[a]	2.20 ± 0.11[c]	0.47 ± 0.05	0.64 ± 0.09[d]

Results are expressed as mean ± SD of three determinations

[a]Significantly higher than control

[b]Significantly lower than crude oil alone treatment

[c]Significantly lower than the control

[d]Significantly higher than crude oil alone treatment

Discussion

Adverse alterations in biochemical parameters have been reported as indices of crude oil and its refined products mediated stress in plants (Achuba 2006; Peretiemo-Clarke and Achuba 2007; Bona et al. 2011; Al-Hawas et al. 2012). The present study showed increases in total sugar, total protein and total amino acid contents of the cowpea seedlings planted in crude oil treated soil relative to the control (Table 2). This increase in these macromolecules indicates metabolic stress (Achuba 2006) and the result correlates with earlier reports of Peretiemo-Clarke and Achuba (2007) and Al-Hawas et al. (2012) on the effect of hydrocarbon treated soil on the seedlings of peanut and Jojoba, respectively. On the other hand, the β-carotene and chlorophyll contents of the cowpea seedlings decreased significantly except at low levels (0.10–0.5%) of crude oil contamination of soil for chlorophyll (Tables 2 and 3). This correlates with the results of Achuba (2006) and Al-Hawas et al. (2012). While both researches found crude oil to cause significant decrease in chlorophyll content in the exposed plants, Al-Hawas et al. (2012) also reported significant decrease in carotenoids and total pigments of Jojoba seedlings with increasing concentration of crude oil. However, treatment with abattoir wastewater (AWW) significantly ($p < 0.05$) increased the β-carotene content at all contamination levels while the chlorophyll content was significantly increased at low levels (0.10–0.5%) of contamination as compared to the corresponding crude oil alone groups. This correlates with the findings of other researchers who employed organic waste products to remediate the adverse effects of soil contamination by crude oil on plant growth and yield (Akujobi et al. 2011; Ekpo and Nya 2012).

Achuba (2006) reported inhibition of activities of α-amylase and starch phosphorylase in the cotyledon of the cowpea seedlings planted in petroleum-treated soil. Treatment of crude oil contaminated soil with AWW enhanced the activities of the starch degrading enzymes, α-amylase and starch phosphorylase (Table 3), although the increase was not significant for starch phosphorylase and was only significant for α-amylase at higher (1.00–2.00%) concentrations of crude oil in soil. Abattoir wastewater enhancement of α-amylase and starch phosphorylase activities in cowpea seedling was reported recently (Achuba and Ekute 2017).

In this study, lipid peroxidation increased as the concentration of crude oil increased (Table 4). This agrees with the findings of Olubodun and Eriyamremu (2015). The increase may have stemmed from petroleum hydrocarbon-induced free radicals generation, with resultant oxidative stress and peroxidation of lipid (Ortega-Villasante 2005; Dietz 2010). Treatment of the crude oil contaminated soil with AWW significantly decreased lipid peroxidation in cowpea seedlings as compared to the corresponding seedlings in crude oil alone treatment.

The activity of xanthine oxidase increased significantly at lower (0.10–0.50%) levels of contamination while it decreased, though not significantly at higher levels (1.00–2.00%) of contamination. Treatment with AWW increased the activity of xanthine oxidase as compared with the corresponding crude oil alone treatment. Earlier report indicated that abattoir wastewater altered sulphite oxidase activity in hydrocarbon contaminated soil (Achuba and Okunbor 2015).

One of the consequences of environmental stress in plants is the imbalance in the generation and removal reactive oxygen species (Al-Hawas et al. 2012). However, living organisms such as plants possess a number of antioxidant enzymes like superoxide dismutase (SOD) and catalase that guide against oxidative damage (Asada 1994). The present study showed that exposure of cowpea seedlings to crude oil contaminated soil negatively altered the activities of catalase, aldehyde oxidase, total superoxide dismutase (SOD), Cu/ZnSOD and MnSOD (Table 5). Al-Hawas et al. (2012) had reported the inhibition of catalase activity as well as increase in SOD activity in Jojoba (*Simmodsia chinensis*) seedling exposed to crude oil contamination levels. On the other hand, Olubodun and Eriyamremu (2015) reported a decreased level of SOD. The inconsistence levels of SOD at different levels of crude oil contamination observed in this present study agrees with previous reports that different levels of SOD activity might occur depending on the living organism involved and environmental conditions (Sgherri et al. 2000; Al-Hawas et al. 2012). Generally, treatment with AWW positively altered the activities of catalase, aldehyde oxidase and various superoxide dismutases. However, this is not concentration dependent. On the whole, these observations indicate that AWW has an ameliorative effect on crude oil mediated oxidative stress.

Abattoir wastewater contains organic substances, hydrocarbon-utilizing microorganisms and a relatively significant amount of plants nutrients (Neboh et al. 2013; Umanu and Owoseni 2013). Therefore, the observed improvement in the metabolic profile of cowpea seedlings in this study may be attributed to the organic matter, hydrocarbon-utilizing microorganisms and essential nutrients present in AWW. Umanu and Owoseni (2013) and Ogbonna (2014) reported that various bacteria strains capable of hydrocarbon degradation were isolated from soil and abattoir effluent and these include Pseudomonas aeruginosa, Bacillus spp, Alcaligenes faecalis, and serratia spp, while the fungi isolated were Aspergillus niger, Penicillium chrysogenum and Fusarium sp. The hydrocarbon-utilizing microorganisms may have enhanced the removal of the toxic hydrocarbons in the contaminated soil (Umanu and Owoseni 2013; Kumar et al. 2014) while the organic matter and essential nutrients may have replenished the soil nutrient status. The promotive effect of biofertilizer was previously reported (Al-Abdallah et al. 2017; Zarei et al. 2017).This may be the basis for enhanced soil fertility (Etuk et al. 2013) and improved metabolic status of cowpea seedlings exposed to AWW pretreated crude oil contaminated soil.

Conclusion

The present study emphasized that the abattoir wastewater is an effective bioremediation resource for crude oil contaminated soil and at the same instance remediated the adverse effects of crude oil exposed plants to crude oil polluted soil. Therefore, abattoir wastewater should be harnessed and used in treating crude oil polluted farmland so as to reduce its toxicity on crop plants.

References

Achuba FI (2006) The effect of sublethal concentrations of crude oil on the growth and metabolism of Cowpea (*Vigna unguiculata*) seedlings. Environmentalist 26:17–20. https://doi.org/10.1007/s10669-006-5354-2

Achuba FI (2014) Petroleum products in soil mediated oxidative stress in cowpea (*Vigna ungiculata*) and Maize (*Zea mays*) seedlings. Open J Soil Sci. 4:417–435. https://doi.org/10.4236/ojss.2014.412042

Achuba FI, Ekute BO (2017) Effect of abattoir wastewater on the metabolic status of cowpea seedlings. In: Proceedings of the fourth Delta State University Faculty of Science International Conference. pp 98–103

Achuba FI, Okoh PN (2014) Effect of petroleum products on soil catalase and dehydrogenase activities. Open J. Soil Sci. 4:399–406. https://doi.org/10.4236/ojss.2014.412040

Achuba FI, Okunbor G (2015) Abattoir waste water attenuates kerosene toxicity on cowpea (*Vigna unguiculuta*) seedlings. Biokemistri 27(4):159–162

Akujobi CO, Onyeagba RA, Nwaugo VO, Odu NN (2011) Effect of nutrient amendments of diesel oil polluted soil on plant growth parameters. Curr Res J Bio Sci. 3(4):421–429

Al-Abdallah NM, Basalah MO, Roushdy SS (2017) The promotive effect of algal biofertilizers on growth and some metabolic activities of *Vigna unguiculata* L. under salt stress conditions Egypt. J Exp Biol 13(2):187–195. https://doi.org/10.5455/egyjebb.20170616083640

Al-Hawas GHS, Shukry WM, Azzoz MM, Al-Moaik RMS (2012) The effect of sublethal concentrations of crude oil on the metabolism of Jojoba (*Simmodsia chinensis*) seedlings. Int J Plant Sci 3(4):54–62

Anon R (1992) Standard methods of water and wastewater examination, 18th edn. American Public Health Association, Washington DC, pp 2–172

Asada K (1994) Production and action of active oxygen species in photosynthetic tissue. In: Foyer CH, Mullineaux PM (eds) Causes of photooxidative stress and amelioration of defense systems in plants. CRC Press, Boca Raton, pp 77–104

Bona C, de Rezende Gedir IM, Santos DO, de Souza LA (2011) Effect of soil contaminated by diesel oil on the germination of seeds and the growth of *Schinus terebinthifolius* Raddi (Anacardiaceae) seedlings. Braz Arch Bio. Technol. 54(6):63–69. https://doi.org/10.1590/S1516-89132011000600025

Crapo JD, McCord JM, Fridovich I (1978) Preparation and assay of superoxide dismutase. Meth Enzymol 53:382–393

Dietz KJ (2010) Redox-dependent regulation, redox control and oxidative damage in plant cells subjected to abiotic stress. In: Sunkar R (ed) Plant stress tolerance: methods and protocols. Humana Press, London, pp 57–70

Duxbury AC, Yentsch CS (1956) Plankton pigment nomographs. J Marine Res 15:92–101

Ekpo FE, Nya EJ (2012) Effect of poultry manure amendments on diesel oil polluted soil on germination and growth performance of forest trees. J Res Environ Sci Toxicol 1(17):195–200

Etuk EA, Ogboi KC, Nwadinigwe CA (2013) Bioremediation of hydrocarbon polluted soil in the lowland forest ecosystem in the niger delta through enhanced natural attenuation process (ENAP). Int J Appl Sci Technol 3(8):129–137

Gupta RP, Gigras H, Mohapatra GV, Kumar A, Chauhan B (2003) Microbial amylases: a biotechnological perspective. Proc Biochem 38:1599–1616

Gutteridge JMC, Wilkins S (1982) Copper-dependent hydroxyl radical damage to ascorbic acid. FEBS Lett 137:327–329

Kumar V, Singh S, Manhas A, Singh J, Singla S, Kaur P, Data S, Negi P, Kalia A (2014) Bioremediation of petroleum hydrocarbon by using *Pseudomonas* species isolated from petroleum contaminated soil. Orient J Chem 30(4):1771–1776. https://doi.org/10.13005/ojc/300436

Lichtenthaler HK (1987) Chlorophylls and carotenoids: pigments of photosynthetic biomembranes. In: Packer L, Douce R (eds) Methods in enzymology. Academic Press, New York, pp 350–382

Lowry OH, Rosebrough NJ, Farr AL, Randal RJ (1951) Protein measurement with the folin phenol reagent. Biol Chem 193:265–275

Misra HP (1972) Fridovich I (1972) The role of superoxide ion in the antioxidation of epinephrine and a simple assay for superoxide dismutase. J Biol Chem 247:3170–3175

Neboh HA, Ilusanya OA, Ezekoye CC, Orji FA (2013) Assessment of Ijebu-Igbo Abattoir effluent and its impact on the ecology of the receiving soil and river. J Environ Sci Toxicol Food Technol 7(5):61–67. https://doi.org/10.9790/2402-0756167

Ogbonna DN (2014) Distribution of microorganisms in water, soils and sediment from abattoir wastes in southern Nigeria. Int J Curr Microbiol App Sci 3(9):1183–1200

Ogbonna DN, Ideriah TJK, Nwachukwu MI (2012) Biodegradation of polycyclic aromatic hydrocarbons by associated microbes from abattoir wastes in the Niger delta, Nigeria. J Microbiol Res 2(6):157–169. https://doi.org/10.5923/j.microbiology.20120206.02

Olubodun SO, Eriyamremu GE (2015) Antioxidant enzymes and crude mitochondria ATPases in the radicle of germinating bean (*Vigna unguiculata*) exposed to different concentrations of crude oil. Int J Biol Food Vet Agric Eng 9(3):220–225

Ortega-Villasante C, Rellán-Álvarez R, del Campo FF, Carpena-Ruiz RO, Hernández LE (2005) Cellular damage induced by cadmium and mercury in *Medicago sativa*. J Expt Bot 56:2239–2251. https://doi.org/10.1093/jxb/eri223

Osakwe SA (2012) Effect of cassava processing mill effluent on physical and chemical properties of soils in Abraka and Environs, Delta State, Nigeria. Chem Mat Res 2(7):27–39

Otitoju O, Udebuani AC, Ebulue MM, Onwurah IN (2017) Enzyme-based assay for toxicological evaluation of soil ecosystem polluted with spent engine oil. Agric Ecol Res Int J 11(3):1–13. https://doi.org/10.9734/jaeri/2017/27605

Peretiemo-Clarke BO, Achuba FI (2007) Phytochemical effect of petroleum on peanut (*Arachis hypogea*) seedlings. Plant Pathol J 6:179–182. https://doi.org/10.3923/ppj.2007.179.182

Rani P, Meena UK, Karthikeyan J (2004) Evaluation of antioxidant properties of berries. India J Clin Biochem 19(2):103–110. https://doi.org/10.1007/BF02894266

Richard JY, Vogel TM (1999) Characterization of a soil bacterial consortium capable of degrading diesel fuel. Int Biodet Biod 44:93–100. https://doi.org/10.1016/s0964-8305(99)00062-1

Romantschuk M, Sarand I, Petanen T, Peltola R, Jonsson-Vihanne M, Koivula T, Yrjala K, Haahtela K (2000) Means to improve the effect of in situ bioremediation of contaminated soil: an over-view of novel approaches. Environ Pollut 107:179–185

Sangeetha J, Thangadurai D (2014) Effect of biologically treated petroleum sludge on seed germination and seedling growth of Vigna unguiculata (L.) Walp. (Fabaceae). Braz Arch Biol Technol 57(3):427–433. https://doi.org/10.1590/s1516-89132014005000011

Seklemova E, Pavlvova A, Kovachenva K (2001) Biostimulation based bioremediation of diesel fuel: field domonstration. Biodegradation 12:311–316

Sgherri CL, Maffei M, Navari-Izzo F (2000) Antioxidative enzymes in wheat subjected to increasing water deficit and rewatering. J Plant Physiol 157:273–279. https://doi.org/10.1016/S0176-1617(00)80048-6

Singh BB, Steinnes H (1976) Uptake of trace element by barley in Zinc polluted soil'. Soil Sci 121(1):38–43

Sivaraj R, Kumar KS, Radhakrishna A (2014) An investigation of growth and yield of *Helianthus annus* on diesel contaminated soil and bioremediated diesel contaminated soil. World J Pharm Res 3(10):1530–1544

Tietz NW (1982) Fundamentals of clinical chemistry. 2nd (ed) W.B. Saunder Co., New York, pp: 994-995

Umanu G, Owoseni RA (2013) Effects of abattoir effluent on microbial degradation of diesel oil in tropical agricultural soil. Pac J Sci Technol 14(1):604–612

UNDP (2006). Niger delta human development report. Pp 229

Wyszkowski M, Wyszkowska J (2005) Effect of enzymatic activity of diesel oil contaminated soil on the chemical composition of oat (*Avena sativa* L.) and maize (*Zea mays* L.). Plant Soil Environ 51(8):360–367

Wyszkowski M, Ziolkowska A (2008) Effect of petrol and diesel oil on content of organic carbon and mineral components in soil. Am Eur J Sust Agric 2(1):54–60

Wyszkowski M, Wyszkowska J, Ziółkowska A (2004) Effect of soil contamination with diesel oil on yellow lupine yield and macroelements content. Plant Soil Environ 50:218–226

Xiao Z, Storms R, Tsang A (2006) A quantitative starch-iodine method for measuring alpha-amylase and glucoamylase activities. Analyt Biochem 351(1):146–148. https://doi.org/10.1016/j.ab.2006.01.036

Zarei M, Abadi VAJM, Moridi A (2017) Comparison of vermiwash and vermicompost tea properties produced from different organic beds under greenhouse conditions. Int J Recycl Org Waste Agric. https://doi.org/10.1007/s40093-017-0186-2

2

Organic mulch sheet formulation as an effort to help plants adapt to climate change

Aniek Iriany[1] · Mochammad Chanan[2] · Gunomo Djoyowasito[3]

Abstract

Purpose This study aimed to discover the precise material composition and thickness (water hyacinth, rice straw, and banana pseudostem) of organic mulch sheet.

Methods This current research was conducted by means of a number of treatments with various material compositions and thickness of organic mulch sheet. Several tests were administered such as chemical analyses on organic mulch sheet, tensile strength, and Vilensky test.

Result Various compositions of materials resulted in various chemical analyses of organic mulch sheet. Organic mulch with the strongest tensile strength was during the treatment of 60% water hyacinth, 20% rice straw, and 20% banana pseudostem, reaching 3.28 N/m^2. The highest sunlight intensity of mulch composition was during the treatment of M4 (50% water hyacinth, 40% rice straw, and 10% banana pseudostem) with no hole and absorbing.

Conclusion The result showed that water hyacinth, rice straw, and banana pseudostem could be used as the materials for organic mulch sheets. They could add organic matters into the soil, have endurance and strength to apply as mulch in crop cultivation, as well as help plants adapt to climate change.

Keywords Organic mulch sheet · Organic matters · Tensile strength · Vilensky · Climate change

Introduction

Mulch is a soil cover that functions to maintain soil temperature and humidity, inhibit the growth of weeds, also lessen soil erosion (Díaz-Pérez and Batal 2002; Kar and Kumar, 2007; Ekinci and Dursun, 2009; Sinkevičienė et al. 2009; Ibarra-Jiménez et al. 2011; Dvořák et al. 2012). In horticulture plantation, both organic mulch and inorganic mulch have been into a common use (Dvořák et al. 2001; George Hochmuth et al. 2002; Kasirajan and Ngouajio 2012; Cowan 2013; Lakkenborg et al. 2014; Azad et al. 2015). Specifically, in this unstable climate and global warming condition (Wai et al. 2007), mulch is utilized to help plants adapt to climate change by modifying microclimate around the growing plants (Dvořák et al. 2001; Ibarra-Jiménez et al. 2011; Kasirajan and Ngouajio 2012). The following concern would be on the environment modification to optimize the growth of plants so as to improve potential production (Scholes et al. 1997; Peng et al. 2004; Widiatningrum and Pukan 2010; Kalra et al. 2013). The growth and productivity of plants generally are affected by rainfall, temperature, humidity, and soil fertility (Pereira and Nova 2008; Ayinde et al. 2011; Mahmood et al. 2012; Iriany et al. 2013; Nakano et al. 2013; Yaghi et al. 2013). Some previous studies have found that mulch usage increases plant productivity (Siwek et al. 2007; Miles et al. 2012; Haapala et al. 2015).

Mulch has been into a common use in crop cultivation, yet with a number of drawbacks, such as the use of environmentally unfriendly synthetic materials that are hard to degrade (transparent plastic, PHP, etc.), apart from their high price that are less affordable for farmers (O'Brine and Thompson 2010; Coolong 2012; Kasirajan and Ngouajio

✉ Aniek Iriany
aniek55@yahoo.co.id

1 The Department of Agrotechnology, Faculty of Agriculture and Animal Husbandry, University of Muhammadiyah Malang, Malang, Indonesia

2 The Department of Forestry, Faculty of Agriculture and Animal Husbandry, University of Muhammadiyah Malang, Malang, Indonesia

3 The Department of Agricultural Engineering, Faculty of Agricultural Technology, Brawijya University of Malang, Malang, Indonesia

2012; Miles et al. 2012). However, the use of organic mulch from rice straw, litter, and others are unstable at the time. Accordingly, it is urgent to come up with alternative organic materials from the accessible surrounding environment to be made for organic mulch for the farming field with simple technology. Organic mulch requires modification into more effective and efficient forms in usage by modifying it into sheet forms. Besides, the use of paper as mulch has been developed in some patents such as using cellulose fibers with additional materials, cellulosic pulp with biodegradation-retarding agent and addition of fertilizer to optimize mulch function (Wright 1936; Pratt and Medford 1955; Yoko et al. 1986; Peter. F et al. 2003). Most of these patents still used additional materials that are inorganic and relatively expensive to increase satisfactory mechanical strength.

An organic material as a renewable resource contains wood or natural fiber called as cellulose. Organic mulch sheet made from fiber can be processed from natural cellulosic source such as kenaf, pineapple leaf fiber, banana fiber, coir, paddy straw, sugarcane, water hyacinth, corn cobs, and many more (Azubuike & Okhamafe 2012; Bhatnagar et al. 2015; Indriyati et al. 2016; Salleh et al. 2015; Main et al. 2014; Sarika et al. 2014; Teygeler 2000). These materials contain fiber waste similar to the materials for this current research. The use of organic fiber waste such as banana pseudostem, straw, and water hyacinth for the organic mulch sheet will improve soil properties as the combination of papermaking and organic fertilizer (Mawlana et al. 2014; Vidya and Girish, 2014). Natural organic mulch eventually breaks down and adds organic materials to the soil. Degradation or decomposition of organic mulch increases the amount of soil organic carbon (Bajorien et al. 2013). Hence, microbial activity in the soil will increase, especially microbial biomass supporting any enzymatic process in the soil (Jodaugienė et al. 2010; Moreno et al. 2011). The issue about the use of water hyacinth that has the ability to absorb heavy metal and will harm soil microorganism and the crop is not true. Metals such as iron, manganese, chrome, and cadmium were highly concentrated at the roots than other parts of the plant like leaves and stem. Commonly, the highest concentration of heavy metal in water hyacinth plant is in root, followed by leaves and stem, respectively (C.N et al. 2014; Das et al. 2016; Saha et al. 2017). Water hyacinth can be used as fertilizer, fish or animal feed and for mushroom cultivation. It means that water hyacinth can be used as an organic matter added into the soil. Previous researches showed that organic fertilizer from contaminated heavy metal water hyacinth and cow dung (mass ratio 1:5 and 1:1) did not exceed the limit of quantities of heavy metal and can be applied. Besides, compost of water hyacinth has a positive effect on the formation of macro-aggregates of the soil, water holding capacity, cation exchange capacity and non-significant changes in pH occurred (Khan and Sarwar 2002; Nyawira 2016; Sindhu et al. 2017).Water hyacinth commonly used as bio-absorption of heavy metal does not contain heavy metal on high proportion that will damage or affect the soil (Gavrilescu 2004).

With simple technology, this organic mulch sheet is cheaper in price as it recycles wastes, more practical and effective in usage, using simpler technology, environmentally friendly, and in long-term increases soil fertility. Accordingly, it is necessary to discover the precise material composition of organic mulch sheet.

Methods

The experiment was conducted in the laboratory of the University of Muhammadiyah Malang and in the farming field in Batu, East Java, in the altitude of 1670 meters above sea level. The experiment was done from March to July 2014.

Material

The making of organic mulch sheet requires the following materials: water hyacinth, rice straw, and banana pseudostem. The tools were: organic mulch sheet molds, a digital scale, scissors, knife, measuring cup, blender, sieve, stove, pan, and caliper.

Experimental design

This research was conducted by employing a completely randomized design. The treatment experiment included compositions of organic mulch sheet materials with six variations and two thickness (twelve combinations), then repeated three times. The variation of organic mulch sheet compositions included M_1 = 40% water hyacinth:40% Rice straw:20% banana pseudostem; M_2 = 50% water hyacinth:20% Rice straw:30% banana pseudostem; M_3 = 50% water hyacinth:30% Rice straw:20% banana pseudostem; M_4 = 50% water hyacinth:40% Rice straw:10% banana pseudostem; M_5 = 60% water hyacinth:20% Rice straw:20% banana pseudostem; M_6 = 0.60%water hyacinth:30% Rice straw:10% banana pseudostem. There are two levels of organic mulch sheet thickness, T_1 = 0.5 mm and T_2 = 1 mm.

Implementation of the study

Making of mold

Preparing Styrofoam and cloth; one hole was made on the Styrofoam with the size of 30 cm × 50 cm; the upper part of Styrofoam was covered by the cloth.

Preparation of tools and materials

Preparing water hyacinth, rice straw, banana pseudostem, and necessary tools.

Cutting and weighing

The total weight of mulch for each combination was 1000 g. Water hyacinth, rice straw, and banana pseudostem were weighted based on the set composition in each treatment before finally being cut into the size of 1 cm.

Pulping

Water hyacinth was blended for 15 min and banana pseudostems were blended for 20 min by the addition of water before being squeezed to take the pulp.

Molding

Molding process started by pouring and flattening the hot pulp into the prepared mold. Then, the pulp in the mold was dried while being steadily pressed to eliminate the water.

Measured variables and data analysis

The following tests were conducted: chemical analysis of the organic materials (C organic, N total, C/N ratio and organic matter), organic mulch sheet tensile strength by Brazilliant test (N/cm^2), Vilensky and sunlight exposure intensity tests. Tabulation was then conducted before statistically being analyzed by means of F Test. To detect the differences among the treatments, BNJ Test in the level of 5% was conducted. The data were analyzed by means of Minitab version 17 software.

Results and discussion

The organic mulch sheet was made of water hyacinth, banana pseudostem, and tannery waste. It produced a compact and strong structure (Li et al. 2010; Sutyasmi 2012; Sahari and Buku 2015). The high cellulose content in the water hyacinth and banana pseudostem (over than 60%) and low lignin content (lower than 5%) through the delignification process are used to remove the lignin contained in the materials (Li et al. 2010; Tumolva et al. 2013; Ramesh et al. 2014).

Chemical analysis

The results are shown in the following Table 1 about chemical analyses on organic materials for organic mulch sheet in various treatments. From various treatments, it could be seen that treatment M3 (50% water hyacinth: 30% rice straw, 20% banana pseudostem) and M1 (40% hyacinth: 40%rice straw, 20% banana pseudostem) showed the highest N total of contents compared to that of other treatments. Previous study showed 50% water hyacinth and 30 and 20% straw organic mulch sheet composition increased C-organic, N-total, and organic matter compared with 60% water hyacinth and 30% straw composition (Iriany et al. 2016). Besides, the nitrogen content of banana pseudostem is the highest than other raw materials. Nitrogen content of water hyacinth, straw, and banana pseudostem in dry matter, respectively, are 0.28%; 0.5–0.8%; and 0.93–1.87% (Dobermann and Fairhurst 2002; Abdullah et al. 2014).

Table 1 The results of chemical analyses on organic materials for organic mulch sheet

Treatment	C-organic (%)	N-total (%)	C/N ratio	Organic matter (%)
M_1 (40:40:20)	27.69	10.16	3	47.90
M_2 (50:20:30)	29.67	7.96	4	51.32
M_3 (50:30:20)	31.41	11.93	3	54.33
M_4 (50:40:10)	32.78	5.73	6	56.71
M_5 (60:20:20)	32.89	7.8	4	56.91
M_6 (60:30:10)	31.03	2.04	15	53.68

Table 2 The results of tensile strength analyses on composition treatment for organic mulch sheet

Treatment	Tensile strength (N/m^2)
M_1 (40:40:20)	28.1
M_2 (50:20:30)	23.2
M_3 (50:30:20)	27.5
M_4 (50:40:10)	18.8
M_5 (60:20:20)	32.8
M_6 (60:30:10)	22.1

Tensile strength

Based on the test of the mulch tensile strength, it resulted in the following numbers as shown in Table 2. Table 2 showed that the highest tensile strength was during the treatment of M5. Previous researches have shown that 60% water hyacinth, 20% straw, and 20% banana pseudostem composition gained the highest tensile strength value than other treatments (Djojowasito et al. 2007; Indriyati et al. 2016). Water hyacinth has a compact, short, and soft fiber than others, therefore, it contributes more effects on the tensile strength of organic mulch sheet. Besides, the tensile strength of organic mulch sheet is higher than biodegradable paper made (patented) by Peter. F et al. (2003)

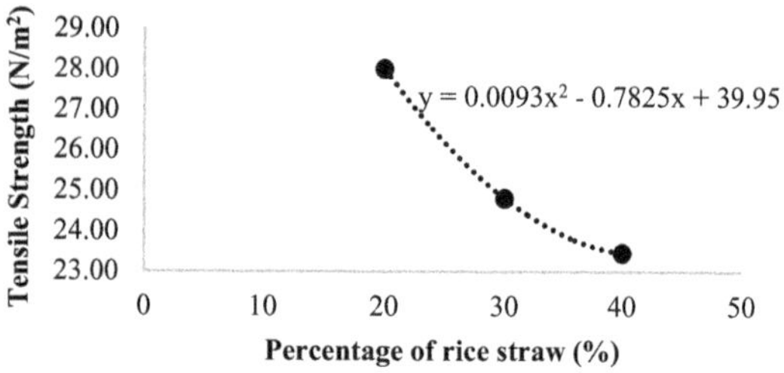

Fig. 1 The correlation between tensile strength and the percentage of rice straw contents in organic mulch sheet

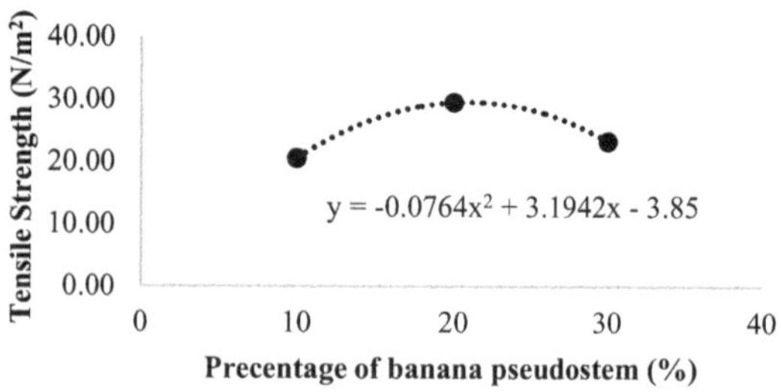

Fig. 2 The correlation between tensile strength and the percentage of banana pseudostem contents in organic mulch sheet

although it is still lower in tensile strength compared to that of fiber composite made from kenaf and pineapple (J, M.K and W.S 2015).

Figure 1 shows that the increase of rice straw proportion resulted in the decrease of tensile strength. Figure 2 shows the parabolic correlation between the proportion of banana pseudostem and tensile strength and the optimum proportion is shown to be 20%. The correlation between tensile strength and water hyacinth proportion displays an inverse manner from the correlation between tensile strength and banana pseudostem proportion (Fig. 3). The optimum proportions of water hyacinth are 40 and 60%. In each material, the maximum tensile strength was gained from rice straw proportion of 20%, banana pseudostem proportion of 20%, and water hyacinth of 60%. It is similar to the previous study revealing that the highest tensile strength has been gained from 20–30% banana pseudostem (Djojowasito et al. 2007); also, the other paper exhibited the highest tensile value of the sample at 15 wt% composition of fiber than 5 and 10 wt% (MdRadzi et al. 2015). Another paper has shown that the proper comparison of hyacinth plant and coconut coir was 88 and 12% (Nugroho et al. 2013) also the additional treatment to increase tensile strength was by alkali and enzyme treatments (Tumolva et al. 2013). For pulp from straw, tensile strength could gain 153.10^5 N/m^2 (Saragi, 2008). Each type of fiber source has its different characteristics as well as its effects on tensile strength of organic mulch sheet. Accordingly, it is crucial to find the proper combination of materials.

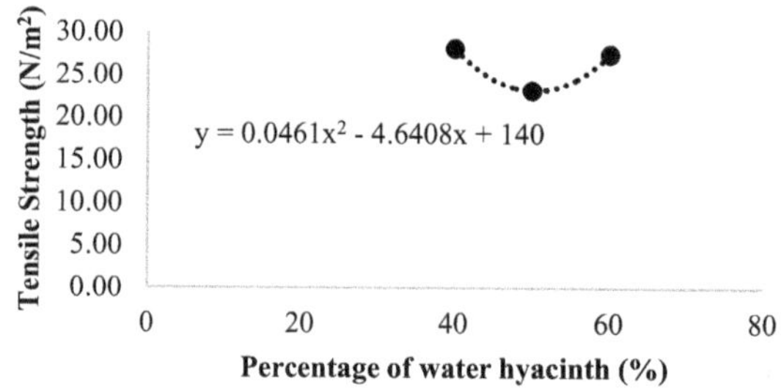

Fig. 3 The correlation between tensile strength and the percentage of water hyacinth contents in organic mulch sheet

Table 3 The result of vilensky test of various organic mulch sheet composition

Treatment	Sunlight intensity			Note
	Before	After		
		Wet	Dry	
M_1 (40:40:20)	15	190	14	Without hole, absorbing
M_2 (50:20:30)	19	45	22	Without hole, non-absorbing
M_3 (50:30:20)	28	238	20	Without hole, absorbing
M_4 (50:40:10)	33	298	40	Without hole, absorbing
M_5 (60:20:20)	8	92	3	Without hole, absorbing
M_6 (60:30:10)	25	194	24	Without hole, absorbing

Vilensky test

Table 3 shows that based on Vilensky test, the highest sunlight intensity of mulch composition was during the treatment of M4 (50% water hyacinth:40% rice straw:10% banana pseudostem). Viewed from the proportion of water hyacinth, the increase in the proportion of water hyacinth could increase sunlight intensity. The picture also portrays that the increase, as well as linear trends, was shown in dry mulch condition. It shows that the increase in the proportion of water hyacinth affects the increase in sunlight intensity (Fig. 4). Fibers from water hyacinth and banana pseudostem that have tied each other prevent the forming of the hole on the mulch (Djojowasito et al. 2007).

Viewed from rice straw proportion used in mulch both with the thickness of 0.5 mm and 1.0 mm, especially for mulch in wet condition, the quadratic curve with the peak/maximum cusp was formed in the proportion of 30%. The same case also occurred in the dry condition of mulch. It indicates that the increase and decrease in rice straw proportion affect the decrease in sunlight intensity (Figs. 5 and 6).

Viewed from banana pseudostem proportion used in mulch both with the thickness of 0.5 mm and 1.0 mm, especially for mulch in wet condition, the quadratic curve tended to decline. In the wet condition, the increase in banana pseudostem proportion affects the decrease in sunlight intensity.

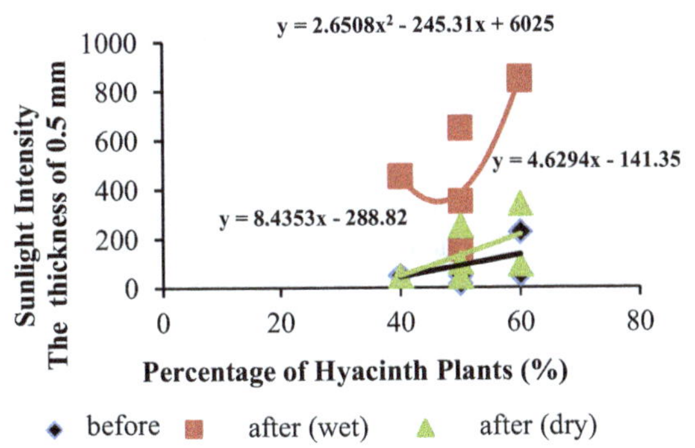

Fig. 4 The correlation between sunlight intensity and the percentage of water hyacinth in 0.5 mm organic mulch sheet

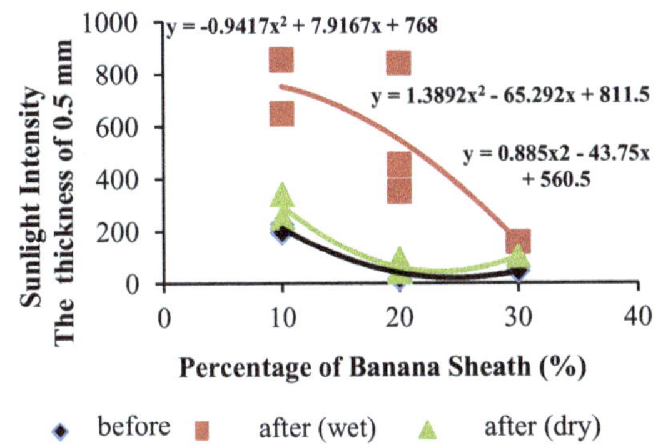

Fig. 7 The correlation between sunlight intensity and the percentage of banana pseudostem in 0.5 mm organic mulch sheet

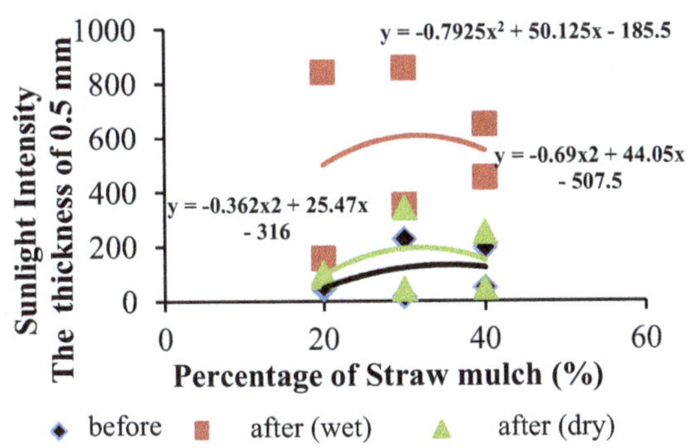

Fig. 5 The correlation between sunlight intensity and the percentage of rice straw in 0.5 mm organic mulch sheet

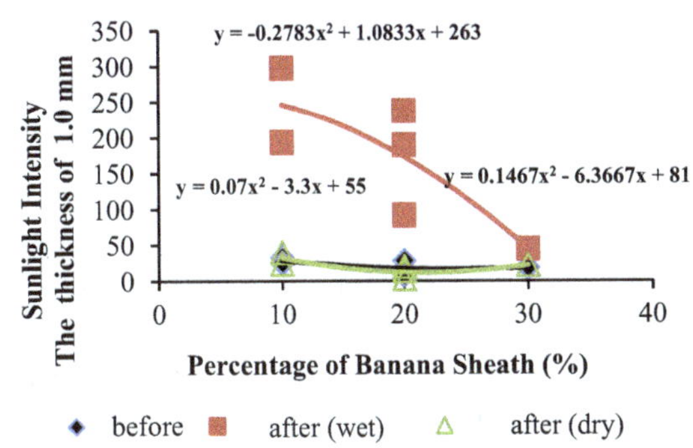

Fig. 8 The correlation between sunlight intensity and the percentage of banana pseudostem in 1.0 mm organic mulch sheet

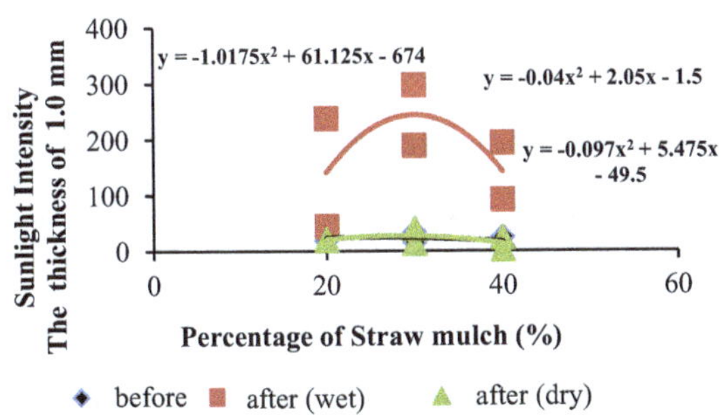

Fig. 6 The correlation between sunlight intensity and the percentage of rice straw in 1.0 mm organic mulch sheet

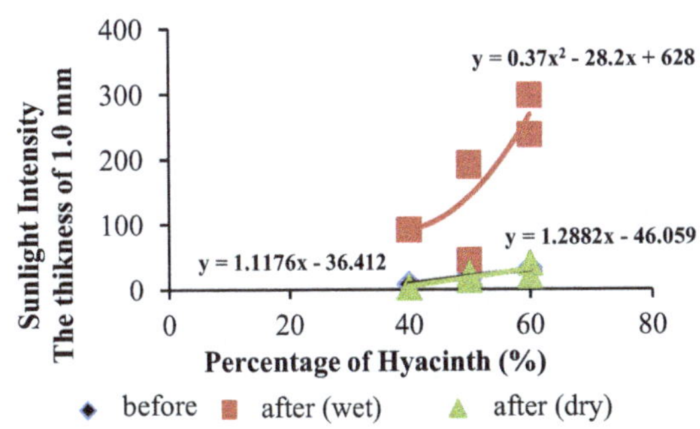

Fig. 9 The correlation between sunlight intensity and the percentage of water hyacinth in 1.0 mm organic mulch sheet

However, in the dry condition, the quadratic curve tended to open upward. It is shown that the minimum cusp of sunlight intensity was formed in the proportion of 20–30%. The higher proportion of banana pseudostem resulted in the increase in sunlight intensity (Figs. 7 and 8).

Viewed from hyacinth proportion used in mulch both with the thickness of 0.5 mm and 1 mm, the quadratic and linear

Table 4 The result of correlation analysis between chemical characteristics of mulch materials and tensile strength of mulch

Chemical characteristics of mulch materials	Correlation	*p* value
C-organic	– 0.113	0.726
N-total	0.072	0.823
C/N	0.032	0.921
Organic materials	– 0.113	0.726

curve tended to ascend along with the increasing proportion of hyacinth. In the wet and dry conditions, the increase in hyacinth proportion affects the increase in sunlight intensity. The higher proportion of hyacinth resulted in the increase in sunlight intensity (Figs. 4 and 9).

The result of correlation test between chemical characteristics of mulch materials and tensile strength of mulch has shown that chemical characteristics of all materials have p value more than 0.05. It proves that there is no significant correlation between C-organic contents, N Total, C/N, and organic materials and tensile strength of mulch (Table 4).

Conclusion

The result of this current research has revealed that water hyacinth, rice straw, and banana pseudostem could be used as materials for organic mulch sheet. The best composition has been shown during the treatment of 60% water hyacinth, 20% rice straw, and 20% banana pseudostem seen from tensile strength and endurance towards the water and normal mulch stretch. The best treatment was the composition of 50% water hyacinth, 40% rice straw, and 10% banana pseudostem. Organic mulch with the strongest tensile strength of 32.8 N resulted during the treatment of 60% water hyacinth, 20% rice straw, and 20% banana pseudostem; also the highest C-organic content was during this treatment that accounted for 32.89%.

Acknowledgement The authors would like to thank University of Muhammadiyah Malang (UMM) for providing necessary facilities to carry out this research. Financial support from DIKTI (Directorate of High Education) for this research is gratefully acknowledged.

References

Abdullah N, Sulaiman F, Miskam MA, Taib RM (2014) Characterization of banana (Musa spp.) plantation wastes as a potential renewable energy source. Int J Biol Biomol 8(8):815–819. https://doi.org/10.1063/1.4803618

Ayinde OE, Muchie M, Olatunji GB (2011) Effect of climate change on agricultural productivity in Nigeria: a co-integration model approach. J Hum Ecol 35(3):189–194

Azad, B., Hassandokht, M. R. and Parvizi, K. (2015) 'Effect of mulch on some characteristics of potato in Asadabad, Hamedan', *IJAAR* , 6(3), pp. 139–147 http://www.innspub.net

Azubuike CP, Okhamafe AO (2012) Physicochemical, spectroscopic and thermal properties of microcrystalline cellulose derived from corn cobs. Int J Recycl Org Waste Agric 1(1):9. https://doi.org/10.1186/2251-7715-1-9

Bajorien K, Jodaugien D, Pupalien R, Sinkevi A (2013) Effect of organic mulches on the content of organic carbon in the soil. Est J Ecol 62(2):100–106. https://doi.org/10.3176/eco.2013.2.02

Bhatnagar R, Gupta G Y, adav, S. (2015) 'A review on composition and properties of banana fibers'. Int J Sci Eng Res 6(5): 143–148 www.ijser.org

Coolong T. (2012) 'Mulches for weed management in vegetable production', Weed Control, Dr. Andrew Price (Ed.) http://cdn.intechopen.com/pdfs/29919/InTech-Mulches_for_weed_management_mulches_for_weed_management.pdf

Cowan J S. (2013) The use of biodegradable mulch for tomato and broccoli production: crop yield and quality, mulch deterioration, and grower's perception. Washington State University

Das S, Goswami S, Talukdar ADAS (2016) Physiological responses of water hyacinth, Eichhornia crassipes (Mart.) Solms, to cadmium and its phytoremediation potential. Turkish J Biol 40:84–94. https://doi.org/10.3906/biy-1411-86

Díaz-Pérez J C, Batal K D (2002) 'Colored plastic film mulches affect tomato growth and yield via changes in root-zone temperature'. J Amer Soc Hort.Sci, 127(1), pp. 127–135. http://journal.ashspublications.org/content/127/1/127.short

Djojowasito G, Ahmad AM, Wijaya K (2007) An evaluation of sheet mulch made from a combination of hyacinth (*Eichornia crassipes* mart. solms) and Musa paradisiaca l. biomass decomposed with either urea or caustic soda. J Teknol Pertan 8(2):110–118

Dobermann A, Fairhurst TH (2002) 'Rice straw management'. Better Crop Int 16:7–11

Dvořák P, Hajšlová J, Hamouz K, Schulzová V, Kuchtová P, Tomášek J (2001) Black polypropylene mulch textile in organic agriculture. Lucr. Ştiinţifice, Prague

Dvořák P, Tomášek J, Kuchtová P, Hamouz K, Hajšlová J, Schulzová V (2012) Effect of mulching materials on potato production in different soil-climatic conditions. Rom Agric Res 29:201–209

Ekinci M, Dursun A (2009) Effects of different mulch materials on plant growth, some quality parameters and yield in melon (Cucumis melo L.) cultivars in high altitude environmental condition. Pak J Bot 41(4):1891–1901

Gavrilescu M (2004) Removal of heavy metals from the environment by biosorption removal of heavy metals from the environment by biosorption. Eng Life Sci. https://doi.org/10.1002/elsc.200420026

George Hochmuth, Robert Hochmuth, Steve Olson (2002) 'New technologies in mulching for vegetable production in Florida'. Citrus Veg 1–4

Haapala T, Palonen P, Tamminen A, Ahokas J (2015) Effects of different paper mulches on soil temperature and yield of cucumber (*Cucumis sativus* L.) in the temperate zone. Agric Food Sci 24:52–58

Ibarra-Jiménez L, Lira-Saldivar RH, Valdez-Aguilar LA, Lozano-Del Río J (2011) 'Colored plastic mulches affect soil temperature and tuber production of potato'. Acta Agric Scand Sect B 61(4):365–371. https://doi.org/10.1080/09064710.2010.495724

Indriyati W, Musfiroh I, Kusmawanti R, Sriwidodo, Hasanah AN (2016) Characterization of carboxymethyl cellulose sodium (na-cmc) from water hyacinth (*Eichhornia crassipes* (mart.) solms) cellulose) growing in Jatinangor and Lembang. IJPST 3(3):100–110

Iriany A, Syekhfani Soemarno, Suprayogo D (2013) Climate change impact to potato farming in the Java of Indonesia in the mountain range of Batu, East Java of Indonesia. J.App Environ Biol Sci 3(6):48–55

Iriany A, Lestari R, Chanan M (2016) Examining organic mulch sheet on the growth and yield of shallot *(Allium ascalonicum* L.). Malang

Jodaugienė D, Pupalienė R, Sinkevičienė A, Marcinkevičienė A, Žebrauskaitė K, Baltaduonytė M, Čepulienė R (2010) The influence of organic mulches on soil biological properties. Zemdirb Agric 97(2):33–40

Kalra N, Suneja P, Mendiratta N, Gupta N (2013) Simulating the impact of climate change and its variability on growth and yield of crops. Clim Chang Environ Sustain 1(1):11–19. https://doi.org/10.5958/j.2320-6411.1.1.002

Kar G, Kumar A (2007) Effects of irrigation and straw mulch on water use and tuber yield of potato in eastern India. Agric Water Manag 94:109–116. https://doi.org/10.1016/j.agwat.2007.08.004

Kasirajan S, Ngouajio M (2012) Polyethylene and biodegradable mulches for agricultural applications: a review. Agro Sustain Dev 32(2):501–529. https://doi.org/10.1007/s13593-011-0068-3

Khan S, Sarwar K (2002) Effect of water hyacinth compost on physical, physico-chemical properties of soil and on rice yield. Pak J Agron 1(2–3):64–65

Lakkenborg H, Campanelli G, Bavec F, Von Fragstein P, Xie Y, Canali S, Titralelli F. (2014) 'Effect of an in-season living mulch on leaching of inorganic nitrogen in cauliflower (*Brassica oleracea* L. var. botrytis) cropping in Slovenia, Germany and Denmark'. In the 4th ISOFAR Scientific Conference 'Building Organic Bridges' 13–15

Li K, Fu S, Zhan H, Zhan Y, Lucia LA (2010) Analysis of the chemical composition and morphological structure of banana pseudostem. BioResources 5(2):576–585. https://doi.org/10.15376/biores.5.2.576-585

Mahmood N, Ahmad B, Hassan S, Bakhsh K (2012) Impact of temperature ADN precipitation on rice productivity in rice-wheat cropping system of Punjab province. J Anim Plant Sci 22(4):993–997

Main NM, Talib RA, Ibrahim R, Rahman RA, Mohamed AZ (2014) Suitability of coir fibers as pulp and paper. In Agric Agric Sci Procedia 2:304–311. https://doi.org/10.1016/j.aaspro.2014.11.043

Matindi CN, Njogu PM, Kinyua R, Nemoto Y (2014) 'Analysis of heavy metal content in water hyacinth (*Eichhornia crassipes*) from Lake Victoria, Kenya'. Int Conf Sustain Res Innov 196–199

Mawlana M, Hossain S, Ferdoushi A (2014) Usefulness of banana (*Musa paradisiaca*) wastes in manufacturing of bio-products: a review. Agri Int J 12(1):148–158. https://doi.org/10.3329/agric.v12i1.19870

MdRadzi M K F, Sulong A B, Muhamad N, MohdLatiff M A, Ismail N F (2015) 'Effect of Filler loading and NaOH addition on mechanical properties of moulded kenaf/polypropylene composite'. Pertanika J Trop Agric Sci 38(4) 583–590. http://www.pertanika.upm.edu.my/

Miles C, Wallace R, Wszelaki A, Martin J, Cowan J, Walters T, Inglis D (2012) Deterioration of potentially biodegradable alternatives to black plastic mulch in three tomato production regions. HortScience 47(9):1270–1277

Moreno MM, Moreno C, Villena J, Mancebo I, Meco R, Campos JA (2011) Effect of different mulches on soil microbiological properties in processing tomato. Res. spain, Geophys

Nakano Y, Higuchi Y, Sumitomo K, Hisamatsu T (2013) Flowering retardation by high temperature in chrysanthemums: involvement of FLOWERING LOCUS T-like 3 gene repression. J Exp Bot 64(4):909–920. https://doi.org/10.1093/jxb/ers370

Nugroho WA, Rahayu FD, Lutfi M (2013) Effect of ingredients formulation of mechanics traits on organic planting bag. J Teknol Pertan 14(2):115–122

Nyawira, M. C. (2016) Analysis of Heavy Metal Content in Water Hyacinth (Eichonia crassipes) from Lake Victoria and Assessment of its Potential as a Feedstock for Biogas Production. University of Nairobi

O'Brine T, Thompson RC (2010) Degradation of plastic carrier bags in the marine environment. Mar Pollut Bull 60(12):2279–2283. https://doi.org/10.1016/j.marpolbul.2010.08.005

Peng S, Huang J, Sheehy JE, Laza RC, Visperas RM, Zhong X, Centeno GS, Khush GS, Cassman KG (2004) Rice yields decline with higher night temperature from global warming. Proc Natl Acad Sci 101(27):9971–9975. https://doi.org/10.1073/pnas.0403720101

Pereira A, Nova NV (2008) Potato maximum yield as affected by crop parameters and climatic factors in Brazil. HortScience 43(5):1611–1614

Peter F L, Cornellius E C, Potnis P S, Cleveland C S, Knauf G H (2003) 'Biodegradable paper-based agricultural substrate'. United States

Pratt, Medford E R, W. (1955) 'Fertilizer Mulch Paper'. United States

Ramesh M, Ananda TS, Aswin US, Eashwar H, Deepa C (2014) Processing and mechanical property evaluation of banana fiber reinforced polymer composites. Procedia Eng. https://doi.org/10.1016/j.proeng.2014.12.284

Saha P, Shinde O, Sarkar S (2017) Phytoremediation of industrial mines wastewater using water hyacinth. Int J Phytoremediation 19(1):87–96. https://doi.org/10.1080/15226514.2016.1216078

Sahari G N A, Buku A (2015) 'Tensile strength of fiber for some type bananas (Ambon, Kepok, Susu)'. Int J Res Eng Technol 4(8) 2319–2322. www.ijret.org

Salleh J, Yusoh M.K, Ruzan WS (2015) 'Tensile strength of some natural-fibre composites'. Pertanika J Trop Agric Sci, 38(4), pp. 575–582. http://www.pertanika.upm.edu.my/

Saragi, D. (2008) Forming and characteristic of paper made of mixture of straw pulp and cement bag trace pulp. Universitas Sumatera Utara

Sarika D, Singh J, Prasad R, Vishan I, Varma VS, Kalamdhad AS (2014) Study of physico-chemical and biochemical parameters during rotary drum composting of water hyacinth. Int J Recycl Org Waste Agric 3(3):63. https://doi.org/10.1007/s40093-014-0063-1

Scholes MC, Powlson D, Tian GL (1997) Input control of organic matter dynamics. Geoderma 79(1–4):25–47. https://doi.org/10.1016/S0016-7061(97)00037-2

Sindhu R, Binod P, Pandey A, Madhavan A, Alphonsa JA, Vivek N, Gnansounou E, Castro E, Faraco V (2017) Bioresource technology water hyacinth a potential source for value addition: an overview. Bioresour Technol 230:152–162. https://doi.org/10.1016/j.biortech.2017.01.035 **Elsevier Ltd**

Sinkevičienė A, Jodaugienė D, Pupalienė R, Urbonienė M (2009) The influence of organic mulches on soil properties and crop yield. Agron Res 7(1):485–491

Siwek P, Kalisz A, Wojciechowska R (2007) Effect of mulching with film of different colours made from original and recycled polyethylene on the yield of butterhead lettuce and celery. Folia Hortic 19(1):25–35

Sutyasmi S (2012) 'Recycling of shaving waste from tanning industry for art paper'. Maj Kulit karet dan Plast 28(2) 113–121 http://dx.doi.org/10.20543/mkkp.v28i2.114

Teygeler R (2000) Water hyacinth paper. Gentenaar and Torley Publisher, Contributions to a sustainable future

Tumolva T, Ortenero J, Kubouchi M, City Q (2013) Characterization and treatment of water hyacinth fibers for NFRP composites. Int Conf Compos Mater 28:1–11

Vidya S, Girish L (2014) 'Water hyacinth as a green manure for organic farming'. Int J Res Applied Nat Soc Sci 2(6) pp. 65–72. www.impactjournals.us

Wai NM, Camerlengo A, Khairi A, Wahab A (2007) A study of global warming in Malaysia. J Teknol 42:1–10

Widiatningrum T, Pukan KK (2010) Growth and production of cauliflower (Brassica oleracea var Botrytis) by organic farming system at lowland climate. Biosaintifika 2(2):115–121

Wright H E (1936) 'Mulch fertilizer paper'. United States

Yaghi T, Arslan A, Naoum F (2013) Cucumber (*Cucumis sativus*, L.) water use efficiency (WUE) under Plastic Mulch and drip irrigation. Agric Water Manag. 128:149–157. https://doi.org/10.1016/j.agwat.2013.06.002

Yoko M, Sugiyama K, Osawa J Hara K. (1986) 'Mulching Paper Sheet'. United States

3

Microbial consortium culture and vermi-composting technologies for recycling on-farm wastes and food production

Girish Chander[1] · Suhas P. Wani[1] · S. Gopalakrishnan[1] · Ankita Mahapatra[1] · Swati Chaudhury[1] · C. S. Pawar[1] · Manoj Kaushal[1] · A. V. R. Kesava Rao[1]

Abstract

Purpose A study was conducted to characterize the 'Madhyam culture' (Excel Crop Care Limited.), an aerobic-composting microbial consortium culture, and understand composting dynamics, product quality and use in crop production vis-à-vis vermi-composting (using earthworms).

Methods 16S rDNA analysis was used to characterize aerobic-composting culture. Aerobic-composting and vermi-composting technologies were evaluated to decompose sorghum straw and dung biomass (80:20 ratio; primed with 0.5% urea and 4% rock phosphate) to study days to maturity and composting dynamics in terms of changes in temperature and microbial population. Compost quality was tested for macro-, micro-nutrients and C:N ratio, and evaluated for food production in on-farm trials.

Results 16S rDNA analysis screened sixteen bacterial isolates—eight related to genus *Bacillus*, three to each *Halobacillus and Staphylococcus*, one to each *Microbacterium* and *Streptomyces*. The population of bacteria was 4.5 cfu ml^{-1} at 10^{-7} dilution. Aerobic- and vermi-composts matured in around 50 and 60 days, respectively. Aerobic-composting throughout recorded relatively higher bacterial population, and higher temperatures during the initial phase. Aerobic-compost tested for high nutrient (1.55% N, 0.93% P, 1.00% K) content and stable C:N ratio (10.3) compared to vermi-compost (1.11% N, 0.43% P, 0.96% K and C:N ratio of 11.7). Field evaluation of both composts showed yield benefit and saving of chemical fertilizers up to 25%.

Conclusions Aerobic-composting (using microbial consortium culture), like vermi-composting, proved to be an effective technology with advantage of no requirement to maintain ambient living conditions in lean periods as is required for earthworms in vermi-composting, but needs more energy/labor for biomass turnings.

Keywords Composting dynamics · Compost quality · Nutrient recycling · Soil carbon building

Introduction

Low soil organic carbon (C) levels and macro-, micro-nutrient deficiencies due to soil mismanagement and misuse are the major challenges for improving productivity in the semi-arid tropics (Chander et al. 2015). Soil organic C is one of the most important parameters governing soil health through influencing soil chemical, physical and biological properties, and thereby influencing crop yields as well (Wani et al. 2003). Moreover, soils are largest terrestrial reservoirs of C (~ 2500 Pg C up to 2 m depth, compared to 600 Pg C in biota and 750 Pg C in atmosphere) (Batjes 1999; Lal and Kimble 1997) and potential C-sink to regulate climate change processes (Lal et al. 2007; Lal 2011). So, practices that add organic matter into soil need to be promoted. In India, ~ 350 million t organic wastes are generated annually from agricultural sources (Pappu et al. 2007), and so recycling C and nutrients contained in such wastes is potential opportunity for soil C-building and cutting costs of chemical fertilizers (Chander et al. 2013; Nagavallemma et al. 2006).

Composting is the technology for conversion of bulky organic wastes into low volume nutrient-enriched and stable product (Yousefi et al. 2013; Kharrazi et al. 2014). Traditional composting (farmers' practices of heaping straw and dung) is

✉ Girish Chander
g.chander@cgiar.org

[1] RP: Asia, International Crops Research Institute for the Semi-Arid Tropics (ICRISAT), Patancheru, Telangana 502324, India

very time consuming and relatively less effective. In such a case, using half decomposed compost/manure/plant residue creates many plant nutrient and pest-related problems, rather than benefits. Vermi-composting is one of the tested technologies to effectively recycle on-farm wastes to produce quality compost for use in crop production (Chander et al. 2013; Wani et al. 2014). Vermi-composting hastens the decomposition process through physical breakdown of the raw biomass coupled with mixing of vast spectrum of microbes with the biomass while passing through earthworm gut. The microbes of earth worm gut are highly potential in digesting the organic materials as well as polysaccharides (Aira et al. 2007; Zhang et al. 2000). However, continuously maintaining ambient living conditions for earthworms is a challenge in vermi-composting and a major cause for many on-farm failures. So, the other option could be to accelerate need-based decomposition of biodegradable materials through the external addition of decomposing microbial consortium culture (aerobic-composting). This may add the convenience of undertaking composting as and when needed without worrying throughout the year about maintaining moisture and feeding material as is needed in vermi-composting. However, relative effectiveness of aerobic-composting via external microbial culture needs to be evaluated vis-à-vis vermi-composting technology, as studies indicate that vermi-composting also renders the advantage of enhancing the diversities of bacterial and fungal communities (Huang et al. 2014). Well-tested scalable composting options are need of the day not only for agricultural wastes, but also for effective disposal of municipal and household wastes (generated to the tune of 125,000 tons per day in a country like India as per the estimates of Central Pollution Control Board (2017) to reduce pollution arising from landfills and open dumps.

Apparently high microbial activity under composting have an indirect role in improving compost nutrient quality by nitrogen fixers, nitrifiers, sulfur oxidizers (Richardson and Simpson 2011) and may also synthesize chemicals which act as plant growth hormones (Pizzeghello et al. 2001; Ghosh et al. 2003; Tomati et al. 1988). Lack of information on standardized practices, processes and products with regard to organic composts is one major drawback in precise use of organics. Therefore, a study was planned to characterize microbial consortia culture (for aerobic-composting) and evaluate its efficacy vis-à-vis vermi-composting in recycling on-farm wastes while understanding microbial dynamics of decomposition process and compost quality and evaluation in field application.

Methodology

Characterization of microbial consortium (Madhyam) culture

'Madhyam' is a culture of micro-organisms developed for accelerated aerobic-composting of organic waste by Excel Crop Care Limited.

Enumeration of microorganisms in microbial consortium (Madhyam) culture

Ten grams of 'Madhyam' sample was homogenized with 90 ml of sterile 0.85% NaCl solution in water and incubated for 1 h at 120 rpm. Then, the samples were serially diluted (up to 10^{-8}) and plated on several medium such as nutrient agar for bacteria (10^{-4} to 10^{-7} dilutions), Bennett's agar for actinobacteria (10^{-3} to 10^{-6} dilutions) and potato dextrose agar for fungi (10^{-2} to 10^{-5} dilutions). The plates were incubated at 28 °C for 2–7 days. The colonies that appeared at the end of incubation were enumerated as colony forming units (CFU) per gram of sample on dry weight basis. Characteristics of the colony were noted and representative single colonies were further cultured and purified using respective medium. The single colonies were stored at 4 °C for subsequent analysis.

Cellulose degradation capability of microbial consortium (Madhyam) culture

Cellulase degrades cellulose in the plant cell wall and cellulase production by bacteria was determined by cellulose congo red agar medium. The test microbes were inoculated on above medium and incubated for 4 days at 28 ± 2 °C. Halo zone around the microbial colonies indicates cellulase production and the observations were documented on a 0–4 rating scale as, 0 = no change; 1 = positive; 2 = halo zone of 1–3 mm; 3 = halo zone of 4–6 mm and 4 = halo zone of 7 mm and above (Hendricks et al. 1995; Gopalakrishnan et al. 2015).

Molecular identification of microbial consortium (Madhyam) culture

The potential cellulose-degrading bacteria identified previously were grown until log phase is reached. The genomic DNA was isolated as per Bazzicalupo and Fani (1995). The 16S rDNA gene was amplified through universal bacterial primer 907R (5′-CCGTCAATTCMTTTRAGTTT-3′) and 785F (5′-GGATTAGATACCCTGGTA-3′) according to the conditions of Pandey et al. (2005). Resulting

PCR products were sequenced at Macrogen Inc. (Seoul, Korea). The sequences obtained were compared with those from the GenBank using BLAST program (Altschul et al. 1990) and aligned through Clustal W software (Thompson et al. 1997). The phylogenetic trees were constructed using neighbor-joining method according to Saitou and Nei (1987).

Evaluation of composting methods

Vermi- and aerobic-composting methods were evaluated for their efficacy in decomposing on-farm wastes in open field conditions at ICRISAT during 2014. The raw biomass was decomposed under the shade by spreading on cement floor base and maintaining moisture through watering the alternate days. Thus, there were 2 treatments: (1) vermi-composting, and (2) aerobic-composting. Under each of the composting method, the raw biomass used was sorghum straw and cow dung (in the ratio 80–20). Sorghum straw was chopped with the help of tractor-run chopper. Urea (0.5%) priming of sorghum biomass was done before 1 week to lower C:N (Carbon:Nitrogen) ratio. In addition, rock phosphate @ 4 kg per 100 kg biomass was also added at start of the experiment. In case of vermi-composting, *Eudrilus eugeniei* and *Eisenia fetida* species of earth worms were used for decomposition of biomass, while in case of aerobic-composting; microbial culture (Madhyam, Excel Crop Care Ltd) @ 1 kg per ton was used.

Raw biomass (~ 1 ton) was filled in layers of sorghum (0.30 m) and cowdung slurry. Vermi-composting was started on a harvested bed of ~ 0.15 m having earthworm culture. In aerobic-composting, microbial culture was spread between the layers and a heap of ~ 0.75 m was prepared on day-1 itself. However, in case of vermi-composting, heap was filled up to 0.75 m height in stages at weekly intervals at days 1, 7 and 14. Intense microbial activity in aerobic-composting leading to high temperatures is required to be regulated through enhanced aeration by regular biomass turnings; so, turnings were done at 10 days interval up to 50 days to maintain the aerobic conditions.

Both composting methods were evaluated for earliness to maturity as judged from color, texture and stabilization of temperature. Other parameters evaluated were microbial dynamics, changes in temperature and chemical analysis of composts (see sections ahead). Bio-washes from vermi- and aerobic-compost were also collected during 30–45 days in plastic containers through outlets for cement floor base at lower end of slope and stored at 4 °C. Bio-washes were also analyzed for nutrient contents and plant growth hormones as described in subsequent sections.

Microbial dynamics during composting

To study microbial dynamics during composting, the samples of vermi- and aerobic-compost were collected at 1, 20 and 55 days of the study in polythene bags and stored at 4 °C. During sampling a treatment at any time, three composite samples of about 100 g each were collected at 0.15–0.20 m depth along length of the heap and thoroughly mixed together for microbial analysis.

For microbial isolation, nutrient agar was used for bacteria, potato dextrose agar for fungi and actinomycetes isolation agar for actinomycetes. All the media were prepared as per standard procedure and sterilized in an autoclave at 121 °C and 15 lb pressure for 20 min. After sterilization, media was dispersed into Petri plates under a sterile atmosphere in laminar air flow and after solidification, these plates were used for the isolation of microorganisms. For enumerating microbial population, dilution plate technique was followed (Ariffin et al. 2008). One gram of compost sample was added to 9 ml of sterile distilled water in a test tube to get 10^{-1} dilution which was serially diluted. Spread plate technique was followed for isolation of microorganisms through spreading 0.1 ml of the desired dilution to the surface of solid agar medium. To check dehydration of medium, the agar plates were wrapped with parafilm, and kept in incubator at 28 + 2 °C. The period of incubation varied for different microorganisms: 24–48 h for bacteria, 72 h for fungi and 72–96 h in case of actinomycetes. The colonies grown in each plate were enumerated with the help of a colony counter as CFU. The total CFU were worked out by taking into account the dilution factor.

Changes in temperature during composting

Daily mean temperature was recorded in both composting types at two depths (0.15 and 0.30 m) using automatic sensors.

Chemical analysis of compost

For estimation of nitrogen (N), phosphorus (P), potassium (K), calcium (Ca) and magnesium (Mg) in compost, 0.5 g of samples was digested using 3.5 ml sulfuric acid–selenium mixture (Sahrawat et al. 2002a). N and P in the digests were estimated with the help of auto-analyzer, and K, Ca and Mg with help of atomic absorption spectrophotometer (AAS). For micronutrients, zinc (Zn), copper (Cu), iron (Fe) and manganese (Mn), 0.2 g compost samples was digested with 2 ml tri-acid and contents in the digests were estimated using AAS (Sahrawat et al. 2002b). S and B in samples were estimated through digesting 0.2 g samples with 2 ml nitric acid plus 0.5 ml hydrogen peroxide and determining the contents in the digest through Inductive Coupled Plasma Atomic

Emission Spectrophotometer (ICP-AES) (Mills and Jones Jr 1996). The pH was measured in 1:2 water suspension (20 g sample and 40 ml water) using glass electrode and total C by Skalar TN/TC analyzer through using 1 g sample (Primacs^{SNC}100 series).

Similar process was followed for macro, micro nutrient chemical analysis in bio-wash samples.

Plant growth hormones analysis in compost bio-washes

Plant growth hormones in compost bio-washes were extracted using standard method by Miezah et al. (2008). A 100 ml of compost washing was taken in 500 ml extraction bottle and then added 200 ml of methanol. This mixture was kept in the shaker for 24 h. The supernatant was then centrifuged for 10 min at 5000 rpm and organic part was stored in freezer. Again, 200 ml methanol was added in aqueous phase and same process followed 4 times to get 800 ml of methanol extract. Then with help of by rotary vacuum evaporator, the methanol extract was concentrated to 2 ml. It was then passed through anhydrous sodium sulfate to make it moisture free and kept in freezer before analysis.

Thin layer chromatography (TLC) technique was used for hormone identification viz. auxins, gibberellins and cytokinins. Each bio-wash extract along with its standards were spotted onto chromatographic plates of 20 cm × 20 cm which were coated with silica gel (60F254) of thickness 0.25 mm. After 4 h of each run, the TLC plates were air dried and then sprayed with coloring reagent, i.e., H_2SO_4. Once the plates were oven dried, R_f value (distance traveled by compound/distance traveled by solvent) of each spot was calculated.

The plant growth hormone standards were used for validating method for auxins (1 ppm), gibberellins (1 ppm) and cytokinins (2 ppm). For isolation of auxins and gibberellins under TLC, iso-propyl alcohol: ammonium hydroxide: water (90:10:10) was used on volume basis, whereas, in case of cytokinins, butyl alcohol: acetic acid: water (80:10:10) was used (Miezah et al. 2008).

On-farm evaluation of composts

The composts prepared following this methodology were evaluated in on-farm studies in India—vermi-compost in Sagar district during 2011–2012 post-rainy season (6 trials with chickpea, 9 trials with wheat) and aerobic-compost in Kadapa district during 2016 rainy season (40 trials with paddy crop). The treatments evaluated were, (1) Balanced nutrition (BN) comprising of soil test-based addition of N, P and K along with deficient micro- and secondary nutrients (i.e. S, Zn and B), and (2) Integrated nutrient management (INM) i.e. 75% of BN inputs + compost. The vermi- or aerobic-compost was added on the basis to replace 25% of N requirement in case of non-legumes and 25% of P requirement in legumes. The on-farm treatment size was 2000 m^2 side by side. All other crop management practices were similar in both the treatments. Composts and all required nutrients except N in non-legumes were added as basal application. In case of N in non-legumes, 50% was added as basal and 25% after 1 month and rest 25% after 2 months. Crop yields were estimated from 3 m × 3 m = 9 m^2 plots, and calculated into kg ha^{-1}.

Results and discussion

Enumeration of microorganisms in microbial consortium (Madhyam) culture

A total of 21 bacterial isolates including one actinomycete were isolated from nutrient agar plates and Bennett's agar plates based on the colony morphology, color and shape. Fungal colonies were not found on the PDA plates while single actinomycete was found on the Bennett's agar. The bacterial population was found to be 4.5 cfu ml^{-1} at 10^{-7} dilution.

Cellulose degradation capability of bacteria isolated from microbial consortium (Madhyam) culture

Of the 21 isolates studied for cellulose degradation, a total of 16 isolates produced the cellulase enzyme. A clear halo zone was found around the colonies in the congo red agar plates. Of the 21 bacteria studied, two (MP-2 and MP-20) were found to produce maximum cellulase (Table 1).

Molecular identification of microbial consortium (Madhyam) culture

The 16 cellulase-producing bacteria were identified through 16S rDNA sequencing. A neighbor-joining dendrogram was generated with the partial sequences of 16 cellulase producing bacteria (1400 bp) and representative sequences from the databases. Phylogenetic analysis of the 16 isolates matched with *Streptomyces* (1), *Microbacterium* (1), *Halobacillus* (3), *Staphylococcus* (3), and *Bacillus* (8 isolates) (Fig. 1; Table 2). *Bacillus* and *Halobacillus* are reported to have various traits including N-fixation, P-solubilization, plant growth promotion, biological control and biofortification (Singh et al. 2008; Richardson and Simpson 2011; Sharma et al. 2013; Gopalakrishnan et al. 2015, 2016). *Streptomyces* sp. are efficient in degrading cellulose and promote crop growth (El-Tarabily 2008; Gopalakrishnan et al. 2011, 2014, 2015, 2016). All the 16 isolates were able to grow at neutral

Table 1 Cellulose degradation capability of bacteria isolated from microbial consortium (Madhyam) culture

S. no.	Isolates	Ratings#
1	MP-1	1
2	MP-2	2
3	MP-3	1
4	MP-4	1
5	MP-5	1
6	MP-6	1
7	MP-7	1
8	MP-8	1
9	MP-9	1
10	MP-10	1
11	MP-11*	–
12	MP-12*	–
13	MP-13*	–
14	MP-14	1
15	MP-15*	–
16	MP-16	1
17	MP-17	1
18	MP-18	1
19	MP-19*	–
20	MP-20	2
21	MA-1	1

*Not identified; # = Observations were recorded on a 0 – 4 rating scale as follows: 0 = no change; 1 = positive; 2 = halo zone of 1 – 3 mm; 3 = halo zone of 4 – 6 mm and 4 = halo zone of 7 mm and above

pH (6.8–7.5) and mesophilic (28–35 °C) temperature (data not shown).

Evaluation of composting methods

Results revealed that aerobic-composting, like vermi-composting, is an effective technology to recycle on-farm wastes into valuable composts in a reasonably short time. Aerobic-compost was ready by around 50th day and vermi-compost by around 60th day. Breakdown of biomass while passing through earthworm gut and mixing of several microbes in coelomic fluid apparently accounted for fast decomposition in vermi-composting. While, high microbial population through added culture and regular turnings at 10-day interval apparently led to fast decomposition in aerobic-composting.

Microbial dynamics during composting

Results revealed that bacteria were the predominant microorganisms found throughout the study followed by actinomycetes and the fungi the least. In general, bacteria occupied 68–92% of the microbial population followed by actinomycetes by 8–32% and fungi the negligible counts (Table 3).

As regards bacterial population, throughout the study, a higher population (21.1–65.3 × 10^{10} CFU/g) was observed under aerobic-composting compared with vermi-composting (19.7–61.6 × 10^{10} CFU/g). The highest population in aerobic-composting was apparently due to external addition of microbial consortium culture dominated by bacterial isolates (see Table 1). In vermi-composting also, there is evidence that different microbes in coelomic fluid get mixed with the biomass while passing through their body and apparently accounts for reasonably higher bacterial number. With passage of time, the bacterial count declined which may be expected due to decomposition of preferable food components of bacteria and also pasteurization as result of high temperatures (Vivas et al. 2009). *Bacillus* spp. was predominant in both the compost types.

Fungi comprised very little component of microbial population (5–9 × 10^3 CFU/g on day 1, 5–10 × 10^3 CFU/g on day 20 and 4–4 × 10^3 CFU/g on day 55) (Table 3). Similar to bacterial count, the fungi population also decreased over the period of time. However, contrary to bacterial population trends, relatively less number of fungal colonies was observed under aerobic-composting apparently due to competition with corresponding higher bacterial population.

Similar to fungi, relatively higher population (71–237 × 10^9 CFU/g) of actinomycete was observed under vermi-composting compared with aerobic-composting (56–75 × 10^9 CFU/g) (Table 3). Mixing of earthworm coelomic fluid with the biomass could be the reason for consistently higher actinomycete population observed during vermi-composting. Further, lower temperature during the vermi-composting also supports the growth of actinomycetes.

Changes in temperature during composting

At 0.15 m depth, up to 30 days, the relatively higher daily mean temperature was recorded under aerobic-composting (38–67 °C) as compared with the vermi-compost (29–48 °C). Under aerobic-composting, it was relatively quite high during the initial 15 days apparently because of high microbial activity, thus indicating rapid decomposition. The temperature reached up to 67 °C by the 9th day (Fig. 2); the turnings at 10th day, however, regulated it to escalate beyond limit which could otherwise negatively affect microbial population and composting process as such and decreased it to 52 °C. By contrast to aerobic-composting, the temperature in vermi-composting remained below 40 °C during 1st week, which may be due to the fact that vermi-composting heap was filled in layers at 0, 7, 14 days. After 1 week, temperature increased and maintained between 40 and 48 °C almost up to 1 month. Between 30 and 60 days (till maturity), the

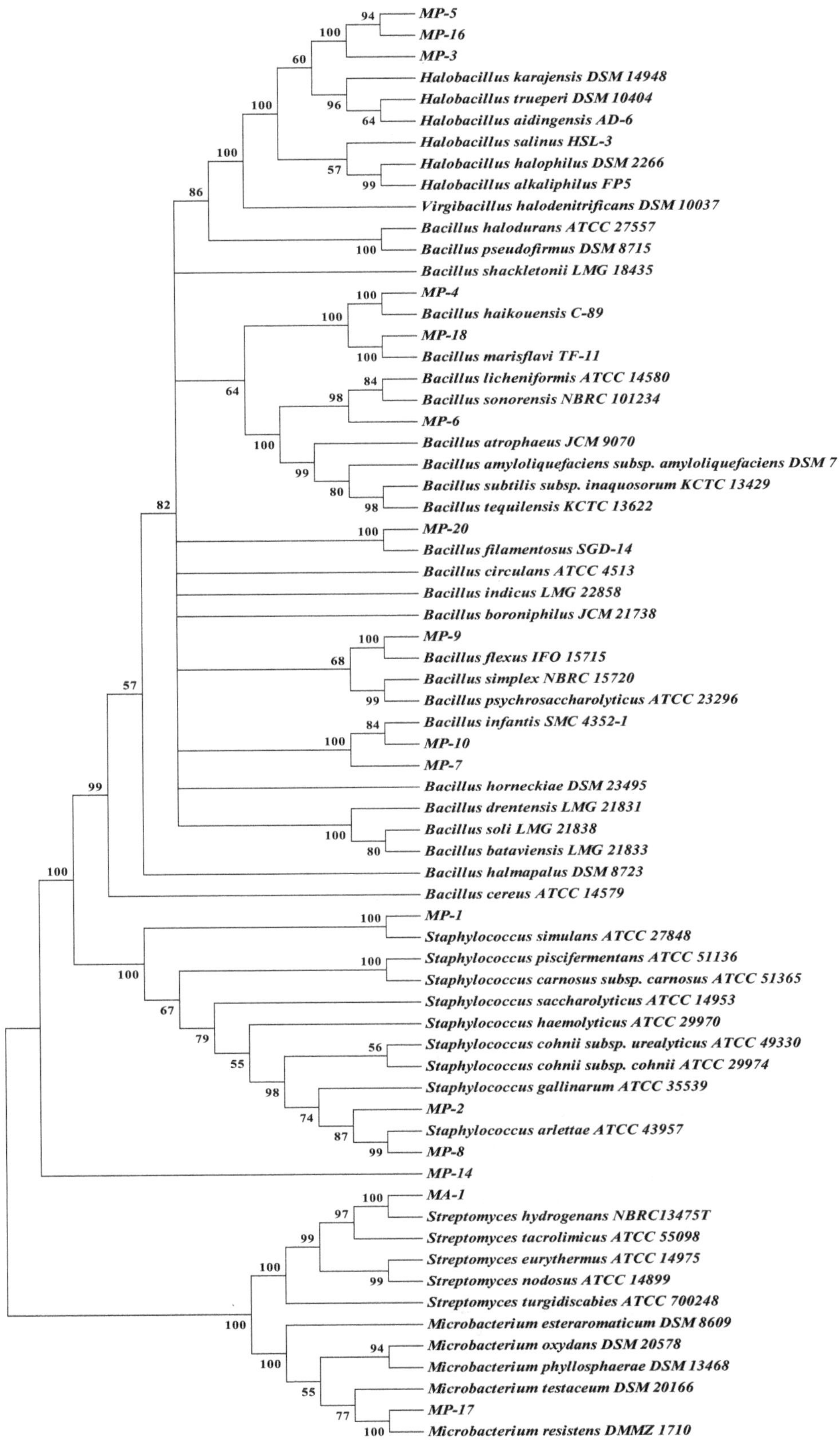

Fig. 1 Phylogenetic relationship between microbial consortium (Madhyam) culture isolates and representative species based on full-length 16S rDNA sequences constructed using the neighbor-joining method

Table 2 Identification of cellulase producing bacteria in microbial consortium (Madhyam) culture

S. no.	Isolates	Scientific name
1	MP-1	*Staphylococcus simulans*
2	MP-2	*Staphylococcus arlettae*
3	MP-3	*Halobacillus* sp.
4	MP-4	*Bacillus aquimaris*
5	MP-5	*Halobacillus* sp.
6	MP-6	*Bacillus licheniformis*
7	MP-7	*Bacillus infantis*
8	MP-8	*Staphylococcus arlettae*
9	MP-9	*Bacillus flexus*
10	MP-10	*Bacillus infantis*
11	MP-14	*Bacillus* sp.
12	MP-16	*Halobacillus*
13	MP-17	*Micro bacterium resistens*
14	MP-18	*Bacillus mariflavas*
15	MP-20	*Bacillus* sp.
16	MA-1	*Streptomyces* sp.

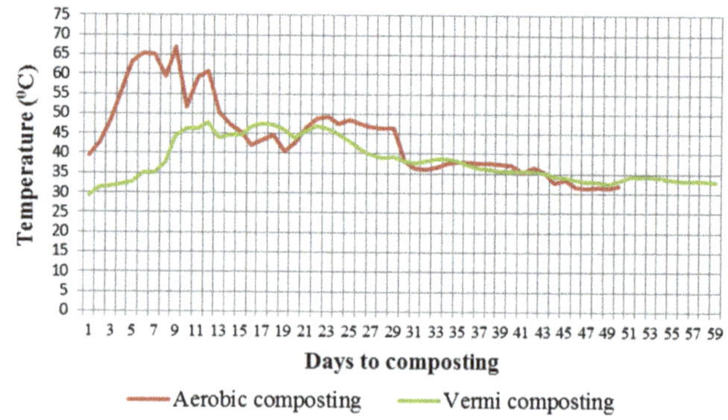

Fig. 2 Changes in daily mean temperature at 0.15 m depth under different composting methods

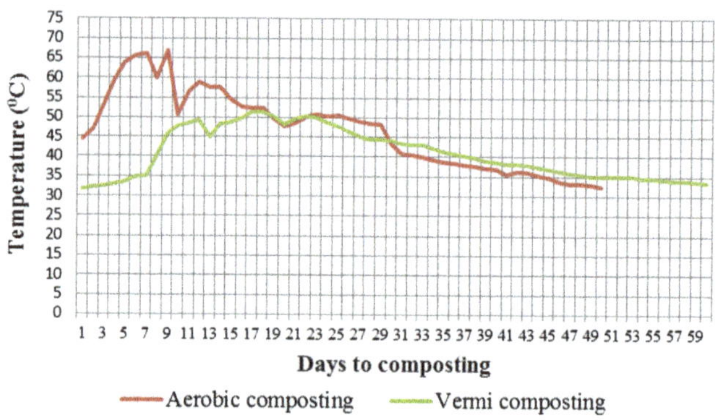

Fig. 3 Changes in daily mean temperature at 0.30 m depth under different composting methods

temperatures were almost similar under both the composting processes and ranged between 31 and 39 °C.

At 0.30 m depth, the trends observed as such were similar to that at 0.15 m depth; however, slightly higher temperature was recorded at 0.30 m depth (Fig. 3). Up to 1 month, a higher temperature was observed under aerobic-composting which varied between 44 and 67 °C compared to vermi-composting between 32 and 51 °C. Similar to depth at 0.15 m, the corresponding temperature at 0.30 m depth beyond 30 days was almost at par in both the composting processes.

Chemical analysis of compost

Results showed aerobic-compost in general having relatively higher nutrient contents w.r.t N, P, K, Ca, S and Zn (Table 4). High nutrient contents in aerobic-compost may be due to better decomposition as a result of higher microbial activity observed in this study. However, vermi-compost was found to contain higher contents of Mg, Cu, Fe, Mn and B. Total C was found to be 160 mg g^{-1} in aerobic-compost and 130 mg g^{-1} in vermi-compost with C:N ratio of 10.3 and 11.7, respectively. The pH was almost same at 6.75 in aerobic-compost and 6.73 in vermi-compost. Reasonably higher nutrient contents in the compost types were probably due to better decomposition (Shah et al. 2015), and enrichment with rock phosphate and urea.

The bio-washes of vermi-compost and aerobic-compost were almost at par in nutritional quality (Table 5). Relatively higher N content in biowash collected from aerobic-compost is apparently due to better decomposition due to higher microbial activity through added culture. Thus,

Table 3 Changes in microbial count under different composting methods

Days to composting	Composting method	Bacteria (× 10^{10} CFU/g)	Fungi (× 10^{3} CFU/g)	Actinomycetes (× 10^{9} CFU/g)
1	Vermi-composting	61.6	9	237
	Aerobic-composting	65.3	5	57
20	Vermi-composting	19.7	10	95
	Aerobic-composting	24.7	5	75
55	Vermi-composting	19.7	4	71
	Aerobic-composting	21.1	4	56

Table 4 Chemical analysis results for mature vermi- and aerobic-compost at 60-day stage of decomposition

S. no.	Compost type	pH	Total C (mg g^{-1})	C:N ratio	Essential plant nutrients (mg kg^{-1})										
					N	P	K	Ca	Mg	S	Zn	Cu	Fe	Mn	B
1	Vermi-compost	6.73	130	11.7	11,084	4275	9576	31,543	5997	1742	88	17.9	7525	395	91
2	Aerobic-compost	6.75	160	10.3	15,519	9323	9982	46,543	4622	1925	101	14.1	4860	305	52

Table 5 Chemical analysis results for compost bio-washes

S. no.	Bio-wash type	Essential plant nutrients (mg kg^{-1})									
		N	P	K	Ca	Mg	S	Zn	Cu	Fe	Mn
1	Vermi-compost bio-wash	227	1	2146	323	117	59	3	1	17	2
2	Aerobic-compost bio-wash	412	1	1988	319	121	72	3	1	12	2

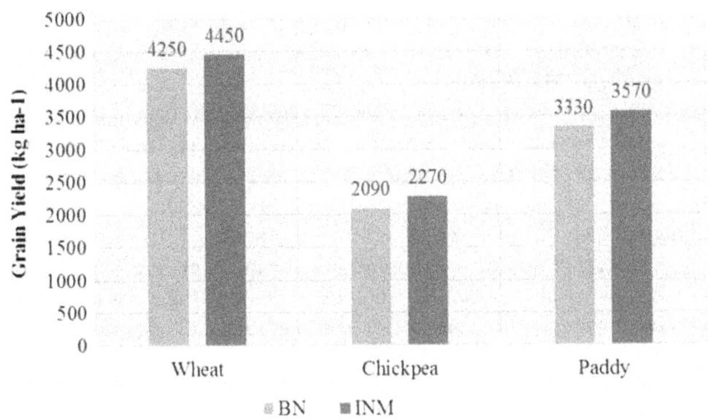

Fig. 4 Effects of soil test-based balanced (BN) and integrated (INM) nutrient management in India—vermi-compost in wheat and chickpea crops in Sagar district, aerobic-compost in paddy in Kadapa district (LSD 5% = 162 in wheat, 37.2 in chickpea and 29.9 in paddy)

these washings may serve as good nutrient source for foliar application to manage nutrient deficiencies in any standing crop.

In standardizing hormones identification method using 25% sulphuric acid spray, a light pink color spot (0.75 R_f value) was observed for auxins, greenish yellow for gibberellins ($R_f = 0.59$) and light brown for cytokinins ($R_f = 0.82$). However, as regards the plant growth hormones in bio-washes, only auxin was detected in vermi-wash, while none of other hormones were detected in any of the bio-washes.

On-farm evaluation of composts

The on-farm evaluation of vermi-compost in wheat and chickpea crops, and aerobic-compost in paddy crop showed that with use of composts, the crop yields are not only maintained or increased over the balanced nutrient management solely through chemical fertilizers (Fig. 4), but also result in savings of chemical fertilizers up to 25%. The integrated use of vermi-compost along with chemical fertilizers recorded higher yield by 5% in wheat (4450 vs 4250 kg ha^{-1}) and 9% in chickpea (2270 vs 2090 kg ha^{-1}). Similarly, the use of aerobic-compost along with chemical fertilizers recorded higher paddy yield by about 7% (3570 vs 3330 kg ha^{-1}) compared to sole use of balanced chemical fertilizers. Other studies have shown similar benefits with INM along with improvement in food nutritional quality through producing nutrient dense food and resilience-building of production systems (Chander et al. 2013; Mengistu et al. 2017).

Thus, aerobic-composting is a sound scalable technology as is vermi-composting for recycling on-farm wastes and suitability of any one is more dependent on farm situations. While continuously feeding the biomass and maintaining moisture is required in vermi-composting to maintain earthworms, which is not the case in aerobic-composting, but may involve recurring cost on the microbial culture and labor for turning the biomass to maintain aerobic conditions.

Conclusions

Aerobic- and vermi-composting proved effective technologies to recycle on-farm wastes into valuable manure. Aerobic-compost matured in around 50 days and vermi-compost in around 60 days. Thus, aerobic-composting may be a good option to handle large quantities of on-farm wastes as and when required without having to constantly maintain ambient living conditions as required for earthworms in vermi-composting. But, it needs more labor or energy for minimum four turnings required during the composting process. Both aerobic- and vermi-composting supported a diverse microbial population of bacteria, fungi and actinomycetes. Growth hormone auxin was detected in vermi-compost bio-wash. Also, both the composts produced had a good nutrient value and high stability. Thus, as per farmers' requirements, both aerobic- and vermi-composting methods have a potential to be used as effective technologies to recycle the plant nutrients in on-farm wastes and help reduce the load of inorganic fertilizers. Its scaling-up will boost the economy and help in maintaining the sustainability of ecosystem.

Acknowledgements Funding support in on-farm scaling-out/evaluation from Government of Andhra Pradesh and Sir Dorabji Tata Trust is gratefully acknowledged. Authors acknowledge help from Mr K Srinivas in assisting to automatically record composting biomass temperatures; and Mr G Pardhasaradhi in chemical analysis of samples. Help from Mr LS Jangawad and Mr Krishna Reddy in facilitating to conduct this study and Dr A Rathore in statistically analyzing yield data is duly acknowledged.

References

Aira M, Monroy F, Dominguez J (2007) Earthworms strongly modify microbial biomass and activity triggering enzymatic activities during vermicomposting independently of the application rates of pig slurry. Sci Total Environ 385:252–261. https://doi.org/10.1016/j.scitotenv.2007.06.031

Altschul SF, Gish W, Miller W, Myers EW, Lipman DJ (1990) Basic local alignment search tool. J Mol Biol 215:403–410. https://doi.org/10.1016/S0022-2836(05)80360-2

Ariffin H, Abdullah N, Umikalsom MS, Shirai Y, Hassan MA (2008) Production of bacterial endoglucanase from pretreated oil palm empty fruit bunch by *Bacillus pumilus* EB3. J Biosci Bioeng 106:231–236. https://doi.org/10.1263/jbb.106.231

Batjes NH (1999) Management options for reducing CO_2 concentrations in the atmosphere by increasing carbon sequestration in the soil. ISRIC, Wageningen, The Netherlands. https://www.researchgate.net/profile/Niels_Batjes/publication/228068233_Management_Options_for_Reducing_CO2_Concentrations_in_the_Atmosphere_by_Increasing_Carbon_Sequestration_in_the_Soil/links/0912f50c0a50b2b348000000/Management-Options-for-Reducing-CO2-Concentrations-in-the-Atmosphere-by-Increasing-Carbon-Sequestration-in-the-Soil.pdf

Bazzicalupo M, Fani R (1995) The use of RAPD for generating specific DNA probes for microorganisms. In: Clapp JP (ed) Methods in molecular biology. Humana, Totowa, pp 155–175. https://doi.org/10.1385/0-89603-323-6:155

Central Pollution Control Board, Delhi (2017) Status report on municipal solid waste management. http://www.cpcb.nic.in/divisionsofheadoffice/pcp/MSW_Report.pdf. Accessed Nov 2017

Chander G, Wani SP, Sahrawat KL, Kamdi PJ, Pal CK, Pal DK, Mathur TP (2013) Balanced and integrated nutrient management for enhanced and economic food production: case study from rainfed semi-arid tropics in India. Arch Agron Soil Sci 59(12):1643–1658. https://doi.org/10.1080/03650340.2012.761336

Chander G, Wani SP, Sahrawat KL, Pardhasaradhi G (2015) Soil test based nutrient management for sustainable intensification and food security: case from Indian semi-arid tropics. Commun Soil Sci Pl Anal 46(S1):20–33. https://doi.org/10.1080/00103624.2014.988087

El-Tarabily KA (2008) Promotion of tomato (*Lycopersicon esculentum* Mill.) plant growth by rhizosphere competent 1-aminocyclopropane-1-carboxylic acid deaminase–producing *Streptomycete actinomycetes*. Plant Soil 308:161–174. https://doi.org/10.1007/s11104-008-9616-2

Ghosh S, Penterman JN, Little RD, Chavez R, Glick BR (2003) Three newly isolated plant growth promoting bacilli facilitate the seedling growth of canola, *Brassica campestris*. Plant Physiol Biochem 41:277–281. https://doi.org/10.1016/S0981-9428(03)00019-6

Gopalakrishnan S, Pande S, Sharma M, Humayun P, Kiran BK, Sandeep D, Vidya MS, Deepthi K, Rupela O (2011) Evaluation of actinomycete isolates obtained from herbal vermicompost for biological control of *Fusarium* wilt of chickpea. Crop Prot 30:1070–1078. https://doi.org/10.1016/j.cropro.2011.03.006

Gopalakrishnan S, Vadlamudi S, Bandikinda P, Sathya A, Vijayabharathi R, Rupela O, Kudapa H, Katta K, Varshney RK (2014) Evaluation of *Streptomyces* strains isolated from herbal vermicompost for their plant growth-promotion traits in rice. Microbiol Res 169:40–48. https://doi.org/10.1016/j.micres.2013.09.008

Gopalakrishnan S, Srinivas V, Alekhya G, Prakash B, Kudapa H, Rathore A, Varshney RK (2015) The extent of grain yield and plant growth enhancement by plant growth-promoting broad-spectrum *Streptomyces* sp. in chickpea. SpringerPlus 4:31. https://doi.org/10.1186/s40064-015-0811-3

Gopalakrishnan S, Srinivas V, Srinivasan S, Sameer Kumar CV (2016) Plant growth-promotion and biofortification of chickpea and pigeonpea through inoculation of biocontrol potential bacteria, isolated from organic soils. SpringerPlus 5:1882. https://doi.org/10.1186/s40064-016-3590-6

Hendricks CW, Doyle JD, Hugley B (1995) A new solid medium for enumerating cellulose-utilizing bacteria in soil. Appl Environ Microbiol 61:2016–2019. https://www.ncbi.nlm.nih.gov/pmc/articles/PMC1388450/

Huang K, Li F, Wei Y, Fu X, Chen X (2014) Effects of earthworms on physicochemical properties and microbial profiles during vermicomposting of fresh fruit and vegetable wastes. Bioresour Technol 170:45–52. https://doi.org/10.1016/j.biortech.2014.07.058

Kharrazi SM, Younesi H, Abedini-Torghabeh J (2014) Microbial biodegradation of waste materials for nutrients enrichment and heavy metals removal: an integrated composting-vermicomposting process. Int Biodeterior Biodegrad 92:41–48. https://doi.org/10.1016/j.ibiod.2014.04.011

Lal R (2011) Sequestring carbon in soils of agro-ecosystems. Food Policy 36:S33–S39. https://doi.org/10.1016/j.foodpol.2010.12.001

Lal R, Kimble JM (1997) Conservation tillage for carbon sequestration. Nutr Cycl Agroecosyst 49:243–253. https://doi.org/10.1023/A:1009794514742

Lal R, Follett RF, Stewart BA, Kimble JM (2007) Soil carbon sequestration to mitigate climate change and advance food security. Soil Sci 172(12):943–956. https://doi.org/10.1097/ss.0b013e31815cc498

Mengistu T, Gebrekidan H, Kibret K, Woldetsadik K, Shimelis B, Yadav H (2017) The integrated use of excreta-based vermicompost and inorganic NP fertilizer on tomato (*Solanum lycopersicum* L.) fruit yield, quality and soil fertility. Int J Recycl Org Waste Agric 6(1):63–77. https://doi.org/10.1007/s40093-017-0153-y

Miezah K, Ofosu-Anim J, Budu GKO, Enu-Kwesi L, Cofie O (2008) Isolation and identification of some plant growth promoting substances in compost and co-compost. Int J Virol 4(2):30–40. https://doi.org/10.3923/ijv.2008.30.40

Mills HA, Jones JB Jr (1996) Plant analysis handbook II: a practical sampling, preparation, analysis and interpretation guide. Micro-Macro Publishing, Athens, GA. https://books.google.co.in/books/about/Plant_Analysis_Handbook_II.html?id=AzorngEACAAJ&redir_esc=y

Nagavallemma KP, Wani SP, Stephane Lacroix, Padmaja VV, Vineela C, Babu Rao M, Sahrawat KL (2006) Vermicomposting: recycling wastes into valuable organic fertiliser. SAT eJ 2(1):1–16. http://ejournal.icrisat.org/agroecosystem/v2i1/v2i1vermi.pdf

Pandey P, Kang SC, Maheshwari DK (2005). Isolation of endophytic plant growth promoting Burkholdeina sp. MSSP from root nodules of *Mimosa pudica*. Curr Sci 89:177–180. http://www.iisc.ernet.in/currsci/jul102005/177.pdf

Pappu A, Saxena M, Asolekar SR (2007) Solid wastes generation in India and their recycling potential in building materials. Build Environ 42(6):2311–2320. https://doi.org/10.1016/j.buildenv.2006.04.015

Sahrawat KL, Kumar Ravi G, Rao JK (2002a) Evaluation of triacid and dry ashing procedures for determining potassium, calcium, magnesium, iron, zinc, manganese, and copper in plant materials. Commun Soil Sci Plant Anal 33:95–102. https://doi.org/10.1081/CSS-120002380

Sahrawat KL, Kumar Ravi G, Murthy KVS (2002b) Sulfuric acid-selenium digestion for multi-element analysis in a single digest. Commun Soil Sci Plant Anal 33:3757–3765. https://doi.org/10.1081/CSS-120015920

Saitou N, Nei M (1987) The neighbor-joining method: a new method for reconstructing phylogenetic trees. Mol Biol Evol 4:406–425. https://doi.org/10.1093/oxfordjournals.molbev.a040454

Shah RU, Abid M, Qayyum MF, Ullah R (2015) Dynamics of chemical changes through production of various composts/vermicompost such as farm manure and sugar industry wastes. Int J Recycl Org Waste Agric 4(1):39–51. https://doi.org/10.1007/s40093-015-0083-5

Sharma A, Shankhdhar D, Shankhdhar SC (2013) Enhancing grain iron content of rice by the application of plant growth-promoting rhizobacteria. Plant Soil Environ 59:89–94. http://www.agriculturejournals.cz/publicFiles/82693.pdf

Singh N, Pandey P, Dubey RC, Maheshwari DK (2008) Biological control of root rot fungus *Macrophomina phaseolina* and growth enhancement of *Pinus roxburghii* by rhizosphere competent *Bacillus subtilis* BN1. World J Microbiol Biotechnol 24:1669–1679. https://doi.org/10.1007/s11274-008-9680-z

Thompson JD, Gibson TJ, Plewniak F, Jeanmougin F, Higgins DG (1997) The CLUSTAL_X windows interface: flexible strategies for multiple sequence alignment aided by quality analysis tools. Nucleic Acids Res 25:4876–4882. https://doi.org/10.1093/nar/25.24.4876

Tomati U, Grappelli A, Galli E (1988) The hormone-like effect of earthworm casts on plant growth. Biol Fertil Soils 5(4):288–294. https://doi.org/10.1007/BF00262133

Vivas A, Moreno B, Garcia-Rodriguez S, Benitez E (2009) Assessing the impact of composting and vermicomposting on bacterial community size and structure, and microbial functional diversity of an olive-mill waste. Bioresour Technol 100(3):1319–1326. https://doi.org/10.1016/j.biortech.2008.08.014

Wani SP, Pathak P, Jangawad LS, Eswaran H, Singh P (2003) Improved management of Vertisols in the semi-arid tropics for increased productivity and soil carbon sequestration. Soil Use Manag 19:217–222. https://doi.org/10.1111/j.1475-2743.2003.tb00307.x

Wani SP, Chander G, Vineela C (2014) Vermicomposting: recycling wastes into valuable manure for sustained crop intensification in the semi-arid tropics. In: Ramesh, Chandra, Raverkar KP (ed) Bioresources for sustainable plant nutrient management. Satish Serial Publishing House, Delhi, India, pp 123–151. https://www.researchgate.net/publication/270528480_Vermicomposting_Recycling_Wastes_into_Valuable_Manure_for_Sustained_Crop_Intensification_in_the_Semi-Arid_Tropics. http://hdl.handle.net/10568/75837

Yousefi J, Younesi H, Ghasempoury SM (2013) Co-composting of municipal solid waste with sawdust: improving compost quality. Clean Soil Air Water 41(2):185–194. https://doi.org/10.1002/clen.201100315

Zhang BG, Li GT, Shen TS, Wang JK, Sun Z (2000) Changes in microbial biomass C, N, and P and enzyme activities in soil incubated with the earthworms *Metaphire guillelmi* or *Eisenia foetida*. Soil Biol Biochem 32:2055–2062. https://doi.org/10.1016/S0038-0717(00)00111-5

4

Germination of several wheat cultivars in desert soil after amendment with raw and digested poultry manure with and without combination with mineral fertilizer

Hassan El-Zeadani[1] · Jamal Abubaker[2] · Mohemed Essalem[1] · Alsanousi Alghali[1]

Abstract

Purpose This study aimed to investigate the effect of time interval between sowing and amendment with non-digested and digested poultry manure and their combinations with urea on the germination of three wheat cultivars (*Slambo*, *Acsad*, and *Karim*) in the desert soil.

Methods The cultivars were sown on four different dates, i.e. directly, 10, 20 and 30 days after amendment. Fertilizers were applied at a rate of 350 kg Tot N ha^{-1}. The germinated seeds were recorded daily starting at day 4 after sowing until day 14 after which the following germination attributes were calculated: time to reach 50% germination ($T_{50\%}$), final germination percentage (FGP) and germination index (GI). Additionally, radicle length, plumule length and seedling dry weight were also determined.

Results Sowing near to the application delayed and reduced seeds germination particularly in the treatments amended with non-digested and digested manure in combination urea. This was confirmed by all germination attributes, i.e. high values of $T_{50\%}$ and low values of FGP and GI. Moreover, sowing near to the application reduced the length of plumule and radicle as well as decreased seedling dry weight of all wheat cultivars.

Conclusion Results suggested that prolonging the period between amendments and sowing to 20 or 30 days would give better germination attributes and increase plumule and radicle length plus increasing seedling dry weight.

Keywords Seed germination · Digestate · Poultry manure · Urea · Wheat cultivars

Introduction

Desert soil is by far the predominant soil in the southern part of Libya, where grain crops such as barley and wheat are cultivated under sprinkler irrigation. Wheat (*Triticum aestivum* L.) occupies the first rank in Libya in terms of cultivated area, i.e. about 165,000 ha with an annual yield of 200,000 tons (Alabasi et al. 2017). In this region, wheat yield is far below the optimal due to the fertility of desert soils is low to provide the crop with the essential plant nutrients. Therefore, the application of fertilizers to this soil is necessary to improve its productivity. Urea is the most widely used chemical nitrogenous fertilizer in Libya, although several studies have shown that it is not good for improving soil properties (Angus et al. 2014; Lungu and Dynoodt 2008). Therefore, replacing chemical fertilizers with organic fertilizers or combine it with organic fertilizers is required to improve soil fertility. Numerous studies have reported that the application of organic fertilizers to soils of arid and semiarid regions enhances significantly soil fertility through improving soil properties compare to application of mineral fertilizers (Diacono and Montemurro 2010; Elhadi et al. 2016; Elnesairy et al. 2016; Hati et al. 2006). Organic fertilizers such as animal manures have a considerable amount of nutrients in organic form and hence is not immediately

✉ Jamal Abubaker
jamal_abubaker@hotmail.com

Hassan El-Zeadani
has64b@gmail.com

Mohemed Essalem
alzain2010@yahoo.com

Alsanousi Alghali
sanoealmosleh@gmail.com

1 Department of Crops, Faculty of Agriculture, Sabha University, P. O. Box 18758, Sabha, Libya

2 Department of Soil and Water, Faculty of Agriculture, Sabha University, P. O. Box 18758, Sabha, Libya

available to the plant until decomposition by heterotrophic soil microorganisms occurs (Ladd and Jackson 1982). In fact, due to the dryness and the absence of soil organic matter, desert soils have a low microbial activity to breakdown organic components and release plant nutrients (Angel et al. 2013; Makhalanyane et al. 2015). It has been reported that microbial communities of the desert soil have higher relative abundances of genes associated with dormancy and lower abundances associated with the nutrient release (Fierer et al. 2012). Therefore, anaerobic digestion of animal manure before application to the soil might be an effective way to increase nutrient availability to the plant. Several studies have shown that fertilizing soils with digested manure results in higher yields compare to compost and mineral fertilizers (Abubaker et al. 2012, 2017; Abubaker 2012). Anaerobic digestion is used worldwide to sustain waste management and produce bioenergy (Angelidaki and Ahring 2000). Moreover, anaerobic digestion leads to the generation of residues called digestate that rich in mineralized plant nutrients (Makádi et al. 2012; Möller and Müller 2012; Odlare et al. 2011). Although, digested and non-digested organic waste has a positive effect on crop yield, however, some other studies have reported an adverse effect of these materials on wheat seeds germination. Abubaker et al. (2017) reported that the application of digestate and non-digested to desert soil, especially in the combination with urea delayed and reduced the germination of wheat seeds. In another study by Bacilio et al. (2003) showed that the composts generated from dairy cow manure have inhibitory effects on wheat seed germination. This negative effect has been attributed to the contaminants that generated during decomposition of organic waste such as ethylene oxide (Wong et al. 1983) and organic acid (Bacilio et al. 2003; Makádi et al. 2012). Moreover, the concentration of ammonia in digestate is high (García-González et al. 2016) and this can reduce and delay wheat seeds germination (Bremner and Krogmeier 1989; Wan et al. 2016). In fact, it is unclear if this adverse effect will last for a short or long period and thus can be avoided by adjusting the cultivation strategy, e.g. prolonging the period between application and sowing. The objective of present study was to investigate the influence of time interval between sowing and application of non-digested and anaerobically digested poultry manure with and without combination with urea on the germination of three wheat cultivars in desert soil.

Materials and methods

Soil sampling and description

Sandy topsoil (0–20 cm) was collected from desert area that has not been cultivated or fertilized before, located at Sabha city, a southern part of Libya (22°30′N, 10°E). After sampling, the soil transported to the laboratory where it was cleaned, sieved (4-mm screen) and stored at lab temperature (20 ± 3 °C). Physical and chemical characteristics of the soil are shown in Table 1. Percentage of sand, silt, and clay was determined according to Bouyoucos (1962). The water-holding capacity of the soil was measured according to Forster (1995). Soil pH was measured in the suspension at a soil-to-deionized water ratio of 1:2 using a pH meter 3030 (Jenway, Ltd., UK). Electrical conductivity (ECe) was determined in a 1:1 (v/v) water-to-soil suspension using the conductivity meter (model 4070, ELE, England). Soil organic matter content (OM) was measured according to Ball (1964). Total soil nitrogen (Tot N) was determined using Kjeldahl technique modified by Bradstreet (1954). Total soil carbon (Tot C) was analysed by a loss-on-ignition method described by Dean (1974). Soil phosphorus (P) was extracted according to the method described by Chapman and Pratt (1961) and determined using Cecil CE 202 spectrophotometer at 420 nm (Super Aquarius, Cecil Instruments, Cambridge, England). Potassium (K) and sodium (Na) were measured using a flame photometer (Jenway, PFP7, UK) according to Protocol: P05-001A and Protocol: P05-002A, respectively. Soil magnesium (Mg) and calcium (Ca) were determined by atomic absorption spectrometer (AAS, Analytik Jena AG 400) according to the methods of Ramakrishna et al. (1966).

Non-digested, digested manure and urea

Approximately 30 kg of poultry manure was collected from broiler farm located in Sabha city, a southern part of Libya. In this farm, wood shavings were used as bedding and grinded grain (i.e. wheat, barley, corn, and soybeans)

Table 1 Physical and chemical properties of the soil used in the germination experiment

Parameters	Values
Physical properties	
Sand (%)	97
Clay (%)	1.3
Silt (%)	1.7
Water-holding capacity (%)	21.5
Chemical properties	
pH_{paste}	8.2
ECe (ds m^{-1})	2.8
OM (%)	0.4
Tot N (%)	0.09
Tot C (%)	0.89
P (g kg^{-1} dw)	0.07
K (g kg^{-1} dw)	0.08
Na (g kg^{-1} dw)	0.25
Mg (g kg^{-1} dw)	0.05
Ca (g kg^{-1} dw)	0.2

as nutriment. The collected manure transported to the laboratory where it was cleaned, portioned and stored at room temperature (20 ± 3 °C) until use. The manure was solid and almost dry contains only 2% of moisture. The digestion of manure was performed anaerobically in a sealable plastic container (30 L volume) provided with a regulator in the lid letting the flow of air from the container and prevents back-flow into it. The digestion started by adding a manure to deionized water at a ratio of 6:10 after which the container sealed. The anaerobic digestion was operated at the mesophilic temperature (42–48 °C) for 30 days retention time, after which the container opened and digested manure portioned into small plastic bags, and stored at − 18 ± 3 °C until use. The physical and chemical characteristics of fertilizers are shown in Table 2. Dry matter (DM) was determined using oven drying at 100 °C for 24 h as described by Windham et al. (1987). The pH was determined at a manure-to-deionized water ratio of 1:6 using a pH meter 3030 (Jenway, Ltd., UK). Tot N was measured by the Kjeldahl method according to Bradstreet (1954). Tot C was measured by a loss-on-ignition method according to Dean (1974). Tot P was extracted by HNO_3 according to Pungor and Horval (1994) and determined using Cecil CE 202 spectrophotometer at 420 nm (Super Aquarius, Cecil Instruments, Cambridge, England). Tot K was extracted by HCl (Protocol: P05-004A) and analysed by the flame photometer (Jenway, PFP7, UK). The mineral fertilizer used in the experiment was urea ($CO(NH_2)_2$) composed of 20% C, 26.6% O, 46.6% N, 6.7% H.

Treatments

The germination experiment was performed in a plastic dishes (7.5 cm diam. × 5 cm height) at the research farm of the faculty of agriculture, Sabha University, Libya. The experiment consisted of five fertilizers, three wheat cultivars (*Slambo*, *Acsad*, and *Karim*) and four sowing dates. Six treatments were used in the experiment showing in Table 3. The experiment was set up in a random block design with four replicates of each treatment, resulting in a total of 252 dishes. The fertilizers were applied at a rate corresponding to 350 kg Tot N ha^{-1}. At the start, 200 g of soil were weighted into each dish and amended with target fertilizer, after that sprayed with 15 ml of distilled water. The combination with mineral fertilizers was applied so that a 50% of the applied N came from the manure and 50% came from the urea. The dishes were irrigated every other day.

Sowing and germination attributes

Four sets of dishes were prepared, each set contains 60 dishes in total 240 dishes (fertilized) + 12 unfertilized dishes were used as control (control was run only at the first sowing). The first set was sowed directly after amendment, the second set sowed 10 days after amendment, the third set sowed 20 days after amendment and the fourth set sowed 30 days after amendment. Each dish was sowed with 15 seeds at the depth of 1 cm (broken and small seeds were avoided), i.e. corresponding to a sown rate of 170 kg seeds ha^{-1}. Before the start of the experiment, all wheat cultivars were tested for their vitalities according to Ellis et al. (1985), which were 99 ± 4, 89 ± 6 and 95 ± 3 for *Slambo*, *Karim* and *Acsad*, respectively. After sowing, all dishes were placed on trolleys under controlled condition (13 h day, 11 h night, temperature 20 ± 2 °C and air humidity 34%). The sowed dishes were checked daily and germinated seeds were counted at day 4, 6, 8, 9, 12 and 14 from sowing. From the germination counts, several germination attributes were calculated including time to reach 50% germination ($T_{50\%}$), final germination percentage (FGP) and germination index (GI). $T_{50\%}$ was calculated according to the formula of Coolbear et al. (1984) modified by Farooq et al. (2005). FGP and GI were calculated according to the formulas described by Dastanpoor et al. (2013). At the end of the experiment (day 14), seedlings were carefully removed from the soil by placing them in flat pans of water where radicles could be freed from soil particles with little injury. After that, the length of plumule and radicle of each seedling was measured. In the

Table 2 Physical and chemical characteristics of non-digested and digested poultry manure used in germination experiment

Parameters	Non-digested manure	Digested manure
DM (%)	98	15.2
pH	9.3	9.7
Tot N (g kg^{-1} dw)	42.6	46.2
Tot C (g kg^{-1} dw)	470	395
C/N	11	9
Tot P (g kg^{-1} dw)	14.2	13.9
Tot K (g kg^{-1} dw)	3.9	18

Table 3 Treatments used in the germination experiment

No.	Abbreviations	Treatments
1	C	No fertilization
2	U	Urea
3	PM	Non-digested poultry manure
4	PM + U	Non-digested poultry manure (50% N) + urea (50% N)
5	DPM	Digested poultry manure
6	DPM + U	Digested poultry manure (50% N) + urea (50% N)

end, seedling dry weight in each treatment was determined after drying in the oven at 75 °C for 24 h.

Statistical analyses

The obtained data were analysed using the SPSS (WIN. Version 17) procedure GLM multiple variables, where two-way ANOVA followed by Tukey (HSD) multiple comparison tests were used for repeated testing of paired differences between treatments regarding $T_{50\%}$, FGP, GI, plumule length, radicle length and seedling dry weight, where time of sowing, fertilizer type, wheat cultivars and their interactions were considered as fixed factors. The dataset was unbalanced as it contained only one set of control running at first sowing. Differences considered significant at $p<0.05$ unless otherwise not stated.

Results

Germination attributes

Application of non-digested and digested poultry manure especially in combination with U had delayed and reduced the germination of the three wheat cultivars considerably in the first and second sowing. However, in the third and fourth sowing, the germination enhanced markedly. This adverse effect was confirmed by all germination attributes i.e. high values of $T_{50\%}$ and low values of FGP and GI. The analysis of variance showed that the germination of wheat seeds was mostly affected by fertilizer type and sowing date rather than wheat cultivars. Statistical analysis showed significant ($p<0.05$) effect of sowing date and fertilizer type on $T_{50\%}$, FGP and GI. Whereas the effect of wheat cultivars was only significant ($p<0.05$) on FGP but not on the other germination attributes. Moreover, the analysis of variance showed significant ($p<0.05$) interaction effect of sowing date × fertilizer type and fertilizer type × wheat cultivars on all germination attributes. Whereas, the interaction effect of sowing date × fertilizer type × wheat cultivars was only significant ($p<0.05$) on FGP but not on the other germination attributes.

Time to reach 50% germination

$T_{50\%}$ represent estimated time for cumulative germination to reach 50% of its maximum (Table 4). In treatments amended with U, $T_{50\%}$ was significantly ($p<0.05$) shorter in the fourth sowing ranged from 3.8 ± 1.2 to 4.7 ± 1.4 for all wheat cultivars compared to other sowing dates where it ranged from 4.4 ± 0.8 to 7.6 ± 2.0 (Table 4). Treatments amended with PM + U, DPM and DPM + U significantly ($p<0.05$) reduced $T_{50\%}$ (ranged from 2.5 ± 0.9 to 3.6 ± 1.0) on the fourth sowing for all wheat cultivars compared to first, second and third sowing (ranged from 3.3 ± 0.6 to 10.3 ± 3.8). Moreover, at the first sowing PM treatments gave lower $T_{50\%}$ by all wheat cultivars ranged from 3 ± 0.0 to 3.7 ± 1.2 compared to treatments amended with U, PM + U, DPM and DPM + U (ranged from 3.8 ± 0.6 to 10.3 ± 3.8).

Final germination percentage

FGP reached its maximum in U treatments at the first and fourth sowing (ranging from 70 ± 10 to $93 \pm 6\%$) while in the second and third sowing was lower (ranging from 60 ± 0 to $77 \pm 6\%$) (Table 5). Treatments amended with PM and DPM showed higher FGP in the first sowing compared to treatments amended with PM + U and DPM + U with exception of that *Karim* cultivar had lower FGP in PM treatment (Table 5).

Moreover, FGP for *Slambo* cultivar was significantly ($p<0.05$) low in the treatment amended with DPM + U compared to treatment amended with PM + U in the first sowing. At the second, third and fourth sowing FGP increased significantly ($p<0.05$) in both PM + U and DPM + U treatments without any significant differences between them. Moreover, in the treatments sowed with *Acsad* and *Karim* cultivars, it seems PM + U had lowered their FGP significantly ($p<0.05$) compared to the application of DPM + U in the first sowing. Overall, FGP had increased significantly ($p<0.05$) in the fourth sowing compared to the other sowing dates, where they have reached their maximum (100%) especially for *Slambo* and *Acsad* cultivars.

Germination index

GI represent germination speed, low values mean slower germination and higher values mean faster germination. GI has increased significantly ($p<0.05$) in U treatments in the fourth sowing date compared to the other sowing dates by all wheat cultivars with exception of *Karim* cultivar where no significant difference was detected between the first and fourth sowing (Table 6). In the third and fourth sowing, GI increased significantly ($p<0.05$) (ranged from 7.5 ± 1.4 to 19.4 ± 1.9) in most treatments amended with PM, PM + U, DPM and DPM + U (Table 6). Furthermore, in the first and second sowing GI had decreased in the treatments amended with PM + U and DPM + U compared with treatments amended with PM and DPM by all wheat cultivars. Moreover, the results demonstrated that *Slambo* cultivar had higher GI compared to other wheat cultivars in most treatments and at most sowing dates.

Table 4 The time to reach 50% germination ($T_{50\%}$) of three wheat cultivars (*Slambo*, *Acsad* and *Karim*) after 14 days growing period sown in desert soil on four sowing dates after amendment with non-digested and anaerobically digested poultry manure with and without combination with urea

Fertilizer type	Wheat cultivars	$T_{50\%}$ (days)			
		First sowing	Second sowing	Third sowing	Fourth sowing
C	*Slambo*	5.7 ± 1.3	–	–	–
C	*Acsad*	5.6 ± 0.6	–	–	–
C	*Karim*	3.4 ± 0.6	–	–	–
Urea	*Slambo*	7.6 ± 2.0	5.0 ± 1.1	4.4 ± 0.8	3.8 ± 1.2
Urea	*Acsad*	5.9 ± 1.4	5.5 ± 1.9	5.0 ± 1.4	3.8 ± 1.4
Urea	*Karim*	6.4 ± 1.1	5.6 ± 3.6	5.4 ± 1.2	4.7 ± 1.4
PM	*Slambo*	3.0 ± 0.0	4.3 ± 2.2	6.6 ± 1.6	3.8 ± 1.4
PM	*Acsad*	3.4 ± 0.8	5.3 ± 2.0	4.6 ± 2.7	4.4 ± 1.3
PM	*Karim*	3.7 ± 1.2	4.5 ± 0.9	4.0 ± 0.0	4.8 ± 1.6
PM + U	*Slambo*	9.1 ± 0.9	3.3 ± 0.6	4.3 ± 0.7	3.0 ± 0.0
PM + U	*Acsad*	10.3 ± 3.8	4.3 ± 2.2	4.4 ± 1.2	3.6 ± 1.0
PM + U	*Karim*	9.2 ± 3.3	6.5 ± 0.5	4.2 ± 1.0	3.0 ± 0.0
DPM	*Slambo*	4.0 ± 0.0	4.1 ± 1.0	4.0 ± 0.9	2.5 ± 0.9
DPM	*Acsad*	3.8 ± 0.6	4.0 ± 0.9	4.3 ± 1.3	3.0 ± 0.0
DPM	*Karim*	5.0 ± 0.9	3.8 ± 1.4	3.6 ± 0.8	3.4 ± 0.8
DPM + U	*Slambo*	7.3 ± 1.2	7.0 ± 0.0	4.7 ± 0.4	3.0 ± 1.4
DPM + U	*Acsad*	7.9 ± 2.6	6.6 ± 2.2	4.2 ± 1.0	3.0 ± 0.0
DPM + U	*Karim*	5.1 ± 2.7	7.0 ± 1.0	4.2 ± 1.0	3.4 ± 1.4

Values represent the average ± standard deviation ($n = 4$)

– no control used in the second, third and fourth sowing

Table 5 Final germination percentage (FGP) of three wheat cultivars (*Slambo*, *Acsad* and *Karim*) after 14 days growing period sown in desert soil on four sowing dates after amendment with non-digested and anaerobically digested poultry manure with and without combination with urea

Fertilizer type	Wheat cultivars	FGP (%)			
		First sowing	Second sowing	Third sowing	Fourth sowing
C	*Slambo*	88 ± 1.4	–	–	–
C	*Acsad*	93 ± 2.5	–	–	–
C	*Karim*	88 ± 6.3	–	–	–
Urea	*Slambo*	93 ± 6	63 ± 15	67 ± 6	93 ± 12
Urea	*Acsad*	70 ± 10	73 ± 6	60 ± 0	83 ± 6
Urea	*Karim*	90 ± 17	70 ± 17	77 ± 6	90 ± 10
PM	*Slambo*	93 ± 12	100 ± 0	93 ± 12	100 ± 0
PM	*Acsad*	100 ± 0	93 ± 12	100 ± 0	100 ± 0
PM	*Karim*	60 ± 0	87 ± 12	80 ± 0	87 ± 12
PM + U	*Slambo*	80 ± 20	93 ± 12	100 ± 0	100 ± 0
PM + U	*Acsad*	60 ± 0	73 ± 23	100 ± 0	100 ± 0
PM + U	*Karim*	47 ± 12	73 ± 12	100 ± 0	87 ± 12
DPM	*Slambo*	93 ± 12	93 ± 12	100 ± 0	100 ± 0
DPM	*Acsad*	100 ± 0	93 ± 12	100 ± 0	100 ± 0
DPM	*Karim*	93 ± 12	87 ± 12	100 ± 0	93 ± 12
DPM + U	*Slambo*	53 ± 23	93 ± 12	100 ± 0	100 ± 0
DPM + U	*Acsad*	73 ± 12	67 ± 31	100 ± 0	93 ± 12
DPM + U	*Karim*	80 ± 0	80 ± 0	87 ± 23	87 ± 23

Values represent the average ± standard deviation ($n = 4$)

– no control used in the second, third and fourth sowing

Table 6 Germination index (GI) of three wheat cultivars (*Slambo*, *Acsad* and *Karim*) after 14 days growing period sown in desert soil on four sowing dates after amendment with non-digested and anaerobically digested poultry manure with and without combination with urea

Fertilizer type	Wheat cultivars	GI (% day^{-1})			
		First sowing	Second sowing	Third sowing	Fourth sowing
C	*Slambo*	11 ± 0.3	–	–	–
C	*Acsad*	14 ± 0.5	–	–	–
C	*Karim*	15 ± 1.5	–	–	–
Urea	*Slambo*	13.9 ± 0.8	11.7 ± 2.7	8.4 ± 2.0	15.3 ± 3.1
Urea	*Acsad*	11.8 ± 1.2	12.5 ± 0.8	7.5 ± 1.4	13.0 ± 2.0
Urea	*Karim*	13.0 ± 4.3	10.8 ± 3.0	10.6 ± 2.1	14.7 ± 2.9
PM	*Slambo*	13.4 ± 2.8	16.7 ± 0.0	15.0 ± 1.7	15.8 ± 0.8
PM	*Acsad*	14.6 ± 3.6	12.9 ± 1.6	14.8 ± 2.5	16.7 ± 1.7
PM	*Karim*	11.7 ± 0.5	13.3 ± 1.4	12.5 ± 0.0	13.2 ± 1.0
PM + U	*Slambo*	7.9 ± 2.4	14.7 ± 1.9	19.4 ± 1.9	16.7 ± 0.0
PM + U	*Acsad*	4.7 ± 0.3	11.3 ± 4.0	17.8 ± 1.0	16.9 ± 1.3
PM + U	*Karim*	3.8 ± 0.6	10.1 ± 1.7	17.8 ± 1.0	14.2 ± 1.4
DPM	*Slambo*	15.0 ± 1.4	16.4 ± 1.7	18.3 ± 1.7	16.0 ± 0.6
DPM	*Acsad*	16.7 ± 0.0	17.2 ± 3.5	17.8 ± 1.9	16.4 ± 0.5
DPM	*Karim*	13.3 ± 1.4	13.5 ± 1.4	16.3 ± 1.8	16.7 ± 3.3
DPM + U	*Slambo*	5.7 ± 1.0	11.0 ± 0.9	16.7 ± 0.0	16.4 ± 1.3
DPM + U	*Acsad*	6.7 ± 0.0	7.6 ± 3.6	17.8 ± 1.0	15.6 ± 1.9
DPM + U	*Karim*	10.3 ± 1.0	10.6 ± 1.0	15.6 ± 4.8	17.5 ± 4.6

Values represent the average ± standard deviation (n = 4)

– no control used in the second, third and fourth sowing

Plumule and radicle length

Analysis of variance showed significant ($p < 0.05$) effect of sowing date and fertilizer type on plumule and radicle lengths. Whereas wheat cultivars have significant ($p < 0.05$) effect only on radicle length. Moreover, there was significant ($p < 0.05$) interaction effect of sowing date × fertilizer type and fertilizer type × wheat cultivars on the plumule and radicle length. Whereas, radicle length was affected significantly ($p < 0.05$) by the interaction of sowing date × wheat cultivars and sowing date × fertilizer type × wheat varieties, but not the plumule length. The results displayed remarkable differences in the plumule and radicle length of seedlings obtained from the control compared to the results found in the other treatments (Fig. 1). In U treatments, plumule and radicle length were reduced significantly ($p < 0.05$) at all sowing dates in comparison with control, with the exception of plumule length of *Slambo* cultivar on the fourth sowing where showed no significant differences between U treatment and the control (Fig. 1). Amendment with PM + U and DPM + U significantly ($p < 0.05$) reduced plumule and radicle length at the first sowing for all wheat cultivars in comparison with treatments of PM, DPM, U and control (Fig. 1). However, in the second, third and fourth sowing plumule and radicle length increased significantly ($p < 0.05$) in the treatments of PM + U and DPM + U in comparison with the first sowing, U treatments and control.

Seedlings dry weight

Analysis of variance displayed significant ($p < 0.05$) effect of sowing date, fertilizer type and wheat cultivars as well as their interaction on seedling dry weight. In the U treatments seedling dry weight was not improved significantly at the first, second and third sowing compared to the control with exception of *Karim*'s which was significantly ($p < 0.05$) higher (Fig. 2). Furthermore, amendment with PM, PM + U, DPM and DPM + U did not improve seedling dry weight significantly on the first and second sowing in comparison with the control and U treatments. However, in the third and fourth sowing seedling dry weight has increased significantly ($p < 0.05$) for most wheat cultivars (Fig. 2).

Discussion

Several studies have shown that application of non-digested and digested manures enhanced wheat growth and yield particularly at combination with U (Abubaker et al. 2017; Kiani et al. 2005). However, in the present study, the germination of wheat cultivars was delayed and reduced when sowing was made directly or after 10 days from amendments with non-digested and digested manures. The adverse effect was more pronounced in the treatments amended with non-digested and digested manures in combination with urea as well as sole urea. This observation

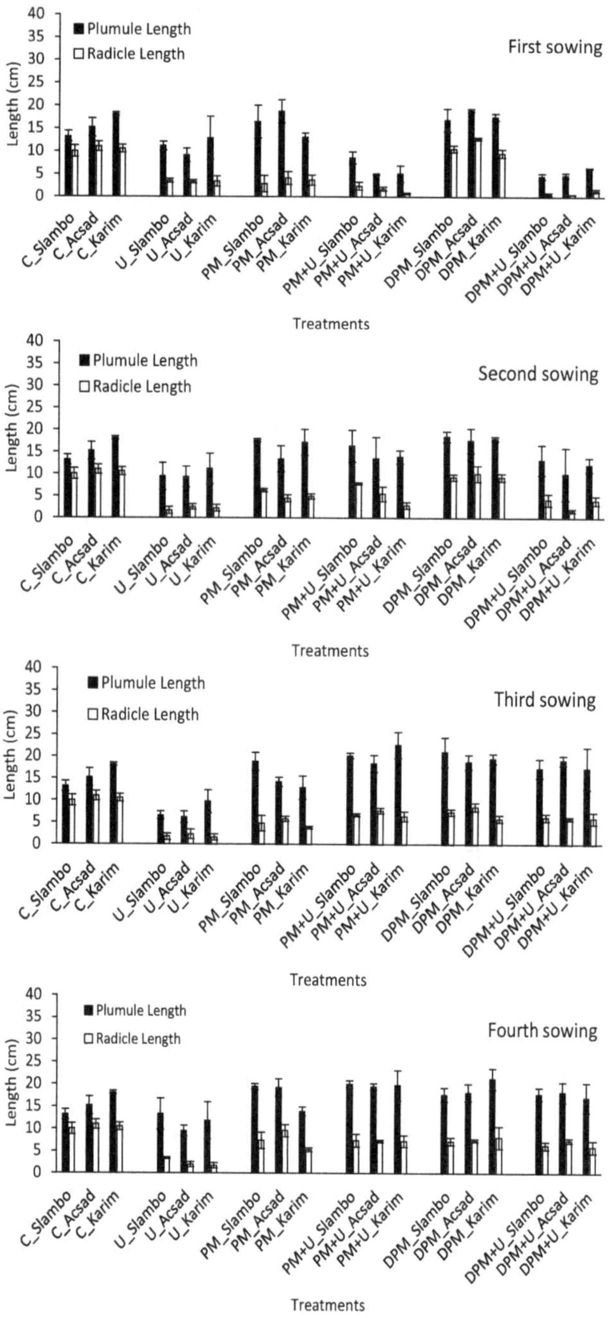

Fig. 1 Plumule and radicle length of three wheat cultivars (*Slambo*, *Acsad* and *Karim*) after 14 days growing period sown in desert soil on four sowing dates after amendment with non-digested and anaerobically digested poultry manure with and without combination with urea. Bars represent mean ± standard deviation ($n = 4$)

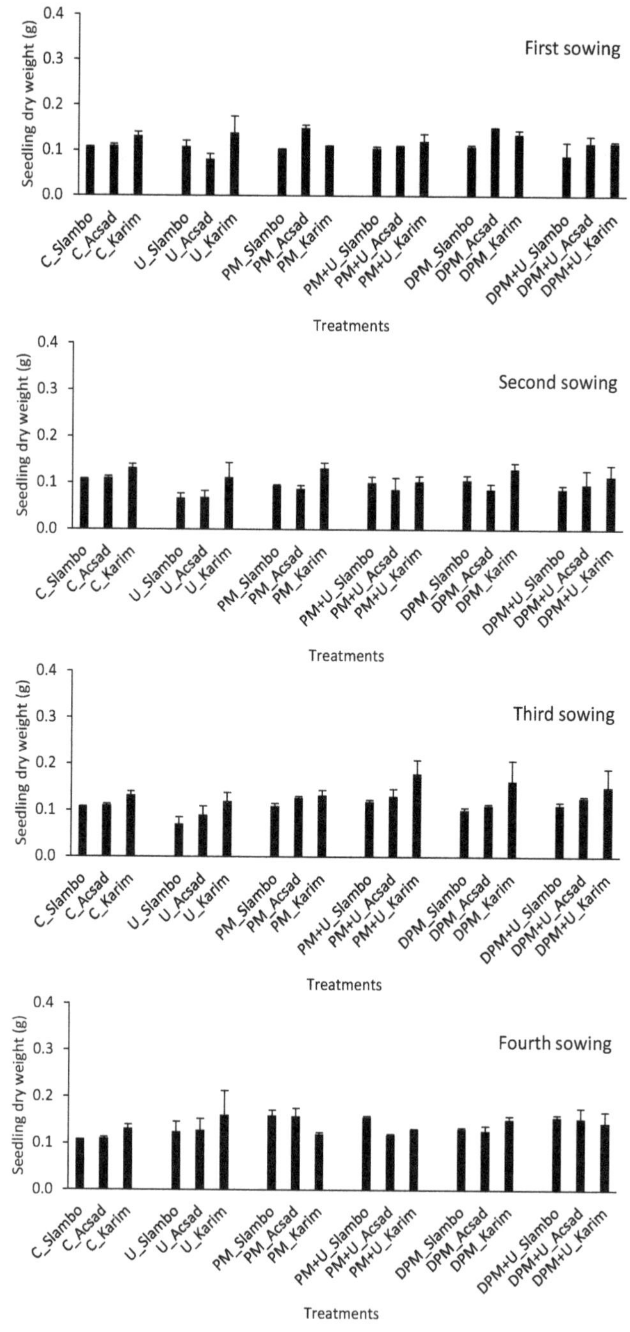

Fig. 2 Seedling dry weight of three wheat cultivars (*Slambo*, *Acsad* and *Karim*) after 14 days growing period sown in desert soil on four sowing dates after amendment with non-digested and anaerobically digested poultry manure with and without combination with urea. Bars represent mean ± standard deviation ($n = 4$)

was in agreement with results reported by Abubaker et al. (2017) who found that sowing wheat seeds two days after amendment with non-digested and digested poultry manure with and without combination with urea reduced germination percentage considerably. Moreover, Gupta and Gupta (2011) reported a similar effect on okra seeds germination when sowing was performed 6 to 18 h after treatment with anaerobically digested poultry manure. This negative effect has been explained by the existence of humic acids in digested manure that formed during anaerobic digestion (Bacilio et al. 2003; Šerá and Novák 2011). In another study, Kaparaju et al. (2012) reported that anaerobically digested orange waste have a toxic effect on seeds germination of

Chinese cabbage and they referred this toxicity to ammonia and organic acids in the digestate. Tunes et al. (2012) have reported that the presence of organic acids which originating from the anaerobic decomposition of organic material can significantly reduce the germination of wheat cultivars, radicle length and dry weight. It is well known that concentration of ammonia in digested manure is higher compared to raw manure (García-González et al. 2016), which was found to reduces wheat seed germination and suppresses root hair formation (Bremner and Krogmeier 1989; Wan et al. 2016). The adverse effect of combining manure with urea on seeds germination and radicle length have been reported in another study where they referred this adverse effect to the presence of ammonia, which formed through hydrolysis of urea by soil urease (Bremner and Krogmeier 1989). Moreover, Wan et al. (2016) have reported an inhibition effect of urea on seed germination and radicle growth in six wheat cultivars, they explained this effect by the formation of ammonia in the soil after application. Soil microorganisms can contribute to ammonia production in the soil by releasing ureases enzyme, which will convert urea to carbon dioxide and ammonia (Maier 2009). However, sowing seeds either 20 or 30 days after amendment enhanced seed germination, radicle length, plumule length and seedling dry weight. This probably was due to the degradation of fertilizers components and the transformation of mineral N by soil microorganisms (Powlson et al. 2001). Kuwatsuka and Shindo (1973) have reported that the concentration of phenolic substances in waste-treated soil decreased rapidly and reached that of the control on the day 30 after waste disposal. In addition, degradation of organic materials will also lead to the release of plant nutrients, which will enhance the growth of radicle and plumule resulted in increasing seedling dry weight. The present study showed that *Acsad* and *Slambo* cultivars had better germination percentages than those of *Karim*. This was probably due to their vitalities, which were higher for *Slambo* (99 ± 4%) and *Acsad* (95 ± 3%) compared to *Karim* (89 ± 6%).

Conclusion

It can be concluded that all tested fertilizers displayed short-term effects on wheat seeds germination, which can be avoided by applying fertilizer at least 20 days prior to the sowing. It seems to fertilize soil 20–30 days before sowing, can reduce the adverse effect of the substances that responsible for delayed germination in non-digested, digested poultry manure and urea. However, the early application of fertilizers, especially digestate, may cause the loss of nutrients such as N in the form of ammonia. Therefore, further work remains to evaluate the effect of prolonging the period between applications and sowing on wheat growth and yield.

Acknowledgements The authors are grateful to Sabha University for providing equipment and chemicals for the research.

References

Abubaker J (2012) Effects of fertilisation with biogas residues on crop yield, soil microbiology and greenhouse gas emissions. Dissertation, Swedish University of Agricultural Sciences

Abubaker J, Risberg K, Pell M (2012) Biogas residues as fertilisers—effects on wheat growth and soil microbial activities. Appl Energy 99:126–134. https://doi.org/10.1016/j.apenergy.2012.04.050

Abubaker J, Elnesairy N, Ahmed S (2017) Effects of non-digested and anaerobically digested farmyard manures on wheat crop cultivated in desert soil. J Aridland Agric 3:1–10. https://doi.org/10.19071/jaa.2017.v3.3127

Alabasi HAM, Mohamed Z, Abd Latif I, Bin Abdullah AM, Iliyas A (2017) Analysis of technical efficiency of traditional wheat farming in Fezzan region, Libya. J Agric Econ Rural Dev 3:251–258

Angel R, Pasternak Z, Soares MIM, Conrad R, Gillor O (2013) Active and total prokaryotic communities in dryland soils. FEMS Microbiol Ecol 86:130–138. https://doi.org/10.1111/1574-6941.12155

Angelidaki I, Ahring BK (2000) Methods for increasing the biogas potential from the recalcitrant organic matter contained in manure. Water Sci Technol 41:189–194

Angus JF, Gupta VVSR, Pitson GD, Good AJ (2014) Effects of banded ammonia and urea fertiliser on soil properties and the growth and yield of wheat. Crop Pasture Sci 65:337–352. https://doi.org/10.1071/CP13337

Bacilio M, Vazquez P, Bashan Y (2003) Alleviation of noxious effects of cattle ranch composts on wheat seed germination by inoculation with *Azospirillum* spp. Biol Fertil Soils 38:261–266. https://doi.org/10.1007/s00374-003-0650-1

Ball DF (1964) Loss-on-ignition as an estimate of organic matter and organic carbon in non-calcareous soils. J Soil Sci 15:84–92. https://doi.org/10.1111/j.1365-2389.1964.tb00247.x

Bouyoucos GJ (1962) Hydrometer method improved for making particle size analyses of soils. Agron J 54:464–465. https://doi.org/10.2134/agronj1962.00021962005400050028x

Bradstreet RB (1954) Kjeldahl method for organic nitrogen. Anal Chem 26:185–187. https://doi.org/10.1021/ac60085a028

Bremner JM, Krogmeier MJ (1989) Evidence that the adverse effect of urea fertilizer on seed germination in soil is due to ammonia formed through hydrolysis of urea by soil urease. Proc Natl Acad Sci USA 86:8185–8188

Chapman HD, Pratt PF (1961) Methods of analysis for soils, plants, and waters. University of California, Department of Agricultural Sciences, Riverside

Coolbear P, Francis A, Grierson D (1984) The effect of low temperature pre-sowing treatment on the germination performance and membrane integrity of artificially aged tomato seeds. J Exp Bot 35:1609–1617. https://doi.org/10.1093/jxb/35.11.1609

Dastanpoor N, Fahimi H, Shariati M, Davazdahemami S, Hashemi SMM (2013) Effects of hydropriming on seed germination and seedling growth in sage (*Salvia officinalis* L.). Afr J Biotechnol 12:1223–1228

Dean WE (1974) Determination of carbonate and organic matter in calcareous sediments and sedimentary rocks by loss on ignition; comparison with other methods. J Sediment Res 44:242–248. https://doi.org/10.1306/74d729d2-2b21-11d7-8648000102c1865d

Diacono M, Montemurro F (2010) Long-term effects of organic amendments on soil fertility. A review. Agron Sustain Dev 30:401–422. https://doi.org/10.1051/agro/2009040

Elhadi EA, Mubarak AR, Rezig FAM (2016) Effects of organic amendments on sand dune fixation. Int J Recycl Org Waste Agric 5:1–8. https://doi.org/10.1007/s40093-015-0111-5

Ellis RH, Hong TD, Roberts EH (1985) Handbook of seed technology for genebanks. Volume I. Principles and methodology. In: Handbooks for genebanks, no. 2, International Board for Plant Genetic Resources, Rome

Elnesairy N, Abubaker J, Mahmod H, Mukhtar N (2016) The impact of *Bradyrhizobium*, farmyard manure and inorganic nitrogen on growth and yield of guar. World J Agric Res 4:56–63. https://doi.org/10.12691/wjar-4-2-4

Farooq M, Basra SMA, Ahmad N, Hafeez K (2005) Thermal hardening: a new seed vigor enhancement tool in rice. J Integr Plant Biol 47:187–193. https://doi.org/10.1111/j.1744-7909.2005.00031.x

Fierer N et al (2012) Cross-biome metagenomic analyses of soil microbial communities and their functional attributes. Proc Natl Acad Sci USA 109:21390–21395. https://doi.org/10.1073/pnas.12152 10110

Forster JC (1995) Soil sampling, handling, storage and analysis A2. In: Alef K, Nannipieri P (eds) Methods in applied soil microbiology and biochemistry. Academic Press, London, pp 49–121. https://doi.org/10.1016/B978-012513840-6/50018-5

García-González MC, Vanotti MB, Szogi AA (2016) Recovery of ammonia from anaerobically digested manure using gas-permeable membranes. Sci Agric 73:434–438. https://doi.org/10.1590/0103-9016-2015-0159

Gupta N, Gupta U (2011) Effect of anaerobically digested slurry of cow dung and kitchen waste on the seed quality in Okra (*Abelmoschus esculentus* L.). J Adv Lab Res Biol 2:1–3

Hati KM, Mandal KG, Misra AK, Ghosh PK, Bandyopadhyay KK (2006) Effect of inorganic fertilizer and farmyard manure on soil physical properties, root distribution, and water-use efficiency of soybean in Vertisols of central India. Bioresour Technol 97:2182–2188. https://doi.org/10.1016/j.biortech.2005.09.033

Kaparaju P, Rintala J, Oikari A (2012) Agricultural potential of anaerobically digested industrial orange waste with and without aerobic post-treatment. Environ Technol 33:85–94. https://doi.org/10.1080/09593330.2011.551839

Kiani MJ, Abbasi MK, Rahim N (2005) Use of organic manure with mineral N fertilizer increases wheat yield at Rawalakot Azad Jammu and Kashmir. Arch Agron Soil Sci 51:299–309. https://doi.org/10.1080/03650340500069114

Kuwatsuka S, Shindo H (1973) Behavior of phenolic substances in the decaying process of plants: I. Identification and quantitative determination of phenolic acids in rice straw and its decayed product by gas chromatography. Soil Sci Plant Nutr 19:219–227. https://doi.org/10.1080/00380768.1973.10432591

Ladd JN, Jackson RB (1982) Biochemistry of ammonification. In: Stevenson FJ (ed) Nitrogen in agricultural soils, vol 22. Agronomy monograph. American Society of Agronomy, Crop Science Society of America, Soil Science Society of America, Madison, pp 173–228. https://doi.org/10.2134/agronmonogr22.c5

Lungu OIM, Dynoodt RFP (2008) Acidification from long-term use of urea and its effect on selected soil properties. Afr J Food Agric Nutr Dev 8:63–76. https://doi.org/10.4314/ajfand.v8i1.19180

Maier RM (2009) Chapter 14—biogeochemical cycling. In: Maier RM, Pepper IL, Gerba CP (eds) Environmental microbiology, 2nd edn. Academic Press, San Diego, pp 287–318. https://doi.org/10.1016/B978-0-12-370519-8.00014-6

Makádi M, Tomócsik A, Orosz V (2012) Digestate: a new nutrient source—review. InTech Eur. https://doi.org/10.5772/1793

Makhalanyane TP, Valverde A, Gunnigle E, Frossard A, Ramond J-B, Cowan DA (2015) Microbial ecology of hot desert edaphic systems. FEMS Microbiol Rev 39:203–221. https://doi.org/10.1093/femsre/fuu011

Möller K, Müller T (2012) Effects of anaerobic digestion on digestate nutrient availability and crop growth: a review. Eng Life Sci 12:242–257. https://doi.org/10.1002/elsc.201100085

Odlare M, Arthurson V, Pell M, Svensson K, Nehrenheim E, Abubaker J (2011) Land application of organic waste—effects on the soil ecosystem. Appl Energy 88:2210–2218. https://doi.org/10.1016/j.apenergy.2010.12.043

Powlson D, Hirsch P, Brookes P (2001) The role of soil microorganisms in soil organic matter conservation in the tropics. Nutr Cycl Agroecosyst 61:41–51. https://doi.org/10.1023/A:1013338028454

Pungor E, Horval G (1994) Practical guide to instrumental analysis. CRC Press, Boca Raton

Ramakrishna TV, Robinson JW, West PW (1966) The determination of calcium and magnesium by atomic absorption spectroscopy. Anal Chim Acta 36:57–64. https://doi.org/10.1016/0003-2670(66)80006-5

Šerá B, Novák F (2011) The effect of humic substances on germination and early growth of Lamb's Quarters (*Chenopodium album* agg.). Biologia 66:470–476. https://doi.org/10.2478/s11756-011-0037-y

Tunes LMd, Avelar SAG, Barros ACSA, Pedroso DC, Muniz MFB, Menezes NLd (2012) Critical levels of organic acids on seed germination and seedling growth of wheat. Rev Bras Sementes 34:366–372. https://doi.org/10.1590/S0101-31222012000300002

Wan X, Wu W, Li C, Liu Y, Wen X, Liao Y (2016) Soil ammonia volatilization following urea application suppresses root hair formation and reduces seed germination in six wheat varieties. Environ Exp Bot 132:130–139. https://doi.org/10.1016/j.envexpbot.2016.08.010

Windham WR, Robertson JA, Leffler RG (1987) A Comparison of methods for moisture determination of forages for near infrared reflectance spectroscopy calibration and validation. Crop Sci 27:777–783. https://doi.org/10.2135/cropsci1987.0011183X002700040034x

Wong MH, Cheung YH, Cheung CL (1983) The effects of ammonia and ethylene oxide in animal manure and sewage sludge on the seed germination and root elongation of *Brassica parachinensis*. Environ Pollut Ecol Biol 30:109–123. https://doi.org/10.1016/0143-1471(83)90008-9

5

Nutrient recovery from anaerobic sludge by membrane filtration: pilot tests at a 2.5 MW_e biogas plant

T. Gienau[1] · U. Brüß[2] · M. Kraume[3] · S. Rosenberger[1]

Abstract

Purpose Membrane filtration is recently applied to recover nutrients and dischargeable water from anaerobic sludge. The purpose of this study is to quantify nutrient separation, membrane performance, and process stability and to increase the economical applicability of the process by energetic optimisation.

Methods At the site of a 2.5 MW_e agricultural biogas plant, a membrane pilot plant was operated over a period of 7 months. It consisted of a screw press separator, a decanter centrifuge, an ultrafiltration unit, and a three-stage reverse osmosis unit. Mass and nutrient balances were generated by sampling and analysing every process stream. Process performance was analysed by monitoring separation efficiencies, membrane flux, cleaning intervals, and energy demand.

Results Solid/liquid separation resulted in separation efficiencies of 70% for total solids and 80% for phosphorus. The solid fraction contained high concentrations of organics and particle-ligated nutrients (20% TS, 8 kg t^{-1} N_{total}, 5.5 kg t^{-1} P_2O_5). The retentate of the reverse osmosis had high concentrations of dissolved ammonia and potassium (4 kg t^{-1} NH_4–N and 10 kg t^{-1} K_2O). 38% of the sludge volume was recovered as clean water.

Conclusion The membrane pilot plant successfully produced a solid N/P-fertiliser, a liquid N/K-fertiliser and clean water. The results contribute to a sound understanding and growing database for future adaption of the process chain. Hydrodynamic optimisation within the pilot plant reduced the energy demand of the ultrafiltration step by 50%, which considerably contributes to the economy of the process.

Keywords Anaerobic digestion · Nutrient recovery · Ultrafiltration and reverse osmosis treatment · Total conditioning process · Fertiliser

Introduction

As a result of the German *Renewable Energies Act*, enacted in 2000, the number and size of biogas plants in Germany has constantly increased. In 2017, more than 9300 biogas plants were operating in Germany with an electrical capacity of $P \approx 4500$ MW_e (German Biogas Association 2017).

✉ S. Rosenberger
s.rosenberger@hs-osnabrueck.de

[1] Faculty of Engineering and Computer Science, Osnabrück University of Applied Science, Albrechtstraße 30, 49076 Osnabrück, Germany

[2] A3 Water Solutions GmbH, Boschstraße 2, 48369 Saerbeck, Germany

[3] Chair of Chemical and Process Engineering, Technische Universität Berlin, Str. des 17. Juni 135, FH 6-1, 10623 Berlin, Germany

Consequently, a considerable amount of anaerobic sludge (digestate) is produced. In 2011, the output of digestate amounted to 65.5 million m^3 containing approximately 390,000 t of nitrogen, 74,000 t of phosphorus, and 330,000 t of potassium (Möller and Müller 2012). In terms of concentration, raw digestate contains about 1.2–9.1 kg_N t^{-1} of total nitrogen with about 44–81% ammonia, 0.4–2.6 kg_P t^{-1} phosphorus and 1.2–11.5 kg_K t^{-1} potassium (Möller and Müller 2012).

Due to its high nitrogen and phosphorus content, anaerobic digestion products are mostly used for local field applications. Anaerobic sludge is characterised by high biological stability and appropriate fertilising quality. Soil and air characteristics are not negatively affected using anaerobic sludge products when applying correct manure procedures with subsurface injection to limit ammonia emissions (Riva et al. 2016). Furthermore, fertilisers based on anaerobic digestion products achieve comparable or even higher biomass production rates compared to synthetic mineral fertiliser for, e.g., the production of wheat (Abubaker et al. 2012) or silage corn (Riva et al. 2016). Local over-fertilisation results in an excess of nutrients in the soil and nutrient leaching into the groundwater (Lee and Oa 2013; Schoumans and Chardon 2014). To guarantee a good fertilising procedure without local over-fertilisation and maintain high groundwater qualities, recent revisions of the fertiliser ordinance in Germany limit field applications of organic residues to 170 kg_N ha^{-1} a^{-1} (DüV 2017). Regarding phosphorus, no surplus application on fields with high or very high phosphate content will be allowed (saldo = 0 kg_P ha^{-1} a^{-1}). Good fertilising practice thus often requires transportation of nutrients between different areas. Given that anaerobic sludge contains more than 90% water, transportation is neither economically nor ecologically sensible. Therefore, a number of different digestate treatment processes emerge which often originate from manure or wastewater treatment (e.g. Fuchs and Drosg 2013; Drosg et al. 2015; Vaneeckhaute et al. 2017; Bolzonella et al. 2017). The combination of mechanical solid/liquid separation and membrane treatment offers the advantage of producing dischargeable or reusable process water on the one hand and concentrated, transport-worthy nutrient products on the other hand.

Solid/liquid separation of organic slurries is achieved by screw press separators and/or decanter centrifuges. Decanter centrifuges achieve high efficiencies of solid and phosphorus separation of about 60–90% for slurries and biogas digestate (Balsari et al. 2006; Lukehurst et al. 2010). According to Lukehurst et al. (2010), separation efficiency for both total solids and nutrients can be improved with the use of stable polymeric flocculants during the centrifugal process. Further treatment of the liquid phase after solid/liquid separation is mostly carried out as a process combination of a micro- or ultrafiltration step (MF, UF) followed by a reverse osmosis step (RO). In Germany, required water specifications for water discharge from biogas plants depend on the quality of receiving water bodies and are defined by the appropriate environmental agency. The requirements for effluents from wastewater treatment plants (size 5) are limited to, i.a., chemical oxygen demand < 75 mg L^{-1}, total nitrogen concentration < 13 mg L^{-1}, and phosphorus concentration < 1 mg L^{-1} (AbwV 2005).

In lab scale experiments, different authors investigated the ultrafiltration performance for anaerobic digester effluents with varying digester feedstock (Waeger et al. 2010; Choo et al. 1992; Zacharof and Lovitt 2014; Meixner et al. 2015; Bhattacharya et al. 2013). The ultrafiltration step has the highest operational energy demand and consequently the highest operational costs (Drosg et al. 2015, Gienau et al. 2018). Membrane material, cross-flow velocity and transmembrane pressure were detected to have the strongest influence on the flux performance of the membrane process, while the flux rates were generally low and the filtration performance limited by organic fouling. This is in accordance with experience from membrane filtration of activated sludges, where organic macromolecules such as proteins and polysaccharides were identified as major membrane foulants (e.g., Rosenberger et al. 2006; Jiang et al. 2008). Waeger et al. (2010) recommended ceramic ultrafiltration membranes with pore sizes of 20–50 nm for biogas digestate filtration applications. Due to pronounced organic fouling, repeated permeate backflushing operations and intensive chemical cleaning strategies were required to ensure stable, high membrane performances.

Only few authors published results of pilot or full scale trials (Chiumenti et al. 2013, Ledda et al. 2013). While Chiumenti et al. (2013) report nutrient balances of a digestate treatment chain at the site of an agricultural biogas plant fed with swine and poultry manure and corn silage, Ledda et al. (2013) report results with digested cattle manure and with digested swine manure. Apart from different feedstock characteristics, the experimental set-ups consisted of different membranes used for the ultrafiltration step. Chiumenti et al. (2013) used ceramic modules with 0.1 μm pore size and Ledda et al. (2013) used 40 kDa grafted polyacrylonitrile membranes. Although the set-up was quite similar, the results of the mass and nutrient balances varied considerably. Ledda et al. (2013) separated the digestates into 38–51% water, 12–32% solid fertiliser, and 30–37% liquid concentrate, while Chiumenti et al. (2013) reported 48% water, 35% solid fertiliser, and 17% liquid concentrate. In the study of Ledda et al. (2013), the different mass distributions occur due to different biogas plant feedstock. Screenings on different biogas digestates reveal the large spectrum of digestate properties (Gienau et al. 2018; Vaneeckhaute et al. 2017). Similar problems have also been reported by Masse et al. (2007) in a review on membrane-based nutrient recovery from manure. The authors conclude that the technology is technically feasible, but that the experimental database is not big enough for a sound technology analysis. Technology reviews on the membrane-based digestate treatment also rely on small available databases. (Drosg et al. 2015, Camilleri-Rumbau et al. 2013; Vaneeckhaute et al. 2017; Bolzonella et al. 2017). When comparing different nutrient recovery technologies, the authors not only emphasize the high technical potential of the membrane treatment chain but also state the relatively high overall costs compared to other technologies such as stripping and drying. Published specific treatment costs vary between 16 and 25 kWh m^3 (Fuchs and Drosg 2013; Chiumenti et al. 2013).

The objective of this paper is to contribute to a sound database on the membrane-based conditioning process for different biogas effluents and to define potential for future reduction of operational costs. Over a period of 7 months, about 1500 m^3 of raw digestate were separated by a screw press separator, a decanter centrifuge, an ultrafiltration unit and a reverse osmosis unit. Besides the characterisation of all process streams, mass and nutrient balances were determined and efficiencies of all process units were calculated. Further focus was on process performance and optimisation of the most energy-intensive process step of the membrane chain, which is the ultrafiltration step.

Materials and methods

Biogas plant

Pilot trials were carried out at the site of an agricultural biogas plant in Bad Bentheim, Lower Saxony (Germany). The biogas plant has a total installed electric power of $P_{total} = 2.5$ MW$_e$. Four combined heat and power units (CHP) are installed on site ($P = 1.01$ MW$_e$) and five further satellite CHP units are located close to local heat-demanding areas in the city ($P = 1.49$ MW$_e$). The biogas plant is fed with about 50 t day^{-1} of cattle manure and dung, with 50 t day^{-1} of corn silage and with 38 t day^{-1} of sugar beet and seasonal crops. The biogas plant is equipped with four fermenters (2×1500 m^3 and 2×1900 m^3), a post fermenter (2200 m^3), and three storage tanks (2700 m^3, 3,400 m^3 and 9700 m^3). The average overall retention time in the fermenters and post fermenter is 60–65 days; the fermenters operate at mesophilic conditions between 40 and 42 °C. The total digestate output of the plant varies from 35,000 to 55,000 t a^{-1}.

Pilot plant for nutrient recovery

The pilot plant for nutrient recovery consists of a two-stage solid/liquid separation unit followed by ultrafiltration (UF) and reverse osmosis (RO) units (Fig. 1).

The digestate (DIG) from the biogas plant is separated with a screw press separator (agriKomp, Quetschprofi, 3 kW) with a mesh size of 500 μm to detach organic fibres and larger particles (Fig. 2, left). A decanter centrifuge dewaters the remaining liquid of the screw press (SP$_L$) (Fig. 2, right, GEA, Westphalia Separator AD 0509, relative centrifugal force: 3,400 g). The liquid/solid separation is

Fig. 2 Installed separation equipment, top: container with decanter centrifuge; left: screw press separator; right: decanter centrifuge

improved by cationic polymeric flocculants with a molecular weight of 3×10^6 to 4×10^6 g mol^{-1}. The applied concentration of polymers was 4–6 $g_{Polymer}$ kg_{TR}^{-1}. These polymers are certified by the manufacturer for 20% biodegradability in 2 years. The solid fractions of the screw press and the decanter centrifuge (SP$_S$ and DEC$_S$) of the digestate were used as solid organic fertiliser. The supernatant from the centrifuge (DEC$_L$) was fed to the ultrafiltration unit (Fig. 3, left). The ultrafiltration module ($\alpha Al_2O_3/TiO_2$, pore size = 50 nm, $A = 7.3$ m^2, Atech, Germany) was operated at transmembrane pressure differences of 4.5–5 bars, cross-flow velocities of 4–5 m s^{-1}, a permeate yield of 60–70%, and temperatures between 40 and 70 °C. The UF permeate (UF$_P$) was further treated by a three-stage reverse osmosis unit (Fig. 3, right, DOW, PA, $A_{total} = 137.2$ m^2) at a transmembrane pressure difference of 12–60 bars, a permeate yield of 70–80% and temperatures between 30 and 35 °C. The retention of ammonia was improved by setting the pH value of the UF permeate to 6.6–6.8 based on the ammonia equilibrium.

The retentate of the RO process (RO$_R$) represents a liquid fertiliser. Clean process water (RO$_P$) is collected and reused for the polymer solution and other applications on site. The screw press separator and the decanter have a capacity of

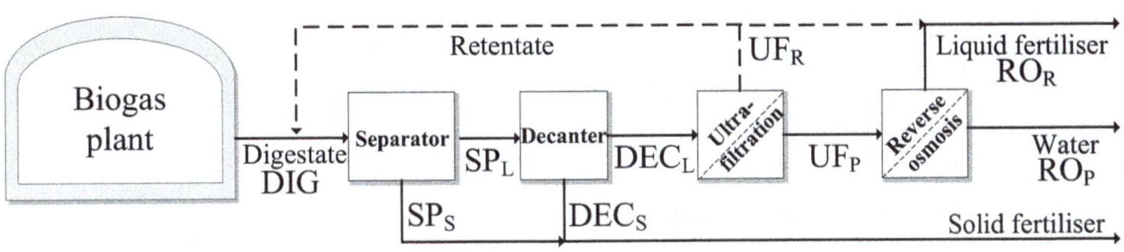

Fig. 1 Process scheme of the total conditioning process with ultrafiltration membrane and reverse osmosis; *DIG* digestate, SP_L liquid screw press separator, SP_S solid screw press separator, DEC_L liquid decanter, DEC_S solid decanter, UF_P permeate ultrafiltration, UF_R retentate ultrafiltration, RO_R retentate reverse osmosis, RO_P permeate reverse osmosis

Fig. 3 Installed membrane equipment, left: ultrafiltration unit; right: reverse osmosis units II and III

5–6 $m^3 h^{-1}$, which represents full-scale concerning a digestate output of 35,000–50,000 t a^{-1}. The membrane process has a capacity of 0.3–0.5 $m^3 h^{-1}$ (scale down factor approx. 1:10), which is depending on temperature and digestate composition. The excess of about 4–5 $m^3 h^{-1}$ from the decanter centrifuge is discharged to the digestate storage tank and/or used as liquid fertiliser.

Analysis and instrumentation

Total solids (TS in weight percent—wt%) and volatile solids (VS in % of TS) were analysed according to the European standards EN 12880 and EN 12879, respectively. Densities were quantified with a pycnometer (25 cm^3, BRAND) or a volumetric flask (500 cm^3, BRAND), depending on the roughness of the material. Measurements were carried out as double determination and have a relative error of < 1% for both TS and density.

Volume flow was recorded with magnetic inductive flow meters for almost all liquid streams. Based on the literature reports on organic fouling on membranes (e.g., Rosenberger et al. 2006, Jiang et al. 2008), polysaccharide and protein concentrations were analysed in the feed and the retentate of the ultrafiltration membrane according to the methods of Dubois (Dubois et al. 1956) and Bradford (Bradford 1976), respectively. All measurements were carried out as double determination and have a relative error of ≤ 5%.

The nutrient and organic carbon concentrations of all liquid streams were measured with standard Hach vial tests LCK 302 (ammonia), LCK 338 (total nitrogen), LCK 228 (potassium), LCK 350 (phosphorus), and LCK 914 (COD). Nutrient concentrations of solid components were analysed by Agrolab institute using standard procedures (Kjeldahl-N: DIN 19684-4, Dumas-N: DIN EN 16168, NH_4–N: DIN 38406-5-2 [E 5-2], P_2O_5: DIN EN ISO 11885 [E 22], K_2O: DIN EN ISO 11885 [E 22]). The relative errors of measurement named by Agrolab institute for nitrogen, ammonia, phosphorus and potassium are < 16.2%, < 15.6%, < 17.2% and < 8.4%, respectively.

Determination of mass balances and separation efficiency

Mass balances

Based on the analysed concentrations of TS, VS and nutrient components, mass balances were set up for both the entire process and each separation unit. As nearly all process streams and intermediates were analysed and each measurement comprises the method's standard deviation, mass balance equations were over-determined, leading to positive or negative balance gaps. Unknown mass flows (e.g., solid fertilisers) were calculated according to TS-concentration or density balances. The mass balances were fitted to minimise balance gaps: mean concentrations were iteratively adapted to fulfil mass balance equations for each single unit operation as well as for the entire process. As boundary condition, each concentration had to remain within the standard deviation of the mean value. The method allowed to almost close the balance gaps (100% of total solids and on average 95.2% of all nutrients).

Separation efficiency

The average values and standard deviations for the separation efficiencies η_i were calculated according to (Eq. 1) to benchmark the efficiency of each unit operation. The efficiency compares the concentration of the component i for each unit operation ($c_{i,\mathrm{unit}}$) with the initial concentration of the digestate ($c_{i,\mathrm{DIG}}$).

$$\eta_i = 1 - \frac{c_{i,\mathrm{unit}}}{c_{i,\mathrm{DIG}}} \quad (1)$$

Results and discussion

Chemical and physical parameters

Visual inspections revealed significant differences between the process fractions. Solid fractions were of dark brown colour, paste-like, and fibrous. Liquid fractions from screw press separator (SP_L), decanter (DEC_L), and retentate of the ultrafiltration step (UF_R) were characterised by pronounced turbidity and brown colour (see Fig. 4). The UF permeate (UF_P) is optically free of organic particles and of lighter colour. Retentate of the reverse osmosis (RO_R) is intensified in colour. The permeate of the last reverse osmosis step (RO_P) optically appears like clean water without organic particles.

Rheological measurements showed that the original digestate is of high viscosity with pseudoplastic properties. The liquid fraction of the decanter is of lower viscosity, but still well above water viscosity, and also shows pseudoplastic rheological behaviour (Gienau et al. 2018). Table 1 shows the chemical and physical parameters of all process streams. The density of all process streams varied between 1000 and 1029 kg m^{-3}. The pH values were in the range of 5.6–8.3. Concentrations of total solids, volatile solids, nutrients and other organic components constantly decreased at each separation unit. TS of the raw digestate (DIG) decreased from 5.8 wt% to no detectable solids in the reverse osmosis permeate (ROP). Most solids were detached by the screw press separator and the decanter, with TS concentrations of 19–21 wt%. Solids from the screw press separator (SPS) had a higher volatile solid content than the decanter solid fraction (DECS). The permeate was particle free after passing the ultrafiltration membrane with a pore size of 50 nm. The remaining COD and TS in the UF permeate can be attributed to salts and low molecular weight organic substances such as saccharides, proteins, and humic substances (Gienau et al. 2018).

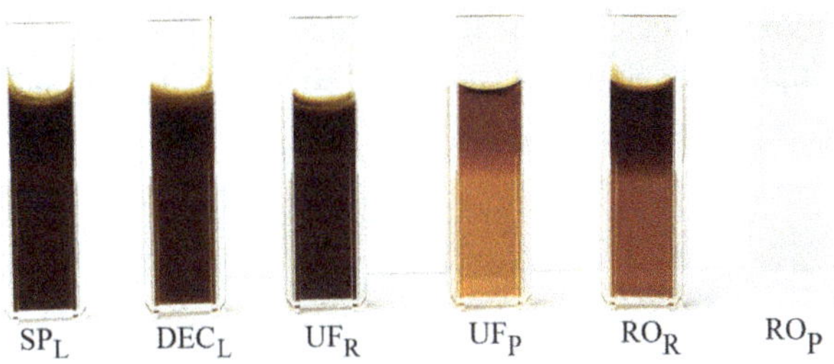

Fig. 4 f.l.t.r: Liquid phase from screw press (SP_L), supernatant from decanter (DEC_L), UF retentate (UF_R), UF permeate (UF_P), RO retentate (RO_R), RO permeate (RO_P)

The nutrient concentrations varied for all process streams. Measured concentrations of input material (i.e., 4.0 kg t^{-1} N_{total}, 1.6 kg t^{-1} P_2O_5, and 4.6 kg t^{-1} K_2O) are well in accordance with published values (Möller and Müller 2012). Highest concentrations of total nitrogen and phosphorus were found in the solid streams from the screw press separator (SP_s) and the decanter centrifuge (DEC_s) with 6.0 kg t^{-1} and 9.7 kg t^{-1}, respectively. The retentate of the reverse osmosis (RO_R) showed the highest values of ammonia, nitrogen and potassium (4.0 kg t^{-1} NH_4–N and 10.1 kg t^{-1} K_2O, respectively). The reverse osmosis membranes retained almost all organic, inorganic and nutrient substances. Chemical oxygen demand of the permeate was reduced from approx. 63 g L^{-1} to < 15 mg L^{-1} in the last reverse osmosis step. The quality of the permeate (RO_P) was characterised by COD < 15 mg L^{-1}, N_{total} = 14 mg L^{-1}, NH_4–N = 10 mg L^{-1} and P_2O_5 < 0.1 mg L^{-1}. The RO permeate thus is of high quality. All concentrations apart from total nitrogen even fulfil WWTP size 5 effluent requirements in Germany (legal limit WWTP size 5: N_{total} = 13 mg L^{-1}). In Germany, there is no legal limit for the discharge of treated digestate effluent. Depending on the quality of the local receiving water body, the local authorities decide on the necessary effluent

Table 1 Characterisation of process streams: digestate (DIG), liquid phase from screw press (SP_L), solid phase from screw press (SP_S), supernatant from decanter (DEC_L), solid phase from decanter (DEC_S), UF retentate (UF_R), UF permeate (UF_P), RO retentate (RO_R), RO permeate (RO_P)

Parameter	Unit	DIG	SP_L	SP_S	DEC_L	DEC_S	UF_P	UF_R	RO_R	RO_P
TS	wt%	5.8 ± 0.4	4.5 ± 0.8	21.1 ± 0.4	1.9 ± 0.0	19.1 ± 0.7	1.0 ± 0.0	3.4 ± 0.3	3.6 ± 0.6	0.002 ± 0.00
VS	%TS	72.8 ± 1.0	70.6 ± 1.7	87.1 ± 0.6	59.5 ± 2.7	75.8 ± 1.8	48.9 ± 14.4	70.9 ± 0.4	31.8 ± 8.7	0.000 ± 0.00
N_{total}	kg t^{-1}	4.0 ± 0.1	3.6 ± 0.2	6.0 ± 0.5	2.4 ± 0.2	9.7 ± 0.5	1.8 ± 0.1	3.4 ± 0.5	4.6 ± 0.4	0.014 ± 0.01
NH_4–N	kg t^{-1}	1.7 ± 0.2	1.8 ± 0.2	2.4 ± 0.2	1.5 ± 0.1	2.4 ± 0.2	1.6 ± 0.1	1.7 ± 0.2	4.0 ± 0.3	0.010 ± 0.01
P_2O_5	kg t^{-1}	1.6 ± 0.1	1.1 ± 0.3	4.7 ± 0.7	0.3 ± 0.0	6.1 ± 0.4	0.0 ± 0.0	0.6 ± 0.1	0.3 ± 0.1	0.000 ± 0.00
K_2O	kg t^{-1}	4.6 ± 0.2	4.8 ± 0.3	4.4 ± 0.2	4.2 ± 0.1	4.7 ± 0.0	4.3 ± 0.2	4.6 ± 0.2	10.1 ± 0.7	0.001 ± 0.00
COD	g L^{-1}	63.0 ± 5.8	51.7 ± 14.3	–	15.7 ± 0.7	–	4.7 ± 1.3	43.8 ± 2.2	7.6 ± 0.6	<0.015
Polys.	g L^{-1}	–	–	–	1.4 ± 0.2	–	0.3 ± 0.0	3.0 ± 0.4	–	–
Protein	g L^{-1}	–	–	–	3.0 ± 0.3	–	0.4 ± 0.1	6.8 ± 0.9	–	–
Density	kg m^{-3}	1029 ± 0.0	1022 ± 0.0	–	1013 ± 0.0	–	1006 ± 0.0	1014 ± 0.0	1028 ± 0.0	1000 ± 0.0
Conduct.	mS cm^{-1}	17.0 ± 0.5	19.5 ± 2.2	–	20.0 ± 0.3	–	20.6 ± 1.0	20.3 ± 0.3	58.3 ± 2.4	0.097 ± 0.1
pH	–	7.7 ± 0.2	7.7 ± 0.1	–	7.8 ± 0.1	–	8.3 ± 0.1	8.0 ± 0.2	7.4 ± 0.4	5.6 ± 1.1

Data given as value ± standard deviation

quality. In a full-scale treatment plant, the RO step can be individually designed to meet the necessary requirements. The obtained effluent concentrations are lower than the values reported in the literature: Ledda et al. (2013) and Chiumenti et al. (2013) report mean effluent concentrations of 20 mg L^{-1} N_{total} and 85 mg L^{-1}, respectively, which could be critical regarding water discharge.

Mass balance

Based on measured mass flow as well as physical and chemical parameters, Fig. 5 shows a Sankey diagram of the complete process, extrapolated to a daily mass flow of 150 t day^{-1} digestate. The input of 150 t day^{-1} digestate contained 94% water (141.3 t day^{-1}). The screw press separator delivered a solid output of 16 t day^{-1}. Before the decanter, an aqueous solution of polymers (7.0 t day^{-1}) was mixed into the liquid stream of the screw press separator to improve separation efficiency of the particularly fine grain fraction. Thus, the decanter separated another 16 t day^{-1} of solids. With significant amounts of total solids, total nitrogen, and phosphorus, both solid fractions represented 32 t day^{-1} of a high-content organic N/P fertiliser. The supernatant of the decanter centrifuge (125 t day^{-1}) was fed to the ultrafiltration unit. The retentate of the ultrafiltration amounted to about 47 t day^{-1}, mainly consisting of water (45.4 t day^{-1}). The UF retentate was reduced in the particulate nutrients phosphorus and organic nitrogen. Ammonia and potassium concentrations are similar to the original digestate. The permeate of the ultrafiltration was nearly free of solids and contained dissolved nutrients/salts and low molecular weight organic components such as proteins and saccharides (compare Table 1). All dissolved components were further transferred to the retentate of the reverse osmosis (22 t day^{-1}). The dischargeable permeate of the reverse osmosis amounted to 56.0 t day^{-1}.

Regarding the complete nutrient separation process, 20% of the input mass flow volume (i.e., from screw press separator and decanter centrifuge) were transferred into a solid N/P-fertiliser (Fig. 6). About 30% of the digestate can be found in the retentate of the ultrafiltration. The retentate of the reverse osmosis amounted to about 14%. The biggest process stream was the treated water after reverse osmosis (35.8%). The UF retentate can be used as liquid N/K-fertiliser. As an alternative, it is often recirculated to the decanter or the digester. Including an internal recirculation loop to the decanter, an RO permeate share of 53.9% of the input mass flow was calculated (Fig. 6). The pilot scale

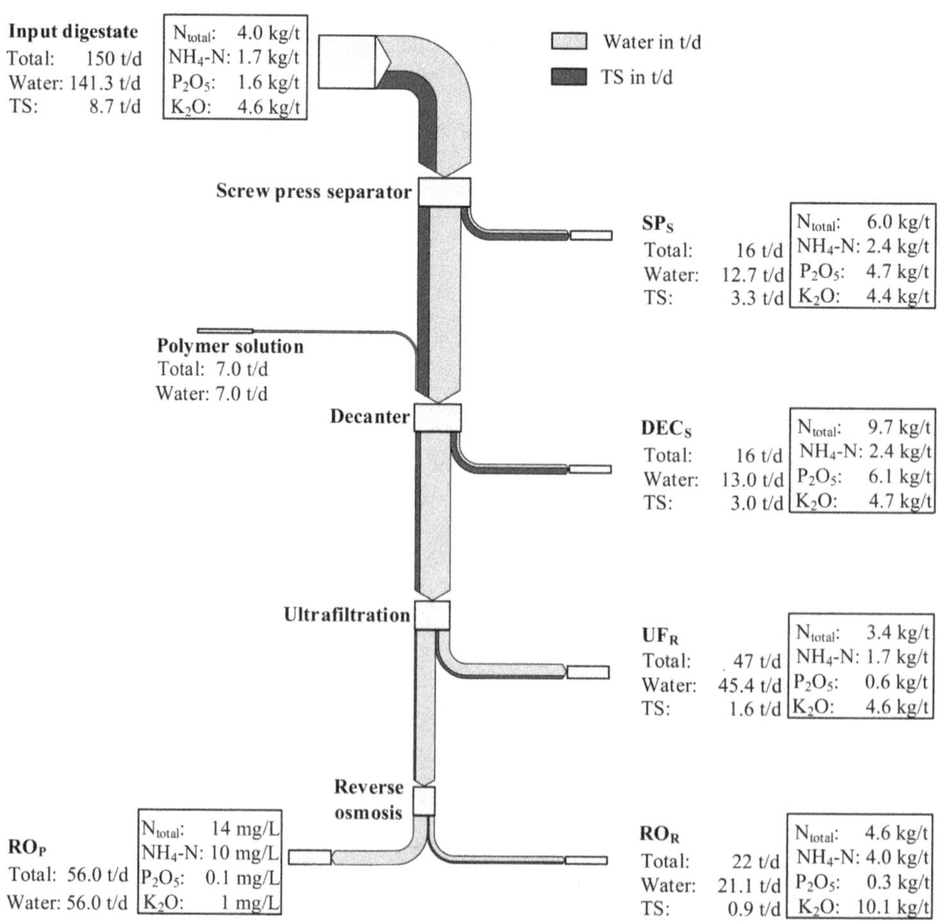

Fig. 5 Sankey diagram of total mass, solids, and water flows, extrapolated to a digestate input mass flow of 150 t day^{-1}

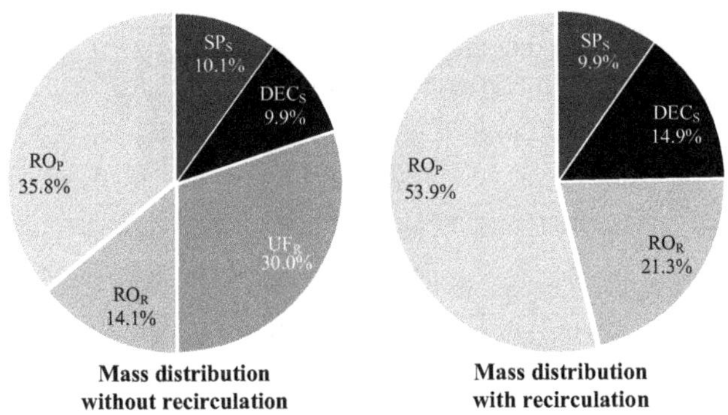

Fig. 6 Mass distribution of the membrane-based nutrient recovery; left: without and right: with recirculation of the UF retentate; the sum of SP_S and DEC_S forms an organic N/P-fertiliser, RO_R represents a liquid N/K-fertiliser and RO_P dischargeable water

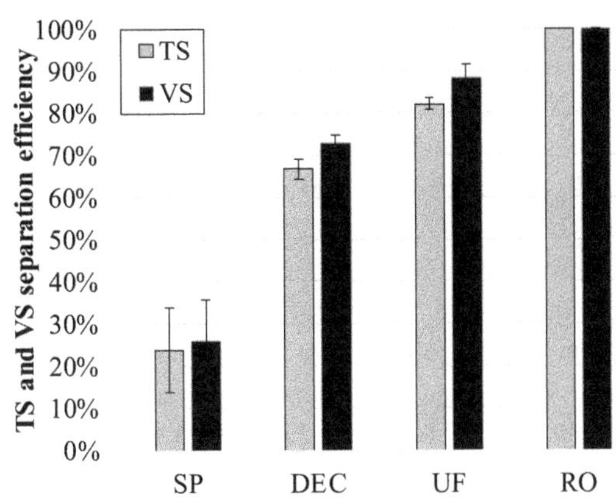

Fig. 7 Cumulated separation efficiencies for total solids (TS) and volatile solids (VS)

results thus support the literature reports [i.e., 46% (Bolzonella et al. 2017); 48% (Chiumenti et al. 2013) and 50% (Drosg et al. 2015)].

Separation efficiencies

The separation efficiency of each process step can be benchmarked by comparing the concentration of each component with the initial concentration of the digestate. Figure 7 shows the cumulated separation efficiency of the organic fractions. The combined liquid/solid separation (SP_S and DEC_S) detached about 67% of TS and nearly 74% of organic matter (VS). For the screw press separator, high standard deviations for the measured parameters occurred. Obviously, these variations were based on fluctuating input quality, composition and TS/VS values of the digestate. The combined liquid/solid separation with supporting flocculants represented an appropriate separation technique for significant TS and VS reduction to enable ultrafiltration processes. The separation of TS and VS after ultrafiltration amounted to 83% and 92%, respectively. After the three-stage reverse osmosis process, no TS and VS were detectable; the separation efficiency was 100%. Total COD reduction was 92.5% after UF and > 99.9% after RO (see Table 1). In comparison, Waeger et al. (2010) report COD removal rates of 80–85% after MF or UF for digestate from a biogas plant fed with organic waste. Chiumenti et al. (2013) report 91.7% COD removal after UF and > 99.9% after RO for a similar biogas plant feedstock to this study.

The cumulated separation efficiency of the nutrients total nitrogen (TN), ammonia (NH_4–N), potassium (K_2O) and phosphorus (P_2O_5) constantly increased at each separation step (Fig. 8). For total nitrogen, a linear separation trend is visible starting with 8% after the screw press separator over 55% after UF to > 98% after reverse osmosis. Ammonia was first separated after the decanter centrifuge (13%) and not significantly affected by the ultrafiltration membrane. As the ammonia equilibrium is not strictly located on the side of ammonia, the centrifugal field can detach gaseous, dissolved components. By shifting the ammonia equilibrium towards ammonia with pH values < 7, the ammonia separation efficiency reached 99.4% after the reverse osmosis step. Waeger et al. (2010) report total nitrogen removal of 5–18%, Ledda et al. (2013) of 60–80%, and Chiumenti et al. (2013) of 61%. The differences can be explained by the percentage of dissolved ammonia nitrogen on total nitrogen due to different feedstock and process conditions. In all studies, removal efficiencies after RO were > 98%.

For potassium, a quite similar separation behaviour can be observed. Potassium concentration in the liquid phase was nearly constant through solid/liquid separation and UF, while separation efficiency reached nearly 100% by reverse osmosis. Phosphorus was constantly separated to 27, 79, 98 and 100% for screw press separator, decanter, ultrafiltration, and reverse osmosis, respectively. Large standard deviations were detected for the screw press separator. As phosphorus is essentially particulate and thus connected to the solid fraction (Lukehurst et al. 2010; Gienau et al. 2018), separation efficiency is linked to the detachment of total solids (compare Fig. 7).

Membrane performance and stability

The ultrafiltration step needs high cross-flow velocities to minimise organic fouling on the membrane surface. UF amounts to more than 50% of total operation costs of the treatment chain (Fuchs and Drosg 2013). Therefore, UF performance was constantly monitored and optimised. During standard operation procedure, UF feed temperature was 42 °C (supported by mesophilic fermenter conditions) and the cross-flow velocity was set to 4–5 m s^{-1} at a transmembrane pressure difference (TMP) of 4 bars (see Test 1 in

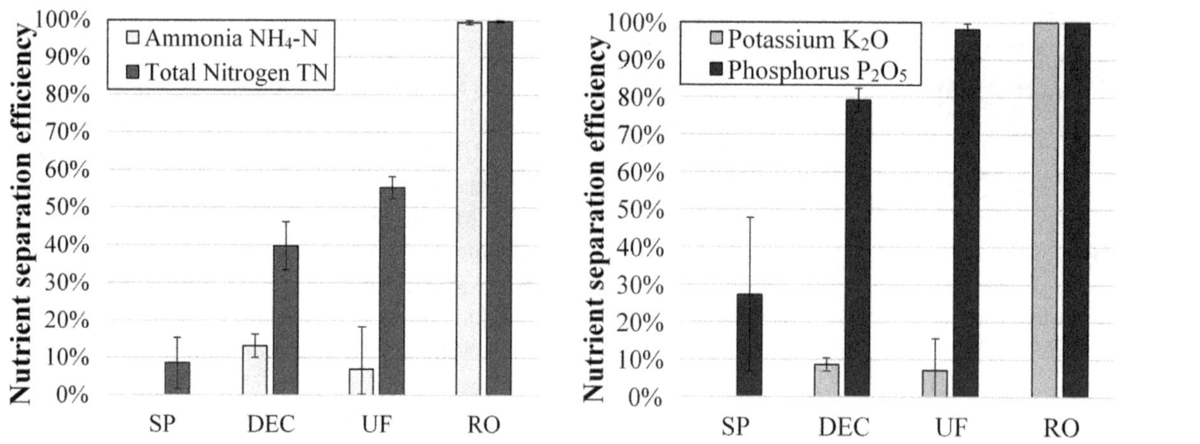

Fig. 8 Cumulated concentration-based separation efficiencies for total nitrogen (N_{total}), ammonia (NH_4-N), potassium (K_2O), and phosphorus (P_2O_5)

Table 2 Ultrafiltration performance of three different optimisation alternatives

Parameter	Unit	Test 1 (Ref.)	Test 2	Test 3
Temperature	°C	42	65	63
Cross-flow velocity	m s^{-1}	4.2	4.8	3.0
Transmembrane pressure	bar	4.0	3.6	3.9
Yield	%	70	70	70
Flux	L m^{-2} h^{-1}	20.5	29.4	23.3
Relative energy	%	100	72	48

Table 2). With a permeate yield of 70%, a mean permeate flux of 20.5 L m^{-2} h^{-1} could be achieved. In comparison, Waeger et al. (2010) report membrane fluxes between 20 and 50 L m^{-2} h^{-1} for digestate from an organic waste biogas plant at TMPs of 1 bar with total permeate recirculation (yield = 0). The differences in flux rates can be explained by the different permeate yield and different feedstock to the biogas plant. Generally, organic waste biogas plants tend to show higher flux rates (Gienau et al. 2018). Besides cross-flow velocity and TMP, the ultrafiltration performance depends on total solid concentration, concentration of biopolymers, and apparent viscosity of the separated anaerobic sludge (Rosenberger et al. 2006; Jiang et al. 2008; Zacharof and Lovitt 2014). For efficient back transport of particles from the membrane surface, the cross-flow velocity has to be sufficiently high to achieve Reynolds numbers (Re) above the laminar flow regime.

The standard operation mode (reference case–100% relative energy demand) was first improved by increasing the feed temperature to 65 °C (Test 2), which reduces fluid viscosity and increases Re. Savings were 28% of process energy due to a significantly improved membrane flux of 29.4 L m^{-2} h^{-1}. In a third test, high temperature was combined with a lower cross-flow velocity of 3 m s^{-1}. As a result of the high temperature, viscosity of the sludge changed to lower values and enabled lower cross-flow velocities for sufficiently turbulent flow regimes. The measured flux of Test 3 was 23.3 L m^{-2} h^{-1}, which is + 14% compared to the reference case (Test 1). By combination of flux improvement and reduction of volume flow rate through the tubular membrane channels, the relative energy demand of UF could be reduced to 48%. Test 3 was found to be the most economic and efficient optimisation case. Long-term studies with technical equipment are planned for the future.

Above results were obtained with frequent chemical membrane cleaning. Cleaning intervals were 1–2 weeks. Chemical cleaning was carried out fully automated with acids and bases for about 2–3 h. The chemicals have no negative impact on the material of the ultrafiltration plant, i.e., stainless steel pipes and ceramic ultrafiltration membranes. Initial membrane flux was achieved after each cleaning step. Availability of the plant was > 90% including maintenance. Life time of the ceramic ultrafiltration membrane and the polymeric reverse osmosis membrane are expected to be 7–9 years and 1–2 years, respectively.

Conclusion

The performed experimental tests demonstrate the successful separation of anaerobic sludge from an agricultural biogas plant over a 7-months operation period. Main products of the process are solid and liquid fertilisers, and dischargeable water, representing about 20, 44 and 36% of total mass flow, respectively. It was shown that an internal recirculation of UF retentate increases dischargeable water yield from 36 to 54%. Solid/liquid separation by screw press separator and decanter centrifuge in combination with polymeric flocculants provided a reliable pre-treatment for the following ultrafiltration unit. The solids from the combined solid/liquid separation were characterised by high total solid concentrations (19–21%), but also by concentrations of 4.7–6.1 kg t^{-1} phosphorus and 6.0–9.7 kg t^{-1} total nitrogen. Thus, the solids represent a high value organic N/P-fertiliser for local-field applications.

The membrane treatment was successfully realised by ceramic ultrafiltration units and a three-stage reverse osmosis unit. For the ultrafiltration membrane, a flux performance of 20–30 L m^{-2} h^{-1}, depending on cross-flow velocity and process temperature, was measured. The retentate of the reverse osmosis was rich in dissolved nutrient salts such as K^+ and NH_4^+ and represented a directly available liquid N/K-fertiliser. The RO permeate in this study almost achieved the German effluent limits for wastewater treatment plants size 5. In Germany, the required effluent quality for discharge of treated digestate is defined for each individual case and depends on the quality of the receiving water body. It is possible to adjust the clear water quality by design of the RO train and pH adaptation. A comparison of obtained nutrient and mass balances with the literature revealed a spread of results, probably depending on feedstock to the biogas plant and on process configuration. For a sound understanding and increasing applicability of the process train, a broader database based on onsite trials is needed. The literature on membrane performance is very limited and mainly focusses on lab-scale trials.

Besides the quality of the nutrient products, the economy of the process depends on its energy consumption. First optimisation schemes resulted in energy reductions of 50% for the ultrafiltration unit. This can be realised by adjusting process temperature and flow conditions in the tubular membrane modules. Future investigations should focus on the long-time behaviour of the ultrafiltration performance and the energy consumption in full scale.

Acknowledgements This research project is supported and financed by the German Federal Environmental Foundation (Deutsche Bundesstiftung Umwelt), Grant no. AZ 31276. The scientific research was organised at Osnabrück, University of Applied Science in cooperation with the company A3 Water Solution GmbH. The pilot plant was provided by A3 Water Solutions GmbH and the local farmer in Bad Bentheim. The authors gratefully acknowledge the entire team of the farm for their practical support.

References

Abubaker J, Risberg K, Pell M (2012) Biogas residues as fertilisers—effects on wheat growth and soil microbial activities. Appl Energy 99:126–134. https://doi.org/10.1016/j.apenergy.2012.04.050

AbwV (2005) Ordinance on requirements for the discharge of waste water into waters (Waste Water Ordinance—AbwV). 1–121

Balsari P, Santoro E, Dinuccio E, Gioelli F (2006) Assessment of the performances of different mechanical solid-liquid separators for pig and cattle slurries. Proc 12th Ramiran Int Conf Technol Recycl Manure Org Residues Whole-Farm Perspect 2:157–159

Bhattacharya P, Ghosh S, Mukhopadhyay A (2013) Combination technology of ceramic microfiltration and biosorbent for treatment and reuse of tannery effluent from different streams: response of defence system in *Euphorbia* sp. J Recycl Org Waste Agric. https://doi.org/10.1186/2251-7715-2-19

Bolzonella D, Fatone F, Gottardo M, Frison N (2017) Nutrients recovery from anaerobic digestate of agro-waste: techno-economic assessment of full scale applications. J Environ Manage. https://doi.org/10.1016/j.jenvman.2017.08.026

Bradford MM (1976) A rapid and sensitive method for the quantitation of microgram quantities of protein utilizing the principle of protein-dye binding. Anal Biochem 72:248–254. https://doi.org/10.1016/0003-2697(76)90527-3

Camilleri-Rumbau MS, Norddahl B, Kjær Nielsen AK, Christensen KV, Fjerbæk Søtoft L (2013) Comparative techno-economical study between membrane technology systems for obtaining concentrated fertilizers from biogas plant effluents. 15th Ramiran Int Conf 8:7–10

Chiumenti A, Da Borso F, Teri F, Chiumente R, Piaia B (2013) Full-scale membrane filtration system for the treatment of digestate from a co-digestion plant. Appl Eng Agric 29:985–990. https://doi.org/10.13031/aea.29.10117

Choo K, Lee C, Pek U, Koh U, Kim S, Koh J (1992) Characteristics of membrane filtration as a post treatment to anaerobic digestion. J Korean Ind Eng Chem 3(4):730–738

Drosg B, Fuchs W, Al Seadi T, Madsen M, Linke B (2015) Nutrient recovery by biogas digestate processing. In: IEA Bioenergy. http://www.iea-biogas.net/files/daten-redaktion/download/TechnicalBrochures/NUTRIENT_RECOVERY_RZ_web1.pdf. Accessed 8 Feb 2018

Dubois M, Gilles K, Hamilton J, Rebers P, Smith F (1956) Colorimetric method for determination of sugars and related substances. Anal Chem 28:350–356. https://doi.org/10.1021/ac60111a017

DüV (2017) Fertiliser Ordinance: (in German) Verordnung über die Anwendung von Düngemitteln, Bodenhilfsstoffen, Kultursubstraten und Pflanzenhilfsmitteln nach den Grundsätzen der guten fachlichen Praxis beim Düngen (Düngeverordnung—DüV)

Fuchs W, Drosg B (2013) Assessment of the state of the art of technologies for the processing of digestate residue from anaerobic digesters. Water Sci Technol. https://doi.org/10.2166/wst.2013.075

German Biogas Association (2017) Biogas sector statistics 2016/2017. https://de.statista.com/statistik/daten/studie/167676/umfrage/branchendaten-fuer-biogas-in-deutschland. Accessed 13 Nov 2017

Gienau T, Brüß U, Kraume M, Rosenberger S (2018) Nutrient recovery from biogas digestate by optimised membrane treatment. Waste Biomass Valori. https://doi.org/10.1007/s12649-018-0231-z

Jiang T, Myngheer S, De Pauw D, Spanjers H, Nopens I, Kennedy MD, Amy G, Vanrolleghem PA (2008) Modelling the production and degradation of soluble microbial products (SMP) in membrane bioreactors (MBR). Water Res 42:4955–4964. https://doi.org/10.1016/j.watres.2008.09.037

Ledda C, Schievano A, Salati S, Adani F (2013) Nitrogen and water recovery from animal slurries by a new integrated ultrafiltration, reverse osmosis and cold stripping process: a case study. Water Res 47:6157–6166. https://doi.org/10.1016/j.watres.2013.07.037

Lee Y, Oa S (2013) Nutrient transport characteristics of livestock manure in a farmland. Int J Recycl Org Waste Agric 2:1–6. https://doi.org/10.1186/2251-7715-2-1

Lukehurst C, Frost P, Al Seadi T (2010) Utilisation of digestate from biogas plants as biofertiliser. In: IEA Bioenergy. http://task37.ieabioenergy.com/files/daten-redaktion/download/publi-task37/Digestate_Brochure_Revised_12-2010.pdf. Accessed 8 Feb 2018

Masse L, Masse DI, Pellerin Y (2007) The use of membranes for the treatment of manure: a critical literature review. Biosyst Eng 98:371–380. https://doi.org/10.1016/j.biosystemseng.2007.09.003

Meixner K, Fuchs W, Valkova T, Svardal K, Lodereer C, Neureiter M, Bochmann G, Drosg B (2015) Effect of precipitating agents on centrifugation and ultrafiltration performance of thin stillage digestate. Sep Purif Technol 145:154–160. https://doi.org/10.1016/j.seppur.2015.03.003

Möller K, Müller T (2012) Effects of anaerobic digestion on digestate nutrient availability and crop growth: a review. Eng Life Sci 12:242–257. https://doi.org/10.1002/elsc.201100085

Riva C, Orzi V, Carozzi M, Acutis M, Boccasile G, Lonati S, Tambone F, D'Imporzano G, Adani F (2016) Short-term experiments in using digestate products as substitutes for mineral (N) fertilizer: agronomic performance, odours, and ammonia emission impacts. Sci Total Environ 547:206–214. https://doi.org/10.1016/j.scitotenv.2015.12.156

Rosenberger S, Laabs C, Lesjean B, Gnirss R, Amy G, Jekel M, Schrotter C (2006) Impact of colloidal and soluble organic material on membrane performance in membrane bioreactors for municipal wastewater treatment. Water Res 40:710–720. https://doi.org/10.1016/j.watres.2005.11.028

Schoumans OF, Chardon WJ (2014) Phosphate saturation degree and accumulation of phosphate in various soil types in The Netherlands. Geoderma 237:325–335. https://doi.org/10.1016/j.geoderma.2014.08.015

Vaneeckhaute C, Lebul V, Michels E, Belia E, Vanrolleghem PA, Tack F, Meers E (2017) Nutrient recovery from digestate: systematic technology review and product classification. Waste Biomass Valori 8:21–40. https://doi.org/10.1007/s12649-016-9642-x

Waeger F, Delhaye T, Fuchs W (2010) The use of ceramic microfiltration and ultrafiltration membranes for particle removal from anaerobic digester effluents. Sep Purif Technol 73:271–278. https://doi.org/10.1016/j.seppur.2010.04.013

Zacharof M, Lovitt RW (2014) The filtration characteristics of anaerobic digester effluents employing cross flow ceramic membrane micro filtration for nutrient recovery. Desalination 341:27–37. https://doi.org/10.1016/j.desal.2014.02.034

6

Assessing the presence of pharmaceuticals in soil and plants irrigated with treated wastewater in Oman

Raya Al-Farsi[1] · Mushtaque Ahmed[1] · Ahmed Al-Busaidi[1] · B. S. Choudri[2]

Abstract

Purpose This study conducted to assess the presence of pharmaceuticals in soils and crops irrigated with treated wastewater in Oman. The study was aimed to evaluate the potential of plant uptake of four commonly used pharmaceuticals in Oman: amoxicillin, sulfamethoxazole, trimethoprim and ibuprofen by radish (*Raphanus raphanistrum subsp. sativius*) in soil culture.

Methods Radish (*Raphanus raphanistrum*) was selected as a test crop. Liquid chromatography mass spectrometry/mass spectrometry (LC–MS/MS) instrument was used to measure concentrations of pharmaceuticals. Groundwater and treated wastewater were used for irrigation. Additionally, there were two spiked treatments; high-spiked concentration (5 mg/l) and low-spiked concentration (1 mg/l).

Results Results of the study showed that trimethoprim and ibuprofen were not detected in any samples of soil and plant. Sulfamethoxazole accumulated in radish roots for low- and high-spiked treatments. Amoxicillin accumulated in roots with low-spiked- as well as high-spiked treatments. Radish showed the ability to translocate some of the selected pharmaceuticals from soil irrigated with treated wastewater if they are present in sufficient concentrations.

Conclusions Radish showed positive uptake of some selected pharmaceuticals from soil irrigated with treated wastewater with spiked (both high and low) concentration of pharmaceuticals and has the ability to accumulate them in the roots. Pharmaceutical usage and disposal must be given attention to prevent their occurrences in the environment. Furthermore, treated wastewater must be given attention, as it is an important source of water in Oman, which suffers from water shortages.

Keywords Pharmaceuticals and personal care products · Treated wastewater · Irrigation · Radish · Translocation · Oman

Introduction

Pharmaceuticals are defined by World Health Organization (WHO) as synthetic or natural chemicals that can be found in human drugs and veterinary drugs. Pharmaceuticals contain active ingredients that have been designed to have positive effects for human and animal health. Many absorbed drugs in tissues of body would undergo metabolic reactions by microorganisms or enzymes, which may produce other compounds. These compounds excreted through urine or feces and transformed by chemical and physico-chemical factors in water (Mompelat et al. 2009). On the other hand, significant fractions of the unmetabolized drugs are excreted from the bodies through urine and feces as active compounds. Wastewater treatment systems are not specifically designed to remove pharmaceuticals, as a consequence, pharmaceuticals are partially removed from wastewater during treatment process which contribute to release into the environment (Carballa et al. 2004; Wang et al. 2005; Joss et al. 2006; Chefetz et al. 2008; Kraigher et al. 2008; Islam, 2013; Weng et al. 2014). Pharmaceuticals have been detected in wastewater effluents, biosolids, biosolid-amended soils, runoff from effluent-irrigated fields, surface and groundwater systems receiving treated wastewater (Kolpin et al. 2002; Pedersen et al. 2005; Kinney et al. 2006; Chefetz et al. 2008; Xu et al. 2009; Miller et al. 2016). Investigation of the potentiality of pharmaceuticals to translocate from soil and water into plant tissues were documented in several studies (Christian et al. 2003; Kumar et al. 2005; Boxall et al. 2006; Ji et al. 2012; Wu et al. 2012a, 2013). Plants have the ability to accumulate pharmaceuticals and personal care products

✉ B. S. Choudri
bchoudri@squ.edu.om

1 College of Agricultural and Marine Sciences, Sultan Qaboos University, Muscat, Oman

2 Center for Environmental Studies and Research, Sultan Qaboos University, Muscat, Oman

(PPCPs) in their tissues through root uptake (Dolliver et al. 2007; Wu et al. 2010, 2012a). Wu et al. (2012b) found that spinach and lettuce can uptake some PPCPs from the soil irrigated with treated wastewater. Further, these authors detected trimethoprim in plant tissues with concentrations of 1.1 ± (0.2–0.4) mg/g in dry weight in both lettuce and spinach, respectively. Sulfamethoxazole was also detected but in lower concentration than trimethoprim. According to Kumar et al. (2005) concentrations of chlortetracycline and sulfamethazine were measured in corn, lettuce, potato, cabbage and green onion and were found in small range of 2–17 μg/kg, while Dolliver et al. (2007) found that the concentration of sulfamethazine in plant tissue was in the range from 0.1 to 1.2 mg/kg.

In a soil–plant system, the fate, behavior and accumulation of PPCPs in plants may be affected due to the chemical, physical and biological processes in the growing system. Soil characteristics such as soil organic carbon content and ion exchange capacity will influence the mobility of pharmaceuticals, and therefore, their availability in the medium (Chefetz et al. 2008; Herklotz et al. 2010; Eggen et al. 2011). There are different processes resulting in elimination of organic compounds in the environment and to reduce plant uptake of PPCPs. These processes include biotic and abiotic. Sorption process is a significant abiotic process to eliminate PPCPs. Organic materials in soil may alter the surface properties and availability of sites for sorption. High organic content can either inhibit or promote sorption of organic compounds to soil (Miller et al. 2016). In addition, microbes may degrade PPCPs in the root zone (Wu et al. 2014).

Several studies worldwide documented the presence of pharmaceuticals in treated wastewater and soils irrigated with treated wastewater or amended by biosolids (Kinney et al. 2006, 2008; Wu et al. 2012a). Recently, studies have shown that pharmaceutical pollutants are of environmental concern and they are increasing with regard to the occurrences and fate. It is very important to note that exposure to pharmaceuticals through number of routes in values higher than the acceptable daily intake will have detrimental effects (Kümmerer 2008). The continuous disposal of pharmaceuticals into the environment through different ways can cause serious health problems and affect plants and aquatic organisms even at lower concentrations. As a consequence, it is a significant activity to evaluate the presence of the pharmaceuticals in water (Wang et al. 2005). However, in case of Oman, there are few studies undertaken by researchers on use of treated wastewater for irrigation and growing crops on experimental basis including the assessment on qualities of treated wastewater (Baawain et al. 2014a, b, 2015; Jaffar Abdul Khaliq et al. 2017; Abdul Khaliq et al. 2017; AlKhamisi et al. 2017). It is pertinent to mention that based on the literature review in the recent past, there are no studies conducted in Oman on understanding the PPCPs in treated wastewater and its translocation into plant root systems.

This study aims to evaluate and measure the accumulation of four most frequent pharmaceuticals consumed in Oman (SQUH 2016, Personal communication) in soil and plants irrigated with treated wastewater. Target compounds are amoxicillin, sulfamethoxazole, trimethoprim and ibuprofen.

Materials and methods

Pharmaceuticals and analytical chemistry

The selected four human drugs (amoxicillin, sulfamethoxazole, trimethoprim and ibuprofen) with purity of > 98% were purchased from Sigma-Aldrich. The pharmaceutical analysis was done in the Central Analytical & Applied Research Unit (CAARU) in college of Sciences, Sultan Qaboos University. Liquid chromatography–mass spectrometry/mass spectrometer (LC–MS/MS) is used for the analysis of pharmaceuticals. Standard stock solutions of selected compounds were prepared and injected in the LC–MS/MS to be identified and added to the instrument library. To calculate detection limits, standards were spiked into the appropriate matrix (water) to a level that would produce a signal < 20 times the baseline noise; detection limits were defined as three times the standard deviation of the signal produced. Each standard was run around six times and the mean was calculated, it was found 0.5 μg/l.

Field experiment

Field experiment and laboratory analysis were conducted to assess the presence of pharmaceuticals in soils and crops irrigated with treated wastewater and to measure their concentration if they are subjected to uptake by plant tissues or accumulate in the soil. The field experiment was conducted at the Agricultural Experimental Station (AES) of Sultan Qaboos University (SQU). Radish (*Raphanus raphanistrum*) was selected as a test crop because it is consumed raw without cooking and both leaves and roots are eaten. It was grown in shadehouse using pots. Compost of sandy loam soil, cow manure and perlite was used as planting media. The soil used for experiment had 14.91% organic matter, 7.9 ± 0.2 (pH) and 1.4 ± 0.01 dS/m (EC). Plants were irrigated according to each plant need with consistent irrigation volume for each plant/replicate set. There were two types of water used for irrigation: groundwater (GW) used as control (pH: 8.0 and EC: 1.4 dS/m) and tertiary-treated wastewater (TWW) (pH: 8.1 and EC: 0.65). Additionally, two spiked treatment were prepared from dS/m TWW and mixed pharmaceuticals; treated wastewater with low concentration of mixed pharmaceuticals (TW + LC) and treated wastewater with high concentration of mixed pharmaceuticals (TW + HC). The treatments were called according to the water type used for irrigation. There were a total of 16 pots (18 cm diameter * 17 cm height) filled with 3 kg of soil. Each treatment contained four replicate containers with three plants each. The spiked water with mix of selected pharmaceuticals was changed for each time of irrigation to minimize the error due to the metabolism of pharmaceuticals in the water and it was used few minutes after spiking. The concentrations of pharmaceuticals were selected to be high considering soil as a planting media, which may reduce the availability of compounds for plant uptake due to the biotic and abiotic process in the soil–plant system. Several process and factors will play role in translocation and uptake of pharmaceuticals and may affect the availability of compounds to the plants. Moreover, the artificially high spiked levels of pharmaceuticals were used in the study to enable analysis. Two concentrations of pharmaceuticals were prepared with TWW. 1 mg of each compound (amoxicillin, ibuprofen, sulfamethoxazole, and trimethoprim) was added to 1 liter of TWW to prepare low concentration of spiked water (TW + LC) mixture. Hence, the concentration of each compound in that case was 1 mg/l. 5 mg of each compound was added to 1 l of TWW to prepare high concentration of spiked water (TW + HC) mixture. The concentration of each compound in that case was 5 mg/l. Plants received total amount of 7.5 mg from each compounds as low concentration (TW + LC) and 37.5 mg as high concentration (TW + HC) for 33 days of growth. All plants were allowed to reach maturity before harvesting. Chlorophyll and leaf area were measured for all plants. The measurements were recorded once a week for each pot. Additionally, radish biomass (roots and shoots) was recorded after harvesting and soil organic matter content was also determined. There were no attempts to collect the leachates from the pots after irrigation.

Field experiment layout

The experimental layout for the radish replicates of the irrigated pots are shown in Fig. 1. Four replicates were set and only three were used to collect the samples. One pot was kept as stock for the replicates.

Sample preparation

For sampling, soil samples were homogenized and crushed with mortar and pestle, then sieved in a 2 mm sieve. Three whole plants were harvested from each pot and combined as one single sample. Samples were transported to the laboratory immediately and separated into roots and leaves. Plant samples were washed using tap water and then rinsed with distilled water to remove the adhering soil. The weights of fresh plants were recorded. They were chopped into small pieces and stored in the freezer under 40 to 80 °C for 2 h to 1 day. Then, they were freeze-dried for 3–4 days to remove residual water using Labconco system. The dried samples were ground into a fine powder using a coffee grinder with stainless steel grinding chamber. All samples were stored under 20 °C until the process of extraction.

Sample extraction

An effective preparation method for plant sample extraction was developed by (Wu et al. 2012a). This method was modified to extract the selected compounds from soil and radish samples. Weight of about 0.2 g of sample was transferred in a 50 ml of glass centrifuge tube. Samples were spiked with deuterated pharmaceutical standards as recovery surrogates. Around 10 ml of MTBE (Methyl tertiary-butyl ether) was added and extracted in a sonication bath for 20 min and centrifuged for 3 min at 3600 rpm. The supernatant was decanted in a 40 ml glass vial. 10 ml of acetonitrile was added to the residual and extracted again in a sonication bath

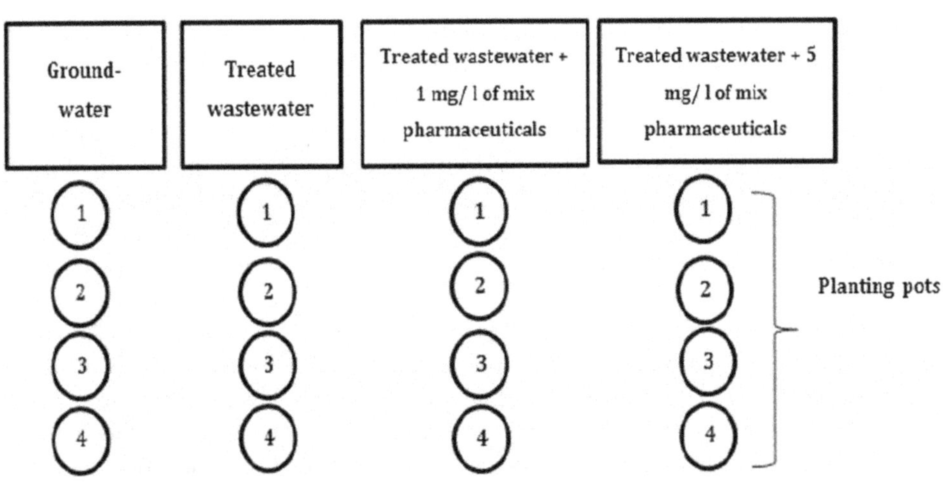

Fig. 1 The layout of the field experiment

for 20 min and centrifuged for 3 min at 3600 rpm, then it was transferred to the same 40 ml glass vial. The combined extract dried under nitrogen gas at 30–40 °C using sample concentrator instrument. Then, it was recovered with 1 ml of methanol, and diluted with 20 ml of ultrapure water. The aqueous sample solutions were loaded under gravity onto a C18 cartridge using a syringe. Air vacuum can also be used to pass the sample through the cartridge by adjusting one drop per second. The cartridge was preconditioned with 3–6 ml of methanol and then with 3–6 ml of ultrapure water before loading the sample. Finally, the analytes were eluted under gravity with 2 ml of methanol and the elution was injected into liquid chromatography–mass spectrometry/mass spectrometry (LC/MS/MS) for detection.

Results and discussion

Soil samples

At the end of the experiment, the values of pH for soil samples were slightly alkaline in the range of 7.9 ± 0.2–7.8 ± 0.1. EC values for all soil samples were low to medium in the range of 0.64 ± 0.17 dS/m to 1.34 ± 0.14 dS/m. The value of pH is negatively correlated to the sorption coefficient of organic compounds and acidic condition promote accumulation in the soil (Zheng et al. 2013).

The organic matter content of soil in all treatments showed similar values in the range of $14.62\% \pm 3.62$ to $14.73\% \pm 1.05$. Therefore, no effects can be assumed on the organic matter content in soils irrigated with treated wastewater even with high concentrations of pharmaceuticals. The organic matter content was found in high levels because the compost containing high amount of cow manure which contributed to increase in organic matter content and other nutrients and minerals in the soil. On the other hand, the organic matter in the soil can affect the availability of pharmaceuticals for plant uptake. According to Miller et al. (2016) organic materials in soil may alter the surface properties and availability of sites for sorption. Then, they can either inhibit or promote sorption of organic compounds to soil. High organic content in soil may result in decreasing sorption of pharmaceuticals to soil particles and making them available for plant uptake or leaching deep in soil layers (Miller et al. 2016).

Pharmaceuticals in soil samples

There was no detection of any of the selected compounds in all soil samples. The disappearance of pharmaceuticals was assumed due to four main reasons; (1) organic compounds can be degraded by biotic and non-biotic process in soil, (2) their degradation was accelerated because of their small biological half-life, (3) high level of organic matter content may play significant role in pharmaceutical sorption to soil particles, (4) complete uptake of pharmaceuticals by plant parts but the non-detectability or low detected levels of those compounds in some plant parts for spiked treatments indicated absence of them from soil. Therefore, there was some uncertainty on the disappearance of those pharmaceuticals from soil. It is also possible that the pharmaceutical leaching beyond the sampling depth but the experiment was not designed to collect the leachates.

The results of this study for soil samples mostly agreed with other studies for the same compounds using soil as planting media but with different crops and experimental conditions. The four selected compounds were not detected in soil samples as found in different studies: trimethoprim and amoxicillin (Boxall et al. 2006), ibuprofen (Winker et al. 2010; Malchi et al. 2014) sulfamethoxazole (Christian et al. 2003). While, Malchi et al. (2014) detected sulfamethoxazole in soil samples.

Radish samples

Plant development and health

In general, a negative effect on plant growth was measured as reduced in the quality and quantity of radish biomass showing the most pronounced effect for treatments spiked with high concentration of mixed pharmaceuticals. Moreover, all plants were attacked by pests such as whiteflies and aphids, the spiked treatments showed less resistance even after using the pesticides to control the pests. The plants in the treatment of high-spiked concentration showed the lowest resistance and some plants lost the mature leaves and few of them did not survive. Plants irrigated with groundwater and treated wastewater were found to be healthy and resisted pests that were exposed to (Fig. 2).

Radish leaves showed normal development for all treatments during growth period except for the spiked treated wastewater treatment. In the last stage of growth, some plants in the spiked treatments declined. In high concentration of spiked pharmaceuticals, some plants died and some of them lost mature leaves. Hence, leaf area was decreasing in that period of growth. The negative effects observed at the high concentrations of pharmaceuticals could be related to plant efficiency in the pharmaceutical uptake at mg/kg level. Pharmaceuticals may inhibit transportation of essential nutrients for growth and toxicity of some organic compounds or elements may cause plant stress and affect the normal growth of plants (Miller et al. 2016). Chlorophyll content showed good and nearly similar values for all treatments in the first week of growth, but it started slightly to decline after reaching maturity. Chlorophyll content of

Fig. 2 Qualitative observation of plant health of the four different treatments, GW: groundwater, TWW: treated wastewater, TW+LC: treated wastewater with low concentration of mixed pharmaceuticals and TW+HC: treated wastewater with high concentration of mixed pharmaceuticals

plants in spiked treatments showed more decrease than other two treatments.

Plants grown using groundwater treatments showed the highest biomass for both roots and shoots. Plants irrigated with treated wastewater also showed good biomass for shoots and roots. Plants irrigated with treated wastewater spiked with low concentration of mixed pharmaceuticals showed similar biomass as treated wastewater treatment. While plants irrigated with treated wastewater of high-spiked concentration showed to have the smallest biomass for both roots and shoots. The loss in leaves of plants and death of other plants in some pots caused the decrease in biomass. Furthermore, the plant roots of high-spiked concentration treatment showed very small biomass.

Pharmaceuticals in plant samples

The complex matrix of plant tissue makes it difficult in the determination of some organic compounds. Two compounds from the selected pharmaceuticals were detected in shoots and roots of radish (*sulfamethoxazole and amoxicillin*). Trimethoprim and ibuprofen were not detected in both plant parts from all treatments, indicating no accumulation or translocation of them into the plant tissues. This was likely because of their small half-lives in the medium (human body, soil, water and other environmental medium) or degradation process by biotic and non-biotic process. Furthermore, the high level of organic matter and alkaline condition in soil decreased their availability in plant soil system. These results did not agree with the study of Boxall et al. (2006), they detected trimethoprim in lettuce and carrot tissue of concentration 5.3–6 μg/kg and Goldstein et al. (2014) detected ibuprofen in tomato and cucumber crops. The accumulation rate in plants of high-concentration-spiked treatment was lower than the plants of low-concentration-spiked treatment and that is because of the small root biomass of plants of high-concentration-spiked treatment; which make small surface area and decrease the available sites for accumulation.

Accumulation in the roots

Analysis of radish roots showed positive detection of pharmaceuticals in both spiked treatments for three compounds of selected pharmaceuticals. Sulfamethoxazole was detected in the roots of both spiked treatments of low (1 mg/l) and high (5 mg/l) concentrations. It was found 1.06 ± 0.63 mg/kg and 1.89 ± 0.68 mg/kg for low- and high-spiked concentrations, respectively. Amoxicillin was also detected in roots of the spiked treatments but in lower concentrations than sulfamethoxazole. It was found to be 0.81 ± 1.41 mg/kg in low spiked treatment and 0.91 ± 1.26 mg/kg in high-spiked treatment (Fig. 2). Accumulation rate could be calculated since to the total amount of plants and soil received was 7.5 and 37.5 mg for low- and high-spiked concentration treatments, respectively. The accumulation and translocation rates ware calculated based on the relationship: [(detected concentration of the compound/the total amount received by plant of the compound) *100]. Around 14% of sulfamethoxazole was accumulated in the roots of low-concentration-spiked treatment, while 5% of it was accumulated in the roots of plants received high concentration of mixed pharmaceuticals. About that 10.8% of amoxicillin was accumulated in plant roots using the treatment of low-spiked concentration, while 2.4% was accumulated on the roots of plants received high concentration of mixed pharmaceuticals.

Translocation to the leaves

The translocation of pharmaceuticals occurred if there was sufficient concentration of them in the media. Sulfamethoxazole translocated from roots to the leaves was found in plants irrigated with treated wastewater spiked with high concentration of mix pharmaceuticals (Fig. 3). The translocation

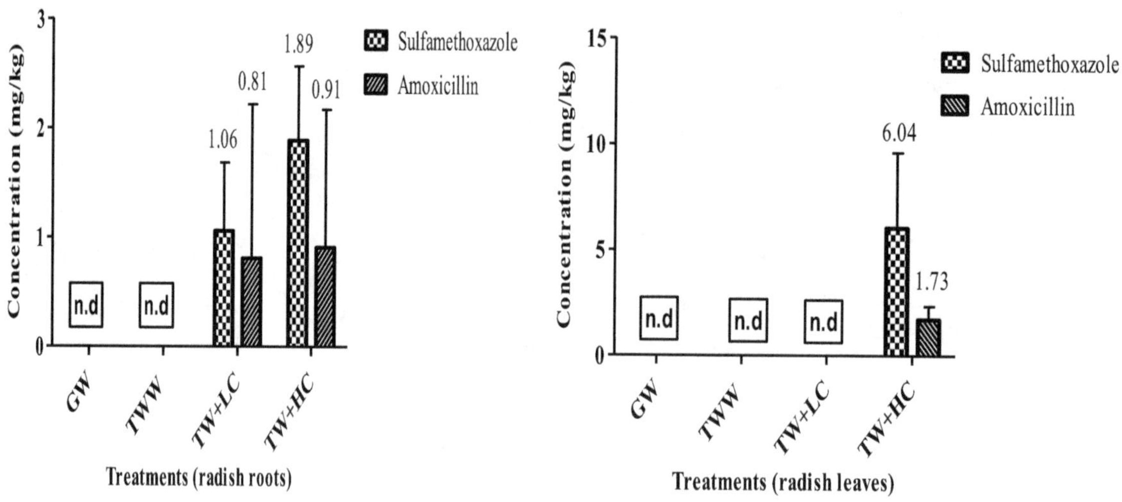

Fig. 3 Detected concentrations of pharmaceuticals accumulated in radish roots and that were translocated to the radish leaves (n.d: not detected), the results were calculated for three replicate samples and presented as mean values, GW: groundwater, TWW: treated wastewater, TW + LC: treated wastewater with low concentration of mixed pharmaceuticals and TW + HC: treated wastewater with high concentration of mixed pharmaceuticals

of sulfamethoxazole and amoxicillin were 6.04 ± 3.56 and 1.73 ± 0.63 mg/kg, respectively. The transferred rate of sulfamethoxazole into the leaves was 16%, while 4.6% of amoxicillin was transferred to the radish leaves.

The translocation and accumulation rate of sulfamethoxazole was higher than amoxicillin in both treatments. Sulfamethoxazole has higher biological half-life in the available medium (human body, soil, water and other environmental medium) by ten times than amoxicillin, which make it available for a longer time, and then it can accumulate and transfer in higher rate. The other percentage of them were assumed to be degraded in the soil or leached deep. The leachate was too small in amount and it was not collected after irrigation. The detected concentrations of selected pharmaceuticals in both roots and shoots showed that radish has the potentiality to uptake and transfer high levels of those pharmaceuticals, which may be available in planting soil. The summary of the results is presented in Table 1.

Results obtained in this experiment did not agree with the findings of Boxall et al. (2006), amoxicillin was not detected in the test crops (lettuce and carrot). While the study of Malchi et al. (2014) agreed with this study for sulfamethoxazole as it was detected in the roots of sweet potatoes and carrots but in small concentrations (0.05 – 0.24 ng/g).

Conclusion

Radish showed positive uptake of some selected pharmaceuticals from soil irrigated by treated wastewater with spiked concentration of pharmaceuticals and had the ability to accumulate them in the roots. It was found that sulfamethoxazole accumulated in radish roots at 1.06 ± 0.63 and 1.89 ± 0.68 mg/kg for low- and high-spiked concentrations, respectively. Amoxicillin accumulated in roots in concentrations of 0.81 ± 1.41 mg/kg in low-spiked treatment and

Table 1 Summary of the detected concentrations of selected pharmaceuticals in radish leaves and roots

Treatment compounds	Accumulation in the roots				Translocation to the leaves			
	Concentrations (mg/kg)		% of total exposed amount		Concentrations (mg/kg)		% of total exposed amount	
	TW + LC	TW + HC	TW + LC	TW + HC	TW + LC	TW + HC	TW + LC	TW + HC
Amoxicillin	0.81 ± 1.41	0.91 ± 1.26	10.8	2.4	n.d	1.73 ± 0.63	n.d	4.6
Sulfamethoxazole	1.06 ± 0.63	1.89 ± 0.68	14	5	n.d	6.04 ± 3.56	n.d	16

The results were calculated for three replicate samples and presented as mean values, groundwater (GW) and treated wastewater (TWW) treatments were free from the four pharmaceuticals. Ibuprofen and trimethoprim were not detected in the spiked treatments. TW + LC: treated wastewater with low concentration of mixed pharmaceuticals and TW + HC: treated wastewater with high concentration of mixed pharmaceuticals, *n.d* not detected

0.91 ± 1.26 mg/kg in high-spiked treatment. Furthermore, radish has the ability to translocate selected pharmaceuticals (sulfamethoxazole and amoxicillin) from soil irrigated with treated wastewater with spiked concentration of pharmaceuticals if they are present in sufficient concentrations. Soil degraded the selected pharmaceuticals by biotic and non-biotic process. Therefore, pharmaceutical usage and disposal must be given attention to prevent their occurrences in the environment. Furthermore, treated wastewater must be given attention, as it is an important source of water in Oman and there are plans to supply treated wastewater for irrigation to cater the shortages in availability of water resources.

References

Abdul Khaliq SJ, Al-Busaidi A, Ahmed M, Al-Wardy M, Agrama H, Choudri BS (2017) The effect of municipal sewage sludge on the quality of soil and crops. Int J Recycl Org Waste Agric 6(4):289–299. https://doi.org/10.1007/s40093-017-0176-4

Alkhamisi SA, Ahmed M, Al-Wardy M, Prathapar SA, Choudri BS (2017) Effect of reclaimed water irrigation on yield attributes and chemical composition of wheat (*Triticum aestivum*), cowpea (*Vigna sinensis*), and maize (*Zea mays*) in rotation. Irrig Sci 35(2):87–98. https://doi.org/10.1007/s00271-016-0522-8

Baawain MS, Al-Omairi A, Choudri BS (2014a) Characterization of domestic wastewater treatment in Oman from three different regions and current implications of treated effluent. Environ Monit Assess 186(5):2701–2716. https://doi.org/10.1007/s10661-013-3572-x

Baawain MS, Al-Jabri M, Choudri BS (2014b) Characterization of domestic wastewater sludge in Oman from three different regions and recommendations for alternative reuse applications Iran. J Public Health 43(2):168–177

Baawain MS, Al-Jabri M, Choudri BS (2015) Characterization of industrial wastewater sludge in Oman from three different regions and recommendations for alternate reuse applications Iran. J Public Health 44(11):1473–1481

Boxall AB, Johnson P, Smith EJ, Sinclair CJ, Stutt E, Levy LS (2006) Uptake of veterinary medicines from soils into plants. J Agric Food Chem 54(6):2288–2297. https://doi.org/10.1021/jf053041t

Carballa M, Omil F, Lema JM, Llompart MA, García-Jares C, Rodríguez I, Gomez M, Ternes T (2004) Behavior of pharmaceuticals, cosmetics and hormones in a sewage treatment plant. Water Res 38(12):2918–2926. https://doi.org/10.1016/j.watres.2004.03.029

Chefetz B, Mualem T, Ben-Ari J (2008) Sorption and mobility of pharmaceutical compounds in soil irrigated with reclaimed wastewater. Chemosp 73(8):1335–1343. https://doi.org/10.1016/j.chemosphere.2008.06.070

Christian T, Schneider RJ, Färber HA, Skutlarek D, Meyer MT, Goldbach HE (2003) Determination of antibiotic residues in manure, soil, and surface waters. Clean Soil Air Water 31(1):36–44

Dolliver H, Kumar K, Gupta S (2007) Sulfamethazine uptake by plants from manure-amended soil. J Environ Qual 36(4):1224–1230. https://doi.org/10.2134/jeq2006.0266

Eggen T, Asp TN, Grave K, Hormazabal V (2011) Uptake and translocation of metformin, ciprofloxacin and narasin in forage-and crop plants. Chemosph 85(1):26–33. https://doi.org/10.1016/j.chemosphere.2011.06.041

Goldstein M, Shenker M, Chefetz B (2014) Insights into the uptake processes of wastewater-borne pharmaceuticals by vegetables. Environ Sci Technol 48(10):5593–5600. https://doi.org/10.1021/es5008615

Herklotz PA, Gurung P, Heuvel BV, Kinney CA (2010) Uptake of human pharmaceuticals by plants grown under hydroponic conditions. Chemosph 78(11):1416–1421

Islam G (2013) Impact of tetracycline on microbial communities in the secondary treatment process of wastewater treatment systems. Master's Thesis, Ryerson University, Toronto, ON, Canada

Jaffar Abdul Khaliq S, Ahmed M, Al-Wardy M, Al-Busaidi A, Choudri BS (2017) Wastewater and sludge management and research in Oman: an overview. J Air Waste Manag Assoc 67(3):267–278. https://doi.org/10.1080/10962247.2016.1243595

Ji X, Shen Q, Liu F, Ma J, Xu G, Wang Y, Wu M (2012) Antibiotic resistance gene abundances associated with antibiotics and heavy metals in animal manures and agricultural soils adjacent to feedlots in Shanghai; China. J Hazard Materials. https://doi.org/10.1016/j.jhazmat.2012.07.040

Joss A, Zabczynski S, Göbel A, Hoffmann B, Löffler D, McArdell CS, Ternes TA, Thomsen A, Siegrist H (2006) Biological degradation of pharmaceuticals in municipal wastewater treatment: proposing a classification scheme. Water Res 40(8):1686–1696. https://doi.org/10.1016/j.watres.2006.02.014

Kinney CA, Furlong ET, Werner SL, Cahill JD (2006) Presence and distribution of wastewater-derived pharmaceuticals in soil irrigated with reclaimed water. Environ Toxicol Chem 25(2):317–326. https://doi.org/10.1897/05-187R.1

Kinney CA, Furlong ET, Kolpin DW, Burkhardt MR, Zaugg SD, Werner SL, Bossio JP, Benotti MJ (2008) Bioaccumulation of pharmaceuticals and other anthropogenic waste indicators in earthworms from agricultural soil amended with biosolid or swine manure. Environ Sci Technol 42(6):1863–1870. https://doi.org/10.1021/es702304c

Kolpin DW, Furlong ET, Meyer MT, Thurman EM, Zaugg SD, Barber LB, Buxton HT (2002) Pharmaceuticals, hormones, and other organic wastewater contaminants in US streams, 1999–2000: a national reconnaissance. Environ Sci Technol 36(6):1202–1211. https://doi.org/10.1021/es011055j

Kraigher B, Kosjek T, Heath E, Kompare B, Mandic-Mulec I (2008) Influence of pharmaceutical residues on the structure of activated sludge bacterial communities in wastewater treatment bioreactors. Water Res 42(17):4578–4588. https://doi.org/10.1016/j.watres.2008.08.006

Kumar K, Gupta S, Baidoo S, Chander Y, Rosen C (2005) Antibiotic uptake by plants from soil fertilized with animal manure. J Environ Qual 34(6):2082–2085. https://doi.org/10.2134/jeq2005.0026

Kümmerer K (2008) Pharmaceuticals in the environment—a brief summary. In: Kümmerer K (ed) Pharmaceuticals in the environment. Springer, Berlin, pp 3–21. https://doi.org/10.1007/978-3-540-74664-5-1

Malchi T, Maor Y, Tadmor G, Shenker M, Chefetz B (2014) Irrigation of root vegetables with treated wastewater: evaluating uptake of pharmaceuticals and the associated human health risks. Environ Sci Technol 48(16):9325–9333. https://doi.org/10.1021/es5017894

Miller EL, Nason SL, Karthikeyan K, Pedersen JA (2016) Root uptake of pharmaceuticals and personal care product ingredients. Environ Sci Technol 50(2):525–541. https://doi.org/10.1021/acs.est.5b01546

Mompelat S, Le Bot B, Thomas O (2009) Occurrence and fate of pharmaceutical products and by-products, from resource to drinking water. Environ Int 35(5):803–814. https://doi.org/10.1016/j.envint.2008.10.008

Pedersen JA, Soliman M, Suffet I (2005) Human pharmaceuticals, hormones, and personal care product ingredients in runoff from agricultural fields irrigated with treated wastewater. J Agric Food Chem 53(5):1625–1632. https://doi.org/10.1021/jf049228m

Wang Y, Hu W, Cao Z, Fu X, Zhu T (2005) Occurrence of endocrine-disrupting compounds in reclaimed water from Tianjin, China. Anal Bioanal Chem 383(5):857–863

Weng S, Sun P, Ben W, Huang CH, Lee LT, Blatchley ER III (2014) The presence of pharmaceuticals and personal care products in swimming pools. Environ Sci Technol Lett 1(12):495–498. https://doi.org/10.1021/ez5003133

Winker M, Clemens J, Reich M, Gulyas H, Otterpohl R (2010) Ryegrass uptake of carbamazepine and ibuprofen applied by urine fertilization. Sci of The Total Environ 408(8):1902–1908. https://doi.org/10.1016/j.scitotenv.2010.01.028

Wu C, Spongberg AL, Witter JD, Fang M, Czajkowski KP (2010) Uptake of pharmaceutical and personal care products by soybean plants from soils applied with biosolids and irrigated with contaminated water. Environ Sci Technol 44(16):6157–6161. https://doi.org/10.1021/es1011115

Wu C, Spongberg AL, Witter JD, Sridhar BM (2012a) Transfer of wastewater associated pharmaceuticals and personal care products to crop plants from biosolids treated soil. Ecotoxicol Environ Saf 85:104–109

Wu X, Conkle JL, Gan J (2012b) Multi-residue determination of pharmaceutical and personal care products in vegetables. J Chromatogr A 1254:78–86. https://doi.org/10.1016/j.chroma.2012.07.041

Wu X, Ernst F, Conkle JL, Gan J (2013) Comparative uptake and translocation of pharmaceutical and personal care products (PPCPs) by common vegetables. Environ Int 60:15–22. https://doi.org/10.1016/j.envint.2013.07.015

Wu X, Conkle JL, Ernst F, Gan J (2014) Treated wastewater irrigation: uptake of pharmaceutical and personal care products by common vegetables under field conditions. Environ Sci Technol 48(19):11286–11293. https://doi.org/10.1021/es502868k

Xu J, Wu L, Chen W, Jiang P, Chang ACS (2009) Pharmaceuticals and personal care products (PPCPs), and endocrine disrupting compounds (EDCs) in runoff from a potato field irrigated with treated wastewater in southern California. J Health Sci 55(2):306–310

Zheng H, Wang Z, Zhao J, Herbert S, Xing B (2013) Sorption of antibiotic sulfamethoxazole varies with biochars produced at different temperatures. Environ Pollut 181:60–67. https://doi.org/10.1016/j.envpol.2013.05.056

7

Poultry feather waste management and effects on plant growth

J. C. Joardar[1] · M. M. Rahman[1]

Abstract

Purpose Poultry industries produce a lot of feathers which are considered as waste and needs to be managed properly. Poultry feathers are rich with keratin protein and therefore they could be a source for good nitrogen fertilizer. Proper treatment of poultry feather waste (PFW) might be an environmental friendly solid waste management tool and a good source of N-rich organic fertilizer.

Methods PFW was treated under a mixture of both aerobic and anaerobic processes. Treated poultry feather waste (TPFW) was analyzed for different nutrient elements. TPFW was applied to the soil at 4, 8, 12 and 16 t/ha along with control. *Ipomoea aquatica* was grown as the test plant to evaluate the growth performance under different rates of TPFW application. Plants were harvested 7 weeks after seed germination.

Results TPFW contained higher amount of organic matter (35.9%) and total nitrogen (4%). Other major nutrient elements were found to be satisfactory. Color of the plants were observed greener in TPFW applied plants than control plants and the green color was pronounced with increasing rate of TPFW application. Plant height (cm/plant), leaf number per plant and weight of plants (g/plant) was also increased significantly due to the application of TPFW at 12 t/ha and above.

Conclusion Proper treatment of PFW might be an environmental friendly, cost effective and sustainable strategy for PFW management that will also play a vital role in nutrient (especially nitrogen) recycling to the soil.

Keywords Poultry · Feather · Waste management · Compost · Nutrient recycling

Introduction

Commercial poultry production has grown rapidly in Bangladesh and it is one of the emerging and important sectors in Bangladesh that has been contributing progressively to the national economy. Poultry sector is playing a very significant role in the reduction of poverty, malnutrition and unemployment problems. At the same time, poultry farms produce a lot of different types of waste products. A mixture of fecal excreta, bedding material, remaining feed, departed chickens, damaged eggs, packing materials and feathers are the major components of this waste. It also includes waste from poultry carrying materials, conveyer belt and cleaning systems (Kelleher et al. 2002).

Poultry feather constitutes up to 10% of the total chicken body weight (Grazziotin et al. 2006) and produce huge amounts in the poultry slaughterhouses as well as in the household during poultry processing. These billions of kilograms of feather have been creating a serious solid waste problem all over the world (Schmidt 1998; Hegedus et al. 1998). Still now, it is very difficult and expensive to dispose of poultry feather and its present uses are not economically valued (Nahm and Nahm 2004). In past and sometimes at present, feathers are roasted or treated at raised temperature and pressure. After drying it is ground to make powder for further use as a feed supplement for livestock. This process is quite expensive and contains protein with low nutritional value (Ekta and Rani 2012). Disposal methods such as burning (cause air pollution) or burying (require a lot of lands and subsequent soil and water pollution) are also occasionally used but they are not environment friendly. Feather meal is produced in developed countries as the management strategy of poultry feather waste by either steam or chemical treatment (Valtcho and Zheljazkov 2005). This poultry feather meal is a readily available and low-cost source of nitrogen (up to 15% total N) and function as a potential biofertilizer (Jeong et al. 2010). Whether it is fresh or dried,

✉ J. C. Joardar
jcjoardar@yahoo.com

[1] Soil, Water and Environment Discipline, Khulna University, Khulna 9208, Bangladesh

feathers are used as fertilizer and also used in landfilling (Pettett and Kurtboke 2004). The consequences of all these processes have some limitations regarding the product quality, cost proficiency, pollution and sustainable environmental quality control (Kim and Patterson 2000). Thus, cost effective and environment friendly poultry feather waste management methods are necessary to mitigate their threat on the environment. Poultry feather, being a rich source of keratin proteins and amino acids, can be converted into valuable N-rich organic fertilizer. Bangladesh being a densely populated country, the soils of the Bangladesh have been subjected of extensive use for cultivation for a long time to meet the future food challenge. Regardless use of high yielding varieties of crops with chemical fertilizer, crop production in Bangladesh is still not enough in comparison to other developed ones due to loss of soils physical health. With the intensification of agriculture, these soils have also witnessed depletion of nutrients and productivity. Composting might be one of the most cost effective and environment friendly approaches of reutilizing the poultry feather waste (Ichida et al. 2001). Composting through anaerobic incorporation is a sustainable and economically viable waste management tool, which drops the greenhouse gases (GHGs) emission, reduces soil and water pollution, lowers the dependence on fossil fuels and produces important nutrient rich organic fertilizer. A good amount of both macro and micro-nutrients retained in organic fertilizers (Gaur 1997; Kalaivanan and Hattab 2016) that improves the physicochemical and biological properties of soil (Adhami et al. 2014; Lim et al. 2015). It is a prerequisite to maintain high agriculture productivity to provide and fulfill the future food demands of fast rising population (Perez-Montano et al. 2014). Chemical fertilizers are being used throughout the world to increase the crop productivity. Excess use of chemical fertilizers enter into the water bodies through rain water causes water pollution too. So, to manage poultry feather waste, composting might be a cheap and easy way of feather-utilization.

The objectives of the present research work were: to manage poultry feather waste through a treatment of a mixture of both aerobic and anaerobic decomposition processes; to examine the influence of treated poultry feather waste on plant growth; and to determine the efficiency of poultry feather waste to be used as nitrogen rich organic fertilizer.

Materials and methods

The research had two parts: (i) poultry feather waste (PFW) was treated as a method of feather waste management, and (ii) pot experiment was carried out in order to evaluate the influence of treated poultry feather waste (TPFW) on growth performance of *gima kalmi* (*Ipomoea aquatica*) as a part of integrated waste management through nutrient recycling.

Poultry feather waste collection

Poultry feather waste was collected from nearby slaughterhouses situated in *Gollamary* green market very close to the Khulna University campus, Bangladesh.

Compost preparation

The collected PFW was first washed well to remove blood and small pieces of meat from feather and the water was removed completely. A small pit (106.7 cm × 76.2 cm × 61 cm) was dug to bury the PFW. The bulk amount of PFW was taken onto the polyethene paper and covered with the polyethene which kept the PFW separated from the soil. Then the PFW which was set inside the polyethene was buried into the soil. The pile was filled with the soil carefully so that soil cannot penetrate to feathers. Again a piece of polyethene was laid on the surface of the soil so that the rain water cannot enter into the pile. The pit was opened at every 10 days interval of time to see the status of PFW decomposition and to control the temperature of feather pile buried into the soil. After 3 months, finally, the pit was opened and the decomposed material was brought out from the pit and was sun dried for 7 days to get final TPFW which was brown in color. Then the TPFW was crushed and passed through 0.5 mm sieve and this sieved material was applied to the soil accordingly.

Soil sample collection

An agricultural field behind the Khulna University campus, Bangladesh was selected for soil sampling. Soil samples at 0–15 cm depth from the surface were collected by following the composite soil sampling method (USDA 2004).

Preparation of soil sample

The collected bulk soils were air-dried spreading on a sheet of paper. The air-dried sample was then broken by crushing it gently using a wooden hammer, mixed thoroughly and were passed through 2-mm sieve. The processed soil was used for plant growth. A tiny portion of the soil was broken again and passed through a 0.5-mm sieve for further laboratory analysis.

Test plant

Gima kalmi (Ipomoea aquatica), a popular vegetable in Bangladesh, was grown as the test plant in the experiment. *Gima kalmi* is a semi-aquatic, tropical plant grown as a vegetable for its tender shoots and leaves. *Gima kalmi* is a leafy vegetable which belongs to the family

Convolvulaceae. The vegetable crop is also known as kangkong, swamp cabbage, water convolvulus, water spinach etc. (Tindal 1983). Use of fertilizer increases the yield of plants, produces high valued foliage and stimulates carbohydrate synthesis (Rai 1981). For successful production, *gima kalmi* requires early and rapidly vegetative growth, which could be influenced by the application of fertilizers. In Bangladesh fertilizer is mostly used as the source of NPK and split application of this fertilizer is commonly practised for leafy vegetable production (Hossain 1990).

Experiment setup

Earthen pots (2-L) containing 1-kg soil in each pot were used. The surface area of the pot was 284 cm^2 and depth was 18 cm. There was a small hole in the bottom of the pots so that the extra water could run away and the roots could free from waterlogging condition. Four rates [4 t/ha (T_1), 8 t/ha (T_2), 12 t/ha (T_3) and 16 t/ha (T_4)] of TPFW were applied along with control (T_0) and there were three replications for each treatment. After mixing TPFW into the soil, the soil was watered well and kept for 2 weeks. About 15–20 seeds of *gima kalmi* were sown in each pot. The pots were arranged in a completely randomized way and the positions were changed every alternative day so that all the pots get equal sunlight. The experiment was carried out under net house condition which is free from birds. The plants were thinned 1 week after seed germination by keeping 10 healthy plants in each pot. As the plant protection measure pesticide (viz. ripcord, bought from local market) was sprayed as needed.

After 7 weeks from the emergence of plants, the plants were harvested by cutting the stems 1-cm above the ground. Visual symptoms were monitored carefully and noted throughout the growing period. After harvest the plant height (cm), number of leaves, fresh weight (g) and dry weight (g) were measured and recorded.

Laboratory analysis of soil and TPFW

Some important properties of soil and TPFW were analyzed in the laboratory by following the recommended protocols (Imamul Huq and Alam 2005).

Agronomic parameters

All the plants in each pot were considered for data collection. Data were collected in respect of the following characters.

Plant height

Plant height was measured in centimeter by using a measuring scale from the ground level to the tip of the stem after harvest. The data was presented as the average height in cm per plant (cm/plant).

Number of leaves

Total leaves of the plants from each pot were counted. The data was recorded and presented as the average number of leaves per plant.

Fresh weight of plant

After harvest the fresh weight (stem plus leaves) of the plants were measured and recorded. The data was presented as the average fresh weight in gram per plant (g/plant).

Dry weight of plant

After taking the fresh weight, the plant samples were dried in an electric oven at 70 °C for 72 h. Then the sample was transferred into desiccators and allowed to cool down to the room temperature. The dry weight of the sample was taken with help of an electric balance and calculated in gram per plant (g/plant).

Statistical analysis

Averages value of three replicated data were expressed as result. A computer built-in statistical software program Minitab-16 was used for data analysis by employing ANOVA technique. Differences between means were statistically analyzed using Fisher method ($p = 0.05$). Computer built-in Microsoft Excel-2010 program was used for all graph preparation.

Results and discussion

Soil analysis

Some basic properties of the soil that was used in the experiment as the medium for plant growth are presented in Table 1. The soil contained low available nutrients including very low total N (0.09%). The soil was silty clay loam in texture and pH of soil was 7.8.

Treated poultry feather waste (TPFW) analysis

The composition of treated poultry feather waste is presented in Table 2. After treatment the poultry feather

Table 1 Some basic properties of soil

Soil textural class	pH	EC (dS/m)	OM (%)	Total N (%)	Available nutrients (μg/g)				
					P	K	S	Zn	B
Silty clay loam	7.8	1.29	1.49	0.09	21.36	0.03	59.39	0.68	0.45

Table 2 Composition of treated poultry feather waste (TPFW)

Nutrients	pH	OM	Total N	P	K	S	Zn	Ca	Mg	C/N ratio
		(%)								
Content	6.75	35.9	4.0	0.50	0.40	0.20	0.02	0.80	0.70	5.21

Fig. 1 Growing plants (*gima kalmi*) treated with different rates of TPFW. [T_0=control, T_1=4 t/ha, T_2=8 t/ha, T_3=12 t/ha and T_4=16 t/ha]

waste reduced waste volume significantly converting waste into valuable organic fertilizer. TPFW contained very high (35.9%) (Table 2) organic matter and total nitrogen (4%) which is the focal point of the present research.

Effect of treated poultry feather waste on plant growth

The effect of TPFW at different rate of application on the growth of the plant (*gima kalmi*) is depicted in the following section.

Visual observation

The color of the plants was greener where TPFW was applied as compared to control plant and the green color of leaves was gradually increased with the application of higher rate of TPFW (Fig. 1). On the other hand, leaves of control plants were yellowish-green in colour. This yellowish colour might be due to the deficiency symptoms of nitrogen in soil. High photosynthetic activity, vigorous vegetative growth and dark green color of the plant is determined by the element nitrogen which is an important component of chlorophyll. Silas et al. (2012) reported that plants show chlorosis due to N deficiency which result in yellowish-green leaves, restricted growth, and fall of lower leaves.

Number of leaves per plant

The presence of the highest number of leaves in a plant could be indicative of higher yield in a particular plant. The number of leaves per plant treated with different rates of TPFW is presented in Table 3. The number of leaves in plants increased with the application of TPFW in soil. The maximum number of leaves per plant was counted in T_2 treatment (8 t/ha) and significantly higher than any other treatment (Table 3).

Height of the plant

Plant height is one of the major indicators that is considered for growth performance. The height of the plant *gima kalmi*

Table 3 Effect of TPFW on the number of leaves, plant height, fresh weight and dry weight of *gima kalmi*

Agronomic attributes	Treatment				
	T_0	T_1	T_2	T_3	T_4
No. of leaves per plant	12.0 ± 0.50^{c}	13.0 ± 0.50^{bc}	15.0 ± 1.0^{a}	14.0 ± 1.53^{abc}	15.0 ± 0.20^{ab}
Plant height (cm/plant)	29.3 ± 1.15^{b}	31.4 ± 1.0^{ab}	32.6 ± 1.25^{ab}	34.1 ± 1.0^{a}	32.5 ± 2.19^{ab}
Fresh weight (g/plant)	3.74 ± 0.60^{b}	3.56 ± 0.39^{b}	4.23 ± 0.07^{ab}	4.94 ± 0.46^{a}	4.93 ± 0.39^{a}
Dry weight (g/plant)	0.56 ± 0.08^{ab}	0.51 ± 0.07^{b}	0.59 ± 0.06^{ab}	0.67 ± 0.02^{ab}	0.70 ± 0.06^{a}

Data represents the average ± standard deviations (n=3). Different letters show the significant differences (p=0.05). [T_0=control, T_1=4 t/ha, T_2=8 t/ha, T_3=12 t/ha and T_4=16 t/ha TPFW application]

treated with different rates of TPFW is shown in Table 3. The highest height (34.06 cm) was recorded at T_3 treatment (12 t/ha) while the shortest (29.33 cm) was in control. Statistical analysis of the result showed that there was a significant difference of the plant height at T_3 (12 t/ha) (Table 3). So, the agronomic parameter showed that the height of the plant was significantly increased when TPFW was applied at 12 t/ha.

Weight of plants

Both the fresh and dry weight (g/plant) of the plants are also presented in the Table 3. The weight of the plants was increased with increasing rate of TPFW application. The weight was significantly increased when the plants were treated with TPFW at 12 t/ha (T_3) and above. The biomass of the plant was significantly increased having promoted canopy and shoot production when TPFW was applied at 12 and 16 t/ha. This implies that superfluous TPFW application would promote canopy and shoot production. This observation is similar to the findings of Myers (1998). In addition, leafy vegetable crops required more nitrogen which is provided by higher application of TPFW, and these higher nitrogen levels increased the yield and biomass content. Our findings also strongly advocated by Verma et al. (1969).

The effects of compost compared to chemical fertilizers are often misunderstood. The nutrients contained in the chemical fertilizers are used rapidly but completely, whereas the nutrients supplied by the compost are used slowly and kept in the soil over a prolonged period (Kumazawa 1984). Treatment of poultry feather waste reduced the volume of solid waste considerably. Farrell and Jones (2009) claimed composting can reduce the mass of solid wastes by about 40% which support the present results of composting from poultry feather waste. Treatment of poultry feather waste enables the recovery of nutrients and is one of the sustainable strategies for dealing with solid waste. The total nitrogen (N) content in treated poultry feather waste is 4% which is relatively higher than that of others (Table 2). Kim et al. (2001) hypothesized that degraded feather which contains keratin is a rich source of nitrogen. The pH value of the TPFW was 6.75 which may support the view of Hubbe et al. (2010) who assumed that pH of final compost drops to neutral. The organic matter (OM) content in the TPFW is very high (Table 2) which illustrated that application of TPFW perhaps one of the best options to reclaim soil organic matter. The carbon and nitrogen ratio of the treated feather was 5.21. Malone (2004) stated that composting is a biological process by which organic material is broken down and decomposed. Treated poultry feather waste could be applied in growing crops instead of and/or with commercial fertilizer as a supplementary source of nutrients for plant growth. The result in the present research showed that the application of TPFW in soil significantly increased the agronomic parameters of *gima kalmi* at 12 t/ha and above. El-Desuki et al. (2001) also stated that plant height, number of leaves and dry weight of plants was significantly increased by increasing the levels of TPFW up to 12 t/ha which support the present results found in the case of TPFW application in *gima kalmi*. The positive effect of compost on plant height, plant growth, yield, leaves color, the number of leaves and dry mass are also reported (El-Raoof 2001; Abdel-Mawgoud 2006; Olaleye et al. 2008).

To produce compost from poultry feather as soil conditioner and organic fertilizer need a large volume of feather waste, thereby it will certainly reduce the solid waste burden from the poultry industries in the environment. In addition to this, it could help in reducing the wide gap between availability and requirement of organic fertilizer for soils in Bangladesh. So, it can be assumed that treatment of poultry feather waste could unlock a possibility for developing a suitable, environment friendly, cost effective and sustainable strategy for poultry feather waste management that will also play a vital role in nutrient recycling to the soil.

Conclusion

Treatment of poultry feather waste under a mixture of both aerobic and anaerobic decomposition processes has been verified in this study. Feather waste generated from poultry industries can be monitored and efficiently managed by employing this strategy. The process adopted is a viable method of converting the useful nutrient present in feathers to soil nutrient that can be used to enhance the crop yield. Feather waste generated from poultry industries could be considered as resource material for nutrient recycling through composting because every day large quantity of feather waste is being generated, it is a good source of nitrogen which is a growth promoter, and the application of TPFW improve the growth performance of crop which has been shown in the present study. It is expected that proper composting of PFW might be an environment friendly, low-cost and sustainable strategy of PFW management. The role of TPFW in nutrient recycling to the soil will increase crop production and will help to improve the food security in meeting the Millennium Development Goals (MDGs) in the country.

Compliance with ethical standards

Conflict of interest The authors would like to declare that there is no conflict of interest with this research.

References

Abdel-Mawgoud AMR (2006) Growth, yield and quality of green bean (*Phaseolus vulgaris*) in response to irrigation and compost applications. J Appl Sci Res 2:443–450

Adhami E, Hosseini S, Owliaie H (2014) Forms of phosphorus of vermicompost produced from leaf compost and sheep dung enriched with rock phosphate. Int J Recycl Org Waste Agric 3:68. https://doi.org/10.1007/s40093-014-0068-9

Ekta T, Rani G (2012) Rapid conversion of chicken feather to feather meal using dimeric keratinase from *Bacillus licheniformis* ER-15. J Bioprocess Biotech 2:4. https://doi.org/10.4172/2155-9821.1000123

El-Desuki M, Amer AH, Omaima M, Sawan Khattab MA (2001) Effect of irrigation and organic fertilization on the growth, bulb yield and quality of sweet fennel under shark El-owinat condition. J Agric Sci Mansoura Univ. 26(7):4465–4481

El-Raoof MA (2001) Productivity of sweet basil cultivated in newly reclaimed land as well as estimation the effect of different organic fertilization levels and plant densities. M.Sc. Dissertation, Faculty of Agriculture, Ain Shams University, Egypt

Farrell M, Jones DL (2009) Critical evaluation of municipal solid waste composting and potential compost markets. Bioresour Technol 100(19):4301–4310. https://doi.org/10.1016/j.biortech.2009.04.029

Gaur AC (1997) Bulky organic manures and crop residues. In: Tandon HLS (ed) Fertilizers, organic manures, recyclable wastes and biofertilizers. FDCO, New Delhi, pp 37–51

Grazziotin A, Pimentem FA, de Jong EV, Brandelli A (2006) Nutritional improvement of feather protein by treatment with microbial keratinase. Anim Feed Sci Tech 126:135–144

Hegedus M, Schmidt J, Rafai P (1998) Állati eredetű melléktermékek hasznosítása (Animal by-product recycle). Mezőgazda Publischer, Budapest

Hossain MZ (1990) Effects of nitrogen and sulphur on growth, yield and nutrient content of rice. M.Sc. Dissertation, Department of Soil Science, Bangladesh Agricultural University, Mymensingh, Bangladesh

Hubbe MA, Nazhad M, Sánchez C (2010) Composting as a way to convert cellulosic biomass and organic waste into high-value soil amendments: a review. BioResources 5(4):2808–2854

Ichida JM, Krizova L, Lefevre CA, Keener HM, Elwell D, Burtt EH (2001) Bacterial inoculum enhances keratin degradation and biofilm formation in poultry compost. J Microbiol Methods 47(2):199–208. https://doi.org/10.1016/S0167-7012(01)00302-5

Imamul Huq SM, Alam MD (2005). A handbook on analyses of soil, plant and water. BACER-DU. University of Dhaka, Bangladesh

Jeong JH, Lee OM, Jeon YD, Kim JD, Lee NR, Lee CY, Son HJ (2010) Production of keratinolytic enzyme by a newly isolated feather-degrading *Stenotrophomonas maltophilia* that produces plant growth-promoting activity. Process Biochem 45:1738–1745. https://doi.org/10.1016/j.procbio.2010.07.020

Kalaivanan D, Hattab KO (2016) Recycling of sugarcane industries byproducts for preparation of enriched pressmud compost and its influence on growth and yield of rice (*Oryza sativa* L.). Int J Recycl Org Waste Agric 5:263–272. https://doi.org/10.1007/s40093-016-0136-4

Kelleher BP, Leahy JJ, Henihan AM, O'Dwyer TF, Sutton D, Leahy MJ (2002) Advances in poultry litter disposal technology. Bioresour Technol 83(1):27–36. https://doi.org/10.1016/S0960-8524(01)00133-X

Kim WK, Patterson PH (2000) Nutritional value of enzyme or sodium hydroxide-treated feathers from dead hens. Poult Sci 79(4):528–534

Kim JM, Lim WJ, Suh HJ (2001) Feather-degrading Bacillus species from poultry waste. Process Biochem 37(3):287–291

Kumazawa K (1984) Beneficial effects of organic matter on rice growth and yield in Japan. Organic matter and rice. International Rice Research Institute, Manila, pp 431–444

Lim SL, Wu TY, Lim PN, Shak KPY (2015) The use of vermicompost in organic farming: overview, effects on soil and economics. J Sci Food Agric 95(6):1143–1156. https://doi.org/10.1002/jsfa.6849

Malone B (2004) Compositing poultry losses. In: Proceedings of poultry information exchange. Queensland, Australia, 19 April 2004. pp 39–42

Myers RL (1998) Nitrogen fertilizer effect on grain. Agron J 90(5):597–602. https://doi.org/10.2134/agronj1998.00021962009000050005x

Nahm KH, Nahm BA (2004) Poultry production and waste management. Yu Han Publishing, Republic of Korea

Olaleye AO, Ndubuaku UM, Dada OA (2008) Comparative study of the performance of jute plant (*Corchorus olitorius*) on home garden soil, farmland and cocoa plantation soils as influenced by varying levels of N-fertilizer. J Trop Agric Food Environ Ext 7(1):78–84

Perez-Montano F, Alias-Vilegas C, Bellogin RA, Del Cerro P, Espunny MR, Jimenez-Guerrero I, Lopez Baena FJ, Ollero J, Cubo T (2014) Plant growth promotion in cereal and plant: from microorganism capabilities to crop production. Microbiol Res 169(5–6):325–336. https://doi.org/10.1016/j.micres.2013.09.011

Pettett AM, Kurtboke DI (2004) Development of an environmentally friendly biofertilizer with keratin degrading and antibiotic producing actinomycetes. Actinomycetologica 18:34–42

Rai MM (1981) Principles of Soil Science. Machillan India Limited, Calcutta, pp 179–182

Schmidt WF (1998) Innovative feather utilization strategies. In: Proceedings national poultry waste management symposium. Auburn University Printing Services, 19–22 October 1998

Silas NE, Murungi JI, Wanjau RN (2012) Levels of macronutrients of leaves of selected plants from highlands east of Mount Kenya. Int J Appl Sci Technol 2(9):105–110

Tindal HD (1983) Vegetables in the tropics. Macmillan Education Ltd., London, pp 98–100

USDA (United States Department of Agriculture) (2004) Soil survey laboratory manual, soil survey investigation report no. 42, version 4.0, USDA-NRCS, Nebraska, USA

Valtcho D, Zheljazkov (2005) Assessment of wool waste and hair waste as soil amendment and nutrient source. J Environ Qual 34:2310–2317. https://doi.org/10.2134/jeq2004.0332

Verma AN, Srivastava DC, Sharma RK (1969) Effect of nitrogen supply on growth, yield, ascorbic acid and oxalic acid content in spinach (*Spinacea oleracea*). Plant Sci 1:189–191

8

Improving the nutritive value of olive pomace by solid state fermentation of *Kluyveromyces marxianus* with simultaneous production of gallic acid

Shadia A. Fathy[1] · Abeer E. Mahmoud[2] · Mona M. Rashad[2] · Magda K. Ezz[1] · Amira T. Mohammed[2]

Abstract

Purpose Olive oil industry is a growing industrial sector in Mediterranean countries. Unfortunately, about 30–35% of total olive weight is discarded as olive pomace (OP), a highly environmental-polluting solid waste.OP is mostly used for composting and production of animal feed; nevertheless, the high oil content, phenolic compounds and fiber contents in addition to its low protein content represent major obstacles for both applications. So, the present study was conducted to evaluate the role of solid state fermentation (SSF) in OP using *Kluyveromyces marxianus* NRRL Y-8281 yeast on the chemical composition and tannin content of OP in a trial for its bioconversion into a value-added animal feed.

Methods Chemical composition, nutritive and energetic values, tannin and gallic acid contents of unfermented and fermented olive pomace were investigated.

Results The fermentation altered the chemical composition of OP, so that crude fiber was decreased by 8.56%, while crude protein, fat and carbohydrate contents were increased by 2.74, 2.63 and 3.57%, respectively. Moreover, the effect of fermentation on cell wall constituents, gross energy, digestible energy, total digestible nutrients and digestible crude protein was evaluated. Furthermore, HPLC analysis revealed that *K. marxianus* mediated fermentation of OP resulted in a sharp decrease in tannin content by 96.75% with 2.8 times increase in gallic acid concentration.

Conclusion SSF of OP by *K. marxianus* does not only eliminate the environmental pollution resulting from its accumulation, but also presents a new eco-friendly valorization technique which leaves OP with an altered chemical composition allowing its use as animal feed or compost.

Keywords Olive pomace · Solid state fermentation · *Kluyveromyces marxianus* · Nutritive value · Gallic acid

Introduction

Olive oil industries play an important economic role in Mediterranean countries. Considering a three phase decanter system (the most widely employed method for virgin olive oil production), two major by-products are generated, a solid residue (30%) known as olive pomace (OP) (also known as orujo and olive cake (OC)) which consists of olive skin, pulp and pit, in addition to significant residual oil content, and a liquid waste (50%) known as olive mill waste water (also known as alpechin) which consists of olive fruit tissue water, washing water and water added for centrifugation (Borja et al. 2006; Dermeche et al. 2013).

It is estimated that the world annual production of OP is 2,881,500 tonnes/year (Nunes et al. 2016). This represents a huge problem for olive oil producing countries, since this vast amount of waste cannot be stored in its original form for extended time due to its high water and significant oil contents that cause rapid deterioration and rancidity often after 4–5 days of its production, as a result, OP is spread on farm lands resulting in serious negative environmental impact regarding soil, underground water and air problems (Brlek et al. 2012).

Three phase OP is a heterogeneous matrix with a chemical composition that differs greatly among different

✉ Abeer E. Mahmoud
ab_essam@yahoo.com

1 Biochemistry Department, Faculty of Science, Ain Shams University, Cairo, Egypt

2 Biochemistry Department, Division of Genetic Engineering and Biotechnology Researches, National Research Centre, Dokki, Giza 12622, Egypt

producers (Christoforou and Fokaides 2016). OP is quite rich in water (30–50%) and contains a substantial amount of residual oil (2–4%) (Azbar et al. 2004). It is also characterized by high fiber (cellulose, hemicellulose, lignin and pectines) and low protein contents. In addition, it is rich in fat, mostly polyunsaturated fatty acids (Ramachandran et al. 2007; Molina-Alcaide and Yáñez-Ruiz 2008; Lafka et al. 2011). The ash content ranges between 1.7 and 4% and major elements present in OP are potassium followed by calcium, magnesium and sodium (Aliakbarian et al. 2011). Despite its variable composition, fibers followed by fat and protein remain the main constituents of OP (Christoforou and Fokaides 2016). OP is a ligno-cellulosic organic material which has low digestibility and energy content in addition to having high tannin content (Al-Masri and Guenther 1999).

Several practices (in particular, composting and animal feed production) have been introduced for disposal and utilization of OP aiming to reduce the environmental-polluting load with concomitant production of value-added bio-products (Cossu et al. 2013).

Low soil organic matter can lead to limited soil fertility and productivity followed by soil degradation. Thus, the periodic addition of organic fertilizers is strongly recommended to restore soil productivity (López-Piñeiro et al. 2011). Generally, natural fertilizers are more preferred than synthetic ones. Having significant contents of organic matter, nitrogen, potassium and phosphorus, OP is considered as a potent organic amendment that improves many soil properties, restores crops productivity and improves the soil nutrient level (Abu-Zreig and Al-Widyan 2002). Raw OP can be applied directly to soil as an organic amendment (Kavdir and Killi 2008). However, the residual oil contained in it results in increased soil hydrophobicity and decreased water retention and infiltration rate. Moreover, due to the high content of phenolic and phytotoxic compounds, utilization of raw OP as soil amendment for long time results in accumulation of toxicants in soil with detectable negative effects on plants in addition to land, ground water and surface water contamination (Lopes et al. 2016).

Livestock production in Mediterranean countries is constricted by the scarcity and fluctuating quantity and quality of the feed supply to animals. Agro-industrial residues can serve as alternative sources for livestock providing an economic and eco-friendly method for their disposal and recycling (Molina-Alcaide and Yáñez-Ruiz 2008). However, utilization of OP as animal feed additive is limited due to its low protein content, high phenolic compounds content and high fiber content which renders OP unpalatable and poorly digestible for ruminants. Moreover, the significant amount of energy contained in OP represented by the cellulose (14–26%) is locked in ligno-cellulosic complex and thus it is inaccessible to ruminal microbes for degradation. Consequently, pre-treatment of OP before animal consumption is necessary to eliminate its toxicity and upgrade its nutritional value (Weinberg et al. 2008). Bioconversion of OP through solid state fermentation (SSF) may represent an excellent pre-treatment strategy (Neifar et al. 2013). SSF is defined as the microbial growth on solid support in the lack or near absence of free water. However, the support must possess enough moisture to sustain microbial growth and metabolism (Pandey 1991). Several researches dealing with SSF of OP using fungal strains have been conducted allowing its use as animal feed additive for both ruminants (cattle, sheep, goats and camels) and poultry (Neifar et al. 2013; Fadel and El-Ghonemy 2015).

The aim of the present work was to evaluate the role of SSF in OP using *Kluyveromyces marxianus* NRRL Y-8281 yeast on the chemical composition and tannin content of OP in a trial for its bioconversion into a value-added animal feed.

Materials and methods

Olive pomace waste

OP was provided during its harvesting season by a local olive-pressing factory (three phase decanter system), located in Al-Arish, North Sinai, Sinai Peninsula, Egypt. It was stored at 4 °C till used.

Chemicals

Gallic acid was obtained from Bio Basic Canada INC (Markham Ontario, Canada). Code No. GB0476. Tannic acid was obtained from Panreac AppliChem (Castellar del Valles, Barcelona, Spain). Code No. 141065. All other chemicals and reagents used in this study were of analytical grade and purchased from Sigma-Aldrich Chemical Co. (St. Louis, MO, USA).

Microorganism

The generally regarded as safe (GRAS) yeast, *Kluyveromyces marxianus* NRRL Y-8281 used in this study was obtained from Agricultural Research Service, Peoria, Illinois, USA.

Adaptation and inoculum preparation of the microorganism

The yeast strain was streaked on YME medium with agar (Wickerham 1951) for 48 h at 30 °C. The stock culture was stored at 4 °C and sub-cultured every 4 weeks, then stored at 4 °C. A loop of the culture was inoculated in 50 ml of sterile inoculum medium (composed of the stock medium without agar), then incubated on controlled incubator shaker (New Brunswick Scientific, USA) at 150 rpm for 24 h at 30 °C.

Solid state fermentation

For SSF, an aliquot of 1 ml of each inoculum containing about 10^8 cells/ml was inoculated in 250 ml Erlenmeyer flasks containing 5 g of sterilized OP (sterilized at 121 °C for 20 min at 15 psi). Incubation was done at static incubator for 48 h at 45 °C (Fathy et al. 2017).

Chemical analysis of unfermented (UFOP) and fermented (FOP) olive pomace

Chemical analysis of unfermented and fermented OP was done as described by AOAC (1997) method.

The moisture, ash, crude protein (CP) and crude fat contents of UFOP and FOP were determined in accordance with the AOAC (1997). Crude protein determination involved the use of routine Kjeldhal nitrogen assay (N × 6.25). Total carbohydrate content was determined using the phenol–sulfuric acid method of Dubois et al. (1956). Cell wall constituents including neutral detergent fiber (NDF), acid detergent fiber and acid detergent lignin were determined according to Goering and Van Soest (1970) and Pv et al. (1991). Hemicellulose was calculated as the difference between NDF and ADF, whereas cellulose was the difference between ADF and ADL. Gross energy (kcal/kg DM) was calculated according to Blaxter (1968) where, each g crude protein = 5.65 kcal, g fat = 9.40 kcal and g (crude fiber and carbohydrate) = 4.15 kcal. Digestible energy (kcal/kg DM) was calculated according to NRC (1977) where, digestible energy (DE) = gross energy × 0.76. Total digestible nutrients (%) was calculated according to NRC (1977) where, total digestible nutrients % = digestible energy/44.3. Digestible crude protein (%) was calculated according to NRC (1977) where, digestible crude protein (%) = $0.85\ X1 - 2.5$. Where $X1$ = crude protein% on DM basis.

High-performance liquid chromatography analysis

Preparation of unfermented and fermented culture filtrates

Culture filtrates were prepared by extracting the samples with eightfold (v/w) acetate buffer (0.02 M, pH 5.5) by shaking (200 rpm) at 30 °C for 60 min. The resultant slurry was centrifuged at 10,070×g for 15 min at 4 °C. Finally, the extracts were collected and considered as a source of tannic and gallic acids.

Analysis of unfermented and fermented culture filtrates

Tannic and gallic acids were estimated in both UFOP and FOP culture filtrates using HPLC technique. Aliquots were withdrawn and filtered through a 0.2 μm membrane then subjected to HPLC analysis.

HPLC system specifications

All samples were analyzed by HPLC system Agilent 1100 series (Waldborn, Germany) equipped with variable wave length detector (G1314A), G1311A quaternary pump, G1322A degasser, G1329A thermostated autosamples and Zorbax 300SB C18 column (Agilent Technology, USA). Ten microliters of each sample and standard were injected and eluted at room temperature with water: methanol (80:20) as a mobile phase at a flow rate of 1 ml/min. Detection was carried out at 272 nm. Standard solutions (3 mg/ml) and (2 mg/ml) for tannic and gallic acid, respectively, were analyzed to compare results.

Data statistical analysis

Data were expressed as mean ± standard error and analyzed statistically using independent sample *t* test for comparison between the fermented and unfermented olive pomace groups. Differences were considered significant at $P < 0.05$.

Results and discussion

The chemical composition of UFOP and FOP was estimated and presented in Table 1. The moisture content of UFOP was found to be 35% which was approximately similar to the ratio of 30% reported by Uribe et al. (2013). Also,

Table 1 Chemical analysis of unfermented (UFOP) and fermented (FOP) olive pomace

Item	Pomace	
	UFOP	FOP
Moisture content (%)	35.00 ± 0.12^b	35.84 ± 0.05^a
Chemical analysis (%) on dry matter basis		
Organic matter	93.52 ± 0.12	93.90 ± 0.19
Crude protein	10.05 ± 0.02^b	12.79 ± 0.11^a
Crude fiber	42.60 ± 0.11^a	34.04 ± 0.06^b
Fat content	13.62 ± 0.05^b	16.25 ± 0.03^a
Carbohydrate content	27.25 ± 0.06^b	30.82 ± 0.01^a
Ash content	6.48 ± 0.05^a	6.10 ± 0.06^b

Values are given as mean ± standard error of three batches

Mean bearing different superscripts in the same raw are significantly different ($P < 0.05$)

the crude protein was found to be 10.05% on dry matter basis which is close to the ratio 9.5% reported by Fadel and El-Ghonemy (2015). Uribe et al. (2014a) stated crude protein ratio of 7.64%, while Neifar et al. (2013) reported lower protein content (6.5%). In addition, the crude fiber was found to be 42.6% on dry matter basis. This result is in accordance with that conducted by Uribe et al. (2013). Crude fiber ratios of 40, 40.72 and 33% were reported by Ramachandran et al. (2007), Brlek et al. (2012) and Fadel and El-Ghonemy (2015), respectively. Besides, the crude fat was found to be 13.62% on dry matter basis which was higher than that reported by Uribe et al. (2013), (2014a) and Fadel and El-Ghonemy (2015) stating crude fat ratio of 8.19, 10.31 and 5.7%, respectively. But, this result was lower than that reported by Vera et al. (2009) who recorded a ratio of 16.5%. Moreover, the carbohydrate content of UFOP was found to be 27.25% on dry matter basis which was lower than the carbohydrate content (77.99%) reported by Uribe et al. (2014b). Finally, the ash content was found to be 6.48% on dry matter basis. This result is in accordance with that reported by Uribe et al. (2014a), (2014b) and (2015), and nearly in the same range of Neifar et al. (2013) and Fadel and El-Ghonemy (2015) who reported 7% ash content, while Uribe et al. (2013) reported 5.47% ash content. Cell wall constituents (NDF, ADF, ADL, hemicellulose and cellulose) values were found to be 56.91, 48.28, 8.74, 8.63 and 39.54%, respectively (Table 2). Neifar et al. (2013) reported ratios of 59, 45, 31, 14.1 and 14.3% for NDF, ADF, ADL, hemicellulose and cellulose, respectively. And, ratios of 62 and 48% were estimated for NDF and ADF, respectively, by Fadel and El-Ghonemy (2015).

The variation in the chemical composition of OP reported by different authors can be attributed to the difference in olive variety, cultivation conditions (cultivation soil, geography of cultivation soil and use of pesticides and fertilizers), climatic conditions, harvesting conditions (the degree of ripening, harvesting time and system) and the applied oil extraction process (Doymaz et al. 2004; Brlek et al. 2012; Uribe et al. 2013, 2014a, 2014b, 2015; Christoforou and Fokaides 2016).

Bioconversion of OP into a better-quality feed using SSF could be an alternative to chemical and physical treatment for enhancing its quality for microbial fermentation in rumen. SSF of OP using *K. marxianus* resulted in an increase in crude protein by 2.74%, changing it from 10.05% for UFOP to 12.79% for FOP. This can be attributed either to the increased biomass or the secretion of enzymes into the fermented medium (Fadel and El-Ghonemy 2015). The increased protein content suggests that the treated substrate is a good protein source for livestock (Neifar et al. 2013; Fadel and El-Ghonemy 2015). Also, a decrease of ash and crude fiber content by 0.38 and 8.56%, respectively, was observed after fermentation. The reduction in the crude fiber content indicates an increased digestibility of the resultant substrate (Neifar et al. 2013). In addition, the biodegradation of OP using *K. marxianus* resulted in an increase in carbohydrate content by 3.57% comparing to UFOP. The increased carbohydrate content can partially be attributed to the degradation of polysaccharide compounds which was associated with an increase in reducing sugars (Neifar et al. 2013). Increased carbohydrate content is an additional advantage besides the reduction of fiber content in the improvement of biomass quality through biological treatment as reported by Jahromi et al. (2011). Moreover, the crude fat was increased by 2.63% over the UFOP which is similar to the finding of Fadel and El-Ghonemy (2015) when growing *Aspergillus oryzae* FK-923 on OP.

Results in Table 2 showed that FOP expressed less cell wall constituents than UFOP as exhibited by the decrease of NDF, ADF, ADL and cellulose by 9.87, 16.16, 16.59 and 16.06%, respectively. Fadel and El-Ghonemy (2015) reported a reduction in NDF and ADF by 28.06 and 32.08%, respectively. While, Neifar et al. (2013) reported a reduction in NDF, ADF and ADL by 23, 14 and 11%, respectively,

Table 2 Cell wall constituents of unfermented (UFOP) and fermented (FOP) olive pomace

Constituent (%)	Pomace	
	UFOP	FOP
Neutral detergent fiber (NDF)	56.91 ± 0.01^a	51.29 ± 0.63^b
Acid detergent fiber (ADF)	48.28 ± 0.13^a	40.48 ± 0.05^b
Acid detergent lignin (ADL)	8.74 ± 0.02^a	7.29 ± 0.11^b
Hemicellulose	8.63 ± 0.08^b	10.81 ± 0.07^a
Cellulose	39.54 ± 0.03^a	33.19 ± 0.06^b

Values are given as mean ± standard error of three batches

Means bearing different superscripts in the same raw are significantly different ($P < 0.05$)

Table 3 Energetic and nutritive values of unfermented (UFOP) and fermented (FOP) olive pomace

Item	Pomace	
	UFOP	FOP
Energetic values		
Gross energy (kcal/kg DM)	4747 ± 3.28^b	4942 ± 7.08^a
Digestible energy (kcal/kg DM)	3608 ± 2.49^b	3756 ± 5.38^a
Nutritive values		
Total digestible nutrients (TDN) (%)	81.44 ± 0.56^b	84.78 ± 0.12^a
Digestible crude protein (DCP) (%)	6.04 ± 0.01^b	8.37 ± 0.10^a

Values are given as mean ± standard error of three batches

Means bearing different superscripts in the same raw are significantly different ($P < 0.05$)

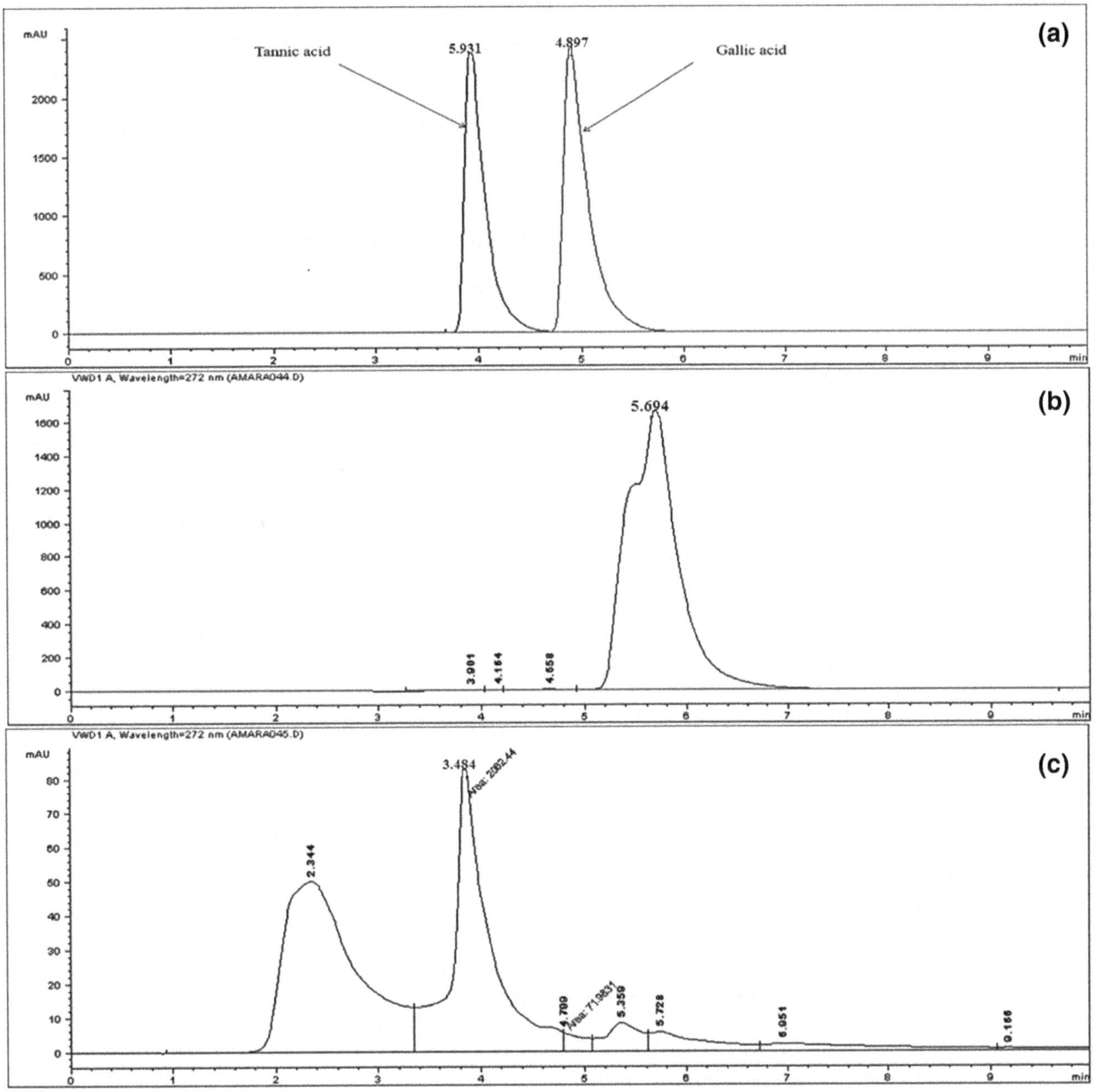

Fig. 1 Typical HPLC chromatograms of tannic acid and gallic acid standards (**a**), UFOP filtrate (**b**) and FOP filtrate (**c**)

when OP was utilized as a solid support for the SSF of *Fomes fomentarius*.

Table 3 shows the energetic and nutritive values of UFOP and FOP. The gross energy of UFOP was found to be 4747 (kcal/kg DM) which was higher than the value 4200 (kcal/kg DM) reported by Fadel and El-Ghonemy (2015). Results showed that fermentation of OP by *K. marxianus* increased the gross energy up to 4942 (kcal/kg DM). While, Fadel and El-Ghonemy (2015) conducted an increase up to 4320 (kcal/kg DM) after fermentation. In addition, results showed that fermentation of OP by *K. marxianus* increased the digestible energy (from 3608 to 3756 (kcal/kg DM), total digestible nutrients (from 81.44 to 84.78%) and digestible crude protein (from 6.04 to 8.37%).

Tannins are the second most abundant group of phenols in nature and they are considered as a plant's secondary metabolites, as they do not have a direct role in plant metabolism. Instead, they have several important biological activities, such as defense against microbial attacks and protection of plant tissues from being attacked by insects and herbivores due to their unflavored taste (Scalbert 1991).

Tannins are uniquely characterized from other phenolic compounds by their ability to precipitate proteins from solutions to other macromolecules such as starch, cellulose, and minerals through formation of strong complexes with them (Lekha and Lonsane 1997; Aguilar et al. 2007). This property is the major source of tannin's undesirable effects in nutrition. Their ability to bind to macromolecules including proteins and minerals prevents their digestion and absorption and leads to inhibition of digestive enzymes (Rodríguez-Durán et al. 2011). Moreover, tannins are responsible for the astringent and bitter taste of fruits, juices and wines which reduces the food intake. In addition, the consumption of large amounts of tannin is associated with development of some forms of cancer (Jana et al. 2014). However, the consumption of low concentrations of tannin in feed is associated with an increase in nitrogen assimilation in ruminants showing higher growth rates and milk production (Belmares et al. 2004; Aguilar et al. 2007). When ingested, tannin can prevent diarrhea, inflammation, diuretics and cancer. In addition, they can be used for the treatment of poisoning caused by heavy metals due to their metal chelating capacity. Moreover, they are considered as potent antimicrobial agents (Jana et al. 2014).

For confirmation of OP tannin degradation by *K. marxianus* fermentation and release of gallic acid in the fermented medium, both UFOP and FOP filtrates were subjected to HPLC analysis. It is well-known that HPLC analysis is very efficient in demonstrating differences in chemical constituents of samples. HPLC profile of OP tannin bioconversion to gallic acid by *K. marxianus* fermentation is shown in Fig. 1a–c. Figure 1a represents the HPLC pattern of tannic acid and gallic acid standards. HPLC profile of UFOP and FOP is shown in (Fig. 1b, c) . Using tannic acid and gallic acid as standards, HPLC data analysis showed revealed that tannic acid concentration in UFOP represented 1154.0 μg/gds, while in FOP, tannic acid represented 37.49 μg/gds. Also, gallic acid concentration in FOP was equal to 61.0 μg/gds comparing to 21.57 μg/gds in UFOP. From these results, it can be concluded that 96.75% of total tannins contained in UFOP were hydrolyzed and degraded producing gallic acid increasing its concentration by 2.8 times as a result of *K. marxianus* fermentation (Fig. 1a–c). The degradation of OP tannin during fermentation by *K. marxianus* can be attributed to the secretion of tannase enzyme produced when OP was used as a sole carbon source for solid state cultivation of *K. marxianus* (Fathy et al. 2017).

Conclusion

Regarding the reduction of tannin content and the increase of gallic acid concentration in addition to the positive effect of fermentation on the chemical composition of OP through increasing protein, carbohydrate and energy contents with decreasing fiber content, SSF of OP by *K. marxianus* is considered as a suitable technique for bioconversion of OP into a value-added animal feed. However, further research work is necessary to optimize the improvement of the chemical composition and to verify the effectiveness of this degraded substrate on the performance of live animals.

Compliance with ethical standards

Conflict of interest The authors declare that they have no conflict of interest.

References

Abu-Zreig M, Al-Widyan M (2002) Influence of olive mills solid waste on soil hydraulic properties. Commun Soil Sci Plant Anal 33:505–517. https://doi.org/10.1081/CSS-120002760

Aguilar CN, Rodríguez R, Gutiérrez-Sánchez G et al (2007) Microbial tannases: advances and perspectives. Appl Microbiol Biotechnol 76:47–59. https://doi.org/10.1007/s00253-007-1000-2

Aliakbarian B, Casazza AA, Perego P (2011) Valorization of olive oil solid waste using high pressure–high temperature reactor. Food Chem 128:704–710. https://doi.org/10.1016/j.foodchem.2011.03.092

Al-Masri M, Guenther K (1999) Changes in digestibility and cell-wall constituents of some agricultural by-products due to gamma irradiation and urea treatments. Radiat Phys Chem 55:323–329. https://doi.org/10.1016/S0969-806X(98)00333-8

AOAC (1997) Analysis of the association of official analytical chemists, 17th edn. AOAC, Washington, D.C.

Azbar N, Bayram A, Filibeli A, Muezzinoglu A, Sengul F, Ozer A (2004) A review of waste management options in olive oil production. Crit Rev Environ Sci Technol 34:209–247. https://doi.org/10.1080/10643380490279932

Belmares R, Contreras-Esquivel JC, Rodríguez-Herrera R, Coronel ARR, Aguilar CN (2004) Microbial production of tannase: an enzyme with potential use in food industry. Food Sci Technol 37:857–864. https://doi.org/10.1016/j.lwt.2004.04.002

Blaxter KL (1968) The energy metabolism of ruminants, 2nd edn. Spring field, Illinois

Borja R, Raposo F, Rincón B (2006) Treatment technologies of liquid and solid wastes from two-phase olive oil mills. Grasas Aceites 57:32–46. https://doi.org/10.3989/gya.2006.v57.i1.20

Brlek T, Voća N, Krička T, Lević J, Vukmirović Đ, Čolović R (2012) Quality of pelleted olive cake for energy generation. Agric Conspec Sci 77:31–35

Christoforou E, Fokaides PA (2016) A review of olive mill solid wastes to energy utilization techniques. Waste Manag 49:346–363. https://doi.org/10.1016/j.wasman.2016.01.012

Cossu A, Degl'Innocenti S, Cristani M, Bedin S, Nuti M (2013) Assessment of the life cycle environmental impact of the olive oil extraction solid wastes in the european union. Open Waste Manag J 6:12–20. https://doi.org/10.2174/1876400220131014001

Dermeche S, Nadour M, Larroche C, Moulti-Mati F, Michaud P (2013) Olive mill wastes: biochemical characterizations and valorization strategies. Process Biochem 48:1532–1552. https://doi.org/10.1016/j.procbio.2013.07.010

Doymaz I, Gorel O, Akgun N (2004) Drying characteristics of the solid by-product of olive oil extraction. Biosyst Eng 88:213–219. https://doi.org/10.1016/j.biosystemseng.2004.03.003

Dubois M, Gilles KA, Hamilton JK, Rebers P, Smith F (1956) Colorimetric method for determination of sugars and related substances. Anal Chem 28:350–356. https://doi.org/10.1021/ac60111a017

Fadel M, El-Ghonemy DH (2015) Biological fungal treatment of olive cake for better utilization in ruminants nutrition in Egypt. Int J Recycl Org Waste Agric 4:261–271. https://doi.org/10.1007/s40093-015-0105-3

Fathy SA, Rashad MM, Ezz MK, Mohammed AT, Mahmoud AE (2017) Enhanced tannase production by *Kluyveromyces marxianus* NRRL Y-8281 under solid state fermentation of olive oil cake. Res J Pharm Biol Chem Sci 8:1698–1708

Goering HK, Van Soest PJ (1970) Forage fiber analysis (apparatus, reagents, procedure and some applications). Agric Handb 379:20

Jahromi M, Liang J, Rosfarizan M, Goh YM, Shokryazdan P, Ho Y (2011) Efficiency of rice straw lignocelluloses degradability by Aspergillus terreus ATCC 74135 in solid state fermentation. Afr J Biotechnol 10:4428–4435. https://doi.org/10.5897/AJB10.2246

Jana A, Halder SK, Banerjee A, Paul T, Pati BR, Mondal KC, Mohapatra PKD (2014) Biosynthesis, structural architecture and biotechnological potential of bacterial tannase: a molecular advancement. Bioresour Technol 157:327–340. https://doi.org/10.1016/j.biortech.2014.02.017

Kavdir Y, Killi D (2008) Influence of olive oil solid waste applications on soil pH, electrical conductivity, soil nitrogen transformations, carbon content and aggregate stability. Bioresour Technol 99:2326–2332. https://doi.org/10.1016/j.biortech.2007.05.034

Lafka T-I, Lazou AE, Sinanoglou VJ, Lazos ES (2011) Phenolic and antioxidant potential of olive oil mill wastes. Food Chem 125:92–98. https://doi.org/10.1016/j.foodchem.2010.08.041

Lekha P, Lonsane B (1997) Production and application of tannin acyl hydrolase: state of the art. In: Neidleman SL, Laskin AI (eds) Advances in applied microbiology, 44th edn. Elsevier Science, Amsterdam, pp 216–260

Lopes D, Pinho I, Santos S, Martins R, Quina M (2016) Germination index as a tool to assess phytotoxicity of olive mill solid wastes. In: 4th international conference on sustainable solid waste management. 23–25 June, Cyprus

López-Piñeiro A, Albarrán A, Nunes JR, Peña D, Cabrera D (2011) Long-term impacts of de-oiled two-phase olive mill waste on soil chemical properties, enzyme activities and productivity in an olive grove. Soil Tillage Res 114:175–182. https://doi.org/10.1016/j.still.2011.05.002

Molina-Alcaide E, Yáñez-Ruiz D (2008) Potential use of olive by-products in ruminant feeding: a review. Anim Feed Sci Technol 147:247–264. https://doi.org/10.1016/j.anifeedsci.2007.09.021

Neifar M, Jaouani A, Ayari A et al (2013) Improving the nutritive value of olive cake by solid state cultivation of the medicinal mushroom Fomes fomentarius. Chemosphere 91:110–114. https://doi.org/10.1016/j.chemosphere.2012.12.015

NRC (1977) National Research Council. Nutrient requirements of rabbits, National Academy of Science, Washington, D.C., USA

Nunes MA, Pimentel FB, Costa AS, Alves RC, Oliveira MBP (2016) Olive by-products for functional and food applications: challenging opportunities to face environmental constraints. Innov Food Sci Emerg Technol 35:139–148. https://doi.org/10.1016/j.ifset.2016.04.016

Pandey A (1991) Aspects of fermenter design for solid-state fermentations. Process Biochem 26:355–361. https://doi.org/10.1016/0032-9592(91)85026-K

Ramachandran S, Singh SK, Larroche C, Soccol CR, Pandey A (2007) Oil cakes and their biotechnological applications–a review. Bioresour Technol 98:2000–2009. https://doi.org/10.1016/j.biortech.2006.08.002

Rodríguez-Durán LV, Valdivia-Urdiales B, Contreras-Esquivel JC, Rodríguez-Herrera R, Aguilar CN (2011) Novel strategies for upstream and downstream processing of tannin acyl hydrolase. Enzyme Res 2011:1–20. https://doi.org/10.4061/2011/823619

Scalbert A (1991) Antimicrobial properties of tannins. Phytochemistry 30:3875–3883. https://doi.org/10.1016/0031-9422(91)83426-L

Uribe E, Lemus-Mondaca R, Vega-Gálvez A et al (2013) Quality characterization of waste olive cake during hot air drying: nutritional aspects and antioxidant activity. Food Bioprocess Technol 6:1207–1217. https://doi.org/10.1007/s11947-012-0802-0

Uribe E, Lemus-Mondaca R, Vega-Gálvez A, Zamorano M, Quispe-Fuentes I, Pasten A, Di Scala K (2014a) Influence of process temperature on drying kinetics, physicochemical properties and antioxidant capacity of the olive-waste cake. Food Chem 147:170–176. https://doi.org/10.1016/j.foodchem.2013.09.121

Uribe E, Lemus-Mondaca R, Pasten A, Astudillo S, Vega-Gálvez A, Puente-Díaz L, Di Scala K (2014b) Dehydrated olive-waste cake as a source of high value-added bioproduct: drying kinetics, physicochemical properties, and bioactive compounds. Chil J Agric Res 74:293–301. https://doi.org/10.4067/S0718-58392014000300007

Uribe E, Pasten A, Lemus-Mondaca R, Vega-Gálvez A, Quispe-Fuentes I, Ortiz J, Di Scala K (2015) Comparison of chemical composition, bioactive compounds and antioxidant activity of three olive-waste cakes. J Food Biochem 39:189–198. https://doi.org/10.1111/jfbc.12120

Van Soest PV, Robertson J, Lewis B (1991) Methods for dietary fiber, neutral detergent fiber, and non-starch polysaccharides in relation to animal nutrition. J Dairy Sci 74:3583–3597. https://doi.org/10.3168/jds.S0022-0302(91)78551-2

Vera R, Aguilar C, Pérez Meléndez P et al (2009) Feeding dry olive cake modifies subcutaneous fat composition in lambs, noting cake resistance to degradation and peroxidation. Chil J Agric Res 69:548–559. https://doi.org/10.4067/S0718-58392009000400010

Weinberg Z, Chen Y, Weinberg P (2008) Ensiling olive cake with and without molasses for ruminant feeding. Bioresour Technol 99:1526–1529. https://doi.org/10.1016/j.biortech.2007.04.022

Wickerham LJ (1951) Taxonomy of yeasts, 1029th edn. United States Department of Agriculture Technical Bulletin, Washington, pp 1–56

9

Vermicomposted tannery wastes in the organic cultivation of sweet pepper: growth, nutritive value and production

Ramom Rachide Nunes[1] · Lívia Botacini Favoretto Pigatin[2] · Ticiane Silva Oliveira[2] · Rhaissa Mecca Bontempi[2] · Maria Olímpia Oliveira Rezende[2]

Abstract

Purpose In this study, vermicomposted tannery wastes were applied in the organic cultivation of sweet pepper in a greenhouse. The effects of this organic matter addition on plant development and on the distributions of Cr(III) and Cr(VI) in plant tissues and organs were assessed.

Methods In a greenhouse, organic sweet peppers were cultivated adding vermicomposted tannery to the substrate (sample VRC) and the results were compared with the samples control (dYL; only soil) and reference (NPK; adding mineral fertilizer). The growth, nutritive value and fruit production were assessed to evaluate the plant development in different types of substrate. Besides evaluating the biostimulant effect of vermicomposts, the distributions of Cr in plant tissues were also studied.

Results Up to three harvested sweet peppers were obtained per plant, compared with only one fruit for the reference treatment with NPK addition. Based on the Cr dynamics in the plants, the highest concentration was found in the fruits, varied as follows: fruits > stem and stalks > leaves = root; however, the Cr concentration in the fruits was statistically the same across all treatments. Additionally, only Cr(III) was detected and quantified in the fruits.

Conclusion The addition of vermicompost was biostimulating to plants and positively influenced their development. Based on the Cr dynamics in the plants, since the Cr concentration was the same across all treatments, it demonstrated that the addition of tannery residues to the vermicompost did not negatively influence the health benefits or food security of the produced fruits.

Keywords Waste management · Tannery wastes · Vermicomposting · Sweet peppers

Introduction

In Brazil, annually an amount of ~ 44.7 thousand metric tons of leather is manufactured. To hold this high production, approximately 450 tanneries plants, located all over the country, make Brazil the fourth leading producer in the world, behind China (209.8), India (52.7), and Italy (51.7 thousand metric tons) (CETESB 2005).

Environmental toxins and demand for safe disposal of industrial wastes are currently the topics of concern. The leather industry is associated with the generation of huge amounts of solid wastes, and waste disposal is a serious problem. Untreated solid waste and wastewater generated by the tanning industry can cause serious environmental damage, particularly to surface water and groundwater (Belay 2010; Vig et al. 2011; Dixit et al. 2015; Scaglia et al. 2016).

Usually, Cr is introduced mainly in the form of $Cr_2(SO_4)_3$ [chromium (III) sulfate] aiming to stabilize the leather by cross linking the collagen fibbers. As final product, the tanned skin by adding Cr salts shows a better texture and quality, resulting in manufactured products with a higher market value. Besides Cr, other contaminants are added in the tanning process, e.g., sodium and ammonium sulfate, thiobenzothiazole, kerosene, etc., but in small

✉ Ramom Rachide Nunes
ramom.rachide@ufrpe.br

1 Academic Unit of Serra Talhada, Federal Rural University of Pernambuco, Serra Talhada, PE, Brazil

2 São Carlos Institute of Chemistry, University of São Paulo, São Carlos, Brazil

concentration. To avoid such pollution, specific chemical, physical and biological treatments have been adopted to reduce waste toxicity and enhance tannery waste use/recycling systems (Aira et al. 2008; Yang et al. 2014; Scaglia et al. 2016; Makkar et al. 2017).

In this context, vermicomposting is an effective technology that can reduce the toxic effects of tannery wastes, based on earthworm activity of *Eisenia fetida* L., which also improves nutrient availability of OM. Earthworms transform organic residues both physically and chemically by mixing and grinding the waste mass, whereas microbes in the guts of worms are responsible for the biochemical transformation of OM (Aira et al. 2008; Yang et al. 2014; Scaglia et al. 2016; Makkar et al. 2017). Moreover, earthworms can accumulate heavy metals in their chloragogen cells, resulting in detoxification of polluted biomass (Huges 1980; Ricci 1996; Nunes et al. 2016; Karimi et al. 2017).

Several forms of industrial waste (i.e., paper mill sludge, sugar industry waste and textile waste) have been successfully vermicomposted to obtain useful and safe organic inputs (Ravindran et al. 2008; Vig et al. 2011). In addition, vermicompost applications increase the yield and growth of several crops, mainly because these organic inputs are rich in plant-available nutrients, such as phosphates, nitrates, and exchangeable soluble potassium and calcium (Arancon et al. 2006; Bachman and Metzger 2008; Gopinath et al. 2009). The use of organic amendments for improving soil productivity and health has been reported previously, usually to provide a source of nutrients in organically managed systems, particularly in organic agriculture practices (Ramesh et al. 2006; Gopinath et al. 2009; Singh et al. 2011; Karimi et al. 2017).

Sweet pepper (*Capsicum annuum* L.) is a highly consumed vegetable that is better adjusted to and, therefore, most frequently produced under greenhouse conditions (Pereira 1990; Scivittaro et al. 1999; Silva et al. 2002; Herman et al. 2008). Greenhouse cultivation of sweet pepper demands the intensive use of agri-inputs and labor; moreover, this intense activity is part of a dynamic market in which seasonal prices fluctuate, demanding rigorous planning by the producer for both crop production and the use of technology to decrease production risks and increase profits (Pereira 1990; Silva et al. 2001; Abul-Soud et al. 2014).

Vermicompost fertilization is well known for its effects on plant nourishment, which result in reduced requirements for (or the elimination of) chemical pesticides and inorganic fertilizers. Additionally, an understanding of the effects of OM addition on sweet pepper plants at each stage of development can provide information to assist with fertilizing programs for crops, even in greenhouses (Bachman and Metzger 2008; Llaven et al. 2008; Gopinath et al. 2009; Abul-Soud et al. 2014).

Therefore, the aim of this work was to determine the effects of application of vermicomposted tannery wastes on organic sweet pepper cultivation in the greenhouse. In particular, the effects of this organic matter addition on plant development, i.e., growth, nutritive value and fruit production, were assessed. Additionally, the distribution of chromium, a heavy metal in tannery wastes, was determined in pepper plant tissues and organs.

Materials and methods

Vermicompost production

This study was conducted using vermicomposted tannery wastes in the form of both sludge and tanned chips mixed with cattle manure and sawdust. The vermicomposters were established in 25 L plastic barrels containing different proportions of fresh wastes (based on dry volume). The proportion of the residues was determined by a combination of their C:N ratios. Designed in triplicate, each barrel received 20 L of the mixture of wastes as follows, based on fresh volume: (1) sludge mix (VRC—S): 14.5 L of cattle manure, 2.5 L of sawdust, and 3 L of sludge; (2) tanned chips mix (VRC—C): 14 L of cattle manure, 2 L of sawdust, and 4 L of tanned chips; and (3) sludge + tanned chips mix (VRC—SC): 14 L of cattle manure, 2 L of sawdust, 1.5 L of sludge, and 2.5 L of tanned chips. For comparison, a vermicomposter with no leather wastes was also prepared (standard treatment, VRC—STD) by mixing fresh-volume 17 L of cattle manure and 3 L of sawdust.

Previously analyzed, the raw materials presented as main characteristics: (1) sludge (S): pH 7.83, OM 36.32%, TOC 17.20%, N 4.63%; (2) tanned chips (C): pH 3.31, OM 84.96%, TOC 48.26%, N 13.30%; (3) cattle manure: pH 8.88, OM 63.02%, TOC 34.30%, N 2.02%; and (4) sawdust pH 4.84, OM 89.11%, TOC 52.69%, N 0.43%.

After the mixtures were prepared, the material stabilized for 1 week; then, all of the contents were turned manually once per week. For vermicompost production, 500 newly born earthworms (*Eisenia fetida* L.) were added to each vermicomposter. The experiment was conducted during the months of April–August 2014 (135 days) in the Laboratory of Environmental Chemistry (LQA), Institute of Chemistry of São Carlos (IQSC), University of São Paulo (USP), Brazil (Nunes et al. 2016).

To characterize the vermicomposts, 0.5 L samples were collected from each vermicomposter. Then, a single combined sample was prepared from the triplicate samples. The material was air-dried, ground and sieved to 0.5 mm. In triplicate, the OM (NEN 5754 1994), total organic carbon (TOC) (ISO 10694 1995), pH (ISO 10390 1994), cation exchange capacity (CEC) (ISO 11260 1994), humic acid content (HA) (Swift 1996), macro and micronutrients (EMBRAPA 2017), and C:N ratio (using a Perkin Elmer 2400; North Billerica, MA, USA) were determined in the vermicomposts.

Sweet pepper organic cultivation

Samples of a dystrophic yellow latosol (dYL) were collected on Campus II of the University of São Paulo (21°59′51″S, 47°56′20″W). Using a tractor, 1.0 ton of soil was drilled at a 40.0 cm depth in the arable fraction. Thereafter, the material was sieved to 2.0 mm and air-dried. Soil collection and preparation were conducted in February 2014, which had an average temperature of 24 °C and precipitation of 198 mm.

To characterize the dYL soil, the following attributes were determined: pH, organic matter (OM), total organic carbon (TOC), cation exchange capacity (CEC), base saturation (BS), macro- and micronutrients, particle-size distribution, and soil texture.

Sweet pepper seedlings (biologically certified) were purchased from SouAgro—Sociedade Agropecuária (São Carlos, SP, Brazil). Each seedling was ~3.0 cm in height and ~30 days from planting. Seedlings were grown on a commercial substrate of coconut fibre and pine bark (Bioplant®, Nova Ponte, MG, Brazil) and were watered manually daily.

When seedlings reached 5.0 cm in height and had at least five permanent leaves, in addition to a healthy appearance, the seedlings were transplanted into 0.5 L plastic pots using the same commercial substrate. At 10 cm in height, the plants were transplanted to 25.0 L vessels for the beginning of the experiment.

Each seedling was grown in a 25.0 L vessel containing 20.0 L of substrate. The cultivation was monitored daily, and to prevent pests, an aqueous extract of orange peel was sprayed on the leaves, a natural repellent that insects avoid. The invasive plants were eliminated manually. Using a watering can (manual watering), 1.0 L of water was added daily to each plant. The volume of 1.0 L was determined from the wilting point, obtained by preliminary tests.

For the organization of the experiment, the vessels were distributed randomly on benches, with 42 vessels distributed on two benches, constructed from two boards on three tripods, with a size of 6.0×0.6 m. In total, 24 vessels were arranged in a 2×12 system (column×row) and the other 14 vessels in a 2×7 system (column×row). Each column corresponded to one board. The distance between the vessels was 10.0 cm and positions were changed weekly at random. Positions were modified to minimize possible external influences, e.g., shade and wind.

Agronomic assays were carried out in triplicate. According to the different types of substrates, the following treatments were studied: SP—dYL: Soil without previous management (control treatment); SP—NPK: Soil and 3.0% (V/V) of mineral NPK fertilizer at 12:5:14 (reference treatment); and SP—VRC: Soil and vermicomposts at concentrations of 1.5, 3.0 and 6.0% (V/V) (Table 1).

Monitoring of sweet pepper cultivation

Sweet pepper growth was monitored biweekly using a metric scale. The height (stalk size) of the plants was measured from the base (substrate) to the culm apex (highest point of the stem). The number of fixed leaves (below the apex) was counted and recorded every 2 weeks for each plant. In the first month, the count was weekly.

The width (w) and length (h) of the five largest leaves of each plant were measured every 15 days. From these data, the average leaf area (A) was estimated (Eq. 1) (Lorenzeni et al. 2015).

$$A = (0.6379 \times w \times h)^{0.9816} \quad (1)$$

Four periods related to flowering and the formation of green and mature fruits were assessed with daily monitoring. The average flowering time was based on initiation of flowering. Fruit formation was measured from shortly after petal fall. Formation of stable green fruits was determined by size and appearance (intense green). The average maturation time was determined by formation of stable, full-red (mature) fruits.

Table 1 Details of samples and different treatments adopted in the study

Sample/treatment	Input	Dosage (V/V) (%)
SP—dYL	–	–
SP—NPK	Mineral fertilizer NPK	3.0
SP—VRC		
SP—1.5%—STD	Vermicompost STD	1.5
SP—3.0%—STD		3.0
SP—6.0%—STD		6.0
SP—1.5%—S	Vermicompost S	1.5
SP—3.0%—S		3.0
SP—6.0%—S		6.0
SP—1.5%—C	Vermicompost C	1.5
SP—3.0%—C		3.0
SP—6.0%—C		6.0
SP—1.5%—SC	Vermicompost SC	1.5
SP—3.0%—SC		3.0
SP—6.0%—SC		6.0

SP sweet pepper, *dYL* dystrophic yellow latosol, *STD* vermicompost control, *S* tannery sludge, *C* tanned chips, *SC* mixture of tannery sludge and tanned chips

The average flowering time was calculated by the difference in days between the appearance of the flowering and the falling of petals. The difference in days between the complete formation of the green fruit and the intense red coloration of the fruit was used to determine the mean maturation time.

The size, circumference and weight were measured of fruits that were well shaped and of appropriate color. Then, the fruits were dehydrated at 40 °C for future determinations. Fruits containing anomalies or that were malformed were counted but not measured and weighed.

The chromium concentration was quantified using an Atomic Absorption Perkin Elmer, model PINAACLE 900T (Waltham, MA, USA). Cr(III) and Cr(VI) were extracted in deionized water using mechanic agitation at 60 rpm for 30 min, followed by vacuum filtration.

At post-harvest, because of the addition of Cr-rich waste, Cr was quantified in all the organs and tissues of the plants to determine possible heavy-metal accumulation in sweet pepper. After being cropped, the plants were separated in roots, stem and stalks, leaves, and fruits. All the analyzed replicates were mixed to form a single sample, which was analyzed in triplicate.

Table 2 Chemical characteristics of the vermicomposted tannery waste mixtures (VRC—S, VRC—C and VRC—SC) and control sample (VRC—STD) (mean, ANOVA bootstrap, $p<0.05$, $n=3$, on a dry matter basis)

	VRC—STD	VRC—S	VRC—C	VRC—SC
pH	8.26b	8.08b	7.67a	8.57c
OM (%)	36.24b	36.98b	33.80a	36.04b
TOC (%)	21.66c	17.65ab	25.47cd	14.64a
HA (%)	2.07b	1.68a	1.46ab	1.07a
CEC (cmol$_c$ kg^{-1})	451ab	459.38ab	478.89ab	424.48a
C:N ratio	8.94bc	8.14bc	6.26b	2.68a
Cr(III) (mg kg^{-1})	<LOQ	4.33b	7.47c	1.09a
N (%)	1.48ab	1.54b	1.32a	2.74c
P (%)	0.70b	0.63b	0.63b	0.59ab
K (%)	3.19 cd	2.09c	1.88b	0.88a
Ca (%)	1.76ab	2.11b	1.49a	1.38a
Mg (%)	0.78b	0.75b	0.71b	0.52a
S (%)	0.45a	0.49a	0.40a	0.36a
B (mg kg^{-1})	4.90a	35.01b	577.67d	145.93c
Cu (mg kg^{-1})	6.40a	33.47b	602.28d	171.93c
Mn (mg kg^{-1})	2.13a	35.90b	570.22d	145.74c
Zn (mg kg^{-1})	0.01a	40.68b	499.17d	160.47c
Fe (mg kg^{-1})	1462.10a	1605.11ab	1455.80a	1614.03ab

VRC vermicompost, *STD* sample control, *S* tannery sludge, *C* tanned chips, *SC* mixture of tannery sludge and tanned chips, *OM* organic matter, *TOC* total organic carbon, *HA* humic acids, *CEC* cation exchange capacity

Values in the same row followed by the same letter are not significantly different at $p<0.05$ according to Duncan's test

Statistical analyses

Following one-way bootstrap ANOVA, the differences among means were evaluated using Duncan's multiple range tests at $p<0.05$. The IBM SPSS statistical software package v. 20 was used for data analyses.

Results and discussion

Substrate characterization

To determine the nutritive potential of the vermicomposts, macro- and micronutrient contents and the chemical characteristics were analyzed, as shown in Table 2. When compared with the control vermicompost (VRC—STD), the attributes of vermicomposts produced from tannery residues (VRC—S, VRC—C and VRC—SC) related to agricultural fertility, e.g., OM, TOC, and CEC, among others, were not significantly different (ANOVA bootstrap, $p>0.05$).

For macronutrient contents, the VRC—STD had higher concentrations of P (0.70%), K (3.19%) and Mg (0.78%) than tannery-based vermicomposts, whereas VRC—S had higher values of Ca (2.11%) and S (0.49%). The lowest N value (1.32%) was in VRC—C, but the concentrations of other macronutrients were not affected. The VRC—SC had the lowest values of all vermicomposts for P (0.59%), K (0.88%), Ca (0.52%) and S (0.36%). Although the determined values were different, the concentrations did not exceed 1.00% (Table 2).

When micronutrient contents were compared, the differences were extreme (Table 2). Except for Fe, the concentrations of micronutrients in VRC—C always greatly exceeded those in the others: B (577.67 mg kg^{-1}), Cu (602.28 mg kg^{-1}), Mn (570.22 mg kg^{-1}) and Zn (499.17 mg kg^{-1}), i.e., values approximately fivefold higher than those determined in VRC—SC. However, the values observed in these two vermicompost types were up to 10,000-fold greater than concentrations determined in the VRC—STD and VRC—S. Therefore, micronutrient quantities were found from high to low as follows: VRC—C > VRC—SC > VRC—S > VRC—STD.

For the content of Cr, as expected, VRC—STD had values of Cr(III) and Cr(VI) below the detectable level (LOQ, 79.80 µg kg^{-1}). In the vermicomposts containing tannery residues, the concentrations of Cr(III) were VRC—C at 7.47 mg kg^{-1}, VRC—S at 4.33 mg kg^{-1} and VRC—SC at 1.09 mg kg^{-1}, with all values significantly different (ANOVA bootstrap, $p<0.05$). For Cr(VI), this form was not detected in any vermicompost (Table 2).

Table 3 Characteristics of dystrophic yellow latosol (dYL) used in sweet pepper organic cultivation (mean, on a dry matter basis)

	pH	OM (%)	TOC (%)	BS (%)	CEC ($cmol_c$ kg^{-1})	Sand (%)	Silt (%)	Clay (%)	Soil texture
dYL	5.74	1.52	0.25	27.99	64.91	58.32	11.07	30.61	Sandy clay loam

	N (%)	P (%)	K ($cmol_c$ kg^{-1})	Ca ($cmol_c$ kg^{-1})	Mg ($cmol_c$ kg^{-1})	Na ($cmol_c$ kg^{-1})	B (mg kg^{-1})	Cu (mg kg^{-1})	Mn (mg kg^{-1})	Zn (mg kg^{-1})	Fe (mg kg^{-1})
dYL	1.23	1.02	0.33	13.03	4.81	<LOQ	<LOQ	179.03	337.21	191.26	59.19

dYL dystrophic yellow latosol, *OM* organic matter, *TOC* total organic carbon, *BS* base saturation, *CEC* cation exchange capacity

Table 4 Sweet pepper plant development on different cultivation substrates (SP—STD, SP—S, SP—C and SP—SC), the reference sample (SP—NPK) and the control sample (SP—dYL) (mean, ANOVA bootstrap, $p<0.05$, $n=3$)

	Height[a] (cm)	Number of leaves[b]	Leaf area[b] (cm^2)
SP—dYL	12.7a	7.6a	8.33a
SP—NPK	19.0a	22.3ab	29.80bc
SP—1.5%—STD	32.0cd	28.3bcd	27.24c
SP—3.0%—STD	34.0cd	35.3bcde	43.12cd
SP—6.0%—STD	41.0de	37.6cde	60.73de
SP—1.5%—S	24.7bc	20.6ab	23.82b
SP—3.0%—S	30.3c	28.0bcd	30.93bc
SP—6.0%—S	43.7e	47.0e	66.43ef
SP—1.5%—C	32.0cd	26.6bcd	31.27bc
SP—3.0%—C	45.0e	39.3de	55.26de
SP—6.0%—C	49.7e	46.6e	79.50f
SP—1.5%—SC	25.7bc	25.6bcd	24.69bc
SP—3.0%—SC	44.0e	48.6ef	67.91ef
SP—6.0%—SC	49.7e	62.3f	128.35g

SP sweet pepper, *dYL* dystrophic yellow latosol, *STD* vermicompost control, *S* tannery sludge, *C* tanned chips, *SC* mixture of tannery sludge and tanned chips

Values in the same column followed by the same letter are not significantly different at $p<0.05$ according to Duncan's test

[a]After 120 days of cultivation (post-harvest)

[b]After 45 days of cultivation

Sweet pepper cultivation

The attributes of the latosol were typical of a dYL soil (EMBRAPA 2018), i.e., acidic with a pH of 5.74, poor OM content of 1.52%, TOC of 0.25%, CEC of 64.91 $cmol_c$ kg^{-1}, and small base saturation of BS 27.99%. For soil particle size distribution, the texture of the dYL was classified as sandy clay loam, with 31.00% clay, 11.00% silt and 58.00% sand (Table 3).

For nutrients, the dYL had low levels of essential elements for plants (Table 3). Therefore, the dYL did not satisfy the nutritional requirements for cultivation for agricultural production, with low levels of the exchangeable bases: K (0.33 $cmol_c$ kg^{-1}), Ca (13.03 $cmol_c$ kg^{-1}) and Mg (4.81 $cmol_c$ kg^{-1}). For micronutrients, the concentrations from high to low were Mn (337.21 mg kg^{-1}), Zn (191.26 mg kg^{-1}), Cu (179.03 mg kg^{-1}) and Fe (59.19 mg kg^{-1}). Levels of Na and some other micronutrients were below the detectable level (LOQ) (Table 3).

Poor in OM and with low CEC and nutrients, the dYL was ideally suited to evaluate the nutritive potential of the vermicomposts. A soil rich in OM could confound interpretation of the results following vermicompost application because the effects of soil OM and OM added from the vermicompost on plant nutrition cannot be discerned individually. Additionally, incorporating OM into soil leads to improved CEC and BS, in addition to increasing TOC levels, which are fundamental attributes for soil quality in agricultural systems (Ndegwa et al. 2000; Canellas and Santos 2005; Tobiašová 2011).

Criteria were established for the transfer of seedlings from the trays to the plastic pots and then to the cultivation vessels: 5- and 10-cm height, respectively. Approximately, 7 and 30 days passed for each transplant, respectively.

In general, plants from the SP—dYL and SP—NPK treatments had distinctly less development compared with those from SP—VRC. Plants grown on substrates containing vermicomposts showed more growth and more fruits were produced, with better appearance and greater weight and dimensions. Generally, as the amount of vermicompost increased, the yield (number of fruits) and plant health (visual analysis) increased.

The development of sweet pepper plants in SP—dYL and SP—NPK (control and reference treatments) was similar; however, fruits were only produced on plants in the SP—NPK treatment. Although vermicomposts with different compositions have been examined in previous studies, similarly satisfactory results have been reported (Bachman and Metzger 2008; Kaur et al. 2010; Vig et al. 2011; Kenyangi and Blok 2013; Nunes et al. 2016). Data related to the development of sweet pepper plants (e.g., growth, nutritive value, and fruit production, among others) obtained in this study are discussed below.

Plant growth in treatments SP—dYL and SP—NPK was not significantly different at 12.7 and 19.0 cm, respectively (ANOVA bootstrap, $p < 0.05$; Table 4). Therefore, the addition of NPK fertilizer did not significantly affect plant growth. Plant growth in the control and reference treatments was significantly different from all treatments in the SP—VRC group at all studied concentrations (ANOVA bootstrap, $p < 0.05$).

Plants cultivated with the control vermicompost (VRC—STD) added at 1.5%, 3.0% and 6.0% did not show significant differences in growth (ANOVA bootstrap, $p < 0.05$), after 120 days of the experiment. The difference in plant growth was ~6–28%, with mean growth between 32.0 and 41.0 cm (Table 4).

Analyzing plant growth under application of the vermicompost S, the mean growth of plants in treatments SP—1.5%—S and SP—3.0%—S was not significantly different at 24.7 and 30.3 cm, respectively. However, both levels of growth were significantly different compared with that of plants in the treatment SP—6.0%—S, with growth of 43.7 cm (ANOVA bootstrap, $p < 0.05$; Table 4).

When vermicompost C was applied, mean plant growth in treatments SP—3.0%—C and SP—6.0%—C was 45.0 and 49.7 cm, respectively, which were not significantly different. Plants in the treatment SP—1.5%—C grew 32.0 cm, which was significantly different from growth in the other levels of vermicompost C (ANOVA bootstrap, $p < 0.05$). At all concentrations, plant growth in the SP—SC treatments was not significantly different compared with that in the SP—C treatments (Table 4).

In general, the growth of all plants that received 1.5% vermicompost was not significantly different (ANOVA bootstrap, $p > 0.05$). Among all treatments, plants with the highest growth were in SP—6.0%—C and SP—6.0%—SC, both at ~49 cm. In samples from the group SP—VRC, plant growth was proportional to vermicompost concentration, with growth at 1.5% < 3.0% < 6.0% (Table 4).

Data related to the leaves (e.g., number, area, and nutritive value) were important to monitor and to qualify in the development of sweet pepper plants in different types of substrate. From the beginning of the experiment (~15 days after planting), the uniqueness in the dimensions of leaves could be observed. After 45 days, whereas the size of leaves from the SP—dYL and SP—NPK treatments remained with the same, leaves in the SP—6.0%—SC treatments had dimensions that were larger than a hand. For the leaf area (A), after 60 days, the area of leaves in SP—dYL was significantly smaller than that in the other treatments (8.33 cm^2). In SP—NPK and SP—1.5%—VRC treatments, the areas of leaves were not significantly different, with values varying between 23.82 and 31.27 cm^2 (ANOVA bootstrap, $p < 0.05$; Table 4).

In general, the nutrient contents in the leaves varied randomly, without tending to the expected. As the sweet pepper continued to develop, the concentrations of nutrients decreased as expected, primarily because of ion dilution in the tissues and organs of the plants (Tables 5 and 6) (Malavolta 1980; Malavolta et al. 1997); however, exceptions were observed.

Nutrients directly influenced plant development, as demonstrated by sweet pepper growth and production. Additionally, visual assessment of the plants, based primarily on the color, shape and texture of leaves, provided important information on the nutritional statuses of the plants (Malavolta 1980; Epstein 1997; Faquin 2005; Epstein and Bloom 2006; Fontes 2006).

Concerning the appearance of leaves, a yellowing was observed in plants of the SP—dYL treatment after 5 days of cultivation that persisted for the entire experiment, with the same occurring after 45 days for plants of the group SP—1.5%—VRC. At concentrations of 3.0% and 6.0% vermicompost, the yellowing occurred at 75 and 90 days of cultivation, respectively. In the SP—NPK treatment, no change was observed in leaf color throughout the experiment. Although the application of mineral fertilizer did not contribute significantly to plant development, nutrition was guaranteed at all plant developmental stages.

Table 5 Leaf analysis (macronutrients) on sweet pepper leaves cultivated on different substrates (SP—STD, SP—S, SP—C and SP—SC), the reference sample (SP—NPK) and the control sample (SP—dYL) (mean, ANOVA bootstrap, $p < 0.05$, $n = 3$)

	N (%)	P (%)	K (%)	Ca (%)	Mg (%)	S (%)
SP—dYL	1.97ab	0.25a	5.71b	1.67a	0.24a	1.03b
SP—NPK	2.29b	0.41a	5.71b	2.99bc	0.73ab	0.79ab
SP—1.5%—STD	2.14b	0.44a	6.63b	2.22ab	0.58ab	0.81ab
SP—3.0%—STD	1.93ab	0.47ab	6.58b	1.85a	0.43a	0.83ab
SP—6.0%—STD	2.18b	0.22a	5.41b	3.57c	0.71ab	1.03b
SP—1.5%—S	1.83ab	0.32a	5.01b	3.11c	0.71ab	1.14bc
SP—3.0%—S	1.80ab	0.48ab	5.42b	3.04c	0.64ab	1.05b
SP—6.0%—S	1.50a	0.67b	5.73b	2.13ab	0.51a	0.92ab
SP—1.5%—C	1.58a	0.20a	5.85b	2.13ab	0.65ab	0.85ab
SP—3.0%—C	1.85ab	0.51a	5.62b	2.48b	0.67ab	0.94ab
SP—6.0%—C	1.99ab	0.65 b	5.71b	2.17ab	0.66ab	0.72a
SP—1.5%—SC	1.54a	0.23 a	2.58a	2.26ab	0.58ab	0.71a
SP—3.0%—SC	1.69a	0.18a	2.56a	2.10ab	0.62ab	0.78a
SP—6.0%—SC	1.74ab	0.24a	2.16a	2.04ab	0.65ab	0.76a

SP sweet pepper, *dYL* dystrophic yellow latosol, *STD* vermicompost control, *S* tannery sludge, *C* tanned chips, *SC* mixture of tannery sludge and tanned chips

Values in the same column followed by the same letter are not significantly different at $p < 0.05$ according to Duncan's test

Table 6 Leaf analysis (micronutrients) on sweet pepper leaves cultivated on different substrates (SP—STD, SP—S, SP—C and SP—SC), the reference sample (SP—NPK) and the control sample (SP—dYL) (mean, ANOVA bootstrap, $p < 0.05$, $n = 3$)

	B (mg kg^{-1})	Cu (mg kg^{-1})	Mn (mg kg^{-1})	Zn (mg kg^{-1})	Fe (mg kg^{-1})
SP—dYL	88.25b	1.36	294.63bc	100.30ab	1223.76c
SP—NPK	89.75b	2.88	353.97c	107.93ab	1002.25c
SP—1.5%—STD	85.04b	0.13	157.44a	99.93ab	738.69a
SP—3.0%—STD	72.65ab	0.23	133.19a	88.09a	577.78a
SP—6.0%—STD	164.12c	1.44	375.32c	145.84b	1316.88d
SP—1.5%—S	116.46c	0.57	211.31b	99.25a	1762.12e
SP—3.0%—S	104.06c	0.54	203.55b	102.22ab	1264.24c
SP—6.0%—S	57.26a	1.97	127.09 a	102.95ab	858.62b
SP—1.5%—C	88.68b	n.d.	277.32b	104.61ab	762.32a
SP—3.0%—C	76.07ab	0.64	100.14a	102.91ab	813.51ab
SP—6.0%—C	50.64a	0.04	245.17b	101.82ab	861.84b
SP—1.5%—SC	78.21ab	0.67	492.57d	143.87b	1091.23
SP—3.0%—SC	58.54a	0.29	356.25c	94.47a	812.90ab
SP—6.0%—SC	67.95a	0.11	305.76c	125.56b	527.06a

SP sweet pepper, *dYL* dystrophic yellow latosol, *STD* vermicompost control, *S* tannery sludge, *C* tanned chips, *SC* mixture of tannery sludge and tanned chips

Values in the same column followed by the same letter are not significantly different at $p < 0.05$ according to Duncan's test

For plants that received the vermicompost control, leaf area in the SP—1.5%—STD treatment (27.24 cm^2) was significantly different from that of the others (43.12 cm^2 in SP—3.0%—STD and 60.73 cm^2 in SP—6.0%—STD), with the difference between the two higher levels not significant. Leaf area of plants from SP—C and SP—SC treatments was not significantly different and their leaves had the largest dimensions and therefore areas (ANOVA bootstrap, $p < 0.05$).

For macronutrients (Table 5), in a comparison between the control and reference treatments (SP—dYL and SP—NPK, respectively), the concentrations in plants from SP—NPK were higher than those in the control, as expected. However, concentrations of K (5.71%) and S (0.79%) were the exceptions and were below those of plants in SP—dYL.

Furthermore, the addition of N, P and K from the mineral fertilizer did not guarantee the highest concentrations of these nutrients among treatments, except for N (2.29%). The highest value of P was determined in SP—6.0%—S (0.67%) and that of K in SP—1.5% —STD (6.63%). The lowest concentrations of macronutrients were in SP—C and SP—SC treatments, with the lowest values of P (0.18% in SP—3.0%—SC), K (2.16% in SP—6.0%—SC) and S (0.71% in SP—1.5%—SC), most likely due to ion dilution in the plant tissues (Malavolta 1980; Malavolta et al. 1997).

Concerning the micronutrients (Table 6), the highest concentrations were in plants from the SP—6.0%—STD treatment, with high levels of B (164.12 mg kg^{-1}), Zn (145.84 mg kg^{-1}) and Fe (1316.88 mg kg^{-1}); and in the plants from SP—SC treatments, with high levels of Cu (29.64 mg kg^{-1} in SP—3.0%—SC) and Mn (492.57 mg kg^{-1} in SP—3.0%—SC). With the exceptions of Zn and Fe, the lowest concentrations of micronutrients were in plants from the SP—C treatments: B (50.64 mg kg^{-1}) and Cu (0.04 mg kg^{-1}) in SP—6.0%—C; and Mn (100.14 mg kg^{-1}) in SP—3.0%—C.

Consistent with our findings, many studies have demonstrated the effects of nutrients on growth, development and fruit production in sweet pepper (Malavolta 1980; Marti and Mills 1991; Aguilera-Gomez et al. 1999; Riga and Anza 2003; Faquin 2005; Albuquerque et al. 2011).

Marti and Mills (1991) studied the effects of Ca uptake by roots on plant growth and development, in addition to its correlation with the N concentration in the substrate. Albuquerque et al. (2011) studied the effect of the soil K concentration on the production of red sweet pepper, in addition to that on biometric fruit data. Riga and Anza (2003) evaluated the effects of Mg deficiency on the physiology of pepper plants and found that it affected plant growth and development. Moreover, Aguilera-Gomez et al. (1999), among other studies, examined the effects of P uptake on the development of sweet pepper plants. Overall, vermicomposts clearly sufficiently nourished the sweet pepper plants to promote their growth, development and fruit production.

Fruit production (red sweet pepper, at the mature stage), in addition to fruit characteristics (dimensions, weight and appearance), are important attributes for evaluating the effects of vermicompost addition on sweet pepper cultivation. In general, the addition of vermicompost was a biostimulant to the plants in this study, increasing the number and quality of cropped fruits (Table 7).

Regarding production, plants subjected to the control treatment (SP—dYL) did not produce any fruit, whereas those exposed to the reference treatment (SP—NPK) produced only one harvested fruit per plant. The optimum results were obtained for the plants exposed to the SP—3.0%—SC and SP—6.0%—SC treatments, which produced three fruits per plant. In addition, the plants subjected to the SP—6.0%—S, SP—3.0%—C, and SP—6.0%—C treatments produced two fruits. For the other treatments, the plants only produced one fruit (Table 7).

Some plants also produced red sweet peppers that did not meet esthetic standards, with irregular shape and coloration, resulting in no commercial value. The worst results were observed for plants in the treatments of group SP—STD

Table 7 Red sweet pepper production on different substrates (SP—STD, SP—S, SP—C and SP—SC), the reference sample (SP—NPK) and the control sample (SP—dYL) (mean, ANOVA bootstrap, $p<0.05$, $n=3$)

	Fruit production (fruit $plant^{-1}$)	Time formation (days)	Fruit characteristics		
			Size (cm)	Circumference (cm)	Weight (g)
SP—dYL	0a	–	–	–	–
SP—NPK	1bcd	94a	11.3bcd	19.1cde	96.8bcd
SP—1.5%—STD	1bcd	121bcde	10.7bcd	13.3abc	46.7abc
SP—3.0%—STD	1cd	105abcd	11.9cd	17.0abcd	93.0bcd
SP—6.0%—STD	1cd	87a	10.9bcd	18.8bcde	134.5d
SP—1.5%—S	1bcd	126e	6.2a	11.2a	23.1a
SP—3.0%—S	1cd	123de	8.6abc	14.7abcd	47.0abc
SP—6.0%—S	1cd	91ab	12.4d	23.0e	107.0cd
SP—1.5%—C	1bcd	110bcde	8.0ab	13.3abc	37.0ab
SP—3.0%—C	2de	92ab	11.7bcd	14.5abcd	84.6bcd
SP—6.0%—C	2de	100abc	12.5d	19.8de	113.3d
SP—1.5%—SC	1bc	120cde	6.0a	13.0ab	24.8a
SP—3.0%—SC	3e	97ab	11.2bcd	17.2bcd	86.7bcd
SP—6.0%—SC	3e	87a	10.5bcd	19.0bcde	100.8cd

SP sweet pepper, *dYL* dystrophic yellow latosol, *STD* vermicompost control, *S* tannery sludge, *C* tanned chips, *SC* mixture of tannery sludge and tanned chips

Values in the same column followed by the same letter are not significantly different at $p<0.05$ according to Duncan's test

because for each two well-formed fruits, one fruit was an anomaly (50% of production). Some treatments showed a significantly higher proportion of well-formed fruits, including in SP—6.0%—SC (5:1 ratio), with only two of ten total fruits showing anomalies, and in SP—3.0%—SC (ratio 8:0), SP—6.0%—C (ratio 6:0), and SP—3.0%—C (ratio 4:1) (ANOVA bootstrap, $p < 0.05$). Plants exposed to some of the treatments showed only well-formed fruits, i.e., SP—6.0%—C, SP—3.0%—SC and the reference treatment, SP—NPK (Table 7).

For fruit formation, less time was required for formation and maturation of red sweet peppers as the vermicompost concentration increased, except for the plants in the SP—C treatments, which all had maturation times between 92 and 110 days that were not significantly different (ANOVA bootstrap, $p < 0.05$). Additionally, plants in all the treatments with 6.0% vermicompost, some treatments with 3.0% (i.e., SP—3.0%—STD, SP—3.0%—C and SP—3.0%—SC) and the SP—NPK reference treatment showed maturation times that were not significantly different at 87–105 days (ANOVA bootstrap, $p < 0.05$, Table 7).

In general, fruits with shorter maturation times had higher weights, varying between 100.8 and 134.5 g, with no significant differences (ANOVA bootstrap, $p < 0.05$). Fruits obtained on plants from treatments with 1.5% vermicompost had fruits that weighed between 23.1 and 46.7 g, with values that were significantly different (ANOVA bootstrap, $p < 0.05$). The difference between fruit weights in the treatments SP—STD and SP—VRC was approximately 480%, indicating the biostimulant effect of vermicompost application on fruit production and attributes.

Concerning the dimensions of fruits, a relation between size and circumference could not be established because larger fruits could have a smaller circumference and vice versa. In general, plants that received 3.0% and 6.0% vermicompost produced fruits with the largest dimensions, varying between 10.5 and 12.5 cm in size and 17.0 and 23.0 cm in circumference, with no significant differences (ANOVA bootstrap, $p < 0.05$). In the treatments that received 1.5% vermicompost, the fruit size and circumference ranged between 6.0 and 8.0 cm and 11.2 and 13.3 cm, respectively (Table 7).

Ribeiro et al. (2000) studied the use of vermicomposts as an alternative for the organic production of sweet pepper in a protected environment. Additionally, the authors compared their results with experiments using a eutrophic red argisol (eRA) and under the application of NPK, imitating a conventional plantation. Although the structure and design of their study were similar to those of this study, the soil selected was different. An eRA (4.26% OM) could mask the results and influence of vermicompost on sweet pepper development because this soil had a certain nutritional potential, different from that of the dYL (EMBRAPA 2018). These findings are similar to those observed in our study. However, in that study, plants that received vermicompost had better results than those planted with NPK and in the control soil based on the following: (1) in organic cultivation (with vermicompost), the production was 16.0 t ha^{-1}, whereas in the conventional planting, production was below 13.1 t ha^{-1}; (2) the production in the control treatment (eRA) was below that obtained in organic cultivation but above that in conventional planting at 13.60 t ha^{-1}; and (3) organic sweet peppers had an average weight of 72.9 g, compared with 67.8 g in the conventional and 65.5 g in eRA treatments.

In a comparison of studies, the fruit weight (84.6–134.5 g), for example, when 3.0% and 6.0% vermicompost were applied in our study, was notably higher than the 72.9 g found by Ribeiro et al. (2000). For fruit size, our organic sweet peppers also had larger dimensions, 9.3 cm in circumference, than the 9.0- and 8.9-cm circumferences found in the conventional cultivation and in eRA, respectively. Although our results were apparently superior, both studies indicated the efficacy and viability of the use of vermicompost in systems of organic agriculture and sweet pepper cultivation.

In general, because of the variety of soils in Brazil and the wide availability of agricultural wastes, scientific works are produced with different aims and goals, making comparisons among studies difficult. However, in all studies reported in the literature, specifically in the areas of environmental chemistry or agronomy and focused on organic agriculture, results and conclusions are consistent with those of this study; organic wastes prepared as inputs for organic agriculture can be applied in the planting of sweet pepper or other crops of economic and social interest.

Chromium

For the dynamics of Cr, the absorption of Cr(III) begins at the root, followed by transport through the tissues of the stem and stalks, reaching the leaves and fruit (Gropper et al. 2009; Thor et al. 2011; Hua et al. 2012). In this study, in some of these stages, Cr accumulation occurred (Table 8).

Initially, the root is expected to act as a type of filter that prevents assimilation of an excessive concentration of Cr by the plant, with accumulation in root tissues (Gropper et al. 2009; Thor et al. 2011; Hua et al. 2012). In the rhizosphere, the Cr(III) levels varied among treatments, although no trend was observed, with values ranging from 23.86 to 65.12 μg kg^{-1}. However, for the treatments SP—dYL, SP—NPK, SP—STD, and SP—1.5%—S, Cr(III) was below the detectable level in roots, indicating metal absorption followed by transport, without accumulation in the root system.

In general, the Cr(III) concentrations varied as follows: fruits > stem and stalks > leaves = root, with exceptions (Table 8). The Cr(III) levels in fruits, stems and stalks were higher than those in the other tissues, most likely because of

Table 8 Chromium distribution in vegetal tissues and organs (roots, stems, stalks, leaves and fruits) of sweet peppers plants

	Roots (μg kg^{-1})	Stem and stalks (μg kg^{-1})	Leaves (μg kg^{-1})	Fruits (μg kg^{-1})
SP—dYL	<LOQ	60.71a	41.65a	–
SP—NPK	<LOQ	61.69a	41.93a	145.06a
SP—1.5%—STD	<LOQ	61.77a	41.64a	145.40ab
SP—3.0%—STD	<LOQ	62.81a	41.60a	156.49ab
SP—6.0%—STD	<LOQ	64.50a	41.57a	159.24ab
SP—1.5%—S	<LOQ	61.74a	41.85a	165.12bc
SP—3.0%—S	44.56cd	82.23b	41.74a	165.20bc
SP—6.0%—S	33.61bc	82.29b	41.64a	164.83bc
SP—1.5%—C	<LOQ	82.19b	41.53a	165.22bc
SP—3.0%—C	44.18cd	92.44c	41.53a	165.12bc
SP—6.0%—C	24.99a	82.28bc	41.69a	164.90bc
SP—1.5%—SC	56.01bcd	82.27b	41.58a	164.88bc
SP—3.0%—SC	23.86ab	82.19b	41.60a	165.03bc
SP—6.0%—SC	65.12e	83.20bc	41.56a	164.69bc

SP sweet pepper, *dYL* dystrophic yellow latosol, *STD* vermicompost control, *S* tannery sludge, *C* tanned chips, *SC* mixture of tannery sludge and tanned chips

Values in the same column followed by the same letter are not significantly different at $p<0.05$ according to Duncan's test

their abilities to accumulate metals in their cells and tissues. In the leaves, the levels of Cr(III) were not significantly different among the treatments, ~41 μg kg^{-1}. In the stems and stalks, the levels of Cr(III) in the treatments SP—dYL, SP—NPK, SP—STD, and SP—1.5%—S were statistically similar, ranging between 60.71 and 64.50 μg kg^{-1}; in the other treatments, levels of Cr(III) were also not significantly different, ranging from 82.19 to 92.44 μg kg^{-1} (ANOVA bootstrap, $p<0.05$; Table 8).

Finally, the fruits are the organs that are of greatest concern because the food security of the consumer is ultimately affected (Gropper et al. 2009; Thor et al. 2011; Hua et al. 2012). In the sweet peppers, Cr(III) concentration varied between 145.06 and 165.20 μg kg^{-1}. Fruits cropped in the treatments SP—NPK and SP—STD, at all concentrations, had levels of Cr(III) that were statistically similar. The Cr(III) levels determined in the treatments involving vermicomposted tannery residues were also statistically similar (SP—S, SP—C and SP—SC; ANOVA bootstrap, $p<0.05$; Table 8). Additionally, tannery residues added in vermicomposting did not increase the Cr contents in fruit because the fruit concentrations of Cr(III) were statistically similar to those quantified in the fruits that received the vermicompost control (VRC—STD) at the same concentration (1.5%, 3.0% or 6.0% V/V).

When comparing the Cr(III) determined in the fruits in this study with other foods from different studies that are commonly reported as 'Cr-rich' or a 'Cr-source', at the level of chromium enrichment in our cropped red sweet peppers, the fruits would be available for human consumption (Table 9). For example, broccoli has 22.0 μg Cr 100 g^{-1}, and brown bread and potatoes have 4.4 and 1.5 μg Cr 100 g^{-1}, respectively. In red and green grapes, the Cr concentrations are 6.5 and 2.1 μg Cr 100 g^{-1}, respectively. In our study, red sweet pepper showed, in general, concentrations ranging from 3.8 to 21.4 μg Cr 100 g^{-1}, which are close to the concentrations present in other foods (Vique et al. 1997; Oliveira and Marchini 2008; NHI 2012).

Conclusions

The present study confirmed the expectation that the vermicomposted tannery wastes can be applied in the organic cultivation of sweet pepper. The vermicompost addition was biostimulating to the plants, influencing development positively, in addition to increasing fruit production. Although the fruits obtained in the SP—NPK (reference treatment) were statistically similar in characteristics to those cultivated with vermicomposted tannery wastes, the production in this treatment was limited to one fruit per plant, whereas in other treatments, up to three red sweet peppers were harvested per plant. Based on the results obtained in the agronomic assays, the suggested dosage to be applied in the organic cultivation of sweet pepper was the 6.0% of vermicompost. Also, mixing both forms of residues, sludge and tanned chips, it is possible to reach a best agronomic efficiency of the vermicompost, by increasing significantly the fruit production. When the levels of Cr were investigated in the plant tissues and organs, only Cr(III) was detected and quantified, as expected. The highest Cr content was found in the fruits; however, the Cr content was the same in fruits in all treatments, which demonstrated that the addition of tannery

Table 9 Cr(III) concentration in the cropped red sweet pepper and comparison with those in other foods found in the literature

	Cr(III) (μg Cr 100 g^{-1}*)	Reference
Sweet pepper		
SP—dYL	14.0bcd	
SP—NPK	6.7abc	
SP—1.5%—STD	14.5bcd	
SP—3.0%—STD	21.4d	
SP—6.0%—STD	3.8a	
SP—1.5%—S	7.7abc	
SP—3.0%—S	17.6cd	
SP—6.0%—S	6.1ab	
SP—1.5%—C	13.9bcd	
SP—3.0%—C	18.6d	
SP—6.0%—C	4.0a	
SP—1.5%—SC	14.3bcd	
SP—3.0%—SC	16.6cd	
SP—6.0%—SC	14.0bcd	
Broccoli	22.0	Oliveira and Marchini (2008)
Potatoes	1.5	NHI (2012)
Brown bread	4.4	NHI (2012)
Green grapes	2.1	Vique et al. (1997)
Red grapes	6.5	Vique et al. (1997)
Apple	0.8	NHI (2012)
Orange juice	1.0	NHI (2012)
Garlic	60.0	NHI (2012)

SP sweet pepper, *dYL* dystrophic yellow latosol, *STD* vermicompost control, *S* tannery sludge, *C* tanned chips, *SC* mixture of tannery sludge and tanned chips

Values in the same column (only for the sweet peppers) followed by the same letter are not significantly different at $p<0.05$ according to Duncan's test

*μg of Cr per 100 g of food

residues in the vermicompost did not negatively influence the quality, health or food security of the produced sweet peppers.

Acknowledgements The authors thank FAPESP - Fundação de Amparo à Pesquisa do Estado de São Paulo (State of São Paulo, Brazil) for providing grants to R. R. Nunes (Process 2013/00744-0) and CNPq - Conselho Nacional de Desenvolvimento Científico e Tecnológico (Brazil) for the support (Process 306715/2013-9).

References

Abul-Soud MA, Emam MSA, Abdrabbo MAA et al (2014) Sustainable urban horticulture of sweet pepper via vermicomposting in summer. J Adv Agric 3:110–122. https://doi.org/10.13140/RG.2.1.1491.9201

Aguilera-Gomez L, Davies FT Jr, Olalde-Portugal V et al (1999) Influence of phosphorus and endomycorrhiza (*Glomus intraradices*) on gas exchange and plant growth of chile ancho pepper (*Capsicum annuum* L. cv. San Luis). Photosynthetica 36:441–449. https://doi.org/10.1023/A:1007032320951

Aira M, Sampedro L, Monroy F, Domínguez J (2008) Detritivorous earthworms directly modify the structure, thus altering the functioning of a microdecomposer food web. Soil Biol Biochem 40:2511–2516. https://doi.org/10.1016/j.soilbio.2008.06.010

Albuquerque FS, Silva EFF, Albuquerque Filho JAC, Nunes MFFN (2011) Growth and yield of sweet pepper fertigated under different irrigation depths and potassium doses. Rev Bras Eng Agrícola e Ambient 15:686–694. https://doi.org/10.1590/S1415-43662011000700006

Arancon NQ, Edwards CA, Lee S, Byrne R (2006) Effects of humic acids from vermicomposts on plant growth. Eur J Soil Biol 42:S65–S69. https://doi.org/10.1016/j.ejsobi.2006.06.004

Bachman GR, Metzger JD (2008) Growth of bedding plants in commercial potting substrate amended with vermicompost. Bioresour Technol 99:3155–3161. https://doi.org/10.1016/j.biortech.2007.05.069

Belay AA (2010) Impacts of chromium from tannery effluent and evaluation of alternative treatment options. J Environ Prot 01:53–58. https://doi.org/10.4236/jep.2010.11007

Canellas LP, Santos GA (2005) Humosfera. CCTA/UENF, Campos dos Goytacazes

CETESB (2005) Curtumes. Companhia Ambiental do Estado de São Paulo, São Paulo

Dixit S, Yadav A, Dwivedi PD, Das M (2015) Toxic hazards of leather industry and technologies to combat threat: a review. J Clean Prod 87:39–49. https://doi.org/10.1016/j.jclepro.2014.10.017

EMBRAPA (2017) Manual de métodos de análise de solo. EMBRAPA Solos, Rio de Janeiro

EMBRAPA (2018) Sistema brasileiro de classificação de solos. EMBRAPA Solos, Rio de Janeiro

Epstein E (1997) The science of composting. Technomic Publishing Company, Lancaster

Epstein E, Bloom AJ (2006) Nutrição mineral de plantas: princípios e perspectivas. Editora Planta, Londrina

Faquin V (2005) Nutrição mineral de plantas. Editora UFLA, Lavras

Fontes PCR (2006) Diagnóstico do estado nutricional das plantas. UFV, Viçosa

Gopinath KA, Saha S, Mina BL et al (2009) Bell pepper yield and soil properties during conversion from conventional to organic production in Indian Himalayas. Sci Hortic (Amst) 122:339–345. https://doi.org/10.1016/j.scienta.2009.05.016

Gropper SS, Smith JL, Groff J (2009) Advanced nutrition and human metabolism. Cengage Learning, Boston

Herman MAB, Nault BA, Smart CD (2008) Effects of plant growth-promoting rhizobacteria on bell pepper production and green peach aphid infestations in New York. Crop Prot 27:996–1002. https://doi.org/10.1016/j.cropro.2007.12.004

Hua Y, Clark S, Ren J, Sreejayan N (2012) Molecular mechanisms of chromium in alleviating insulin resistance. J Nutr Biochem 23:313–319. https://doi.org/10.1016/j.jnutbio.2011.11.001

Huges HBN (1980) The biology of polluted waters. University Press, Liverpool

ISO 10390 (1994) Soil quality: determination of pH. International Organization for Standardization, Genebra

ISO 10694 (1995) Soil quality: determination of organic and total carbon after dry combustion (elementary analysis). International Organization for Standardization, Genebra

ISO 11260 (1994) Soil quality: determination of effective cation exchange capacity and base saturation level using barium chloride solution. International Organization for Standardization, Genebra

Karimi H, Mokhtari M, Salehi F et al (2017) Changes in microbial pathogen dynamics during vermicomposting mixture of cow manure–organic solid waste and cow manure–sewage sludge. Int J Recycl Org Waste Agric 6:57–61. https://doi.org/10.1007/s40093-016-0152-4

Kaur A, Singh J, Vig AP et al (2010) Cocomposting with and without *Eisenia fetida* for conversion of toxic paper mill sludge to a soil conditioner. Bioresour Technol 101:8192–8198. https://doi.org/10.1016/j.biortech.2010.05.041

Kenyangi A, Blok W (2013) Vermicompost as a component in potting mixes for growth promotion in ornamental plants. Rwanda J 28:53–63. https://doi.org/10.4314/rj.v28i0.4

Llaven MAO, Jimenez JLG, Coro BIC et al (2008) Fruit characteristics of bell pepper cultivated in sheep manure vermicompost substituted soil. J Plant Nutr 31:1585–1598. https://doi.org/10.1080/01904160802244738

Lorenzeni MZ, Rezende R, Souza AHCS et al (2015) Estimativa da área foliar em função das dimensões lineares da folha de pimentão. In: XLIV Congresso Brasileiro de Engenharia Agrícola, São Pedro

Makkar C, Singh J, Parkash C (2017) Vermicompost and vermiwash as supplement to improve seedling, plant growth and yield in *Linum usitassimum* L. for organic agriculture. Int J Recycl Org Waste Agric 6:203–218. https://doi.org/10.1007/s40093-017-0168-4

Malavolta E (1980) Elementos de nutrição mineral de plantas. Ceres, São Paulo

Malavolta E, Vitti GC, Oliveira SA (1997) Avaliação do estado nutricional das plantas: princípios e aplicações. POTAFOS, Piracicaba

Marti HR, Mills HA (1991) Calcium uptake and concentration in bell pepper plants as influenced by nitrogen form and stages of development. J Plant Nutr 14:1177–1185. https://doi.org/10.1080/01904169109364276

Ndegwa PM, Thompson SA, Das KC (2000) Effects of stocking density and feeding rate on vermicomposting of biosolids. Bioresour Technol 71:5–12. https://doi.org/10.1016/S0960-8524(99)00055-3

NEN 5754 (1994) Determination of organic matter content in soil as loss-on-ignition. Netherlands Normalisation Institute, Delft

NHI (2012) Office of dietary supplements. Dietary supplement fact sheet: chromium. In: National Institutes of Health. http://ods.od.nih.gov. Accessed 30 Nov 2017

Nunes RR, Bontempi RM, Mendonça G et al (2016) Vermicomposting as an advanced biological treatment for industrial waste from the leather industry (tanneries). J Environ Sci Health Part B 51:271–277. https://doi.org/10.1080/03601234.2015.1128737

Oliveira JED, Marchini JS (2008) Ciências nutricionais. Sarvier, São Paulo

Pereira AL (1990) Cultura do pimentão. DNOCS, Fortaleza

Ramesh P, Singh M, Panwar NR et al (2006) Response of pigeonpea (*Cajanus cajan*) varieties to organic manures and their influence on fertility and enzyme activity of soil. Indian J Agric Sci 76:252–254

Ravindran B, Dinesh SL, Kennedy LJ, Sekaran G (2008) Vermicomposting of solid waste generated from leather industries using epigeic earthworm *Eisenia foetida*. Appl Biochem Biotechnol 151:480–488. https://doi.org/10.1007/s12010-008-8222-3

Ribeiro LG, Lopes JC, Martins Filho S, Ramalho SS (2000) Adubação orgânica na produção de pimentão. Hortic Bras 18:134–137. https://doi.org/10.1590/S0102-05362000000200012

Ricci MSF (1996) Manual de vermicompostagem. EMBRAPA Rondônia, Porto Velho

Riga P, Anza M (2003) Effect of magnesium deficiency on pepper growth parameters: implications for determination of magnesium critical value. J Plant Nutr 26:1581–1593. https://doi.org/10.1081/PLN-120022367

Scaglia B, Nunes RR, Rezende MOO et al (2016) Investigating organic molecules responsible of auxin-like activity of humic acid fraction extracted from vermicompost. Sci Total Environ 562:289–295. https://doi.org/10.1016/j.scitotenv.2016.03.212

Scivittaro WB, Melo AMTA, Tavares M et al (1999) Caracterização de híbridos de pimentão em cultivo protegido. Hortic Bras 17:147–150. https://doi.org/10.1590/S0102-05361999000200014

Silva MAG, Boaretto AE, Muraoka T et al (2001) Efeito do nitrogênio e potássio na nutrição do pimentão cultivado em ambiente protegido. Rev Bras Ciência do Solo 25:913–922. https://doi.org/10.1590/S0100-06832001000400014

Silva MAA, Escobedo JF, Galvani E (2002) Influência da cultura do pimentão (*Capsicum annuum* L.) nos elementos ambientais em ambiente protegido. Irriga 7:230–240. https://doi.org/10.15809/irriga.2002v7n3p230

Singh RP, Singh P, Araujo ASF et al (2011) Management of urban solid waste: vermicomposting a sustainable option. Resour Conserv Recycl 55:719–729. https://doi.org/10.1016/j.resconrec.2011.02.005

Swift RS (1996) Organic matter characterization. In: Sparks DL, Page AL, Helmke PA et al (eds) Methods of soil analysis: chemical methods. Soil Science Society of America, Madison, pp 1018–1020

Thor MY, Harnack L, King D et al (2011) Evaluation of the comprehensiveness and reliability of the chromium composition of foods in the literature. J Food Compost Anal 24:1147–1152. https://doi.org/10.1016/j.jfca.2011.04.006

Tobiašová E (2011) The effect of organic matter on the structure of soils of different land uses. Soil Tillage Res 114:183–192. https://doi.org/10.1016/j.still.2011.05.003

Vig AP, Singh J, Wani SH, Dhaliwal SS (2011) Vermicomposting of tannery sludge mixed with cattle dung into valuable manure using earthworm *Eisenia fetida* (Savigny). Bioresour Technol 102:7941–7945. https://doi.org/10.1016/j.biortech.2011.05.056

Vique CC, Teissedre PL, Cabanis MT, Cabanis JC (1997) Determination and levels of chromium in French wine and grapes by graphite furnace atomic absorption spectrometry. J Agric Food Chem 45:1808–1811. https://doi.org/10.1021/jf960691b

Yang J, Lv B, Zhang J, Xing M (2014) Insight into the roles of earthworm in vermicomposting of sewage sludge by determining the water-extracts through chemical and spectroscopic methods. Bioresour Technol 154:94–100. https://doi.org/10.1016/j.biortech.2013.12.023

10

The challenge of using date branch waste as a peat substitute in container nursery production of lettuce (*Lactuca sativa* L.)

Najla Dhen[1] · **Safa ben Abed**[2] · **Amin Zouba**[1] · **Faouzi Haouala**[3] · **Bouthaina AlMohandes Dridi**[1]

Abstract

Purpose Peat-based substrates constitute the preferred media in conventional and organic nursery production. Nevertheless, in recent years, there has been a growing interest in environmental impacts associated with peat extraction that has increased with the demand of these non-renewable substrates. The re-use of organic wastes as substrate seems to be good solution to substitute commercial peat. This study evaluates date-palm peat (wastes of date-palm branches base locally known as "Kornef") as an alternative nursery substrate.

Methods The research was conducted in a completely block randomized design with a linear substitution (0, 25, 50, 75 and 100%) of peat with date-palm waste peat for transplant production of lettuce (*Lactuca sativa* L.). Different physicochemical parameters (porosity, bulk density, pH, and CEC) of cultivation substrates were measured. The growth parameters (leaves number, leaves area, length, chlorophyll values, fresh and dry weight of seedlings) were evaluated at the end of growth period (when the seedling reached the commercial transplanting size).

Results The results showed that date-palm waste peat is an appropriate media for nursery production, showing similar properties with commercial peat and best plant response with 25 and 50% substitution.

Conclusions In view of low cost, availability and large area of date-palm cultivation in Tunisia and over the world, it seems that peat can be replaced with substrate of date-palm wastes in the horticulture sector.

Keywords Nursery · Kornef · Substrate · Commercial peat · Plantlet growth · Tunisia

Introduction

Even if greenhouse horticulture occupies a small portion of agricultural land in the world, it has recently expanded considerably in many areas because of its contribution to local or national economy (Massa et al. 2011). Greenhouse production is considered as an effective technique to increase production in many countries (Rostami et al. 2014) due to the control of production factors (fertilizing, luminosity, watering) or the nursery substrates quality.

Compared to direct sowing, nursery production is considered as the most consistent method to ensure the proper establishment of a range of commercial horticultural crops with significant economic value (Herrera et al. 2008). One of the most important cultural considerations to produce a successful greenhouse or nursery crop is to identify the appropriate substrate components or pre-formulated substrates (known also as growing mixes or media). In fact, substrates could have direct or indirect effects on transplant growth and development. Traditionally, greenhouse and nursery operations have primarily utilized soilless substrates (Nelson 2003:692). They used organic materials such as peat moss or tree bark blended with other organic or inorganic components such as vermiculite, perlite and sand (Bilderback et al. 2005; Vaughn et al. 2011).

Sphagnum peat-based substrates are the most standard growing media used in container crops. However, their use causes serious environmental issues since they are

✉ Najla Dhen
dhen.najla@yahoo.fr

1 Laboratory of Biochemistry and Environmental Toxicology, High Institute of Agronomy of Chott-Mariem, University of Sousse, 4042 Chott Mariem, Tunisia

2 Laboratory of Dry Land Farming and Oasis Cropping, Arid Regions Institute, Medenine, Tunisia

3 Department of Biology, College of Sciences, Al Imam Mohammad Ibn Saud Islamic University (IMSIU), Riyadh 11623, Saudi Arabia

considered as a non-renewable resource and because of the negative impacts on wetland ecosystems associated with peat extraction (Ceglie et al. 2015). In fact, peat mining is increasing throughout the world (Basirat 2011). Its extraction exacerbates climate change because of the release of stable and sequestered carbon into the active carbon cycle (Cleary et al. 2005; Dunn and Freeman 2011; Barrett et al. 2016). This has generated a global movement to develop and utilize sustainable growing media for container-grown plant production.

Awareness of the pollution associated with intensive agriculture forces greenhouse growers to adopt more environment-friendly cultivation methods (Massa et al. 2011). The use of an eco-friendly material (rapidly renewable, locally available with a low cost and able to perform as well as commercial products) as growing medium appears to be a viable alternative crop production system (Abdelrahman et al. 2012; Rostami et al. 2014; Abdelrahman et al. 2016).

For the partial or complete replacement of commercial peat in seedling production, many researchers have studied the use of organic materials derived from primary wastes and transformed wastes of agricultural, agro-industrial, animal manure and municipal solid waste streams (Vaughn et al. 2011; Ceglie et al. 2015; Barrett et al. 2016). The use of switch grass (*Panicum virgatum* L.), *Miscanthus* (Altland 2010), rice (*Oryza sativa* L.) hulls (Locke et al. 2013), poultry manure (Flynn et al. 1995), poultry feather fiber (Evans and Vance 2007), olive mill waste (Raviv et al. 2009), coconut (*Cocos nucifera* L.) husk fiber (Barrett et al. 2016), corn/sweatcorn waste, kenaf (*Hibiscus cannabinum* L.) stem core (Vaughn et al. 2011), municipal solid waste (Herrera et al. 2008) has been reported. Yet, despite this research, few of these materials have been largely adopted in container crops production. Results concerning their potentiality have varied significantly and were not always satisfactory (Carmona et al. 2012). New growing media should be environmentally sustainable, economically feasible and as efficient as those it is replacing.

In recent years, cocopeat (also known as coir dust, coir meal and coir pith), a waste product of the coconut industry, has been adopted as a renewable sphagnum peat substitute (Pickering 1997; Yau and Murphy 2000). Cocopeat has sufficient physical properties similar to peat. It provides plant roots high water content, high total pore space; it has low bulk density, low shrinkage and slow biodegradation (Meena et al. 2017). In contrast to peat, cocopeat is characterized by high re-wetting capacity (Blok and Wever 2008). Given that coco peat is obtained from coconut tree, its industry is geographically limited to America, tropical Africa and Asia (Barrett et al. 2016). Every year, huge amounts of money are spent to import coco peat from other countries (Shirani 2013). On the other hand, palm, which belongs to the same family as coconut, presents high similarity with fiber of coconut fruit hull (Basirat 2011; Ghehsareh et al. 2011) and is considered an important product in Tunisia and Mediterranean basin. These findings suggest that date-palm peat may be considered as a proper substitute for commercial peat in the future.

In the history of humanity, the date-palm (*Phoenix dactylifera* L.) cultivation and industry have been considered important in day-to-day life. Worldwide, there are approximately 105 million palm trees, an estimated number of over 3 675 000 tons of residues are discarded annually (Almi et al. 2015) leading to environmental problems (Chandrasekaran and Bahkali 2013; Ghehsareh 2013). Date-palm tree waste includes leaves, branches, stem barks and fronds. They are the most commonly generated waste. They are obtained by seasonal pruning of palm trees which is an essentially agricultural practice and discarded with no valorization. Therefore, from the economic and environmental considerations, the utilization of fronds from the date-palm wastes is a promising project (Almi et al. 2015).

Palm wastes have higher water holding capability than coco peat and can absorb water 8.5 times its dry weight (Chandrasekaran and Bahkali 2013; Shirani 2013). Wastes of date-palm trees seem to be an innovative material in Horticulture industry, to be used as growing media (Ghehsareh 2013) or as an organic fertilizer when used as a biochar (Mahdi et al. 2015). The performance of palm-date wastes peat for plant growth may be leveraged in the production of potted plants but only a few studies have used palm-date wastes as a substitute for peat in potting substrates.

Therefore, the aim of this study was to evaluate the use of fronds palm-date wastes, locally known in Tunisia as "kornef", as an alternative material for peat substitution for the transplant production of lettuce (*Lactuca sativa* L.) in nursery production.

Materials and methods

This experiment was performed to identify the effectiveness of date-palm branches wastes to substitute commercial peat and which is the best and most appropriate percentage for producing healthy, strong and homogeneous lettuce plantlets in greenhouse conditions.

Vegetal material and sowing

Lettuce (*Lactuca sativa* L.), was chosen as plant model to study the attitude of date-palm branch wastes peat as potting media.

One seed was sowed and grown per cell of foamed polystyrene plug trays with 48 cells of 55 mL per treatment. The experiment was conducted in an unheated polyethylene-covered greenhouse under natural daylight conditions at the

Higher Institute of Agronomy of Chott-Mariem (Sousse, Tunisia).

Growing media

Two growing media were evaluated. Commercial peat (*Sphagunum* peat) (P) (KLASMANN Potgrond H80) without previous fertilization was used as control and mixed with date-palm fronds wastes peat (K), locally known as "Kornef".

This was prepared by chopping the frond palm-date wastes into small sizes (2–3 cm and then to 0.2–4 mm). Because of salinity, the material can present phytotoxic levels of sodium and potassium. Thereby several washes in fresh water and a buffering treatment (in which calcium nitrate ($CaNO_3$) is added to leach out and remove any excess sodium and potassium from the material). They were kept in 1.5 m^3 plastic bags to control the moisture and temperature. Some amounts of animal fertilizer and nitrogen fertilizer (0.2 to 0.4 g/l) were added as a fermentation starter. Air holes were made on the bags to ensure respiration and the moisture was adjusted to 50%. During 2 months of incubation, these materials were mixed together every week and put into the bags again.

Physico-chemical and chemical characteristics of growing media

Prepared substrates properties were assessed according to the methods described by Ceglie et al. (2011). In the growing media, pH (ph mètre), electrical conductivity (conductimètre Mettler Toledo), porosity and bulk density were analyzed according to the standard European norms described by Tittarelli et al. (2009).

A 1:5 (V:V) material: water suspension and a filtrate obtained from the diluted materials were used for measurement of pH and electrical conductivity (EC) using a Jenway combined pH and Conductivity Meter 3540 (Belda et al. 2016).

Experimental design

A linear substitution (0, 25, 50, 75 and 100%) of the peat with date-palm fronds wastes peat was considered in a completely randomized plot design, with five treatments and five replications. The treatments were irrigated, according to the environmental conditions, using enough water close to the field capacity to avoid stress in the lettuce seedlings (Ceglie et al. 2015). Two weeks after sowing, a Long Ashton nutrient solution (Hewitt 1966) (1.5 mM $MgSO_4$,$7H_2O$; 2 mM K_2SO_4; 4 mM $CaCl_2.2H_2O$; 1.87 Mm $NaH_2PO_4.2H_2O$; 0.13 Mm $Na_2HPO_4.12H_2O$; 4 mM $NaNO_3$; 4 mM NH_4Cl 0,1388 mM H_3BO_3;0.0208 mM $MnSO_4$,$4H_20$; 2.3 μM $ZnSO_4.7H_2O$; 3.3 μM $CuSO_4.5H_2O$; 0.2 μM $Na_2MoO_4.2H_2O$; 89 μM Fe-EDTA) was applied at the same rate for all the treatments once a week.

The substrate treatments were as follows:

*S*1) Commercial peat as control.

*S*2) A mixture of Date-palm wastes (25%) with Commercial peat (75%).

*S*3) A mixture Date-palm wastes (50%) with Commercial peat (50%).

*S*4) A mixture Date-palm wastes (75%) with Commercial peat (25%).

*S*5) Date-palm wastes (100%).

Forty days after sowing, seedlings reached the commercial transplanting size. They were harvested at random from each experimental unit, avoiding those placed next to the edges. In the stems, seedling length (L) was measured from the root collar to the tip of the shoot; Seedlings' numbers of leaves (NL) were counted; also, seedling's leaf area (LA) was determined by leaf area meter.

To determine chlorophyll content, fresh tissue (1.0 g) was sampled from the youngest fully expanded leaf, extracted with 80% acetone and read using a UV/visible spectrophotometer 4201/50 at 663 and 645 nm. Acetone was used to correct for any turbidity in the extract before chlorophyll concentrations were calculated as follows (Patel and Thethwar 2015).

$$\text{Chl a (mg ml}^{-1}) = 12.7\ A_{663} — 2.69\ A_{645.}$$

$$\text{Chl b (mg ml}^{-1}) = 22.9\ A_{645} — 4.68\ A_{663.}$$

$$\text{Chla} + \text{Chlb} = 8.02\ A_{663} + 80.20\ A_{645.}$$

A_{663} and A_{645} represent absorbance values read at 663 and 645 nm wavelengths, respectively.

Mutual shoot and root's fresh (FW) and dry (DW) (60 °C, 3 days) matter were weighted and their ratio were calculated.

Statistical analysis

The mean values of experimental data were statistically analyzed by ANOVA to evaluate the significant effect of the nursery substrate factor. Means were compared by Duncan's test at the 5% level using as data analysis software system SPSS 20.

Results and discussion

Physico-chemical properties of the growing media

The main hydro-physical and chemical properties of the different growing media compared to the values established for an "ideal" substrate (Abad et al. 2001; Noguera et al. 2003) are shown in Table 1. The peat substitution with date-palm fronds wastes or "Kornef" generated an increase in the pH values and in the salinity contents of the mixtures. The

Table 1 Pysico-chemical properties of the growing media

	pH	EC (mS cm^{-1})	TPS (% vol)	AV (% vol)	TAW (% vol)	BD (g cm^{-3})
Acceptable/optimum range[a]	5.3–6.5	<0.5	>85	20–30	–	<0.4
*S*1 (control)	5.67c	0.93d	82.50b	22.17c	60.33a	0.29a
*S*2	5.71c	2.41c	83.83ab	25.83bc	58.00a	0.212b
*S*3	6.03b	3.67b	85.00ab	29.83ab	55.17a	0.184c
*S*4	6.06b	4.78a	86.83ab	31.50a	55.33a	0.146d
*S*5	6.46a	4.88a	88.50a	33.67a	54.83a	0.116e
Significance	**	**	NS	**	NS	**

*S*1 = 100% Peat, *S*2 = 75% Peat + 25% palm-date wastes, *S*3 = 50% Peat + 50% palm-date wastes, *S*4 = 25% Peat + 75% palm-date wastes, *S*5 = 100% palm-date wastes

Mean values in columns followed by the same letter are not statistically different according to the Duncan test at $P<0.05$. Statistical significance: *NS* not significant, ** statistically significant at $P<0.01$, * statistically significant at $P<0.05$

BD bulk density, *EC* electrical conductivity, *TPS* total porosity space, *AV* air volume, *TAW* Total available water

[a]According to Abad et al. (2001) and Noguera et al. (2003)

date-palm fronds wastes-based substrate (*S*5) manifested the highest pH and EC values (6.46 and 4.88 mS/cm, respectively). EC values being higher than those suggested as optimum values (EC < 0.5 mS m^{-1}). This fact was also revealed by other researchers when investigating peat substitution using date-palm wastes (Ghehsareh et al. 2012; Rostami et al. 2014; Ceglie et al. 2015; Radhouani et al. 2015).

Analyzing the physical properties, all substrates showed suitable values of the bulk density (BD) (<0.4 g cm^{-3}). BD decreased with the incorporation of date-palm wastes giving sufficiently low values so as not to present a problem in transportation or handling. The amount of BD in date-palm wastes peat (*S*5) was minima (0.116 g cm3); therefore, root media aeration in this treatment should be better than others. When root media aeration is sufficient, supplying of water and nutrient elements for plants is better (Ghehsareh 2013). Generally, BD and total pore space (TPS) are inversely correlated parameters which could improve plant root growth (Ceglie et al. 2015). This result was in accordance with the values of TPS obtained for different substrates. Higher proportion of date-palm wastes in growing media resulted in significantly greater values of TPS. TPS values of *S*1 and *S*2 were slightly under the 'ideal' substrate threshold for the commercial peat substrate but were ideal for other substitutions with date-palm wastes. The replacement of peat moss by 50% of date-palm fronds wastes (*S*3) provided the best air volume (AV) value (29.83%) which promotes plant growth. High aeration is important in plug trays with small container [low height (5–7 cm)], where the poor drainage after watering is a limiting factor in nursery production (Carmona et al. 2012). Significant differences can be observed in all treatments AV values but were within the acceptable range (20–30%). Generally, organic wastes have suitable bulk density, porosity and water holding capacity (Borji et al. 2010; Ghehsareh 2013). Whilst, in this study, mixtures with a higher proportion of date-palm wastes showed the lowest amount of total available water (54.83%). The highest amounts were observed with the control treatment (*S*1) (TAW; 60.33%). The findings of Samiei et al. (2005) support our results when investigating the effectiveness of peat moss and date-palm wastes as substrates for Aglaonema (*Aglaonema Commutatum*) growing.

A suitable growing media would provide sufficient anchorage to the plant, represent a reservoir for needed water and nutrients, allow oxygen diffusion to the roots and promote gaseous exchange between roots and atmosphere (Bunt 1988: 309; Ghehsareh 2013; Barrett et al. 2016). To be classed as a good growing media, the materials must provide an appropriate balance of air and water based on its physical structure, for healthy root development. Hence, both mixture of date-palm wastes and peat, *S*2 and *S*3, seem to have the greatest characteristics compared to peat-based substrate used as control.

Plant response with the different substrates

A classical method to evaluating novel nursery substrates is to conduct experiments comparing the response of a model plant species grown in this substrate with those grown in standard medium. Plant performance is determined as quantitative measures of plant growth, e.g., growth index and biomass (Barrett et al. 2016).

The effectiveness of peat substitution with different rates of branch date-palm wastes (kornef) peat affected some characteristics of lettuce growth in nursery (Table 2). Interestingly, the peat substitution with 25–50% of this organic

Table 2 Lettuce growth attitude on a linear substitution of commercial peat

	SL (cm)	SNL	SLA (cm^2)	FW (g)		DW (g)	
				Roots	Shoot	Roots	Shoot
Mixtures							
*S*1	8.8a	6.8a	21.74a	0.24b	1.3ab	0.075a	0.120ab
*S*2	6.6a	6.4a	19.53ab	0.218b	0.962b	0.072a	0.086abc
*S*3	9.5a	7a	19.02ab	0.618a	2.33a	0.112a	0.148a
*S*4	5.8a	6.2a	10.23c	0.058b	0.53b	0.022b	0.024c
*S*5	9.6a	6.6a	13.61bc	0.052b	0.84b	0.012b	0.080bc
Significance	NS	NS	*	*	*	**	**

*S*1 = 100% Peat, *S*2 = 75% Peat + 25% palm-date wastes, *S*3 = 50% Peat + 50% palm-date wastes, *S*4 = 25% Peat + 75% palm-date wastes, *S*5 = 100% palm-date wastes

Mean values in columns followed by the same letter are not statistically different according to the Duncan test at $P < 0.05$. Statistical significance: *NS* not significant, ** statistically significant at $P < 0.01$, * statistically significant at $P < 0.05$

SL seedling length, *SNL* seedlings' numbers of leaves, *SLA* seedling leaf area, *FW* fresh weight, *DW* dry weight

waste peat stimulated foliar expansion (seedling leaf area, leaves numbers and seedling high) and thereby the accumulation of the fresh and dry matter. Seedlings grown in *S*3 [(Peat: Kornef) (v:v) (1:1)] revealed significant differences at 5% level and showed better fresh and dry weight as compared with the commercial peat-based substrate (*S*1) and others substrates. However, fresh and dry matters of plantlets grown in *S*4 and *S*5 have lower properties.

Seedling leaf area in different substrates had significant difference at 5% level. Media amended with 25 and 50% date-palm wastes showed great leaf area values compared to the control. All treatments offer high number of leaves (SNL) that had not any significant difference with peat media used as control. Plantlets grown in commercial peat-based substrate (*S*1), 50% of date-palm wastes (*S*3) and 100% of date-palm wastes (*S*5) were longer (seedling length up to 9.6 cm) than plants grown in other substrates. These findings can be justified, essentially, by the substrates more favorable aeration that enhances the root oxygen demand and eventually permits more absorption of water and nutrients (Borji et al. 2010). Sufficient properties with a view to low bulk density and high porosity in palm peat media lead to better support of water and nutrient for plant and then better growth.

The highest amount of root and shoot weight in both dry and fresh conditions was related to 50% peat substitution (*S*3) with date-palm wastes that had significant difference at 5% level as compared with other substrates. Roots had more growth in this media's mixture because of its good porosity (85%) and so its low bulk density (0.184 g cm^{-3}) compared with values established for an ideal substrate. Accordingly, resistance to root motion was minimal in this media. By contrast, the lowest root growth of lettuce plantlets was obtained with 100% peat substation with date-palm wastes (*S*5) despite its good physicochemical properties. These results show that, in general, increasing rates of this organic peat offered excellent growing properties compared to commercial peat but complete substitution produced a decrease in the plant growth (estimated by the root dry and fresh matter in *S*5).

Greatest plantlets biomass (fresh and dry weights) yielded from lettuce grown in such formulations (20–50% of date-palm wastes) may have been affected by pH of the materials. The recommended pH range for lettuce production is 6.0–6.8 (Flynn et al. 1995). The pH values of all treatments were within the recommended range, averaging 6.03 for 50% of incorporating this organic waste peat (Kornef) into the commercial peat.

Changes in pigment content are linked to visual symptoms of plant stress and photosynthetic productivity. So that, chlorophyll content has long been relied upon as an indicator of environmental stress impact on plants (Parekh et al. 1990). There are no major significant differences regarding the levels of chlorophyll content of transplants leaves grown in *S*2 and *S*3, but are significantly different from other treatments (Table 3). The color intensity of the seedlings (Chla, Chlb, Chla + b) grown in these two (*S*2 and *S*3) substrates was higher than in control peat and other mixtures. The

Table 3 The effect of different substrates on chlorophyll content (mg. g^{-1} fresh matter) of lettuce transplants

	*S*1	*S*2	*S*3	*S*4	*S*5
Chl a	14.434b	20.115a	20.439a	8.775c	4.458c
Chl b	4.685ab	6.316a	8.388a	1.148b	3.761ab
Chl (a + b)	19.116b	26.426a	28.820a	9.921c	8.217c

*S*1 = 100% Peat, *S*2 = 75% Peat + 25% palm-date wastes, *S*3 = 50% Peat + 50% palm-date wastes, *S*4 = 25% Peat + 75% palm-date wastes, *S*5 = 100% palm-date wastes, Values followed by the same letter are statistically similar referring to the test of Duncan at 5%

chlorophylls can provide a more thorough understanding of physiological responses to growth conditions and plant adaptations to its environment.

During the examination of the morphological responses of lettuce plantlets (Fig. 1) grown in different nursery substrates, satisfactory seedling growth was noted by *S*2 and *S*3 substrates. Incorporating date-palm wastes peat to Sphagnum peat-based substrate (control) in a range of 25–50% significantly improved lettuce plantlets growth.

Date-palm waste properties seem to be enhancer factors for their use as growing medium especially when blended with other substrates. This fact was in accordance with Ghehsareh et al. (2011) and Soltani and Naderi (2016) who found that physiochemical properties of growing media affected the seedling growth. Results of Flynn et al. (1995) on lettuce plants demonstrated that the greatest lettuce biomass yields were from broiler litter composted with peanut hulls mixed with Promix (commercial substrate with a high proportion of peat moss) at a ratio of 3:1 compost: promix and they supposed that it may have been affected by pH of the materials. The results obtained by Ceglie et al. (2015) revealed that green waste compost (GC) and palm fibre trunk (PF) waste-derived substrates were suitable for their use in peat (SP) substitution. Using a mixture design, authors have reported that a substrate formulation with 20% GC, 39% PF and 31% SP offers the best lettuce, tomato and melon plantlets growth in nursery stage. This formulation includes a high amount of palm fibre waste which confirms our results about date-palm wastes effectiveness in peat substitution. This effectiveness was reported by different authors, such as Ghehsareh (2013), in a study using different mixtures of date-palm wastes and rice hull. He observed that the highest cucumber yield in greenhouse culture and plant height was related to palm waste. To evaluate the possibility of replacing Coco peat as cultivation substrate with date-palm wastes (with different particular size), a study was conducted by Rostami et al. (2014), which indicates that the highest growth and yield of strawberry production belonged to date-palm culture media with 0–5 mm particular size. Rahbarian and Sardoei (2014) evaluated the substrate effect on the herb *Padanus* Spp. Some growth indices were greatest on 50% peat palm + 25% sand + 25% perlite and 100% peat palm substrates, respectively. They reported that palm peat substrates perform as well as peat moss for most plant growth which proves that palm peat can serve as an alternative media for peat.

Although agricultural reuse of plant wastes is well established, only few studies have reported the suitability of date-palm wastes as an effective potting media. However, their use as substrates can involve issues of toxicity (Radhouani et al. 2015). Serious consequences can be caused by direct contact with theses substrates in concentrated form (Christoulaki et al. 2014).Thereby, a mixture of date-palm waste and peat-based substrate seems to be the best formulation to use in nursery production of lettuce. Indeed, fresh wood waste is rarely used as a stand-alone growth medium, and it usually forms a constituent (normally less than 50%) in mixtures (Christoulaki et al. 2014; Barrett et al. 2016). These findings support our results indicating a peat substitution with a range of 25–50% of palm-dates wastes or kornef.

Conclusion

The present study reveals that a supply of 25–50% date-palm fronds wastes peat in commercial peat may affect commercial substrate properties and improve plant performance for greenhouse cultured lettuce in nursery stage. This encouraging rate is estimated by good seedling growth (high length, number of leaves and biomass) when compared to commercial peat and other substrates. Results demonstrate that date-palm wastes have a great potential to be used in a mixture as a substrate for plant growth in soilless systems. Properties (pH, EC, TPS, BD, etc.) of date-palm wastes peat and its mixture with commercial peat appear to be suitable and were within the optimum range.

Fig. 1 Morphology of lettuce plantlets grown in different nursery substrates. *S*1 = 100% Peat (Control), *S*2 = 75% Peat + 25% palm-date wastes, *S*3 = 50% Peat + 50% palm-date wastes, *S*4 = 25% Peat + 75% palm-date wastes, *S*5 = 100% palm-date wastes

From the obtained results, it can be concluded that date-palm wastes peat material can perform as well as commercial materials it is replacing, giving an environmentally sustainable and commercially viable substrate.

With considering that the continuous use of *Sphagnum* peat as commercial growing media is being questioned, the use of fiber from date-palm wood waste is a promising project. Switching to an alternative material can be extremely costly and complicated if it does not possess very similar properties to peat. So, a mixture of a new organic media with commercial peat may convince transplant producers.

Future studies should be aimed at improvement and optimizing the proportion of used palm-date waste and an investigation on obtained substrate properties using advanced analysis such as morphology and surface chemistry analysis would be interesting.

Author contribution Conceived and designed the experiments, analyzed the data and wrote the paper: ND and BAD. Performed the experiments: ND and AZ. Contributed materials/analysis tools: ND and SBA. Revision of the manuscript: FH and BAD.

References

Abad M, Noguera P, Bures S (2001) National inventory of organic wastes for use as growing media for ornamental potted plant production: case study in Spain. Bioresour Technol 77:197–200. https://doi.org/10.1016/S0960-8524(00)00152-8

Abdelrahman HM, Ceglie FG, Erriquens FG, Verrastro V, Rivera CM, Tittarelli F (2012) Compost based growing media for organic melon seedlings production. Acta Hort (ISHS) 933:99–106. https://doi.org/10.17660/ActaHortic.2012.933.10

Abdelrahman HM, Ceglie FG, Awad FA, Tittarelli F (2016) Growth responses of organic tomato seedlings to N liquid fertilizers and compost-amended growing media. Compost Sci Util 1065-657X:2326–2397. https://doi.org/10.1080/1065657X.2016.1183533

Almi K, Lakel S, Benchabane A, Kriker A (2015) Characterization of date palm wood used as composites reinforcement. Acta PhysPol A 127(4):1072–1074. https://doi.org/10.12693/APhysPolA.127.1072

Altland JA (2010) Use of processed biofuel crops for nursery substrates. J Environ Hortic 28(3):129–134. https://doi.org/10.24266/0738-2898-28.3.129

Barrett GE, Alexander PD, Robinson JS, Bragg NC (2016) Achieving environmentally sustainable growing media for soilless plant cultivation systems—a review. Sci Hortic 212:220–234. https://doi.org/10.1016/j.scienta.2016.09.030

Basirat M (2011) Use of palm waste cellulose as a substitute for common growing media in Aglaonema growing. J Ornament Hort Pl 1(1) 1–11. http://jornamental.iaurasht.ac.ir/article_513685.html

Belda RM, Lidon A, Fornes F (2016) Biochars and hydrochars as substrate constituents for soiless growth of myrtle and mastic. Ind Crop Prod 94:132–142. https://doi.org/10.1016/j.indcrop.2016.08.024

Bilderback TE, Warren SL, Owen JR, Albano JP (2005) Healthy substrates need physicals too. Hort Technol 15: 747–751. http://horttech.ashspublications.org/content/15/4/747.abstract?sid=9097cdbd-3240-439c-846e-7fa5385dcfe9

Blok C, Wever G (2008) Experience with selected physical methods to characterize the suitability of growing media for plant growth. Acta Hortic 779: 239–249. http://www.actahort.org/books/779/779_29.htm

Borji H, Ghahsareh AM, Jafarpour M (2010) Effects of the substrate on tomato in soilless culture. Res J Agric Biol Sci 6(6):923–927

Bunt BR (1988) Media mixes for container grown plants: a manual on the preparation and use of growing media for pot plants, 2nd edn. Springer, Netherlands, p 309p

Carmona E, Moreno MT, Avilés M, Ordovás J (2012) Use of grape marc compost as substrate for vegetable seedlings. Sci Hortic 137:69–74. https://doi.org/10.1002/cad.20018

Ceglie FG, Elshafie H, Verrastro V, Tittarelli F (2011) Evaluation of olive pomace and green waste composts as peat substitutes for organic tomato seedling production. Compost Sci Util 19(4):293–300. https://doi.org/10.1080/1065657x.2011.10737011

Ceglie FG, Bustamante MA, Ben Amara M, Tittarelli F (2015) The challenge of peat substitution in organic seedling production: optimization of growing media formulation through mixture design and response surface analysis. PLoS One 10(6):e0128600. https://doi.org/10.1371/journal.pone.0128600

Chandrasekaran M, Bahkali AH (2013) Valorization of date palm (Phoenix dactylifera) fruit processing by-products and wastes using bioprocess technology—review. Saudi J Biol Sci 20:105–120. https://doi.org/10.1016/j.sjbs.2012.12.004

Christoulaki M, Gouma S, Manios T, Tzortzakis N (2014) Deployment of sawdust as substrate medium in hydroponically grown lettuce. J Plant Nutr 37(8):1304–1315. https://doi.org/10.1080/01904167.2014.881870

Cleary J, Roulet NT, Moore TR (2005) Greenhouse gas emissions from Canadian peat extraction, 1990–2000: a life cycle analysis. J Hum Environ 34(6):456–461. https://doi.org/10.1579/0044-7447-34.6.456

Dunn C, Freeman C (2011) Peatlands: our greatest source of carbon credits. Carbon Manag 2(3):289–301. https://doi.org/10.4155/cmt.11.23

Evans MR and Vance L (2007) Physical properties of processed poultry feather fiber-containing greenhouse root substrates. Horttechnology 17 (3): 201–304. http://horttech.ashspublications.org/content/17/3/301.full?sid=e0b6b87d-b1a3-44ad-bec6-e14e11cee390

Flynn RP, Wood CW, Guertal EA (1995) Lettuce response to composted broiler litter as a potting substrate component. J Am Soc Hortic Sci 120(6):964–970

Ghehsareh AM (2013) Effect of date palm wastes and rice hull mixed with soil on growth and yield of cucumber in greenhouse culture. Int J Recycl Org Waste Agricult. 2–17. https://doi.org/10.1186/2251-7715-2-17

Ghehsareh AM, Borji H, Jafarpour M (2011) Effect of some culture substrates (date-palm peat, cocopeat and perlite) on some growing indices and nutrient elements uptake in greenhouse. Afr J Microbiol Res 5(12):1437–1442. https://doi.org/10.5897/AJMR10.786

Ghehsareh AM, Hematian M, Kalbasi K (2012) Comparison of date-palm wastes and perlite as culture substrates on growing indices in greenhouse cucumber. Int J Recycl Org Waste Agricult 1:5. https://doi.org/10.1186/2251-7715-1-5

Herrera F, Castillo JE, Chica AF, López Bellido L (2008) Use of municipal solid waste compost (MSWC) as a growing medium in the nursery production of tomato plants. Bioresour Technol 99:287–296. https://doi.org/10.1016/j.biortech.2006.12.042

Hewitt EJ (1966) Sand and water culture methods used in the study of plant nutrition, 2nd revised ed. Commonwealth Agricultural Bureau, Farnham Royal, Bucks, pp 547

Locke J, Altland J and Ford C (2013) Gasified rice hull biochar affects nutrition and growth of horticultural crops in container substrates. J Environ Hortic. 31(4): 195–202. http://hrijournal.org/doi/full/10.24266/0738-2898.31.4.195

Mahdi Z, El Hanandeh A and Yu QJ (2015) Date palm (*Phoenix Dactylifera L.*) seed characterization for biochar preparation. In: The 6th international conference on engineering, project, and production management (EPPM), 2–4 September 2015, Gold Coast

Massa D, Incrocci L, Maggini R, Bibbiani C, Carmassi G, Malorgio F, Pardossi A (2011) Simulation of crop water and mineral relations in greenhouse soilless culture. Environ Model Softw 26:711–722. https://doi.org/10.1016/j.envsoft.2011.01.004

Meena AK, Garhwal OP, Mahawar AK, Singh SP (2017) Effect of different growing media on seedling growth parameters and economics of papaya (Carica papaya L) cv. Pusa delicious. Int J Curr Microbiol Appl Sci 6(6):2964–2972. https://doi.org/10.20546/ijcmas.2017.606.353

Nelson PV (2003) Greenhouse operation and management, 6th edn. Department of Horticultural Science, North Carolina State University, Raleigh, p 692

Noguera P, Abad M, Puchades R, Maquieira A, Noguera V (2003) Influence of particle size on physical and chemical properties of coconut coir dust as a container medium. Commun Soil Sci Plant Anal 34:593–605. https://doi.org/10.1081/CSS-120017842

Parekh D, Puranik RM and Srivastava HS (1990) Inhibition of chlorophyll biosynthesis by cadmium in greening maize leaf segments. Biochem Physiol Pflanz 186: 239–242. https://www.cabdirect.org/cabdirect/abstract/19910742075

Patel K and LK Thethwar (2015) Study of plants for their chlorophyll content, phenolic content and antioxidant properties. Baharat J Scien Technol 1(1): 65–67. http://www.cvruresearch.org

Pickering JS (1997) An alternative to peat. The Garden 122:428–429

Radhouani C, Leonardi, B, Lechaieb L, Ben Yahya A, Ferchichi A (2015) Assessment of the potential of two local composts for seeds germination and vegetable production. J New Sci 20(1):780–787

Rahbarian P, Sardoei AS (2014) Effect of different media on growth, sucker and chlorophyll of padanus spp in under system mist. Intl J Farm Alli Sci 3(3):285–288

Raviv M, Medina S, Krassnovsky A, Laor Y, Aviani I (2009) Composting olive mill waste and assessment of its horticultural value. Acta Hortic 819: 353–359. http://www.actahort.org/books/819/819_42.htm

Rostami Z, Ghahsare AM, Kavoos B (2014) Date Palm waste application as culture media for strawberry and its impact on some growth indices and yield components. Agri Communi 2(3):15–21. https://www.cabdirect.org/cabdirect/abstract/20143287046

Samiei L, Halighi KA, Kafi M, Samavat S, Arghavani M (2005) An investigation of substitution of peat moss with palm tree celluloid wastes in growing aglaonema (Aglaonema Commutatum Cv. Silver Queen). Iran J Agric Sci 36(2):503–510

Shirani M (2013) Use of date-palm wastes as a substrate on tomato yield and number. Biol J Armenia 1(65):63–67

Soltani M, Naderi D (2016) Yield compounds and nutrient elements of carnation (*Dianthus caryophyllus* L.) under different growing media. Open J Ecol 6:184–191. https://doi.org/10.4236/oje.2016.64019

Tittarelli F, Rea E, Verrastro V, Pascual JA, Canali S, Ceglie FG (2009) Compost-based nursery substrates: effect of peat substitution on organic melon seedlings. Compost Sci Util 17(4):220–228. https://doi.org/10.1080/1065657X.2009.10702427

Vaughn SF, Deppe NA, Palmquist DE, Berhow MA (2011) Extracted sweet corn tassels as a renewable alternative to peat in greenhouse substrates. Ind Crops Prod 33:514–517. https://doi.org/10.1016/j.indcrop.2010.10.034

Yau PY and Murphy RJ (2000) Biodegraded cocopeat as a horticultural substrate. Acta Hortic 517:275–278. http://www.actahort.org/books/517/517_33.htm

11

Substitution of manure source and aerator in nursery media on sandy loam topsoil and their fertility indices 4 months after formulation

C. V. Adubasim[1] · C. M. Igwenagu[2] · G. O. Josiah[2] · S. E. Obalum[2] · U. M. Okonkwo[2] · I. M. Uzoh[2] · S. Sato[3]

Abstract

Purpose In soil-based nursery media, topsoil, poultry droppings and sawdust conventionally provide anchorage, nutrients and aeration, respectively. Considering poultry droppings' scarcity and sawdust's inertness nutrient-wise, more readily available organic wastes should be explored as substitutes. Here, we evaluated the effect of such substitution on media fertility, aimed at seeking alternatives to the conventional practice.

Methods In a topsoil-manure-aerator volume ratio of 3:2:1, poultry droppings was substituted with pig slurry (slurry) or cattle dung (dung) as manure and sawdust with rice-husk dust (huskdust) as aerator, giving seven soil-based media including reference medium (topsoil+droppings+sawdust) and the control (topsoil+topsoil+topsoil). They were watered regularly and analysed for fertility parameters 4 months later.

Results Reference had the highest pH (8.60) and topsoil + dung + huskdust/control the lowest (6.83). Substituting sawdust with huskdust enhanced pH, organic matter and Mg^{2+} in droppings/dung-amended media (topsoil+droppings+huskdust/topsoil+dung+huskdust) unlike slurry-amended ones where it too reduced total nitrogen (0.19 vs 0.11%). The substitution also enhanced available phosphorus in topsoil+droppings+huskdust (117.50 mg kg^{-1}) and topsoil+dung+huskdust (71.50 mg kg^{-1}) but reduced K^+ in the latter where it too had moderating effects on Na^+. Reference surpassed topsoil+slurry+huskdust for Ca^{2+}, but was surpassed by topsoil+droppings+huskdust for Mg^{2+}. Reference/topsoil+droppings+huskdust and topsoil+slurry+huskdust/control showed highest and lowest CEC, respectively. Excluding pH, topsoil+dung+huskdust and topsoil+slurry+sawdust were, notably, consistently similar. Overall, droppings-amended > dung-amended > slurry-amended media and, for available phosphorus only, sawdust-aerated < huskdust-aerated media.

Conclusion Based on fertility status 4 months after blending, topsoil+droppings+huskdust could serve as alternative to the conventional nursery medium, or topsoil+dung+huskdust where near-neutral pH is preferred to increased phosphorus/cations release.

Keywords Soil-based media · Organic manure · Pig slurry · Cattle dung · Rice-husk dust · Nutrient release

Introduction

The formulation of growth media for raising seedlings in the nursery influences the development and quality of the seedlings which in turn greatly defines the success of their re-establishment and biomass productivity in the field (Bhardwaj 2014). For soil-based media, the materials used to complement the formulation influence their physicochemical properties (Wilson et al. 2001; Sahin et al. 2005), the reason for which the physical and chemical components of such media affect percentage germination, seedling emergence and growth (Ede et al. 2015). When the proportion of silt and clay in a soil-based medium is high, the porosity of the medium may decrease, which implies poor aeration

✉ S. E. Obalum
sunday.obalum@unn.edu.ng; ijewelle@yahoo.com

[1] Department of Plant Biology and Biotechnology, University of Benin, Benin City, Nigeria

[2] Department of Soil Science, University of Nigeria, Nsukka 410001, Nigeria

[3] Department of Environmental Engineering for Symbiosis, Soka University, 1-236 Tangi-machi, Hachioji-shi, Tokyo 192-8577, Japan

and seed respiration (Baiyeri and Mbah 2006). Poor seed respiration results in lowered metabolic energy required for seed germination, thereby leading to seed decay (Baiyeri and Mbah 2006). Therefore, a favourable balance of water and air within the pore system of nursery media is needed to ensure seed survival as well as to support plant growth (Bruckner 1997; Caron and Nkongolo 1999).

Research on nursery media formulation in Nigeria has been traditionally devoted to horticultural concerns, focusing mostly on seedlings performance. Although there have been indications that the physicochemical fertility status of nursery media could influence seedlings performance (Baiyeri and Aba 2013), such status is often not first ascertained empirically before the adoption of media for raising seedlings. A good growing medium should contain the right proportion of materials such that it provides anchorage to seedlings while serving as reservoir of nutrients and water as well as allowing for adequate aeration to particularly ensure the diffusion of oxygen to the roots (Abad et al. 2002; Villagra and Cavagnaro 2006). Sandy loam and/or gravelly topsoils, when used in soil-based nursery media for anchorage, have been found to engender good moisture retention in such media (Ngulube 1996).

In Nigeria, the conventional practice is the use of topsoil mainly to provide anchorage, poultry droppings as manure and river sand as aerator in the volume ratio of 3:2:1 to formulate nursery media for seedlings (Ede et al. 2015; Ngwuta et al. 2016). However, nursery media blended using coarse-textured topsoil and poultry droppings (droppings) benefit from inclusion of sawdust in terms of enhanced total porosity (Baiyeri and Aba 2013). Such inclusion of sawdust thus promotes the aeration of soil-based nursery media. This, coupled with the high density and chemical inertness of river sand, informed the use of sawdust in place of river sand and hence the tagging of topsoil + droppings + sawdust as the reference medium in this fertility-oriented study of soil-based nursery media. In agronomic terms, this topsoil + droppings + sawdust has been shown to be similar to the conventional practice (Nwauzoma and Moses 2013), and to be the best among some soil-based/soilless media without river sand (Alaba et al. 2015). Besides, we are averse to the inclusion of river sand in nursery media, considering the pedo-agronomic implications of the unethical after-use disposal of such media back onto the soil surface.

However, poultry droppings are in high demand for its other uses in agricultural and ecological domains, whereas sawdust contains lignin and is poor nutrient-wise, limiting its role in nursery media to aeration. Amidst these challenges, there is need to explore the use of apparently less nutrient-rich sources of organic manure and more nutrient-rich aerator as substitutes. For instance, cattle dung (dung) and pig slurry (slurry) are more available in larger amounts in the agro-industry but are rather poor compared to poultry droppings not in terms of nutrient reserve but mineralization rate (Uzoh et al. 2015). On the other hand, rice-husk dust (huskdust) is generated in large quantities daily, but mostly has low agronomic utility, often times to the detriment of the environment. It is a valuable organic input owing to its small size which allows excellent soil-litter contact as well as its low polyphenol content (Schulz et al. 2003). Besides being of high nutrient content and moisture retention capacity, rice-husk dust is a good aerator (Baiyeri and Mbah 2006).

To our knowledge, virtually no study has evaluated the effect of replacing poultry droppings with pig slurry or cattle dung in soil-based nursery media. Again, despite the potential of rice-husk dust in such media, we are aware of only two studies that evaluated its effect. One is the study by Baiyeri and Mbah (2006) where it was used not necessarily as aerator but to blend soilless media by replacing topsoil in a mixture of topsoil, poultry droppings and river sand in varying ratios. These authors reported enhanced growth of African breadfruit (*Treculia Africana*) seedlings due to the replacement. In the other study, topsoil + droppings + huskdust and topsoil + droppings + sand were found to outperform topsoil + droppings + sawdust and other soil-based media in terms of seedling performance (Hofeleña and Martinez 2008). Regarding fertility assessment of soil-based media containing poultry droppings or other manures, the use of rice-husk dust to serve both as aerator and for nutrient supply is poorly researched. Use of these organic wastes in blending soil-based nursery media could be a way of beneficially disposing them.

This study was conceptualized out of the present and the future challenge of procuring poultry droppings and the greater availability of the duo of pig slurry and cattle dung as well as the inertness of sawdust vis-à-vis the abundance and great nutrient reserve in rice-husk dust. The study was initiated to investigate, under situations of retaining topsoil in nursery media formulation, the effect of substituting poultry droppings with pig slurry or cattle dung, on one hand, and sawdust with rice-husk dust, on the other, while still maintaining the topsoil–manure–aerator volume ratio of 3:2:1, on their chemical fertility indices. The aim was to seek alternative or even superior formulations to the conventional practice.

Materials and methods

Location of experiment

The research was established at the University of Nigeria Teaching and Research Farm in Nsukka campus of the University. The Farm is located at 06°52′N and 07°24′E, and is on an elevation of about 447 m asl. The mean annual rainfall is 1600 mm, while the mean values for minimum and maximum temperatures are 21 and 31 °C, respectively. The area also records variable relative humidity throughout the year in the range of 55–90%.

Media formulation and experimental design

The experiment involved the formulation of soil-based nursery media from topsoil, three sources of organic manures (poultry droppings, pig slurry and cattle dung) and two aerators (sawdust and rice-husk dust). A small portion of the farm under fallow was identified and cleared, and the needed topsoil collected from the surface layer (0–20 cm) of the soil. The soil, deeply weathered brownish red coarse-textured (sandy loam) Ultisols, is underlain by false-bedded sandstone. The topsoil was air-dried and homogenised by thorough mixing. Two of the manures, poultry droppings and pig slurry, were procured from the poultry and piggery units, respectively, of the animal section of the farm while the third, cattle dung, was sourced from the local abattoir. All the manures were cured by sun-drying and then crushed and sieved. The sawdust was collected from the 'Timber Shed' at Nsukka and the rice-husk dust from its dumpsite around a rice mill at nearby Adani, both in southeastern Nigeria.

Treatments consist of various combinations of topsoil with each of the three manure sources and each of the two aerators in turn. Because the aim was to contest the conventional formulation, all media had topsoil but poultry droppings was replaced with pig slurry or cattle dung and sawdust with rice-husk dust, maintaining the normal topsoil + manure + aerator ratio of 3:2:1 (vol./vol./vol.). With this substitution pattern, there were six variously formulated nursery media, including the conventional formulation adopted in this study (topsoil–droppings–sawdust), regarded here as the reference medium. A control (topsoil without manure and aerator such that the topsoil + manure + aerator ratio was 6:0:0 on volume basis, translating into topsoil + topsoil + topsoil volume ratio of 3:2:1) was also included, bringing the number of treatments to seven. The treatments are summarized in Table 1.

Treatments were thoroughly mixed. Potting was done using 0.15-L black polyethylene bags with perforations at the bottom for drainage. All treatments were replicated three times in a completely randomised design (CRD) under an 'open' tree shade. Watering of the nursery media was done regularly, the need of which was occasionally obviated or supplemented by rainfall. Sampling was scheduled to take place at 4 months after formulation and potting, when decomposition/mineralization of the various organic materials used in the blending must have attained optimum stage (Schulz et al. 2003; Uzoh et al. 2015). Sub-samples of the soil-based media were taken and air-dried to constant weight. Thereafter, they were crushed and passed through a 2-mm mesh sieve before analyses for some chemical fertility indices.

Laboratory analyses

Sub-samples of the organic wastes were analysed using standard procedures, to determine their chemical composition shown in Table 2. The pH of the various media was determined on deionised water and 0.1 N KCl suspensions in a 1.0:2.5 soil-liquid ratio (McLean 1982), and the values

Table 1 Summary of the variously formulated soil-based nursery media

Formulations/treatments (soil-based nursery media)	Topsoil (three parts by volume)	Organic manure (two parts by volume)	Aerator (one part by volume)
Topsoil + droppings + sawdust (reference medium)	Topsoil	Poultry droppings	Sawdust
Topsoil + droppings + huskdust	Topsoil	Poultry droppings	Rice-husk dust
Topsoil + slurry + sawdust	Topsoil	Pig slurry	Sawdust
Topsoil + slurry + huskdust	Topsoil	Pig slurry	Rice-husk dust
Topsoil + dung + sawdust	Topsoil	Cattle dung	Sawdust
Topsoil + dung + huskdust	Topsoil	Cattle dung	Rice-husk dust
Topsoil only (control)	Topsoil	Topsoil	Topsoil

Table 2 Chemical composition of the different organic wastes used in the formulation of soil-based nursery media (chemical properties are expressed in percentages)

Organic waste	Organic Carbon	Nitrogen	Phosphorus	Calcium	Magnesium	C:N ratio
Poultry droppings	23.94	2.31	0.88	0.19	0.17	10.36
Pig slurry	13.97	1.19	0.32	0.18	0.17	11.74
Cattle dung	28.33	1.12	0.28	0.20	0.19	25.29
Sawdust	9.11	0.30	0.36	0.29	0.12	30.37
Rice-husk dust	37.91	0.80	0.41	2.20	1.20	47.39

measured potentiometrically using a Beckman's zeromatic glass electrode pH metre. Organic matter was determined by the Walkley–Black wet dichromate oxidation method (Nelson and Sommers 1996). Total nitrogen was determined using micro-Kjeldahl digestion and distillation procedure as described by Bremner (1996). Available phosphorus was extracted with Bray 2 solution; thereafter, values were measured colorimetrically using the method described by Olsen and Sommers (1982).

To determine the cation exchange properties of the nursery media, the samples were leached with neutral (pH 7) 1 N NH_4OAc. Two of the exchangeable bases (K^+ and Na^+) in the leachate were measured with flame photometer while the other two (Ca^{2+} and Mg^{2+}) were determined by the complexometric titration method (Thomas 1982). The soil exchangeable acidity was determined by displacement with 1 N KCl followed by titration with NaOH (McLean 1982). Apparent cation exchange capacity (CEC) was determined by the NH_4OAc displacement method (Rhoades 1982). Thereafter, base saturation of the nursery media was calculated as the ratio of the four exchangeable bases to the effective cation exchange capacity (e-CEC), being the sum of the exchangeable bases and the exchangeable acidity.

Statistical analysis

One-way analysis of variance for a CRD was done on the data using the software GENSTAT Discovery Edition 1 Release 4.23 (VSN Int. Ltd., Hemel Hempstead, UK). In this analysis, the least significant difference (LSD) was used in mean separation at $p \leq 0.05$.

Results and discussion

The pH and organic matter content of the 4-month-old soil-based nursery media

There were differences in the pH of the seven nursery media 4 months after formulation (Table 3). The values were high, with the highest values of 8.60 recorded for topsoil + droppings + sawdust, the reference medium of this study. This was followed by topsoil + droppings + huskdust and topsoil + dung + sawdust. The lowest pH values of 6.83 were recorded for both topsoil + dung + huskdust and the control. These two media thus showed pH values closest to the range of 5.50–6.50 usually considered as optimum for seedlings in nursery media (Goh and Haynes 1977; Miller and Jones 1995). The pH is an important index of chemical quality and horticultural value of nursery media. For instance, Salifu et al. (2006) reported that pH alone explained 54% of variation in height of *Quercus rubra* seedlings in three variously formulated soilless media. The favourable pH of the topsoil + dung + huskdust may thus be considered its advantage over the other media formulations.

The presence of rice-husk dust in topsoil + dung + huskdust would partially explain the near-neutral pH found in this nursery medium (Baiyeri and Mbah 2006). From the present results, it appears that adding cattle dung and rice-husk dust to topsoil to serve as manure and aerator, respectively, in nursery media is ineffectual in increasing the soil pH. Considering that the pH shown by the control is a 'near-neutral' value usually preferred by seedlings of most plants (Goh and Haynes 1977), it could be that the combination of cattle dung and rice-husk dust has the potential to buffer the pH of the nursery medium against increases beyond this optimal value. Notably, the replacement of sawdust known to be the conventional aerator with rice-husk dust caused a 'desirable' reduction in the pH of all media except the ones with pig slurry as organic manure where the opposite prevailed.

All the formulated nursery media showed increases in organic matter content compared with the control; topsoil + slurry + huskdust was the only one where the increase over the control was marginal (Table 3). The highest values were recorded for topsoil + slurry + sawdust and topsoil + dung + huskdust. Therefore, the similarity between topsoil + slurry + huskdust and the control is rather surprising. It could be, however, that pig slurry is more compatible with sawdust and cattle dung with rice-husk dust regarding providing nursery media with aeration that favours certain

Table 3 The pH, organic matter, total nitrogen and available phosphorus contents of the various soil-based nursery media 4 months after formulation

Formulations/treatments (soil-based nursery media)	PH_{water}	PH_{KCl}	% Organic matter	% Total nitrogen	Available phosphorus (mg/kg)
Topsoil + droppings + sawdust (reference medium)	8.60	7.57	2.80	0.17	87.70
Topsoil + droppings + huskdust	8.07	7.33	3.39	0.15	117.50
Topsoil + slurry + sawdust	7.17	6.60	4.82	0.19	57.80
Topsoil + slurry + huskdust	7.50	6.73	2.36	0.11	41.10
Topsoil + dung + sawdust	8.23	7.10	3.12	0.15	37.60
Topsoil + dung + huskdust	6.83	6.30	4.33	0.19	71.50
Topsoil only (control)	6.83	6.33	1.42	0.06	10.3
$LSD_{0.05}$	0.29	0.23	1.14	0.05	25.55

biochemical processes that enhance organic matter status in manure-amended soils. In what seems like an inverse of the trend for pH, the replacement of sawdust with rice-husk dust caused organic matter to increase in all those soil-based media amended with poultry droppings (marginally though) and cattle dung; the opposite prevailed in those amended with pig slurry.

Total nitrogen and available phosphorus in the 4-month-old soil-based nursery media

The results for percent total nitrogen somewhat mirrored those for organic matter content (Table 3). Values were highest in topsoil + slurry + sawdust and topsoil + dung + huskdust and lowest in the control, while the increment in topsoil + slurry + huskdust over the control was marginal. Nursery media content of total nitrogen has, however, been shown not to be a good measure of their fertility status as evaluated by seedling performance; the topsoil + slurry + sawdust and/or topsoil + dung + huskdust would be deemed superior to topsoil + slurry + huskdust showing lower values only if they also show higher contents of other macronutrients including available phosphorus (Asiah et al. 2004). Also, the five nursery media formulations showing similar contents of total nitrogen here may differ when evaluated by seedling performance because both net nitrogen mineralization and net nitrogen immobilization are possible (Clark and Cavigelli 2005). The replacement of sawdust with rice-husk dust caused marginal increases in percent total nitrogen of all media except those amended with pig slurry where the replacement caused a significant reduction instead.

Also, there were differences in available phosphorus content of the various nursery media (Table 3). The topsoil + droppings + huskdust recorded the highest values while the control recorded the lowest values. Low levels of other plant nutrients in nursery media from germination through the early weeks of seedling growth could be tolerated because nutrient requirements during this stage are minimal, but not for phosphorus (Miller and Jones 1995). Available phosphorus being highest in topsoil + droppings + huskdust is thus an edge this formulation has over the rest of the nursery media. The trend due to the replacement of sawdust with rice-husk dust (higher values in the nursery media amended with poultry droppings and cattle dung and corresponding lower values in those amended with pig slurry) subsisted; however, the replacement induced marginal reductions this time around.

Cation exchange properties of the 4-month-old soil-based nursery media

The data for the exchangeable bases and acidity of the various soil-based nursery media 4 months after formulation are shown (Table 4). The content of K^+ was higher in all the formulated nursery media relative to the control. There were no differences in either of the droppings-amended two media and the slurry-amended two media due to replacement of sawdust with rice-husk dust, but the dung-amended counterparts differed, with topsoil + dung + sawdust showing higher K^+ than topsoil + dung + huskdust. Perhaps, aerator replacement had no effect on aeration porosity of the droppings-amended and the slurry-amended media (data not yet processed, hence not shown); otherwise, this observation implies that the capacity of the concerned soil-based media to supply potassium is not influenced by their degree of aeration.

The Ca^{2+}, apart from its being higher in the reference medium compared to topsoil + slurry + huskdust, indicated similar values among the seven soil-based media. Hence, the replacements done in the reference medium (i.e, replacing poultry droppings with pig slurry or cattle dung as organic manure and sawdust with rice-husk dust

Table 4 The exchangeable bases and acidity including the cation exchange capacity of the various soil-based nursery media 4 months after formulation

Formulations/treatments (soil-based nursery media)	K^+	Ca^{2+}	Mg^{2+}	Na^+	H^+	e-CEC	CEC	BS
			$cmol_c\ kg^{-1}$				$cmol\ kg^{-1}$	
Topsoil + droppings + sawdust (Reference medium)	0.32	9.33	10.00	0.19	1.00	20.85	30.53	0.95
Topsoil + droppings + huskdust	0.27	6.93	12.27	0.24	0.80	20.51	24.80	0.96
Topsoil + slurry + sawdust	0.20	7.47	6.80	0.14	0.80	15.41	29.60	0.95
Topsoil + slurry + huskdust	0.25	6.00	3.40	0.23	0.80	10.68	19.47	0.93
Topsoil + dung + sawdust	0.34	7.73	2.87	0.33	0.73	12.00	23.07	0.94
Topsoil + dung + huskdust	0.23	7.60	6.53	0.15	0.73	15.24	34.93	0.95
Topsoil only (Control)	0.04	6.80	2.67	0.03	0.67	10.21	12.67	0.93
$LSD_{0.05}$	0.10	2.63	1.77	0.04	0.19	1.60	7.16	0.02

e-CEC effective cation exchange capacity, *CEC* apparent cation exchange capacity, *BS* base saturation

as aerator), when considered individually, produced no effect on the Ca^{2+} content of the soil-based media; differences were due to simultaneous replacement of poultry droppings with pig slurry and sawdust with rice-husk dust.

Treatment also affected Mg^{2+} of the nursery media with highest and lowest values in topsoil + droppings + husk-dust and the control, respectively (Table 4). Contrary to the observation for pH, replacement of sawdust with rice-husk dust caused increases in Mg^{2+} content of droppings-amended and dung-amended soil-based media but decreases in that of slurry-amended ones. Because of the positive role of Mg^{2+} in seedling performance in the nursery (Baiyeri and Aba 2013), it can be considered an important fertility index of nursery media. For the droppings-amended media, therefore, the significant improvement in Mg^{2+} over the reference medium by topsoil + droppings + huskdust may have some agronomic relevance. Notably, topsoil + slurry + huskdust and topsoil + dung + sawdust were so low in Mg^{2+} that they had similar values as the control, thus pointing to the aforementioned compatibility of pig slurry with sawdust and not rice-husk dust and of cattle dung with rice-husk dust and not sawdust in the formulation of soil-based nursery media.

The Na^{+} content of the nursery media was also affected by treatment. All the formulated media recorded higher values compared with the control. There were also differences among them, with the highest values in topsoil + dung + sawdust, plausibly due to the use of soapy materials for washing in abattoirs. Both the replacement of manure and aerator contributed to these differences. For the aerator replacement, increases in Na^{+} due to the change from sawdust to rice-husk dust were recorded for the droppings- and slurry-amended media; decreases occurred only for the dung-amended media. The decreases were such significant that, among the variously blended media, the highest values in topsoil + dung + sawdust dropped to lowest in topsoil + dung + huskdust. Considering the widely known dispersive tendency of sodium in soils and the fact that it is not an essential but functional plant nutrient (Maathuis 2014), the results just stated favour the adoption of topsoil + dung + huskdust in nursery practice. The topsoil + slurry + sawdust was similar to topsoil + dung + huskdust in Na^{+} content. In the context of doubtful utility of excessive sodium in soils, this observation lends credence to the aforementioned compatibility of pig slurry with sawdust and of cattle dung with rice-husk dust in the formulation of soil-based nursery media.

The exchangeable acidity of the soil-based media was dominated by H^{+}, with Al^{3+} occurring in traces. Higher values of the H^{+} were recorded in the reference medium relative to the rest of the nursery media. However, the H^{+} tended to decrease in the order droppings-amended media, slurry-amended media, dung-amended media, and the control.

The aggregated effect of the differences in the exchangeable bases and H^{+} in the e-CEC of the soil-based media shows that all the formulated media but the topsoil + slurry + huskdust were superior to the control. However, all treatments showed e-CEC of above 10 cmol kg^{-1} regarded as the minimum for nursery media (Miller and Jones 1995). Reference medium and topsoil + droppings + huskdust had similar e-CEC, values of which were higher than the rest of the treatments. By contrast, the replacement of sawdust with rice-husk dust as aerator caused an increase in the e-CEC of both the slurry-amended and the dung-amended media. That the change of aerator from sawdust to rice-husk dust affected not the e-CEC of droppings-amended media but that of slurry- and dung-amended media suggests that the complementary role of rice-husk dust in enhancing the fertility of nursery media could be masked in the presence of such nutrient-rich sources of manure as poultry droppings.

But for the higher values in the reference medium than topsoil + droppings + huskdust, the results for the apparent CEC of the soil-based media were somewhat congruent to those for the e-CEC. Generally, the media formulations of this study were of low cation exchange capacities. This is despite the increases in their organic matter content due to the organic manures involved in the formulation. It could be that the formulations were yet to attain full decomposition and mineralization at the time of sampling which took place 4 months after formulations. Another plausible reason is that texture seems to have stronger influence than organic matter on CEC of the soil used for the formulations (Obalum et al. 2013).

The change of the aerator from sawdust to rice-husk dust resulted in the percent base saturation being higher in droppings- and dung-amended media compared to slurry-amended media which together with the control showed the lowest values. Also, neither of the droppings-amended and the dung-amended media differed in percent base saturation due to the change of the aerator from sawdust to rice-husk dust, unlike the slurry-amended media in which the replacement lowered the percent base saturation, hence the lowest values recorded in topsoil + slurry + huskdust.

Conclusions

Substitution of organic manure and aerator used in complementing coarse-textured topsoil in the blending of nursery media could lead to considerable variations in their chemical fertility indices some 4 months after formulation/blending.

In the conventional medium (topsoil + droppings + sawdust), the organic manure poultry droppings could be retained or substituted with cattle dung; this substitution is efficacious only when the aerator sawdust is also substituted with rice-husk dust. Thus, topsoil + droppings + huskdust could serve as an alternative to the conventional medium; however, topsoil + dung + huskdust would be a better one if near-neutral pH is priced more than phosphorus availability and cation exchange.

The above conclusions are drawn from the preliminary results on fertility status of the variously formulated soil-based nursery media. The study on substitution of manure source and aerator in such media is yet terminated. Presently, we are validating the soil test results via agronomic evaluation of the nursery media commencing from 12 months after formulation, at the end of which the nursery media shall be analysed once again to check for any change in fertility status and to test the observed differences in fertility indices for consistency.

Acknowledgements Special thanks to some students in the B.Agric. Degree Programme of the Faculty of Agriculture of the University of Nigeria Nsukka (UNN) who were in their 4th year in the 2016/2017 Session for their co-operation while initiating and managing the experiment. The support of the staff of the Laboratory Unit of the Department of Soil Science of UNN, particularly Ms Ngozi Ojionuka, in the processing/analyses of the samples is also gratefully acknowledged.

References

Abad M, Noguera P, Puchades R, Maquieira A, Noguera V (2002) Physico-chemical and chemical properties of some coconut dusts for use as peat substitute for containerized ornamental plants. Bioresour Technol 82:241–245. https://doi.org/10.1016/S0960-8524(01)00189-4

Alaba FT, Motolani AO, Adeola AC (2015) Efficiency of seed treatments and different growth media on the germination of seeds of *Xylopia aethiopica* (Dunal) A. Rich J Biol Chem Res 32:849–853

Asiah A, Mohd Razi I, Mohd Khanif Y, Marziah M, Shaharuddin M et al (2004) Physical and chemical properties of coconut coir dust and oil palm empty fruit bunch and the growth of hybrid heat tolerant cauliflower plant. Pertanika J Trop Agric Sci 27:121–133

Baiyeri KP, Aba SC (2013) Physicochemical characteristics of nursery media formulated from organic wastes influenced growth and dry matter yield of banana (Musa AA var. Pisang lilin) plantlets. Niger J Hort Sci 17:135–145

Baiyeri K, Mbah BN (2006) Effects of soilless and soil-based nursery media on seedling emergence, growth and response to water stress of African breadfruit (*Treculia Africana* Decne). Afr J Biotechnol 5:1405–1410

Bhardwaj RL (2014) Effect of growing media on seed germination and seedling growth of papaya cv. 'Red Lady'. Afr J Plant Sci 8:178–184. https://doi.org/10.5897/AJPS11.265

Bremner JM (1996) Total nitrogen. In: Sparks DL (ed) Methods of soil analysis part 3: chemical methods, vol 5. Soil Science Society of America, Madison, pp 1085–1122

Bruckner U (1997) Physical properties of different potting media and substrate mixture—especially air- and water-capacity. Acta Hortic 450:263–270. https://doi.org/10.17660/ActaHortic.1997.450.31

Caron J, Nkongolo VKN (1999) Aeration in growing media: recent developments. Acta Hort 481:545–551

Clark S, Cavigelli M (2005) Suitability of composts as potting media for production of organic vegetable transplants. Compost Sci Util 13:150–156. https://doi.org/10.1080/1065657X.2005.10702232

Ede AE, Ndubuaku UM, Baiyeri KP (2015) Media effects on emergence and growth of Moringa (*Moringa oleifera* Lam) seedlings in the nursery. Am J Exp Agric 7:181–189. https://doi.org/10.9734/AJEA/2015/13602

Goh KM, Haynes RJ (1977) Evaluation of potting media for commercial nursery production of container-grown plants. N Z J Agric Res 20:363–370. https://doi.org/10.1080/00288233.1977.10427348

Hofeleña LT, Martinez CP (2008) Factors influencing growth performance of *Jatropha curcas* L. (Tuba-Tuba) seedlings. Southeast Philipp J Res Dev 16:89–101

Maathuis FJM (2014) Sodium in plants: perception, signalling, and regulation of sodium fluxes. J Exp Bot 65:849–858. https://doi.org/10.1093/jxb/ert326

McLean EO (1982) Soil pH and lime requirement. In: Page AL, Miller RH, Kenney (eds) Methods of soil analysis, part 2: chemical and microbial properties. Agronomy monograph no. 9. American Society of Agronomy, Madison, pp 199–224

Miller JH, Jones N (1995) Organic and compost-based growing media for tree seedling nurseries. World bank technical paper no. 264—forestry series. The World Bank, Washington, p 75

Nelson DW, Sommers LE (1996) Total carbon, organic carbon and organic matter. In: Sparks DL (ed) Methods of soil analysis, part 3: chemical and microbial properties. American Society of Agronomy, Madison, pp 539–580

Ngulube MR (1996) Ecology and management of Uapaca kirkiana in Southern Africa. PhD Thesis, University of Wales

Ngwuta AA, Peter-Onoh CA, Obiefuna JC, Chigbundu NI, Nwokeji EM, Chris-Emenyonu CM, Metu CN (2016) Juvenile phenology of *Jatropha curcas* as influenced by selected nursery growth media. Int J Agric Rural Dev 19:2506–2510

Nwauzoma AB, Moses K (2013) Factors affecting seedling emergence and dry matter characteristics in *Musa balbisiana* Colla. ISRN Bot 2013:5. https://doi.org/10.1155/2013/582581

Obalum SE, Watanabe Y, Igwe CA, Obi ME, Wakatsuki T (2013) Improving on the prediction of cation exchange capacity for highly weathered and structurally contrasting tropical soils from their fine-earth fractions. Commun Soil Sci Plant Anal 44:1831–1848. https://doi.org/10.1080/00103624.2013.790401

Olsen SR, LE Sommers (1982) Phosphorus. In: Page AL, Miller RH, Keeny (eds) Methods of soil analysis, part 2: chemical properties, 2nd edn. Agronomy monograph no. 9. American Society of Agronomy, Madison, pp 15–72

Rhoades JD (1982) Cation exchange capacity. In: Page AL (ed) Methods of soil analysis, part 2: chemical methods. Agronomy monograph no. 9. American Society of Agronomy, Madison, pp 149–157

Sahin U, Ors S, Ercisli S, Anapali O, Esitken A (2005) Effect of pumice amendment on physical soil properties and strawberry plant growth. J Central Europ Agric 6:361–366

Salifu KF, Nicodemus MA, Jacobs DF, Davis AS (2006) Evaluating chemical indices of growing media for nursery production of quercus rubra seedlings. HortScience 41:1342–1346

Schulz S, Tian G, Oyewole B, Bako S (2003) Rice mill waste as organic manure on a degraded Alfisol. Agric Ecosyst Environ 100:221–230. https://doi.org/10.1016/S0167-8809(03)00198-1

Thomas GW (1982) Exchangeable cations. In: Page AL (ed) Methods of soil analysis, part 2: chemical methods. Agronomy monograph no. 9. American Society of Agronomy, Madison, pp 159–165

Uzoh IM, Obalum SE, Ene J (2015) Mineralization rate constants, half-lives and effects of two organic amendments on maize yield and carbon-nitrogen status of loamy Ultisol in southeastern Nigeria. Agro-Sci 14:35–40. https://doi.org/10.4314/as.v14i3.7

Villagra PE, Cavagnaro JB (2006) Water stress effects on the seedling growth of *Prosopis argentina* and *Prosopis alpataco*. J Arid Environ 64:390–400. https://doi.org/10.1016/j.jaridenv.2005.06.008

Wilson SB, Stoffella PJ, Graetz DA (2001) Use of compost as a media amendment for containerized production of two subtropical perennials. J Environ Hort 19:37–42

12

Direct and residual effect of various vermicompost on soil nutrient and nutrient uptake dynamics and productivity of four mustard Pak-Coi (*Brassica rapa* L.) sequences in organic farming system

N. Nurhidayati[1] · Masyhuri Machfudz[2] · Indiyah Murwani[1]

Abstract

Purpose To study the direct effect on the first mustard cropping and the residual effect of three kinds of vermicompost on the productivity of the second, third, fourth Pak-Coi mustard sequential cropping.

Method A field experiment was conducted for four sequential planting periods of mustard Pak-Coi in silty clay Inceptisol. The experiment used a factorial randomized block design with two factors. Factor I comprised three kinds of vermicompost with three types of bedding materials; V1—spent mushroom waste, V2—coconut husk, and V3—sugarcane trash, while factor II comprised four different application rates: 5, 10, 15, and 20 t ha^{-1}. From these two factors, there were 12 treatments plus one control treatment (without vermicompost application).

Results Application of vermicompost increased soil NPK content. The highest nutrient uptake was also found in the residual effect for the second cropping. Nutrient uptake of the third and fourth mustard cropping had a decreasing trend. The highest yield on the first cropping was found in the application of vermicompost V2 and V1 with the application rates of 10–15 t ha^{-1}. In the second cropping, the productivity increased, whereas the third and fourth sequential cropping decreased in the vermicompost V1 and V2, whereas in the vermicompost V3 the productivity of the third and fourth cropping was slightly increased.

Conclusion The application of vermicompost to organic farming provides nutrient availability in four mustard subsequent planting periods. Differences in the dynamics of nutrient availability, nutrient uptake and crop yields are affected by the vermicompost materials used.

Keywords Vermicompost · Soil nutrient · Nutrient uptake · Plant productivity

Introduction

The long-term overcropping and overuse of inorganic fertilizers without organic input degrade soil quality and health and cause environmental pollution as well (Albiach et al. 2000). Decreasing soil quality can be avoided by proper and careful soil management as practiced by organic farmers. The application of organic fertilizer is an important practice in organic farming to improve soil quality, enhance microbial activity and nutrient recycle to produce high-quality crops. Thus, organic farming system produces healthy plant (Knapp et al. 2010; Dimitri et al. 2012). Increase in microbial biomass activity also results changes in soil enzymes that play an important role in soil biochemical reactions (Garcia-Gil et al. 2000; Ros et al. 2006). Organic matter input into the soil gives a long-term positive effect due to humus production that can increase the soil cation exchange capacity and water retention (Weber et al. 2007).

One of the organic fertilizers often used by organic farmers is vermicompost. Vermicompost is a compost produced by earthworm activity. During the vermicomposting process, earthworms play an important role in degrading organic materials. This process gradually decreases C:N ratio and increases the surface area of organic matter, hence making it easier for microorganisms to decompose it. Castings produced by earthworms will be biochemically

✉ N. Nurhidayati
nurhidayati@unisma.ac.id

1 Department of Agrotechnology, Faculty of Agriculture, University of Islam Malang, East Java, Indonesia

2 Department of Agribusiness, Faculty of Agriculture, University of Islam Malang, East Java, Indonesia

decomposed by microorganisms (Domínguez et al. 2010). Vermicompost is widely used in the cultivation of horticultural crops. Vermicompost has an extensive potential for use on agricultural crops to replace or, alternatively, combine with inorganic fertilizers. The yield of some horticultural crops on treatment using vermicompost and inorganic fertilizers was not significantly different, but the quality of yield was higher than that of the inorganic fertilizers (Nurhidayati et al. 2015, 2016). Other researchers have evaluated the effects of vermicompost on the growth of other plants through various greenhouse and field studies. Vermicompost have a positive effect on physicochemical properties of soil hence improving growth and yield of tomato (Atiyeh et al. 2000a, b; Atiyeh et al. 2001; Hashemimajd et al. 2004; Gutiérrez-Miceli et al. 2007). Application of vermicomposts to the soil also increased soil humic and plant growth hormone hence enhanced the growth and yield of pepper (Arancon et al. 2004a, 2005). Soil improvement with vermicompost can increase growth and yield of the other crops such as garlic (Argüello et al. 2006), aubergine (Gajalakshmi and Abbasi 2004), strawberry (Arancon et al. 2004b), and sweet corn (Lazcano et al. 2011). The combined effects of vermicompost and vermiwash can also increase the yield of organic linseed crop (Makkar et al. 2017). The improvement of soil physicochemical and the addition of plant growth substances in the soil stimulated the development of roots and growth of shoots and plant leaf area (Edwards et al. 2004; Lazcano et al. 2009). Roy et al. (2010) showed that the plots using vermicompost treatment had higher plant height, shoot and root weight as compared to compost or soil plots.

Vermicompost is a natural organic fertilizer product by earthworm. It releases the nutrients gradually into the soil (Chaoui et al. 2003). Therefore, it has the residual effect on the succeeding plant. Jat and Ahlawat (2006) reported that vermicompost application enhanced soil nutrient status (N and P) in subsequent crops compared to no vermicompost application. However, the residual value was determined by the vermicompost quality and the application rate. The effect of vermicompost on plants was different. It is dependent on crop species and genotype (Lazcano et al. 2011). Nutrient dynamics in the soil and the availability of nutrients in the soil, especially the total N and P available in the soil due to vermicompost applications have been studied by Jat and Ahlawat (2006) on chickpea–maize cropping system. Where the dynamics of nutrient availability in soil and nutrient uptake in vermicompost application after chickpea increased and decreased after maize. However, little is known about the direct and residual effects of various vermicompost on one plant species with several planting periods associated with soil nutrient, nutrient uptake dynamics and crop yields. The main aim of this study was to determine the direct and residual effects of various vermicompost with different rates on soil nutrient, nutrient uptake dynamics and productivity of mustard Pak-Coi for four planting periods.

Materials and methods

Study site and soil characteristics

This study was a field experiment conducted at Landungsari village, Dau district, Malang regency in March–July 2017 with an altitude of 544 m above sea level, 07°55′42.1″S latitude, and 112°35′55.0″E longitude. The average temperature was 22–28 °C while the rainfall is 1750 mm/year. The field had a semi-technical irrigation system. The type of soil was an Inceptisol. The soil characteristic was silty clay comprising 46% clay, 46% silt, and 8% sand. The soil had bulk density of 1.23 g cm^{-3}. The soil was low in organic carbon (0,96%), with pH 5.7, low in total N (0.15%), high in phosphorus (228.26 mg kg^{-1}), and low in available K(0.33 cmol kg^{-1} soil).

Vermicompost preparation

Making of vermicompost was done in vermicomposting bin having a size of 80 × 120 × 30 cm. Materials used in making of the vermicompost were the spent mushroom waste, coconut husk, sugar cane trash, cow dung, dry leaf litter, and fresh vegetable waste. The vegetable wastes were collected from the traditional market and were cut into the size of vegetable wastes pieces of ± 5 cm. The dry leaf litter and sugar cane trash were grounded by a grinder. The spent mushroom waste, coconut husk and sugarcane trash served as bedding materials, while the mixture of cow dung, leaf litter and the vegetable waste were used as worm feeds. The amount of material needed to fill the bin was 40 kg of bedding materials and 20 kg of worm feed. This amount was adapted to the capacity of the worm to decompose organic matter on a daily basis. Vermicomposting process lasted for 28 days using the earthworm species, *Lumbricus rubellus*. Thereafter, composting was done for 14 days. The vermicompost is also added with fish meal and eggshell flour with a composition of 5% as the additive materials. The addition of eggshell flour and fish meal was done after vermicomposting process. During the vermicomposting process, moisture content was maintained at 80%. The vermicomposting bin is covered with a black cloth to avoid the sun. The chemical compositions of the resulting vermicomposts were analyzed (Table 1).

Experimental design

This experiment used a factorial randomized block design (RBD) which consisted of two factors. Factor I comprised

Table 1 Chemical compositions of three kinds of vermicompost on dry weight basis

No.	Chemical properties	V1	V2	V3
1	Organic-C (%)	17.39	34.66	19.09
2	Polyphenol (%)	0.79	0.67	1.58
3	Cellulose (%)	26.75	27.86	27.34
4	Lignin (%)	25.08	24.15	15.8
5	Total N (%)	2.04	2.28	2.22
6	C:N ratio	8.52	15.22	8.60
7	P (%)	10.63	0.73	8.08
8	K (%)	0.23	1.05	0.52
9	pH	7.4	7.2	7.1
10.	Ash (%)	4.66	4.73	3.86

three kinds of vermicompost (V) with three different bedding materials; V1: mixture of spent mushroom waste, cow dung, vegetable waste, and leaf litter, V2: mixture of coconut husk, cow dung, vegetable waste and leaf litter, and V3: sugarcane trash, cow dung, vegetable waste and leaf litter. Factor II comprised 4 different rates of vermicompost application: D1 = 5 t ha^{-1}, D2 = 10 t ha^{-1}, D3 = 15 t ha^{-1}, and D4 = 20 t ha^{-1}. From these two factors, there were 12 treatment combinations plus one control treatment (without vermicompost application). Each treatment was replicated 3 times to obtain 39 units of experimental plot, and each experimental plot had 5 samples of the observed plants.

Experimental procedures

Pak-Coi mustard seeds were planted in the medium mixture of cow dung and soil with a ratio of 1:1. This seedling takes 3 weeks (21 days) to be ready for transplanting. Land preparation was done by tilling the soil with a hoe and then making a bed of 2.5 m × 1.5 m. All experimental plots received the application of cow dung with a rate of 5 t ha^{-1}. Three kinds of vermicompost were weighed in accordance with the application rate. Then, the vermicompost was incorporated into the well-prepared beds evenly using the hoe 1 week before planting. Mustard seedlings were planted at a spacing of 20 cm × 20 cm, 21 days after seeding and a week after vermicompost application. Watering and manual weeding was done. The mustard was harvested 28 days after transplanting. After harvesting the first crop of mustard, the land was cleared for 1 week and planting the second crop of mustard and so on for up to the fourth mustard cropping, to test the residual effects on the second, third and fourth cropping. The observed variables were soil nutrient, nutrient uptake, fresh and dry weights of total biomass, and weight of marketable yield for the fourth mustard cropping period.

The nitrogen (N) content in leaves of these crops was measured using modified Kjeldahl's method (Jackson 1973). The P content was calculated by vanado-molybdo-phosphoric acid yellow colour method (Jackson 1973) and the K content was measured with a flame photometer. The N, P and K content in leaves was multiplied with the respective dry matter of total biomass to get N, P and K uptake in crop. The soil samples from the experimental plots were analyzed for total N (modified Kjeldahl's method, Jackson 1973) and available-P (Bray I method, Jackson 1973) and extractable-K content (flame photometer) at 2 weeks after transplanting during the first mustard cropping and after harvesting the first, second and the third mustard cropping.

Statistical analysis

The collected data were statistically confirmed using analysis of variance (*F* Test) at $P \leq 0.05$ and the differences between each treatment were determined by Tukey test ($P \leq 0.05$) using Minitab Version 14.12. Dunnett's test at 5% level was used to compare all treatments with control. For statistical analysis of data, Microsoft Excel was used.

Results and discussion

Soil nutrient content

Based on the results of the analysis of variance at $P = 5\%$, variables of the soil nutrients content at the fourth soil sampling showed that the tested treatment significantly ($P < 0.05$) influenced the soil nutrient content of soil. The soil nutrient dynamics of each treatment on the fourth soil sampling is presented in Fig. 1. The dynamics of N and K nutrients in soils up to the fourth soil sampling had a similar pattern, where the increased availability of N and K occurs on the second soil sampling (after harvesting mustard 1). On the third and fourth soil sampling, soil N and K decreased (Fig. 1a, c). Nutrient dynamics of P in soils differ from N and K, where the highest P nutrient availability was found in the first soil sampling (2 weeks after transplanting). The succeeding soil sampling P tended to decrease (Fig. 1b).

Figure 1a also illustrated that the direct effects of application of three vermicompost treatments, on soil N was found highest in vermicomposts V2 and V3 at application rates of 15–20 t ha^{-1}. On the second and third soil sampling (after harvesting mustard 1 and 2) the same pattern of nutrient availability was still observed. However, at the fourth soil sampling (after harvesting mustard 3), the soil nutrients in the plots treated with vermicompost V3 at an application rate of 20 t ha^{-1} were higher than the other treatments. This suggests that the different bedding materials affected the release of nutrients into the soil. Bedding derived from the spent mushroom waste had the highest lignin content providing a low nutrient release although the vermicompost had a high

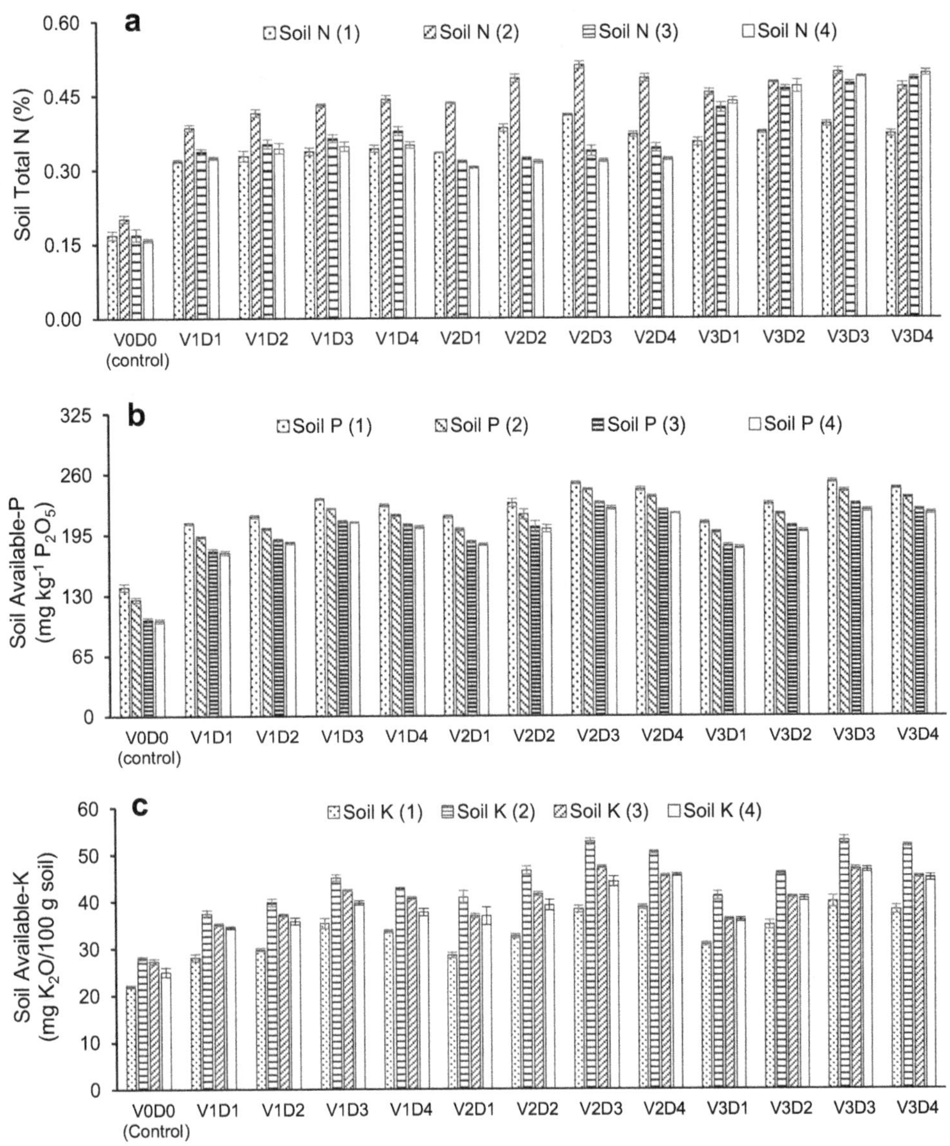

Fig. 1 The dynamics of soil N, P and K nutrient (**a–c**) due to direct and residual effect of three kinds of vermicompost on various application rates and the fourth soil sampling. (remarks: V1, V2, and V3 = kind of vermicompost, D1, D2, D3, and D4 = application rate (5, 10, 15, and 20 t ha^{-1}), (1), (2), (3), and (4) = time of soil sampling)

P content. Vermicompost V2 and V3 had a higher soil N content than V1. Thus, it released a higher N (Fig. 1b).

The highest soil N and K were found in the second soil sampling (after harvesting mustard 1), while the highest soil P was found at the first soil sampling (2 weeks after transplanting). In the next soil sampling, the soil P content decreased until the fourth soil sampling. The highest soil K was found on the plot treated with vermicompost V2. Vermicompost V2 contains bedding derived from coconut husk that has the highest K content. The chemical composition of vermicompost V2 had the highest K content (Table 1).

The dynamics of nutrient release for the fourth mustard cropping showed different patterns due to the application of three kinds of vermicompost. These results indicated that nutrient release is influenced by nutrient content in vermicompost. Yadav et al. (2010) showed that the characteristics of the raw material of vermicompost influenced the quality of vermicompost such as the total nutrient contents. Vermicompost V2 and V3 showed the highest nutrient availability especially N and K content on the second soil sampling. While vermicompost V3 still showed the highest N nutrient availability on the last soil sampling. Vermicompost V3 contained bedding derived from sugarcane trash which has a high N and the lowest lignin content (Table 1), which can increase the availability of N in the soil, while Vermicompost V1 contained the highest lignin. Differences in chemical composition of raw materials of organic matters such as organic-C, total N and lignin content as well as C:N and Lignin:N ratio affected the rate of decomposition and release of nutrients into the soil (Wardle et al. 2009; Moore et al. 2011).

De Araújo et al. (2010)reported that organic materials are mostly composed of carbon. The release of nutrient vermicompost was determined by the carbon quality and C:N ratio of the organic matters used. Thus, C:N ratio was an indicator of vermicompost maturity (Pattnaik and Reddy 2010). In addition to C:N ratio, the lignin content of organic matter affected the dynamics of N in the soil. Lignin was one of the most slowly decomposing components (Rahman et al. 2013). The three kinds of vermicompost had a C/N ratio < 20. This suggested that vermicomposts have high maturity and stability. An organic material with C:N ratio < 20 will be mineralized in soil releasing nutrients that can be absorbed by the plant. Among the three kinds of vermicompost, vermicompost V2 (bedding derived from coconut husk) had the highest C/N ratio of 15.22. Vermicompost V2 has the highest K content, because the coconut husk has a high K content.

Plant nutrient uptake

N, P and K uptake of the first mustard cropping was directly affected by three vermicompost applications, where vermicompost V1 and V2 provided the highest N uptake at an application rate of 10 t ha^{-1}, while the residual effect on the second plant, the highest N uptake was found in vermicompost V2 with application rate of 20 t ha^{-1}. Residual effects on the third and fourth mustards, vermicompost V2 and V3 at application rates of 15–20 t ha^{-1} gave the highest N uptake (Table 2).

The P uptake of mustard was not influenced by the direct effects of three vermicompost applications. However, the different rate significantly affect P uptake of the mustard where the highest P uptake was found at an application rate of 10 t ha^{-1}. The residual effect of three kinds of vermicompost on the P uptake of the second mustard was found on the plot treated with vermicompost V1 at application rates of 15–20 t ha^{-1}, while on the third and fourth mustard cropping, the highest P uptake was found in vermicompost 3 with an application rate of 15–20 t ha^{-1} (Table 2).

K uptake of the first mustard cropping was influenced by the direct effects of three vermicompost applications where the vermicompost V2 provided the highest K uptake at an application rate of 10–15 t ha^{-1}. The highest residual effect of the three kinds of vermicompost on K uptake of the second mustard was found in vermicompost V1 with application rate of 15–20 t ha^{-1}, while the highest residual effect on the third and fourth mustard was found in the plot treated with vermicompost V3 with application rate of 15–20 t ha^{-1}. These results demonstrated that the vermicompost derived from bedding of sugarcane trash gave the best residual effect on nutrient uptake until the fourth mustard cropping (Table 2).

Plant nutrient uptake highly depends on the nutrient release from the soil solid phase in the form of mineral and organic materials to the soil solution. The release of nutrients from organic matter occurs biochemically through the balance of mineralization and immobilization processes, while from mineral materials occurs physicochemically through adsorption and desorption, and precipitation and dissolution. The mineralization controls the soil solution concentration directly when the soil mineral does not release nutrient into the soil solution (Comerford 2005). Vermicompost as an organic fertilizer also undergoes mineralization releasing nutrients into the soil solution. The three kinds of vermicompost used in this study had a C/N ratio < 20. Thus, the vermicomposts had a high mineralization rate. Chaoui et al. (2003) reported that the vermicompost had a lower C/N ratio and indicated that it is more suitable for use as a soil amendment. It is due to the high NH_4^+ content released and converted into NO_3^- through the nitrification process during vermicomposting (Prabha 2009). Vermicompost contains high available nutrients, allowing not only a short supply of plant nutrient but also increasing reserves for the succeeding crops (Jat and Ahlawat 2006).

Crop yield

The fresh weight of total biomass differed significantly ($P < 0.05$) between the tested treatment combinations from either the direct effect for the first mustard cropping or the residual effect for the second, third and fourth mustard cropping (Table 3). The plots treated with vermicompost V2 at a rate of 10 t ha^{-1} had higher fresh weight of total biomass and weight of marketable yield than the other treatments though not significantly different with the treatment of vermicompost V1 at a rate 10–15 t ha^{-1} for direct effect of vermicompost (Tables 3 and 4). For the second mustard cropping, the treatments with vermicompost V1 at the rate of 10–20 t ha^{-1} significantly had the highest yield, but not significantly different with the vermicompost V3 at the rate of 20 t ha^{-1}. For the third mustard cropping, the treatments with vermicompost V3 at the rate of 15–20 t ha^{-1} significantly had the highest yield, but not significantly different with the vermicompost V1 at the rate of 20 t ha^{-1}. The fourth mustard cropping treated by vermicompost V3 at the rate of 15–20 t ha^{-1} significantly had the highest fresh weight of total biomass and weight of marketable yield (Tables 2, 3). All of the plots treated with three kinds of vermicompost significantly ($P < 0.05$) had higher yield than control (without vermicompost). Application of vermicomposts improved soil physicochemical and biological properties and added plant growth substance. Hence, it had a positive effect on the vegetative growth of plants and improved the yield and quality of horticultural crops (Atiyeh et al. 1999, 2000a, b, 2001; Hashemimajd et al. 2004; Gutiérrez-Miceli et al. 2007; Arancon et al. 2004b; Nurhidayati et al. 2016; Nurhidayati et al. 2017).

There was a trend of increased yield on the plots treated with three kinds of vermicompost for the residual effect

Table 2 Nutrient uptake of the first, second, third, and fourth mustard cropping as influenced by direct and residual effect of three kinds of vermicompost at various application rate

Treatments	Plant N-uptake (kg ha^{-1})			
	Mustard 1	Mustard 2	Mustard 3	Mustard 4
Kinds of vermicompost				
Vermicompost 1	160.31b	195.43b	100.63a	88.37a
Vermicompost 2	168.20b	166.29a	104.58ab	96.44a
Vermicompost 3	142.46a	169.37a	117.70b	131.59b
HSD 5%	14.15	20.63	14.49	12.56
Application rate				
5 t ha^{-1}	147.14a	150.40a	80.48a	84.09a
10 t ha^{-1}	184.79c	165.55a	105.62b	103.66b
15 t ha^{-1}	159.24b	193.17a	118.58bc	116.11b
20 t ha^{-1}	136.79a	199.00b	125.86c	118.01b
HSD 5%	19.76	28.80	20.22	17.53
Treatments	Plant P-uptake (kg ha^{-1})			
	Mustard 1	Mustard 2	Mustard 3	Mustard 4
Kinds of vermicompost				
Vermicompost 1	33.39a	34.96b	11.86a	10.65a
Vermicompost 2	32.69a	28.92a	12.18a	11.50a
Vermicompost 3	31.13a	30.01a	13.90b	14.55b
HSD 5%	3.06	3.64	1.60	1.37
Application rate				
5 t ha^{-1}	29.04a	26.09a	9.09a	9.40a
10 t ha^{-1}	37.57b	29.28a	12.20b	11.81b
15 t ha^{-1}	33.18ab	34.39b	14.13b	13.71c
20 t ha^{-1}	29.83a	35.43b	15.16c	14.00c
HSD 5%	4.27	5.08	2.23	1.91
Treatments	Plant K-uptake (kg ha^{-1})			
	Mustard 1	Mustard 2	Mustard 3	Mustard 4
Kinds of vermicompost				
Vermicompost 1	102.00ab	145.62b	72.82a	31.29a
Vermicompost 2	109.38b	123.11a	74.92ab	33.78a
Vermicompost 3	91.85a	131.92a	84.75b	43.06b
HSD 5%	11.33	17.14	9.90	4.04
Application rate				
5 t ha^{-1}	78.93a	100.07a	57.70a	26.72a
10 t ha^{-1}	120.68b	120.91a	75.31b	34.60b
15 t ha^{-1}	112.93b	159.52b	85.54bc	40.74c
20 t ha^{-1}	91.77a	153.70b	91.45c	42.12c
HSD 5%	15.81	23.92	13.83	5.64

Means followed by different letters in the same column are statistically significant different at Tukey test, $P=0.05$

of the second mustard cropping. The highest increase was found on the vermicompost V1 as 41% and followed by V3 as 26% and V2 as 0.5% as compared with the same treatments on the direct effect of the first mustard cropping. The next residual effect for the third and fourth mustard cropping had a trend to decrease yield on the plots treated with the vermicompost V1 and V2, while on the plots treated with the vermicompost V3 had a trend to increase yield slightly (Fig. 2a, b).

The results of this study corroborated with those reported by Jat and Ahlawat (2006) that the application of vermicompost had a positive effect on dry matter production directly

Table 3 Crop yield of the first, second, third, and fourth mustard cropping as influenced by direct and residual effect of interaction of three kinds of vermicompost and at various application rate

Treatments	Fresh weight of total biomass (t ha^{-1})			
	Mustard 1	Mustard 2	Mustard 3	Mustard 4
V0D0 (control)	37.31	47.65	42.48	45.07
V1D1	*47.81a	*60.35bcd	^{ns}44.00ab	^{ns}40.10a
V1D2	*54.00ab	*71.11ef	*51.97bc	^{ns}47.67abc
V1D3	*54.34ab	*79.96f	*54.77bcd	^{ns}50.54bcd
V1D4	*51.01a	*73.11ef	*58.50cde	^{ns}46.25abc
V2D1	*44.13a	^{ns}49.21a	^{ns}37.74a	^{ns}42.10abc
V2D2	*64.53b	*55.78abc	*51.59bc	^{ns}45.64abc
V2D3	*54.49ab	*59.96bc	*57.39cd	*50.98cd
V2D4	*52.80ab	^{ns}52.50ab	^{ns}48.05abc	^{ns}40.99ab
V3D1	*48.16a	^{ns}52.73ab	^{ns}47.84abc	^{ns}48.35bc
V3D2	*50.04a	*55.89abc	*57.56cd	*59.11de
V3D3	*51.62a	*64.79cde	*65.43de	*68.61ef
V3D4	*46.38a	*70.90def	*71.01e	*73.24f
HSD 5%	11.247	10.72	13.39	9.1
Dunnet 5%	6.09	5.80	7.24	5.20

Means followed by different letters in the same column are statistically significant different at Tukey test, $P=0.05$; * = significant; ns = non-significant at Dunnet test, $P=0.05$

Table 4 Marketable yield of the first, second, third, and fourth mustard cropping as influenced by direct and residual effect of interaction of three kinds of vermicompost and at various application rate

Treatments	Marketable yield (t ha^{-1})			
	Mustard 1	Mustard 2	Mustard 3	Mustard 4
V0D0 (control)	33.92	45.10	28.52	27.88
V1D1	*43.15ab	*57.55ab	*40.76ab	*37.01a
V1D2	*47.51ab	*68.39bcd	*48.92bc	*44.51abc
V1D3	*50.64abc	*77.48d	*51.58bcd	*47.40cd
V1D4	*47.95ab	*70.22cd	*55.09cde	*43.24abc
V2D1	*41.17a	*46.33a	*35.10a	*39.28abc
V2D2	*60.20c	*47.98a	*47.79bc	*42.65abc
V2D3	*51.90bc	*57.67ab	*54.44cde	*47.83cd
V2D4	*48.00ab	*50.25ab	*44.84abc	*37.54ab
V3D1	*45.20ab	*49.99ab	*44.65abc	*45.53bc
V3D2	*46.26ab	*53.32ab	*54.10cd	*55.96d
V3D3	*47.78ab	*59.67bc	*62.08de	*65.78e
V3D4	*42.56ab	*68.42bcd	*67.26e	*70.50e
HSD 5%	9.85	11.32	13.13	9.81
Dunnet 5%	5.33	6.13	7.10	5.31

Means followed by different letters in the same column are statistically significant different at Tukey test, $P=0.05$; * = significant; ns = non-significant at Dunnett test, $P=0.05$

on the first crop and gave residual effects on the succeeding crop. Chaoui et al. (2003) showed application of slow release fertilizer increased the synchrony between the plant's nutrient requirements and nutrient availability hence to increase plant biomass. The favorable response of vermicompost might be due to relatively higher nutrient content of vermicompost, thereby enriching the supply of the entire essential macro- and micronutrients and increasing plant nutrient uptake (Gupta and Garg 2008; Nath et al. 2009; Chauhan and Joshi 2010; Shak et al. 2014). Increase in crop yield, however, was not only caused by the availability of adequate nutrients and synchrony of plant's requirements, but also due to the improvement of soil physical properties affected by vermicompost application. Azarmi et al. (2008) reported that vermicompost applications can improve the bulk density and total porosity of the soil. Horticultural crops require soil conditions that are crumb due to shallow rooting. Vermicompost could also improve the soil aeration and infiltration (Arancon et al. 2008), thus providing a better medium for root growth (Chaoui et al. 2003). Improvement of soil physical conditions can create a good growth medium for crops hence the growth and crop yields increase (Manivannan et al. 2009).

Conclusion

Application of three kinds of vermicompost significantly increased soil nutrient availability, nutrient uptake and crop yield as compared with the control (without vermicompost application) for four mustard cropping sequences. The application of vermicompost provided the highest availability of N and K in soils on the second soil sampling (after harvesting mustard 1). While the highest P nutrient availability was found on the first soil sampling (2 weeks after transplanting the first mustard cropping). Application of vermicompost directly affected nutrient uptake and mustard yield, where the highest was found in vermicompost with bedding derived from coconut husk with dose of 10 t ha^{-1} but not significantly different with vermicompost using bedding of spent mushroom waste. The residual effect of vermicompost on the second mustard gave an increase in nutrient uptake and yield as compared with the direct effect, where the highest was found in vermicompost with bedding derived from the spent mushroom waste at rates of 10–15 t ha^{-1} and sugarcane trash at a rate of 20 t ha^{-1}. Vermicompost with bedding sugarcane trash with a rate of 20 t ha^{-1} provided the best residual effect on the third and fourth mustard sequential cropping. These results suggest that vermicompost application is required in organic farming systems to provide the best direct and residual effect on the availability of soil nutrient, plant nutrient uptake and crop yields.

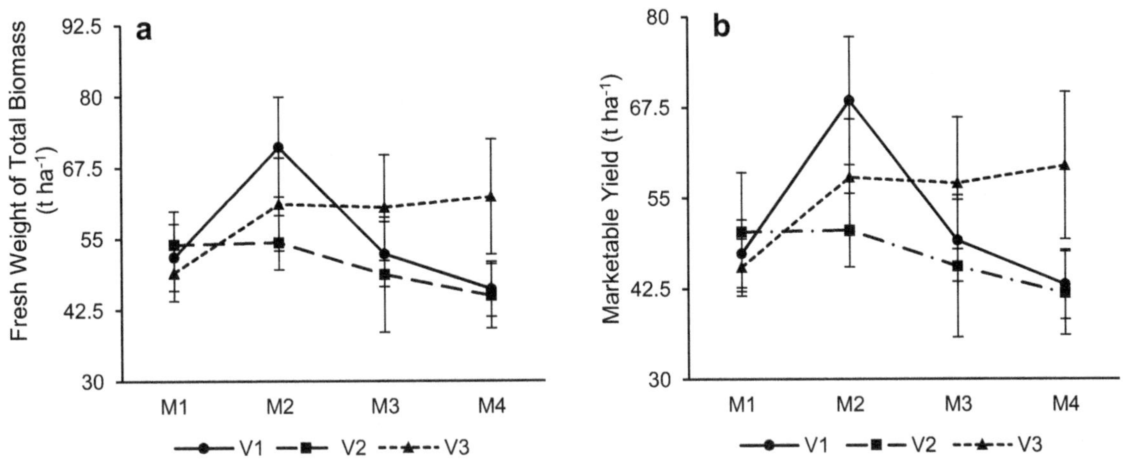

Fig. 2 The productivity of mustard due to direct and residual effect of three kinds of vermicompost at various application rates and the fourth planting periods of mustard. (remarks: M1, M2, M3 and M4 = The first, second, third and fourth mustard cropping)

Acknowledgements The authors would like to thank Directorate of Higher Education, Ministry of Research and Technology and High Education, Indonesia for their financial support through the research grant scheme of the University Excellent Research 2017.

References

Albiach R, Canet R, Pomares F, Ingelmo F (2000) Microbial biomass content and enzymatic activities after the application of organic amendments to a horticultural soil. Bioresour Technol 75(1):43–48. https://doi.org/10.1016/S0960-8524(00)00030-4

Arancon NQ, Edwards CE, Atiyeh RM, Metzger JD (2004a) Effects of vermicompost produced from food waste on the growth and yields of greenhouse peppers. Bioresour Technol 93(2):139–144. https://doi.org/10.1016/j.biortech.2003.10.015

Arancon NQ, Edwards CA, Bierman P, Welch C, Metzger JD (2004b) Influences of vermicomposts on field strawberries: 1. Effects on growth and yields. Bioresour Technol 93(2):145–153. https://doi.org/10.1016/j.biortech.2003.10.014

Arancon NQ, Edwards CA, Bierman P, Metzger JD, Lucht C (2005) Effects of vermicomposts produced from cattle manure, food waste and paper waste on the growth and yield of peppers in the field. Pedobiologia 49(4):297–306. https://doi.org/10.1016/j.pedobi.2005.02.001

Arancon NQ, Edwards CA, Babenko A, Cannon J, Galvis P, Metzger JD (2008) Influences of vermicomposts, produced by earthworms and microorganisms from cattle manure, food waste and paper waste, on the germination, growth and flowering of petunias in the greenhouse. Appl Soil Ecol 39:91–99. https://doi.org/10.1016/j.apsoil.2007.11.010

Atiyeh RM, Subler S, Edwards CA, Metzger J (1999) Growth of tomato plants in horticultural potting media amended with vermicompost. Pedobiologia 43:1–5

Atiyeh RM, Subler S, Edwards CA, Bachman G, Metzger JD, Shuster W (2000a) Effects of vermicomposts and compost on plant growth in horticultural container media and soil. Pedobiologia 44(5):579–590. https://doi.org/10.1078/S0031-4056(04)70073-6

Atiyeh RM, Arancon NQ, Edwards CA, Metzger JD (2000b) Influence of earthworm-processed pig manure on the growth and yield of greenhouse tomatoes. Bioresour Technol 75(3):175–180. https://doi.org/10.1016/S0960-8524(00)00064-X

Atiyeh RM, Edwards CA, Subler S, Metzger JD (2001) Pig manure vermicompost as a component of a horticultural bedding plant medium: effects on physicochemical properties and plant growth. Bioresour Technol 78(1):11–20. https://doi.org/10.1016/S0960-8524(00)00172-3

Azarmi R, Giglou MT, Taleshmikail RD (2008) Influence of vermicompost on soil chemical and physical properties in tomato (*Lycopersicum esculentum*) field. Afr J Biotechnol 7(14):2397–2401

Chaoui HI, Zibilske LM, Ohno T (2003) Effects of earthworm casts and compost on soil microbial activity and plant nutrient availability. Soil Biol Biochem 35:295–302. https://doi.org/10.1016/S0038-0717(02)00279-1

Chauhan A, Joshi PC (2010) Composting of some dangerous and toxic weeds using *Eisenia foetida*. J. Am Sci 6 (3):1–6. http://www.americanscience.org

Comerford NB (2005) Soil factors affecting nutrient bioavailability. In: Bassiri Rad H (ed), Ecological studies, vol. 181. Nutrient acquisition by plants an ecological perspective. ISBN: 978-3-540-24186-7. Springer, Berlin, 7 pages

De Araújo ASF, De Melo WJ, Singh RP (2010) Municipal solid waste compost amendment in agricultural soil: changes in soil microbial biomass. Rev Environ Sci Biotechnol 9(1):41–49. https://doi.org/10.1007/s11157-009-9179-6

Dimitri C, Kemp L, Sooby J, Sullivan E (2012) Organic farming for health and prosperity. Organic Farming Research Foundation. Washington, D.C. Ofrf.org. 75 pages. https://www.ofrf.org/sites/ofrf.org/files/docs/pdf/HP-report-web.pdf

Domínguez J, Aira M, Gómez Brandón M (2010) Vermicomposting: earthworms enhance the work of microbes. In: Insam H, Franke-Whittle I, Goberna M (eds) Microbes at work: from wastes to resources. Springer, Berlin Heidelberg, pp 93–114

Edwards CA, Domínguez J, Arancon NQ (2004) The influence of vermicomposts on plant growth and pest incidence. In: Shakir SH, Mikhaïl WZA (eds) Soil zoology for sustainable development in the 21st century, Cairo, pp 397–420

Garcia-Gil JC, Plaz C, Soler-Rovira P, Polo A (2000) Long-term effects of municipal solid waste compost application on soil enzyme activities and microbial biomass. Soil Biol Biochem 32(13):1907–1913. https://doi.org/10.1016/S0038-0717(00)00165-6

Gajalakshmi S, Abbasi SA (2004) Neem leaves as a source of fertilizer-cum-pesticide vermicompost. Bioresour Technol 92(3):291–296. https://doi.org/10.1016/j.biortech.2003.09.012

Gupta R, Garg VK (2008) Stabilization of primary sewage sludge during vermicomposting. J Hazard Mater 153(3):1023–1030. https://doi.org/10.1016/j.jhazmat.2007.09.055

Gutiérrez-Miceli FA, Santiago-Borraz J, Molina JAM, Nafate CC, Abdud-Archila M, Llaven MAO, Rincón-Rosales R, Dendooven L (2007) Vermicompost as a soil supplement to improve growth, yield and fruit quality of tomato (*Lycopersicum esculentum*). Bioresour Technol 98(15):2781–2786. https://doi.org/10.1016/j.biortech.2006.02.032

Hashemimajd K, Kalbasi M, Golchin A, Shariatmadari H (2004) Comparison of vermicompost and composts as potting media for growth of tomatoes. J Plant Nutr 27(6):1107–1123. https://doi.org/10.1081/PLN-120037538

Jackson ML (1973) Soil chemistry analysis, 2nd edn. Prentice Hall of India Pvt. Ltd, New Delhi

Jat RS, Ahlawat IPS (2006) Direct and residual effect of vermicompost, biofertilizers and phosphorus on soil nutrient dynamics and productivity of chickpea-fodder maize sequence. J Sustain Agric 28(1):41–54. https://doi.org/10.1300/J064v28n01_05

Knapp BA, Ros M, Insam H (2010) Do composts affect the soil microbial community? In: Insam H, Franke-Whittle I, Goberna M (eds) Microbes at work: from wastes to resources. Springer, Berlin, pp 93–114

Lazcano C, Arnold J, Tato A, Zaller JG, Domínguez J (2009) Compost and vermicompost as nursery pot components: effects on tomato plant growth and morphology. Span J Agric Res 7(4):944–951

Lazcano C, Revilla P, Malvar RA, Domínguez J (2011) Yield and fruit quality of four sweet corn hybrids (*Zea mays*) under conventional and integrated fertilization with vermicompost. J Sci Food Agric 91(7):1244–1253. https://doi.org/10.1002/jsfa.4306

Makkar C, Singh J, Parkashi C (2017) Vermicompost leachate reduces some negative effects of salt stress in pomegranate. Int J Recycl Org Waste Agric 6:255–263. https://doi.org/10.1007/s40093-017-0168-4

Manivannan S, Balamurugan M, Parthasarathi K, Gunasekaran, G, Ranganathan LS (2009) Effect of vermicompost on soil fertility and crop productivity – beans (*Phaseolus vulgaris*). J Environ Biol 30(2):275–281

Moore TR, Trofymow JA, Prescott CE, Titus BD (2011) Nature and nurture in the dynamics of C, N and P during litter decomposition in Canadian forest. Plant Soil 339:163–175. https://doi.org/10.1007/s11104-010-0563-3

Nath G, Singh K, Singh DK (2009) Chemical analysis of vermicomposts/vermiwash of different combinations of animal, agro and kitchen wastes. Aust J Basic App Sci 3(4):3671–3676. http://www.ajbasweb.com/old/ajbas_oct-dec_2009.html

Nurhidayati, Ali U, Murwani I (2015) Influence of the kind of vermicompost material and earthworm *Pontoscolex corethrurus* population on the yield and quality of phak-coi mustard (*Brassica rapa* L.) with organic potting media. In: Proceeding of the first international conference on life science and biotechnology exploration and conservation of biodiversity. ISBN: 978-602-9030-98-3.p.168-176

Nurhidayati, Machfudz M, Murwani I (2017) Combined effect of vermicompost and earthworm Pontoscolex corethrurus inoculation on the yield and quality of broccoli (*Brassica oleracea L*) using organic media. JOBARI 22(4):148–156

Nurhidayati N, Ali U, Murwani I (2016) Yield and quality of cabbage (*Brassica oleracea* L.var.Capitata) under organic growing media using vermicompost and earthworm *Pontoscolex corethrurus* inoculation. Agric Agric Sci Proc 11:5–13. https://doi.org/10.1016/j.aaspro.2016.12.002

Pattnaik S, Reddy MV (2010) Nutrient status of vermicompost of urban green waste processed by three earthworm species: *Eisenia fetida*, *Eudrilus eugeniae*, and *Perionyx excavatus*. App Environ Soil Sci article ID 967526 (2010). 13 pages. http://dx.doi.org/10.1155/2010/967526

Prabha ML (2009) Waste management by vermitechnology. Indian J Environ Prot 29(9):795–800

Rahman MM, Tsukamoto J, Rahman MM, Yoneyama A, Mostafa KM (2013) Lignin and its effects on litter decomposition in forest ecosystems. Chem Ecol 29(6):540–553. https://doi.org/10.1080/02757540.2013.790380

Ros M, Pascual JA, García C, Hernández MT, Insam H (2006) Hydrolase activities, microbial biomass and bacterial community in a soil after long-term amendment with different composts. Soil Biol Biochem 38(12):3443–3452. https://doi.org/10.1016/j.soilbio.2006.05.017

Roy S, Arunachalam K, Dutta BK, Arunachalam A (2010) Effect of organic amendments of soil on growth and productivity of three common crops viz. *Zea mays*, *Phaseolus vulgaris* and *Abelmoschus esculentus*. App Soil Ecol 45:78–84. https://doi.org/10.1016/j.apsoil.2010.02.004

Shak KPY, Wu TY, Lim SL, Lee CA (2014) Sustainable reuse of rice residues as feedstocks in vermicomposting for organic fertilizer production. Environ Sci Pollut Res 21(2):1349–1359. https://doi.org/10.1007/s11356-013-1995-0

Wardle DA, Bardgett RD, Walker LR, Bonner KI (2009) Among- and within-species variation in plant litter decomposition in contrasting long-term chronosequences. Funct Ecol 23:442–453. https://doi.org/10.1111/j.1365-2435.2008.01513.x

Weber J, Karczewska A, Drozd J, Licznar M, Licznar S, Jamroz E, Kocowicz A (2007) Agricultural and ecological aspects of a sandy soil as affected by the application of municipal solid waste composts. Soil Biol Biochem 39(6):1294–1302. https://doi.org/10.1016/j.soilbio.2006.12.005

Yadav KD, Tare V, Ahammed MM (2010) Vermicomposting of source separated human faeces for nutrient recycling. Waste Manag 30(1):50–56. https://doi.org/10.1016/j.wasman.2009.09.034

13

Application of rock phosphate enriched composts increases nodulation, growth and yield of chickpea

Allah Ditta[1] · Juma Muhammad[1] · Muhammad Imtiaz[2] · Sajid Mehmood[3] · Zheng Qian[3] · Shuxin Tu[3]

Abstract

Purpose Phosphorus (P) deficiency in agricultural land is one of the main factors which reduce the crop production and yield. To increase the crop growth, the availability of P using the rock phosphate (RP) can be enhanced using organic waste with the addition of phosphate solubilizing microorganisms (PSMs). Moreover, the combination along with time and dose application of RP and compost can also affect the P-availability through mineralization and/immobilization.

Methods To investigate the effect of different combinations of RP and compost (0:100, 25:75, 50:50, 75:25 and 100:0) and time of application (30, 15, 7, 3 and 0 days before sowing), each was conducted in two consecutive pot experiments while, the rate of application (100–1000 kg ha^{-1}) was investigated under field conditions, in comparison to control (recommended P fertilizer).

Results Rock phosphate enriched compost (RP-EC) with a combination ratio of 50:50; RP and compost applied before 7 days of sowing in pot experiments resulted in the maximum nodulation, growth and productivity of chickpea. Under field conditions, the maximum increase of 35.3% in no. nodules $plant^{-1}$, 26.7% in dry wt. of nodules $plant^{-1}$ and 20.8% in grain yield (t ha^{-1}) compared to control was obtained by RP-EC @ 1000 kg ha^{-1}. The same treatment indicated an increase of 12.9 and 4.3% in P contents in straw and grains, respectively, compared to control. However, most results were non-significant when RP-EC applied at the rate of 1000 kg ha^{-1}.

Conclusion Application of RP-EC, with the ratio of 50:50 (RP:Compost) and application rate of 800 kg ha^{-1} before 7 days of sowing, exhibited maximum growth and development, and can be highly recommended for optimum production of chickpea.

Keywords Optimum combination ratio · Rock phosphate · Time and rate of application · Compost · Chickpea · Phosphorus

Introduction

Globally, the agricultural soils are getting phosphorus (P) deficient after nitrogen (Vance et al. 2003) and hence, impairs the various physiological and biochemical processes within plants (Wu et al. 2005). Generally, on dry weight basis, it constitutes ~ 0.2% of the plant biomass (Schachtman et al. 1998). Phosphorus, being an immobile nutrient in soil necessitates the application of chemical phosphatic fertilizers as its available sources for optimum production of crop plants. However, nutrient availability from chemical fertilizers is not more than 20% and has forced the poor farmers to add two times more than the optimum application rate of P-fertilizers. The addition of extra P in the soil led to consume more resources for the crop production, and hence, the costs of production of crop plants per unit area increase (Aziz et al. 2006). Moreover, there is expected depletion of high quality rock phosphate (RP) which is the major sources of P-fertilizers by the year 2050 (Vance et al. 2003). These circumstances necessitate to find out an economical and eco-friendly way to increase the availability of P, and at the same time, reduce the losses of P, and RP enriched with compost and phosphate solubilizing microorganisms (PSMs) can be a good solution.

✉ Allah Ditta
ad_abs@yahoo.com; allah.ditta@sbbu.edu.pk

[1] Department of Environmental Sciences, Shaheed Benazir Bhutto University Sheringal, Dir (Upper), Khyber Pakhtunkhwa 18000, Pakistan

[2] School of Environmental Science and Engineering, Guangzhou University, Guangzhou 510006, China

[3] Microelement Research Center, College of Resources and Environment, Huazhong Agricultural University, Wuhan 430070, China

Rock phosphate (RP) is an important natural source of P and is used as raw material for the production of chemical phosphatic fertilizers (Reddy et al. 2002). The direct application of RP has been found suitable for acidic soils as low pH helps to solubilize the RP and increases available form of P to the plants. However, this approach does not work for alkaline/or calcareous soils due to high pH (Caravaca et al. 2004). The application of enriched-RP with compost to alkaline/or calcareous soils could be a wise approach for the solubilization of RP. Furthermore, addition of organic fertilizers also improves the physicochemical as well as biological properties of soil (Adhami et al. 2014; Lim et al. 2015). This approach also had positive effects on crop because organic fertilizers contained plenty of macro and micro-nutrients (Gaur 1997; Kalaivanan and Hattab 2016). Typically, most of the organic fertilizers are produced from biodegradable solid wastes using compost or/and vermicompost (Wu et al. 2014; Lim et al. 2016).

Besides, some rhizobacteria have also been used to improve the P-bank in the soil such as with P-solubilizing and ACC-deaminase activities. Ultimately, these approaches also help to increase P availability to the crop plants and alleviate the stress induced by ethylene (Glick et al. 1998). Ethylene, a gaseous phytohormones, is produced under biotic and abiotic stresses (Arshad and Frakenberger 2002). Its higher concentration (≥ 25 µg L^{-1}) inhibits root growth and nodulation in legumes (Mattoo and Suttle 1991; Arshad and Frakenberger 2002). Inoculation of rhizobacteria with ACC-deaminase activity significantly improved growth parameters of maize and mung bean (Shaharoona et al. 2006, 2007; Davari et al. 2012), root proliferation (Dey et al. 2004), and transformed soil P (Chen et al. 2006). So, the use of rhizobacteria with both of the above-mentioned activities could be a good strategy to decrease ethylene production during nodulation and also improve P availability especially from RP.

Most of the organic P is mineralized and certain organic acids are produced during composting. These organic acids lower the pH of microclimate and hence, increase the solubility of RP, ultimately the bioavailability of P. On the other hand, combined application of rhizobacteria carrying P-solubilizing as well as ACC-deaminase activities with RP-EC can reduce the ecological issues due to overuse of fertilizers and enhance the nutrient use efficiency. In literature, there are few reports of combined use of RP, compost and/PSM (Biswas and Narayanasamy 2006; Saleem et al. 2013). Moreover, no research work reports about the optimum combination ratio of RP and compost, its time and rate of application except on lentil (Ditta et al. 2015; Ditta and Khalid, 2016). Here in this study, we have investigated the effects of the optimum ratio of combination between RP and compost in RP-EC, its time and rate of application with PSM strain (*Bacillus* spp.) on nodulation, growth and productivity of chickpea.

Materials and methods

Pot experiments

Effect of different combinations of RP and compost, i.e., 0:100, 25:75, 50:50, 75:25 and 100:0 (W/W) on nodulation, growth and yield parameters of chickpea was investigated in the first pot experiment under wire house conditions in comparison to control (recommended NPK). The different combination ratios of RP-EC were applied at 600 kg ha^{-1} (0.3 g kg^{-1} soil), thoroughly mixed with soil and the pots were arranged in completely randomized design (CRD) with six replications. Remaining N and P were applied through urea and diammonium phosphate (DAP) after analysis and calculations for available N and P in each combination. RP used in the studies was chemically analyzed for P using standard methods and it contained 22% P_2O_5 as total P and 0.022% P_2O_5 as Olson P. The *Bacillus strain* (PSM) had high ACC-deaminase and P-solubilizing activities, 1.53 µmol mL^{-1}, 753.0 µg P mL^{-1}, respectively (Baig et al. 2012) was taken from the Laboratory of Environmental Sciences. The inoculum was prepared separately for PSM and rhizobia using National Botanical Research Institute's Phosphate (Nautiyal 1999) and N-free medium, respectively, in Erlenmeyer flasks (250 ml) which were incubated for 48 h in orbital shaking incubator (Model OSI-503 LD; Firstek Scientific, Japan), at 28 ± 1 °C and 100 rev min^{-1}. For surface sterilization, chickpea seeds (var. Noor-2009) were soaked in ethanol (70% v/v) for 2 min and then in NaClO (5% v/v) for 5 min. The seeds were thoroughly washed 3 times using sterile distilled water. The inoculum was mixed with sugar solution (15% *w*/*v*) and sterilized peat + clay (1:1 w/w) and surface sterilized seeds were dipped. Seeds treated with sterilized broth, sugar solution and peat plus clay were used as control.

After finding out optimum combination ratio of RP and compost, another pot study was conducted under wire house conditions to investigate the effect of different application times (30, 15, 7, 3 and 0 days before sowing) of RP-EC on nodulation, growth and yield contributing parameters comparison to control (60 kg P ha^{-1} using DAP). Chickpea (var. Noor-2009) was used as test crop in the experiment. RP-EC with an optimum combination ratio of RP and compost, i.e., 50:50, applied at the rate of 600 kg ha^{-1} as described in the earlier pot experiment. The inoculum preparation and seed inoculation were performed using standard methods as described earlier in the above paragraph.

In both pot experiments, inoculated seeds were sown in plastic pots with 25 cm diameter and 23 cm of height. Each pot contained 8 kg dried and sieved (2 mm) soil collected from the University research area. Respective treatment of RP-EC in both the trials was thoroughly mixed in the soil manually before sowing of inoculated chickpea seeds. Remaining N and P were applied through urea and diammonium phosphate (DAP) after analysis and calculations for available N and P in each combination. Physicochemical properties of soil used in the experiment were determined using standard methods (Ryan et al. 2001). The soil was sandy clay loam in texture with alkaline pH 7.6 and $EC_{extract}$ 3.4 dS m^{-1}, low in organic matter (0.67%). The nutrient contents of the soil were as follows: total N = 0.056%, available P = 8.6 mg kg^{-1} and extractable K = 110 mg kg^{-1}. The pots were arranged in the CRD with six replications. Each pot contained two plants in the beginning while after proper germination, thinning was done to select one uniform germinated plant/pot. Standard procedures were adopted for the estimation of growth and yield parameters such as root dry weight, fresh biomass and grain yield at maturity. The roots were excised from the shoot part and were dried in oven at 105 °C for 24 h till constant dry weight was obtained using a digital balance. P contents in grain and straw samples were also determined (Olsen et al. 1954; Ryan et al. 2001). At the flowering stage, three replicates were randomly selected from each treatment for nodulation parameters, washed to remove soil; nodules were excised manually, placed on filter paper and dried. Their number was counted manually and electrical balance was used for dry weight of nodules.

Field experiment

Under field conditions, the effect of different application rates (100–1000 kg ha^{-1}) of RP-EC with PSM and with a combination ratio of RP and compost (50:50 W/W), applied 7 days before sowing, on nodulation, growth and productivity of chickpea compared to control (60 kg P ha^{-1} using DAP) was investigated at the University research farm. The treatments were arranged using randomized complete block design (RCBD) in triplicate. For sowing, hand drill was used to maintain row to row distance of 40 cm. After a week, plant to plant distance (30 cm) was maintained by thinning. The recommended rate of N and K (25 and 25 kg ha^{-1}) were applied in control using urea and sulfate of potash (SOP). Remaining N and P were applied through urea and diammonium phosphate (DAP) after analysis and calculations for available N and P in treatments where RP-EC was applied. Rhizobia and PSM inoculum were prepared using standard methods as described above and were applied in all the treatments. The inoculated seeds were sown in respective plots (1.5 × 1.5 m^2). Similar standard methods as described in pot experiments were adopted for the estimation of agronomic and yield parameters at maturity and P contents from grain and straw samples (Olsen et al. 1954; Ryan et al. 2001). Each treatment in triplicate was randomly selected using a square meter ring and plants were harvested at the flowering stage for nodulation parameters. Number of nodules $plant^{-1}$ were counted manually and their dry wt. using digital balance.

Preparation of RP-EC

RP-EC was prepared following the procedures reported in our previous paper (Ditta et al. 2015). Briefly, fruit peels and vegetable wastes (mainly consisted of green peas waste) were collected during January–March, 2013, from the local market, oven dried (70 °C) overnight, ground, sieved (< 2.0 mm) and was mixed with RP in different combinations as described earlier (0:100, 25:75, 50:50, 75:25 and 100:0), respectively. After mixing, the mixture was composted for 7 days with moisture contents (40% *v/w*) maintained manually in a locally manufactured composter with a load capacity of 500 kg. For the preparation of each RP-EC with specific ratio, the composter was thoroughly cleaned each time to remove the previous one. Simple compost without RP was also prepared using similar methods as described earlier.

Chemical analysis of compost, RP-EC

For chemical analysis of simple compost and RP-EC with specific combinations of RP and compost, standard methods were used with four replications. Loss-on-ignition method was employed for carbon contents (Nelson and Sommers 1996). The available N ($NH_4 + NO_3$) was determined by the MgO Devarda alloy method (Ryan et al. 2001). Sodium bicarbonate ($NaHCO_3$) method was used for available P determination (Olsen et al. 1954) while total P was determined through vanadomolybdophosphoric yellow color method (Jackson 1973). The results of chemical analysis are given in Table 1.

Statistical analysis

Data collected in each treatment were subjected to statistical scrutiny through analysis of variance (ANOVA) using Statistix 8.1 and to compare each treatment mean, least significant difference (LSD) test at $\alpha = 0.05$ was applied (Steel et al. 1997).

Table 1 The chemical analysis of the compost and the RP- enriched composts

Parameters	Raw organic material	Combination ratios of RP and compost				
		0:100	25:75	50:50	75:25	100:0
Carbon (g kg^{-1})	345 ± 3.2†	255 ± 6.6	237 ± 4.4	215 ± 5.5	174 ± 3.3	–
Nitrogen (g kg^{-1})	13.8 ± 0.1	22.1 ± 0.3	17.1 ± 0.2	15.3 ± 0.3	12.4 ± 0.2	–
Total P (g kg^{-1})	3.5 ± 0.06	3.7 ± 0.9	59.7 ± 2.1	123.0 ± 4.4	175.0 ± 2.1	224 ± 2.6
Olson P (mg kg^{-1})	340 ± 10	450 ± 20	2070 ± 20	2380 ± 30	2150 ± 20	220 ± 10
C:N	25.00 ± 0.34	11.53 ± 0.18	13.85 ± 0.28	14.05 ± 0.63	14.03 ± 0.45	–
C:P	98.57 ± 1.63	54.26 ± 0.96	12.03 ± 0.12	9.19 ± 0.18	7.73 ± 0.07	–

† Shows the standard error of means (SEM) where $n = 4$

Results

Effects of RP-EC with different ratios of combination on nodulation parameters, growth and yield parameters and P contents (g kg^{-1})

The formation of nodules in leguminous crops is very important because it helps to increase the uptake of nitrogen to the plants and in this process, availability of P is very critical. As clear from Table 1, the availability of P was increased with different ratios of RP and compost, and the maximum availability was observed with 50:50 ratio of combination between RP and compost. The same ratio produced maximum no. of nodules plant^{-1} and dry wt. compared to other combination ratios (Table 2). The maximum no. of nodules plant^{-1} were 86.1% more than that treated with 0:100 combination ratio of RP and compost, respectively.

The obtained results show that the application of different combination ratios of RP and compost exhibited a significant increase in grain yield and fresh biomass (g pot^{-1}) as clear from the data presented in Table 2. The results also show that about 82.1% more grain yield was obtained when plants treated with 50:50 as compared to 0:100 ratio of RP and compost. The same treatment resulted in the maximum fresh biomass (43.8 g pot^{-1}).

Moreover, the straw and grain of chickpea accumulated higher P contents about 2.8 and 5.0 g kg^{-1}, respectively, when treated 50:50 ratio of RP and compost (Table 2).

Effects of Application time of RP-EC on nodulation parameters, growth and yield parameters and P contents (g kg^{-1})

Application time of an organic amendment is critical with respect to the availability of nutrients through mineralization/immobilization and it has an indirect impact on nodulation, growth and yield in plants. The results indicated that the application of RP-EC before 7 days of sowing enhanced the production of nodules plant^{-1} as well as their dry biomass. (Table 3). This application time exhibited 57 nodules plant^{-1} which were 119.2% more as compared to that applied before 30 days of sowing. The same treatment resulted in the maximum dry wt. of nodules (0.74 g plant^{-1}).

The dry weight of root, grain yield and fresh biomass were also increased with the application of RP-EC before 7 days. The results show that chickpea exhibited about 2.47 , 43.7 g pot^{-1}, and 80% more dry weight, fresh biomass and

Table 2 Impact of RP-EC with different ratios of combination on nodule, growth, yield and P uptake in chickpea under soil pot culture conditions

Combination ratios of RP and compost	Nodules (No. plant^{-1})	Dry wt. of nodules (g plant^{-1})	Root dry wt. (g plant^{-1})	Grain yield (g pot^{-1})	Fresh biomass (g pot^{-1})	P in straw (g kg^{-1})	P in grains (g kg^{-1})
0:100	36 c†	0.46 c	1.34 d	6.23 b	27.27 d	1.4 d	3.8 b
25:75	45 bc	0.57 b	2.25 b	11.07 a	37.58 c	2.4 ab	4.5 ab
50:50	67 a	0.79 a	2.47 a	11.37 a	43.83 a	2.8 a	5.0 a
75:25	47 b	0.58 b	1.78 c	10.23 a	38.61 b	2.0 bc	4.2 ab
100:0	38 bc	0.48 bc	1.03 e	7.40 b	27.04 d	1.5 cd	3.7 b
LSD	9.77***	0.11***	0.30***	2.10***	1.79***	0.51***	0.99***

*** Significant at $p \leq 0.001$

† Means sharing the same letter, for a parameter, do not differ significantly at $p \leq 0.05$ using least significance difference (LSD) test

Table 3 Impact of application time of RP-EC with optimum combination ratio (50:50) on nodule, growth, yield and P uptake in chickpea under soil pot culture conditions

Application time (Days before sowing)	Nodules (No. plant^{-1})	Dry wt. of nodules (g plant^{-1})	Root dry wt. (g plant^{-1})	Grain yield (g pot^{-1})	Fresh biomass (g pot^{-1})	P in straw (g kg^{-1})	P in grains (g kg^{-1})
30	26 c†	0.41 c	1.34 d	6.40 d	27.2 d	1.5 c	3.6 b
15	35 b	0.52 b	2.25 b	11.24 ab	35.8 c	2.6 ab	4.3 ab
7	57 a	0.74 a	2.47 a	11.52 a	43.7 a	3.1 a	4.8 a
3	37 b	0.53 b	1.78 c	10.40 b	38.5 b	2.3 b	4.1 ab
0	28 c	0.43 c	1.03 e	7.56 c	26.9 d	1.7 c	3.5 b
LSD value	4.74**	0.05**	0.14**	1.02**	0.87**	0.25**	0.47**

*** Significant at $p \leq 0.001$

†Means sharing the same letter, for a parameter, do not differ significantly at $p \leq 0.05$ using least significance difference (LSD) test

grain yield, respectively, than the application of RP-EC before 30 days of sowing.

The accumulation of P contents in the straw and grains of chickpea was also increased when RP-EC was applied before 7 days of sowing. With the application of RP-EC before 7 days of sowing, chickpea accumulated P contents about 3.1 and 4.8 g kg^{-1} in straw and grains, respectively, and this P accumulation was higher than other application times of RP-EC (Table 3).

Effect of field application of RP-EC on nodulation parameters, growth and yield parameters and P contents (g kg^{-1})

Under the field conditions, application of RP-EC at the rate of 1000 kg ha^{-1} with phosphate solubilizing microorganisms (PSMs) resulted in the maximum number of nodules plant^{-1} (about 69 nodules) (Table 4). A similar trend was observed for the dry weight of nodules (g). However, the results about nodulation parameters: number of nodules plant^{-1} and dry weight of nodules showed non-significant difference when exposed to RP-EC at the rate of 800 and 1000 kg ha^{-1}.

The obtained results indicated that chickpea plants treated with the application of RP-EC at the rate of 1000 kg ha^{-1} with PSM exhibited more grain yield and fresh biomass, which were about 3.66 and 13.59 t ha^{-1}, respectively, as compared to control where recommended NPK was applied (Table 4).

Under field conditions, the P accumulated more in straw and grains was about 3.5 and 4.8 g kg^{-1}, respectively, when RP-EC applied at the rate of 1000 kg ha^{-1} with PSM. Furthermore, the application of RP-EC at the rate of 800 and 1000 kg ha^{-1} showed statistically non-significant differences in P accumulation in chickpea plants (Table 4).

Table 4 Impact of field application of RP-EC with optimum combination ratio (50:50) on nodule, growth, yield and P uptake in chickpea

Rate of application (kg ha^{-1})	Nodules (No. plant^{-1})	Dry wt. of nodules (g plant^{-1})	Root dry wt. (g plant^{-1})	Grain yield (t ha^{-1})	Fresh biomass (t ha^{-1})	P in straw (g kg^{-1})	P in grains (g kg^{-1})
100	29 e†	0.52 d	1.35 e	1.87 f	6.93 e	1.7 f	3.7 c
200	31 de	0.54 d	1.57 d	2.44 e	8.93 d	2.0 e	3.9 bc
300	39 d	0.63 c	1.88 c	2.63 de	10.62 c	2.3 d	4.1 b
400	58 bc	0.72 b	2.13 b	2.93 cd	10.73 bc	2.7 c	4.6 a
500	63 ab	0.85 a	2.44 a	3.23 bc	11.86 bc	3.2 b	4.8 a
800	68 a	0.89 a	2.55 a	3.35 ab	11.97 b	3.4 a	4.8 a
1000	69 a	0.90 a	2.57 a	3.66 a	13.59 a	3.5 a	4.8 a
NPK (recommended)	51 c	0.71 bc	2.44 a	3.03 bc	11.70 bc	3.1 b	4.6 a
LSD value	9.49***	0.07***	0.21***	0.35***	1.33***	0.23***	0.32***

*** Significant at $p \leq 0.001$

† Means sharing the same letter, for a parameter, do not differ significantly at $p \leq 0.05$ using least significance difference (LSD) test

Discussion

Chemical analysis of the composted finished product, i.e., RP-EC with different combination ratios of RP and compost revealed that the maximum improvement in the P availability (2380 mg kg^{-1}) was exhibited with 50:50 combination ratio between RP and compost (Table 1). The C:N is an important indicator of compost maturity as it indicates the extent of decomposition of biological wastes being used during composting and its stabilization, ranges from 15 to 25 of matured compost materials (Watson 2003). The production of organic acids such as acetate, lactate, oxalate, tartrate, succinate, citrate, gluconate, ketogluconate, glycolate, etc., during the composting process by the microbes might have resulted in increasing the availability of nutrients especially P as observed the chemical analysis after composting (Table 1). Due to mineralization of nutrients present, availability of carbon and other nutrients is increased during composting, which increases the microbial population and ultimately the production of organic acids (Rashid et al. 2004; Biswas and Narayanasamy 2006; Chakraborty et al. 2011). The presence of various organic acids in the composted materials helps to accelerate the break down the RP by pushing the dissolution reaction to the right. The increase in production of CO_2 during the composting process occurred due to low pH. This CO_2 helps to produce carbonic acids in the presence of water and have been reported to increase the available P in the soil and from RP (Chien 1979). Previous studies indicated that production of microbial extracellular products such as enzymes and organic acids can increase the mineralization of nutrients (Chien 1979; Jansson et al. 1988).

Another factor that influences the nodulation in legumes is the source of carbon and energy which increases the population of N-fixing bacteria (Hill and Patriquin 1996). In our studies, fruit and vegetable compost was used as it is easily mineralized thereby, increasing the availability of C and energy to diazotrophic bacteria. Similarly, any biotic or abiotic stress also induces ethylene production (Shaharoona et al. 2006), including rhizobial infection during nodulation (Suganuma et al. 1995), which may ultimately reduce root growth (van Workum et al. 1995). The application of microbes with ACC-deaminase activity in growth medium help in reducing this ethylene stress. In the present study, addition of phosphate solubilizing microorganisms (PSMs) contained both ACC- and P-solubilizing activities; the former activity might have caused a reduction in ethylene levels and the P-solubilizing activity increased P availability which helped to increase nodulation (Table 2, 3 and 4). Moreover, the compost material used in the present study supplied micro- as well as macro nutrients which promoted better root growth and nodulation (Cheuk et al. 2003; Ahmad et al. 2008a, b). Other researchers have also reported similar results (Shaharoona et al. 2006, 2007; Ditta et al. 2015).

The results about to find the optimum ratio and then its time of application to release the nutrients from RP-EC are presented in Table 1. Moreover, the effect was more pronounced when microbial strain with high P-solubilization and ACC-deaminase activity was applied. The addition of P-solubilization in RP-EC maximally increased the available P while, ACC-deaminase activity enhanced the root growth by reducing ethylene stress. The Tables 2, 3 and 4 contained the results about releasing of nutrients and availability of nutrients, which revealed that growth and yield parameters were increased using applied treatments. Furthermore, the improvement in growth and production of chickpea was more pronounced with the inoculation of bacterial strain with ACC-deaminase and P-solubilizing activities. The application of organic fertilizers alone may not help to attain the standard yield of crops due to their low and slow nutrient availability (Fageria et al. 2010). According to Roy et al. (2001), organic inputs are beneficial; however, the inclusion of additional P with phosphate solubilizing microorganisms may produce yields at par with inorganic fertilizers. The obtained results show an agreement with the results of previous studies (Mattoo and Suttle 1991; Surange et al. 1995; Chang and Janzen 1996; Nevens and Reheul 2003; Verma et al. 2013).

With the application of RP-EC and PSM, an increase in P contents was observed in straw and grain samples which might be due to the higher P-solubilizing activity of PSM strain, the better root growth via alleviating ethylene stress through ACC-deaminase activity, and/or better solubilization of P in RP during the composting process. The addition organic matter increases the dissolution of RP by complexing Ca^{2+} ions in the soil solution (Sanyal and De Datta 1991) and ultimately increases the available P for the crop plants grown in calcareous soils. Singh and Amberger (1995) found that amount of loosely bound P in the soil increases while fraction held as Ca–P decreases with the addition of phosphor-compost as compared to RP application only. Other researchers have also reported an increase in P availability which ultimately resulted in higher growth, productivity and nutrient contents of various crops (Belimov et al. 2002; Vassilev et al. 2006; Glick et al. 2007; Salimpour et al. 2010).

Conclusions

In conclusion, RP-EC with a combination ratio of 50:50 (RP:Compost), applied before 7 days of sowing under pot culture conditions and at the rate of 800 kg ha^{-1} with PSM under field conditions produced maximum nodulation, growth and yield parameters of chickpea in comparison to conventional chemical phosphatic fertilizers was observed with the application of. Therefore, RP-EC can be used as an alternate source of P for maximum production of leguminous crops. In future, there is need to evaluate and ensure conditions.

Acknowledgements Financial support for this study was provided by the Higher Education Commission (HEC) of Pakistan, under Indigenous Ph.D. Fellowship Program (PIN no. 074-2995-AV4-029). We are also thankful to Dr. Khurram S. Baig for giving us bacterial strain (*Bacillus strain*) used in the study. We are also thankful to Microbiology Section at Ayub Agricultural Research Institute (AARI), Faisalabad for providing rhizobia specific to chickpea.

References

Adhami E, Hosseini S, Owliaie H (2014) Forms of phosphorus of vermicompost produced from leaf compost and sheep dung enriched with rock phosphate. Int J Recycl Org Waste Agric 3:68. https://doi.org/10.1007/s40093-014-0068-9

Ahmad R, Arshad M, Khalid A, Zahir ZA (2008a) Effectiveness of organic-/bio-fertilizer supplemented with chemical fertilizers for improving soil water retention, aggregate stability, growth and nutrients uptake of maize (*Zea mays* L.). J Sustain Agric 31:57–77. https://doi.org/10.1300/J064v31n04_05

Ahmad R, Khalid A, Arshad M, Zahir ZA, Mahmood T (2008b) Effect of composted organic waste enriched with N and L-tryptophan on soil and maize. Agron Sustain Dev 28:299–305. https://doi.org/10.1051/agro:2007058

Arshad M, Frakenberger WT Jr (2002) Ethylene: agricultural sources and applications. Kluwer/Academic Publishers, New York

Aziz T, Rahmatullah Maqsood MA, Tahir IA, Cheema MA (2006) Phosphorus utilization by six brassica cultivars (*Brassica juncea* L.) from tri-calcium phosphate; a relatively insoluble P compound. Pak J Bot 38:1529–1538

Baig KS, Arshad M, Shaharoona B, Khalid A, Ahmed I (2012) Comparative effectiveness of Bacillus spp. Possessing either dual or single growth-promoting traits for improving phosphorus uptake, growth and yield of wheat (*Triticum aestivum* L.). Ann Microbiol 62:1321–1330. https://doi.org/10.1007/s13213-011-0352-0

Belimov AA, Safronova VI, Mimura T (2002) Response of spring rape (*Brassica napus* L. var. Oleifera) to inoculation with plant growth promoting rhizobacteria containing 1-aminocyclopropane-1-carboxylate deaminase depends on nutrient status of the plant. Can J Microbiol 48:189–199. https://doi.org/10.1139/w02-007

Biswas D, Narayanasamy G (2006) Rock phosphate enriched compost: an approach to improve low-grade Indian rock phosphate. Bioresour Technol 97:2243–2251. https://doi.org/10.1016/j.biortech.2006.02.004

Caravaca F, Alguacil MM, Azcon R, Diaz G, Roldan A (2004) Comparing the effectiveness of mycorrhizal inoculum and amendment with sugar beet, rock phosphate and *Aspergillus niger* to enhance field performance of the leguminous shrub *Dorycnium pentaphyllum* L. Appl Soil Ecol 25:169–180. https://doi.org/10.1016/j.apsoil.2003.08.002

Chakraborty A, Chakrabarti K, Chakraborty A, Ghosh S (2011) Effect of long-term fertilizers and manure application on microbial biomass and microbial activity of a tropical agricultural soil. Biol Fertil Soils 47:227–233. https://doi.org/10.1007/s00374-010-0509-1

Chang C, Janzen HH (1996) Long-term fate of nitrogen from annual feedlot manure application. J Environ Qual 25:785–790. https://doi.org/10.2134/jeq1996.00472425002500040019x

Chen Y, Rekha P, Arun A, Shen F, Lai W-A, Young C (2006) Phosphate solubilizing bacteria from subtropical soil and their tricalcium phosphate solubilizing abilities. Appl Soil Ecol 34:33–41. https://doi.org/10.1016/j.apsoil.2005.12.002

Cheuk W, Lo KV, Branion RMR, Fraser B (2003) Benefits of sustainable waste management in the vegetable greenhouse industry. J Environ Sci Health 38:855–863. https://doi.org/10.1081/PFC-120025565

Chien SH (1979) Dissolution of phosphate rock in acid soils as influenced by nitrogen and potassium fertilizers. Soil Sci 127:371–376. https://doi.org/10.1097/00010694-197906000-00008

Davari M, Sharma SN, Mirzakhani M (2012) Residual influence of organic materials, crop residues, and biofertilizers on performance of succeeding mung bean in an organic rice-based cropping system. Int J Recycl Org Waste Agric 1:14. http://www.ijrowa.com/content/1/1/14. Accessed 13 Nov 2017

Dey R, Pal KK, Bhatt DM, Chauhan SM (2004) Growth promotion and yield enhancement of peanut (*Arachis hypogaea* L.) by application of plant growth promoting rhizobacteria. Microbiol Res 159:371–394. https://doi.org/10.1016/j.micres.2004.08.004

Ditta A, Khalid A (2016) Bio-organo-phos: a sustainable approach for managing phosphorus deficiency in agricultural soils. In: Tandon HLS (ed) Organic fertilizers— from basic concepts to applied outcomes (ISBN 978-953-51-4701-5). InTech, Rijeka, pp 109–136. https://doi.org/10.5772/62473

Ditta A, Arshad M, Zahir ZA, Jamil A (2015) Comparative efficacy of rock phosphate enriched organic fertilizer vs. mineral phosphatic fertilizer for nodulation, growth and yield of lentil. Inter J Agric Biol 17:589–595. https://doi.org/10.17957/IJAB/17.3.14.954

Fageria NK, Baligar VC, Jones CA (2010) Growth and mineral nutrition of field crops. CRC Press, Boca Roton, pp 156–157

Gaur AC (1997) Bulky organic manures and crop residues. In: Tandon HLS (ed) Fertilizers, organic manures, recyclable wastes and biofertilizers. FDCO, New Delhi, pp 37–51

Glick BR, Penrose DM, Li J (1998) A model for lowering plant ethylene concentration by plant growth promoting rhizobacteria. J Theor Biol 190:63–68. https://doi.org/10.1006/jtbi.1997.0532

Glick BR, Cheng Z, Czarny J, Duan J (2007) Promotion of plant growth by ACC deaminase-producing soil bacteria. Eur J Plant Pathol 119:329–339. https://doi.org/10.1007/s10658-007-9162-4

Hill NM, Patriquin DG (1996) Maximizing N_2 fixation in sugarcane litter. In: International symposium on sustainable agriculture for the tropics—the role of biological nitrogen fixation, programme and abstracts. Serope´dica, Embrapa-CNPAB, pp 59–60

Jackson ML (1973) Methods of chemical analysis. Prentice Hall of India (Pvt.) Ltd, New Delhi

Jansson M, Olsson H, Pettersson K (1988) Phosphatases; origin, characteristics and function in lakes. Hydrobiol 170:157–175. https://doi.org/10.1007/BF00024903

Kalaivanan D, Hattab KO (2016) Recycling of sugarcane industries byproducts for preparation of enriched pressmud compost and its influence on growth and yield of rice (*Oryza sativa* L.). Int J Recycl Org Waste Agric 5:263–272. https://doi.org/10.1007/s40093-016-0136-4

Lim SL, Wu TY, Lim PN, Shak KPY (2015) The use of vermicompost in organic farming: overview, effects on soil and economics. J Sci Food Agric 95(6):1143–1156. https://doi.org/10.1002/jsfa.6849

Lim SL, Lee LH, Wu TY (2016) Sustainability of using composting and vermicomposting technologies for organic solid waste biotransformation: recent overview, greenhouse gases emissions and economic analysis. J Clean Prod 111:262–278. https://doi.org/10.1016/j.jclepro.2015.08.083

Mattoo AK, Suttle JC (1991) The plant hormone ethylene. CRC Press Inc, Boca Raton

Nautiyal CS (1999) An efficient microbiological growth medium for screening phosphate solubilizing microorganisms. FEMS Microbiol Lett 170:265–270. https://doi.org/10.1111/j.1574-6968.1999.tb13383.x

Nelson DW, Sommers LE (1996) Total carbon, organic carbon and organic matter. In: Bigham JM (ed) Methods of soil analysis: part 3-Chemical methods. Soil Science Society of America Society of Agronomy, Inc., Madison, pp 961–1010

Nevens F, Reheul D (2003) The application of vegetable, fruit and garden waste (VFG) compost in addition to cattle slurry in a silage maize monoculture: nitrogen availability and use. Eur J Agron 19:189–203. https://doi.org/10.1016/S1161-0301(02)00036-9

Olsen SR, Cole CV, Watanabe FS, Dean LA (1954) Estimation of available phosphorus in soils by extraction with sodium bicarbonate. vol 939. U.S. Dept. of Agriculture, Washington, DC

Rashid M, Khalil S, Ayub N, Alam S, Latif F (2004) Organic acids production and phosphate solubilization by phosphate solubilizing microorganisms (PSM) under in vitro conditions. Pak J Biol Sci 7:187–196. https://doi.org/10.3923/pjbs.2004.187.196

Reddy MS, Kumar S, Babita K, Reddy MS (2002) Biosolubilization of poorly soluble rock phosphates by *Aspergillus tubingensis* and *Aspergillus niger*. Bioresour Technol 84:187–189. https://doi.org/10.1016/S0960-8524(02)00040-8

Roy SK, Sharma RC, Trehan SP (2001) Integrated nutrient management by using farmyard manure and fertilizers in potato-sunflower-paddy rice rotation in the Punjab. J Agric Sci 137:271–278. https://doi.org/10.1017/S0021859601001472

Ryan J, Estefan G, Rashid A (2001) Soil and plant analysis: laboratory manual. ICARDA, Aleppo

Saleem MM, Arshad M, Yaseen M (2013) Effectiveness of various approaches to use rock phosphate as a potential source of plant available P for sustainable wheat production. Inter J Agric Biol 15:223–230

Salimpour S, Khavazi K, Nadian H, Besharati H, Miransari M (2010) Enhancing phosphorous availability to canola (*Brassica napus* L.) using P solubilizing and sulfur oxidizing bacteria. Aus J Crop Sci 4:330–334

Sanyal SK, De Datta SK (1991) Chemistry of phosphorus transformation in soils. Adv Soil Sci 16:1–120. https://doi.org/10.1007/978-1-4612-3144-8_1

Schachtman DP, Reid RJ, Ayling SM (1998) Phosphorus uptake by plants: from soil to cell. Plant Physiol 116:447–453. https://doi.org/10.1104/pp.116.2.447

Shaharoona B, Arshad M, Zahir ZA (2006) Effect of plant growth promoting rhizobacteria containing ACC-deaminase on maize (*Zea mays* L.) growth under axenic conditions and on nodulation in mung bean (*Vigna radiata* L.). Lett Appl Microbiol 42:155–159. https://doi.org/10.1111/j.1472-765X.2005.01827.x

Shaharoona B, Jamro GM, Zahir ZA, Arshad M, Memon KS (2007) Effectiveness of various *Pseudomonas* spp. and *Burkholderia caryophylli* containing ACC-deaminase for improving growth and yield of wheat (*Triticum aestivum* L.). J Microbiol Biotechnol 17:1300–1307

Singh CP, Amberger A (1995) The effect of rock phosphate-enriched compost on the yield and phosphorus nutrition of rye grass. Am J Alternative Agri 10:82–87. https://doi.org/10.1017/S0889189300006196

Steel RGD, Torrie JH, Dicky DA (1997) Principles and procedures of statistics—a biometrical approach. McGraw-Hill, Singapore, pp 204–227

Suganuma N, Yamauchi H, Yamamoto K (1995) Enhanced production of ethylene by soybean roots after inoculation with *Bradyrhizobium japonicum*. Plant Sci 111:163–168. https://doi.org/10.1016/0168-9452(95)04239-Q

Surange S, Wollum AG, Kumar N, Nautiyal CS (1995) Characterization of Rhizobium from root nodules of leguminous trees growing in alkaline soils. Can J Microbiol 43:891–894. https://doi.org/10.1139/m97-130

van Workum WAT, Van Brussel AAN, Tak T, Wijffelman CA, Kijne WJ (1995) Ethylene prevents nodulation of *Vicia sativa* ssp. nigra by exopolysaccharides deficient mutants of *Rhizobium leguminosarum* bv viciae. Mol Plant Microb Int 8:278–285. https://doi.org/10.1094/MPMI-8-0278

Vance CP, Uhde-Stone C, Allan DL (2003) Phosphorus acquisition and use: critical adaptations by plants for securing a non-renewable resource. N Phytol 157:423–447. https://doi.org/10.1046/j.1469-8137.2003.00695.x

Vassilev N, Vassileva M, Nicolaeva I (2006) Simultaneous P-solubilizing and biocontrol activity of microorganisms: potentials and future trends. Appl Microbiol Biotechnol 71:137–144. https://doi.org/10.1007/s00253-006-0380-z

Verma SL, Penfold C, Marschner P (2013) Mobilization of rock phosphate by surface application of compost. Biol Fertil Soils 49:287–294. https://doi.org/10.1007/s00374-012-0717-y

Watson ME (2003) Testing Compost. Extension Fact Sheet ANR-15-03. Ohio State University. http://ohioline.osu.edu/anr-fact/0015.htmlS. Accessed 21 Nov 2005

Wu C, Wei X, Sun HL, Wang ZQ (2005) Phosphate availability alters lateral root anatomy and root architecture of *Fraxinus mandshurica* Rupr. seedlings. J Integr Plant Biol 47:292–301. https://doi.org/10.1111/j.1744-7909.2005.00021.x

Wu TY, Lim SL, Lim PN, Shak KPY (2014) Biotransformation of biodegradable solid wastes into organic fertilizers using composting or/and vermicomposting. Chem Eng Trans 39:1579–1584. https://doi.org/10.3303/CET1439264

14

Mobility of phosphorus from organic and inorganic source materials in a sandy soil

Raphael Passaglia Azevedo[1] · Ignacio Hernán Salcedo[2] · Priscila Alves Lima[3] · Vânia da Silva Fraga[4] · Regina Maria Quintão Lana[5]

Abstract

Purpose The long-term use of manure as a source of nutrients can promote phosphorus (P) leaching, especially in sandy soils. The aim was to evaluate P mobility from organic and mineral sources in columns, linked with the Dystric Xeropsamments adsorption characteristics with long-term organic fertilization regime.

Methods The mineral, chemical, and physical properties of the samples were characterized, including Langmuir adsorption and desorption kinetics. The P mobility was determined in one leaching experiment, in 20 cm soil columns. The topsoil (0–10 cm) layer was treated with organic (cattle, swine, goat, and hen manure) and inorganic fertilizers. Leaching was corresponding to 10 pore volumes (PV), 1 PV day^{-1}. Aliquots of the leachate were collected to analyze P concentrations. After the leaching, the columns were sliced into 5-cm sections for the analysis of water-extractable P (WEP).

Results The mineral source obtained higher leaching of P and between the organic sources the bovine and swine manure. The latter were the ones that had higher value of P_w in the soil after the leaching. Due to the increase of the adsorption capacity of P with depth, there was a reduction in the mobility of P, and an unbalance of P_w in the soil was found.

Conclusions The mobility of P depends on the concentration of the soluble P soil or added material; moreover, the presence of Fe and Al oxides, even in small amounts, reduces the mobility of P in sandy soil.

Keywords Leaching · Desorption · Adsorption · Manure

Introduction

Low-income small farming in Agrest is characterized by limited commercial use of nitrogen or phosphorus fertilizers. Soil fertility, therefore, solely relies on organic matter management (Nascimento et al. 2003), and that is the reason why organic fertilizers, especially manure, are widely used by farmers in that region. In the organic fertilizers being considered as a main source of nutrients to the plants, the soil conditioner effects have not been studied so far.

Furthermore, the use of manure in agriculture varies according to the availability, amount, and type of manure (cattle, swine, goat, and poultry). In the northeastern parts of Brazil, solid cattle manure and goat manure are frequently used in doses ranging from 4 to 20 Mg ha^{-1} (Sabourin et al. 2000; Silva and Menezes 2007; Galvão et al. 2008; Galvão

In memoriam of Ignacio Hernán Salcedo.

✉ Raphael Passaglia Azevedo
raphaelpassaglia@hotmail.com

Priscila Alves Lima
priscilaea@hotmail.com

Vânia da Silva Fraga
vfraga@cca.ufpb.br

Regina Maria Quintão Lana
rmqlana@terra.com.br

[1] Agronomy, Universidade Federal de Uberlândia, Av. Para, 1720, Uberlândia, MG, Brazil

[2] Instituto Nacional do Semiárido (INSA), Av. Francisco Lopes de Almeida, S/N, Serrotão, Campina Grande, PB 58434-700, Brazil

[3] Instituto Federal do Norte de Minas, Higway BR 367 Almenara/Jequitinhonha, km 111, Rural Zone, Almenara, MG 39900-000, Brazil

[4] Dpto. de Solos e Engenharia Rural, Universidade Federal da Paraíba, Higway PB 049 km 12, Areia, PB, Brazil

[5] Instituto de Ciências Agrárias-ICIAG, Universidade Federal de Uberlândia, Av. Amazonas s/n Blok 2E and 01, Uberlândia, MG, Brazil

and Salcedo 2009). In the south of the country, on the other hand, because the animal concentration of swine and cattle slurry is higher, doses around 80 m^3 ha^{-1} are often used (Piovesan et al. 2009; Bertol et al. 2010; Ceretta et al. 2010).

Nutrients in animal manure are lower in concentration and unbalanced compared to commercial fertilizers, which hinder an even supply of nutrients to crops. The doses are usually based on demand for N of cultures, the size of the area or in the load capacity of the vehicle of transport common in the region and are generally multiples of 4 tons. Moreover, the wide variability of the chemical composition of manure may cause accumulation of one or more nutrients in the soil (Galvão et al. 2008), as the P, for example.

Accumulation of P, in the topsoil layers, has been reported in the literature (Ojekami et al. 2011; Kang et al. 2011; Rubæk et al. 2013). In Brazil, annual applications of manure in areas with Dystric Xeropsamments of sandy texture in sub-humid region in the northeastern Brazil (Agreste) have caused accumulation of nutrients in the surface layer (Silva and Menezes 2007; Galvão et al. 2008; Oliveira et al. 2011). Consequently, these nutrients move down the soil profile, including phosphorus (Galvão and Salcedo 2009; Xavier et al. 2009), which has little mobility in clayey soils. The accumulation of P in the topsoil increases the risk of its vertical displacement beyond the reach of the roots. This fact, in addition to representing an economic loss for the farmer, can become an environmental problem due to the contamination of subsurface water (Sharpley et al. 2004; Nelson et al. 2005; Galvão et al. 2008; Oliveira et al. 2011; Lee and Oa 2013).

To assess the risk, it is necessary to analyze the chemical composition of the soil, especially the content of poorly crystalline minerals represented by the oxalate-soluble Fe and Al fraction. Even sandy soils may contain small amounts of Fe and Al in their clay fraction, but that is enough to reduce the mobility of P (Sims et al. 1998). These elements are also components of oxides and hydroxides which are mainly responsible for the adsorption of P (Novais et al. 2007; Tarkalson and Leytem 2009; Fink et al. 2016).

For these reasons, the objective of this study was to evaluate phosphorus mobility in Dystric Xeropsamments fertilized with organic (cattle, swine, goat, and hen manure) and inorganic source materials for a long period by balancing phosphorus inputs and outputs and relating mobility to soil adsorption characteristics.

Methods

Soil collection and analysis

The soil samples were collected in municipality of Remigio, PB (6°59′14.9″ latitude S; 35°48′14.8″ longitude W) because of its fertilization regime. The chosen area has been fertilized exclusively with cattle manure for nearly 40 years. Part of the manure is produced locally, while the rest is purchased elsewhere in the area. However, little is known about the exact quantities used each year.

The climate of the region is characterized according to the scale of Koppen as As', hot and humid, with a rainy season between February and August, with average rainfall of the last 20 years of 917.6 mm (AESA 2017). The soil was classified as Dystric Xeropsamments (Soil Survey Staff 2014). The samples were collected from an area of 2 × 3 m at two depths: 0–10 and 10–20 cm. Once collected, the samples were air dried, crushed, and sieved through 4-mm mesh for mounting the columns, and through a 2-mm mesh for mineralogical, chemical, and physical analysis.

The mineralogical analysis was performed using X-ray with Cu Kα radiation, 30 kV and 30 mA, with scan from 5° to 55° (2θ) on samples initially sieved through the 2-mm mesh, which after grinding were sieved through a 200 mesh (Table 1).

The following soil chemical attributes were determined: pH in water 1:2.5 (m/v), phosphorus extracted with Mehlich-1 (MEP) (EMBRAPA 2011), remaining P (P-rem) (Alvarez et al. 2000), and water-extractable phosphorus (WEP) which was extracted with deionized water and the supernatant filtered through a membrane of 0.45 μm, adapted by del Campillo et al. (1999). The concentrations of P in the extracts were determined according to Murphy and Riley (1962). The total organic carbon (TOC) was determined by wet oxidation–diffusion (Snyder and Trofymow 1984) and the water-extractable carbon (WEC) by the method proposed by Bartlett and Ross (1988) and adapted by Mendonça and Matos (2005). In addition to this, iron (Fe) and aluminum (Al) were extracted with ammonium oxalate and determined by atomic absorption (Schwertmann 1964) (Table 2). The following physical attributes were analyzed: particle size (densitometer method), particle density (Pd), bulk density (Bd), and total porosity (TP) (EMBRAPA 2011) (Table 3).

Phosphorus desorption kinetics was determined using paper strips impregnated with iron (Fe strips) in triplicate (van der Zee et al. 1987) for 1, 7, 22, 28, 43, 48, and 72 h. The results were processed by the SigmaPlot 10 program (Systat Software 2006) to estimate phosphorus surface adsorption (Q_{in}) and the rate of desorption (K_d). The P-adsorption test was performed by the "batch" method

Table 1 Main mineral components of Dystric Xeropsamments with long-term organic fertilization

Depth (cm)	Quartz (%)	Feldspar (%)	Hematite (%)
0–10	97.0	2.29	0.69
10–20	82.5	16.5	1.03

Table 2 Chemical analysis of Dystric Xeropsamments with long-term organic fertilization

Attribute	Unit	Depth (cm)	
		0–10	10–20
pH	1:2.5 (H_2O)	6.07	6.06
MEP	mg kg^{-1}	8.25	2.88
P-rem	mg kg^{-1}	53.2	49.3
WEP	mg kg^{-1}	1.15	0.79
Al^{3+}	mg kg^{-1}	0	0
WEC	mg kg^{-1}	79.1	42.7
TOC	g kg^{-1}	6.59	4.45
OM	g kg^{-1}	11.4	7.7
Fe-ox	mg g^{-1}	0.11	0.17
Al-ox	mg g^{-1}	0.12	0.22

pH determined in water, *MEP* extracted with Mehlich-1, *WEP and WEC* extracted with water, *P-rem* remaining P (Alvarez et al. 2000), *TOC* sulfuric–phosphoric digestion, *OM* TOC × 1.724, *Fe-ox, Al-ox* extracted with ammonium oxalate

with increasing concentrations of P (0, 5, 10, 20, 40, 60, and 100 mg L^{-1}) with three replications and agitation for 24 h. This method was an adaptation of the method described by Farias et al. (2009). The amount of P in the solid phase was the difference between the P added to and recovered from the equilibrium solution. The P added to the solid phase (Q_{in}) was found in the desorption experiment with Fe strips (van der Zee et al. 1987). Experimental data were adjusted using the adsorption equation proposed by Soares and Casagrande (2009):

$$[\mathrm{P}]_{ads} = \frac{V_{solution}[(C_0 - C_e)]}{M_{soil}} + Q_{in}, \quad (1)$$

where $[\mathrm{P}]_{ads}$ is the concentration of adsorbed P (mg kg^{-1}), $V_{solution}$ is the volume of solution (L), C_0 is the initial concentration of P in the solution (mg L^{-1}), C_e is the remaining concentration of P in the equilibrium solution (mg L^{-1}), M_{soil} is the soil mass (kg), and Q_{in} is the adsorbed phosphorus extracted with iron strips (mg kg^{-1}).

The results of P_{ads} and C_e were processed using the Langmuir equation with the aid of the SigmaPlot 10 program (Systat Software 2006).

P-leaching experiment with soil columns

To evaluate the vertical displacement of P for the organic or inorganic sources, polyvinyl chloride (PVC) columns were sequentially filled with soil layers. The experiment was conducted at the Departamento de Solos e Engenharia Rural do Centro de Ciências Agrárias—CCA. One assay was conducted with 20-cm columns containing soil from 0 to 10 and 10 to 20-cm layers.

The organic source materials were cattle, swine, goat, or hen manure, while the inorganic source material was a mixture of ammonium sulfate [$(NH_4)_2SO_4$], monosodium phosphate (PO_4H_2Na), and potassium chloride (KCl).

After composting, the organic sources were dried, crushed, sieved through 2-mm mesh, and stored in airtight containers for chemical analysis. The analysis was determined: pH in water 1:10 (m/v); electric conductivity (EC) 1:10 (m/v); total P (P_t) obtained by digestion with H_2SO_4 and H_2O_2 (Thomas et al. 1967); P extracted with Mehlich-1 solution (MEP); P extracted with water (WEP) and the extracts determined by the method of Murphy and Riley (1962); WEC determined by the method of Mendonça and Matos (2005) adapted from Bartlett and Ross (1988); TOC obtained by wet oxidation–diffusion (Snyder and Trofymow

Table 4 Chemical analysis of organic sources used in experiments with soil columns

Attribute	Units	Organic source materials (manure)			
		Cattle	Swine	Goat	Hen
pH	(1:10)	8.02	7.70	9.07	9.78
EC	dS m^{-1}	2.71	4.05	5.06	5.48
Pt	g kg^{-1}	4.42	14.3	11.6	20.1
WEP	g kg^{-1}	1.23	3.66	0.89	1.99
MEP	g kg^{-1}	3.43	10.7	2.50	3.75
TOC	g kg^{-1}	315	319	277	113
WEC	g kg^{-1}	1.15	1.30	1.30	0.93
C/P		71.4	22.4	23.8	5.60
Ashes	%	33.3	25.1	38.3	64.4

pH analyzed in water 1:10 (m/v), *Pt* sulphuric–perchloric acid digestion, *MEP* extracted with Mehlich-1, *WEP, WEC* extracted with water, *TOC* sulfuric–phosphoric acid digestion 3:2 (H_2SO_4/H_3PO_4), *C/P* relation between Ct and Pt

Table 3 Physical analysis of Dystric Xeropsamments with long-term organic fertilization

Depth (cm)	Bd (g cm^{-3})	Pd (g cm^{-3})	TP (%)	Sand (g kg^{-1})	Silt (g kg^{-1})	Clay (g kg^{-1})	Textural class (g kg^{-1})
0–10	1.62	2.6	37.7	852	96	52	Loamy sand
10–20	1.63	2.62	37.8	808	96	96	Loamy sand

Bd bulk density, *Pd* particle density *TP* total porosity = [(Pd – Bd)/Pd] × 100

1984); and ash obtained by the method of EMBRAPA (2011) (Table 4).

The manure dose was calculated based on the content of P_t in the cattle manure, considering the application of 16 Mg ha^{-1} in the first 20 cm of soil, which is the mean annual dose applied in the region (Galvão and Salcedo 2009; Oliveira et al. 2011). However, we decided to apply 8 Mg ha^{-1} of cattle manure to the 0–10 cm layer to better observe P displacement. Based on this dose, an amount of P was set for other fertilizers, adjusting them to the same P input with cattle manure (35.3 kg P ha^{-1}) (Table 5). The soil volume mixed with the sources was 1767 cm^3, having a mass of 2650 g (0–10 cm).

The experimental design was completely randomized blocks with six treatments. Four organic source materials (cattle, swine, goat, and hen manure), one inorganic source, and a control were evaluated with three repetitions.

The columns were made of PVC pipes (150-mm diameter and 25-cm height), maintaining a 5-cm border from the top. First, the inner surface of each column was sanded. Second, the sand was washed with an acidic solution of 10% HCl. The sand was then glued with polychloroprene glue and synthetic resins to the inner wall of the tubes to increase soil adhesion to the wall and to avoid artificial preferential flow paths by the internal walls during the percolation of the solution. Finally, the bottoms of the columns were closed with a nylon screen of 77-μm mesh to prevent soil from escaping during the percolation of the solution and another nylon screen (50% shading) was used to maintain the soil in the column. Both were fastened with wire and PVC glue.

The organic sources were mixed with soil in the top layer (0–10 cm). The treated soil was moistened with distilled water until it reached 40% of the total pore volume (PV). Then, the setup was incubated for 15 days to stabilize the microbial activity, keeping moisture constant with daily weighings. After the incubation, the soil was air dried. This procedure was not carried out for the inorganic source.

Soil in the columns was prepared manually, assuming a mean bulk density of 1.5 g cm^{-3} found for this type of soil (Santos et al. 2012; Kang et al. 2011; Oliveira et al. 2011; Santos et al. 2010). The layer of 0–10 cm was gently set, because the soil when plowed is less dense. To dissipate the impact of the drops and evenly distribute the solution on the surface of the column, and to avoid the induction of preferential paths, three discs with a mesh of 50% shading capacity were placed on the soil with the leaning nozzle. After the assembly, six columns were placed on a wooden structure with funnels to collect the leachate into containers.

Table 5 Amounts of dry matter from organic sources added to the soil column

Unit	Organic source materials (manure)				Inorganic source material
	Cattle	Swine	Goat	Hen	$NaH_2PO_4H_2O$
Mg ha^{-1}	8	2.48	3.03	1.76	0.16
mg g^{-1}	4.40[a]	14.3	11.6	20.1	999.9
g $column^{-1}$	14.1	4.40	5.40	3.10	0.278

[a]Fixed dose

Rounds and volume of percolated solution

A hospital infusion set with tubing and flow regulators to supply $CaCl_2$ at 0.001 mol L^{-1} was mounted on the columns (Roy and Dzomback 1996). The flow rate was 10 ± 1 cm^3 min^{-1}.

The volume of the applied solution was calculated using the pore volume (PV) of the columns (Eq. 2), which was 1500 cm^3:

$$PV = \pi r^2 h \left[1 - \left(\frac{\text{Ds}}{\text{Dp}}\right)\right]. \tag{2}$$

In total, ten pore volumes were applied to each column—1 PV per day. The displacement of the saline solution was vertical and downward. The collection of the column effluents started with the beginning of drainage, splitting each PV into three equal aliquots using calibrated containers. The aliquots were filtered through a 0.45-μm membrane. The concentration of P was determined according to Murphy and Riley (1962). Once the leaching stopped, the columns with the soil were dried in an oven at 50 °C and sliced evenly into 5-cm sections. The soil of each section was crushed, homogenized, and sieved through a 2-mm mesh. Finally, WEP values were obtained.

The balance of WEP in the columns was carried out by calculating its input and output. For this, the amounts of WEP already in the soil (soil WEPi) were added to the amounts in the source (source WEPi), and then subtracted from the leached amount, thus generating the final WEP (final. WEP). This value was compared with the final observed WEP (final obs. WEP) extracted with saline solution from the columns after leaching.

Statistical analysis

We performed analysis of variance (ANOVA) for total leachate and the comparison of means by Tukey test at 10% probability.

To evaluate the influence of the sources at each depth, we performed ANOVA and Tukey test ($p < 0.1$). In addition to ANOVA and Tukey ($p < 0.1$), a linear mixed model was created to compare the depths for each source. It estimates an additional error in analysis of variance to compensate for P variations between depths due to non-independence of the data, the lack of randomness, and autocorrelation of P content between depths. The program used for statistical analysis for SAS software 9.3 (2011) and for the construction of the figures the SigmaPlot 12.0.

Results and discussion

Leaching columns

The leaching of P was higher with the inorganic source starting at 9.67 mg of P for the first PV and decreasing to 0.26 mg for the tenth PV (Fig. 1a). However, P from the organic sources behaved differently, starting with a mean ($n = 12$) of 0.15 mg of leached P and increased to 0.27 mg at the tenth PV. Among the organic sources, P from swine manure leached the most, with increasing levels until the sixth PV, stabilization until the ninth PV, and increasing again after the tenth PV (Fig. 1b).

The leaching of P for the inorganic source had a logarithmic trend ($r^2 = 0.99$) (Fig. 2a), while the organic sources and the control demonstrated a linear trend ($r^2 = 0.99$) and did not differ ($p < 0.1$) (Fig. 2b).

Considering that the input of P was the same for all sources (35.3 kg ha^{-1} or 62.5 mg column^{-1} of P), the different behaviors of P from the organic and inorganic sources were due to the water-extractable P in the manures (Kang et al. 2011; Conti et al. 2015): 17.4, 16.0, 4.75, 6.18, and 62.5 mg of WEP column^{-1} added with cattle, swine, goat and hen manure, and the inorganic source, respectively.

Leached P for the inorganic source, which reached 22.3 mg or 34.4% of added P, was 18 times higher than the average of leached P for the organic materials (Fig. 2a). However, as the organic sources contained much less WEP, the percentage of leached P ranged between 6.6 and 18.8% of WEPi. On the other hand, the leached P showed relations with the WEP content in the sources, because the cattle and swine manure lost nearly double the value compared to the goat and hen manure (Fig. 2b). The physicochemical characteristics of the organic sources, such as lignin (Galvão et al. 2008) and the chemical forms of P in manure (soluble phosphate ion, nucleotides, phytin, Ca phosphates, and polyphosphates) certainly contributed to the variations of soluble P in the source materials.

Higher P losses for an inorganic source were also found by Kang et al. (2011). These authors also analyzed P leaching for organic and inorganic sources in columns with sandy soil. They observed higher P leachability for inorganic sources with 63 and 71% of P lost added, respectively, via triple superphosphate and potassium phosphate (KH_2PO_4) using approximately 23 PV. However, in this study, 10 PV were used and the losses reached 34.4% of P added via the inorganic source.

Perhaps, in our experiment, P from different sources was partially dragged down the profile with the percolating solution through artificial preferential flow paths within the soil column (Djodjic et al. 1999; Sharpley et al. 2003). We tried to prevent or reduce this effect by coating the inner surface of the columns with washed sand. Furthermore, there is consistency of data regarding the difference between the soluble inorganic and the organic source materials with

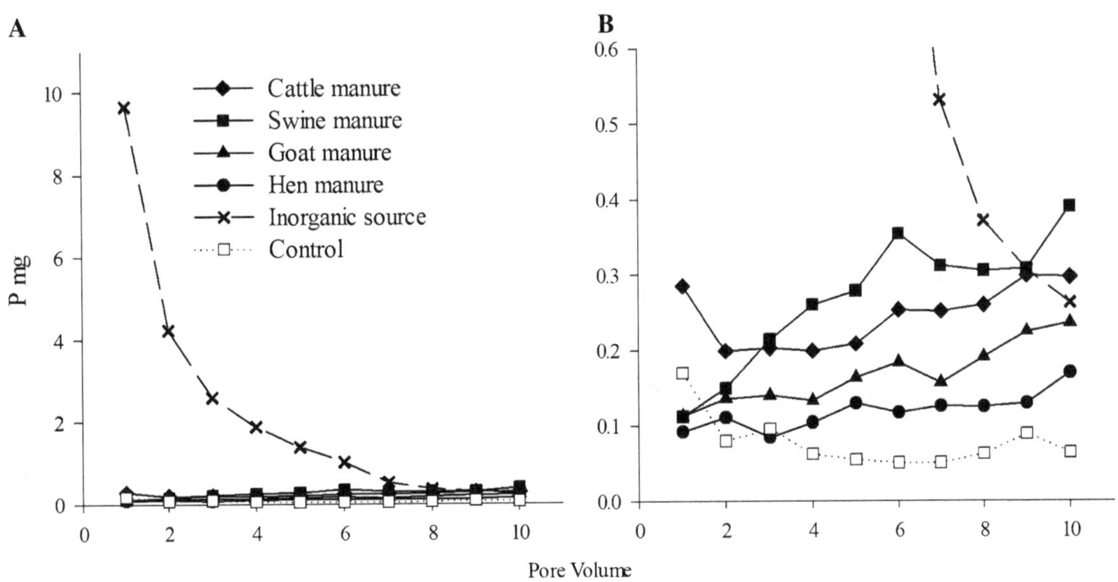

Fig. 1 Leaching of P (mg) for different sources in Dystric Xeropsamments in 20-cm columns as function of pore volume. **a** Leaching of P (mg) in all treatments; **b** the same **a** with cut in the "*y*"-axis at 0.6 mg of P

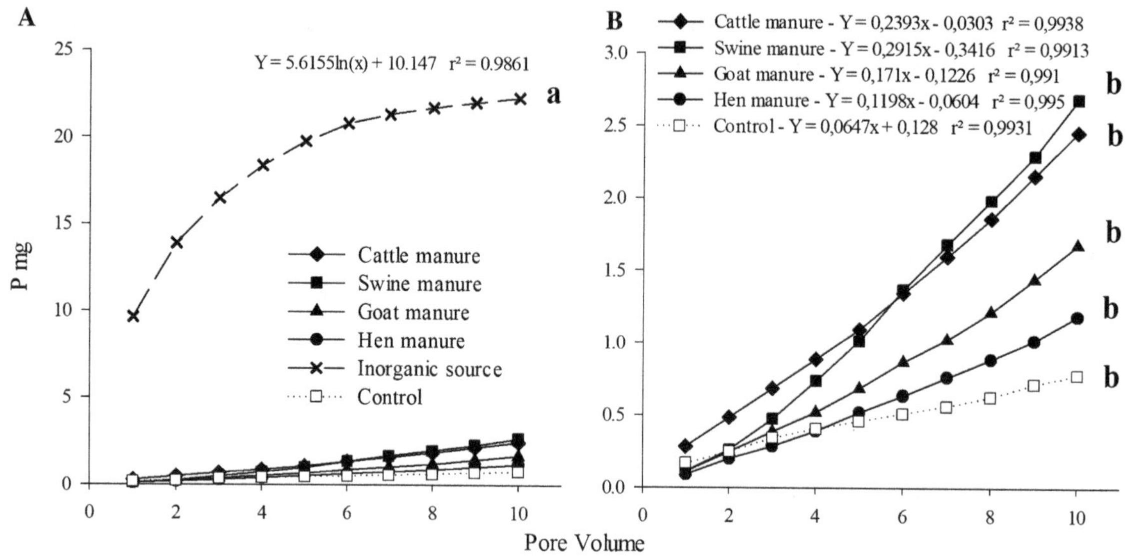

Fig. 2 Leaching of P (mg) for different sources in Dystric Xeropsamments in 20-cm columns depending on the volume of pores. **a** P (mg) in all treatments; **b** the same **a** with cut in the "y"-axis at 3 mg of P. Treatments followed by the same lower case letter do not differ by Tukey test ($p < 0.1$)

lower volume of WEP. As already mentioned leached P for the organic sources shows correlation with WEPi. For these reasons, although the possibility of preferential paths cannot be totally ruled out, it was considered unlikely.

Another process in soils that may influence the leaching of P is the competition of P with organic matter compounds (organic acids and fulvic acid) for the same adsorption sites. As a result, P may remain in soil solution and be subject to leaching (Guppy et al. 2005a, b). In soils with low content of clay, organic and humic acids effectively block the adsorption sites, promoting lower P adsorption than in soils with high clay content (Andrade 2001). To reinforce the hypothesis of competition, non-tabulated results indicate that the soil placed in the columns contained, on average, 52.6 mg kg^{-1} of C per layer in water-soluble form. It is also possible that the percolation of organic P occurred (Galvão and Salcedo 2009) due to phosphatase-catalyzed hydrolysis during the collection of the percolated solution.

The concentrations of WEP remaining in soil columns after leaching varied depending on the P sources (Fig. 3). The lowest WEP concentrations were recorded for the control and for the soils in the columns treated with the inorganic source. The concentrations were uniform at different depths and lower in the case of the inorganic source, which is consistent with the high loss of P caused by leaching for this source (Figs. 1, 2). This is due to the mobility of this

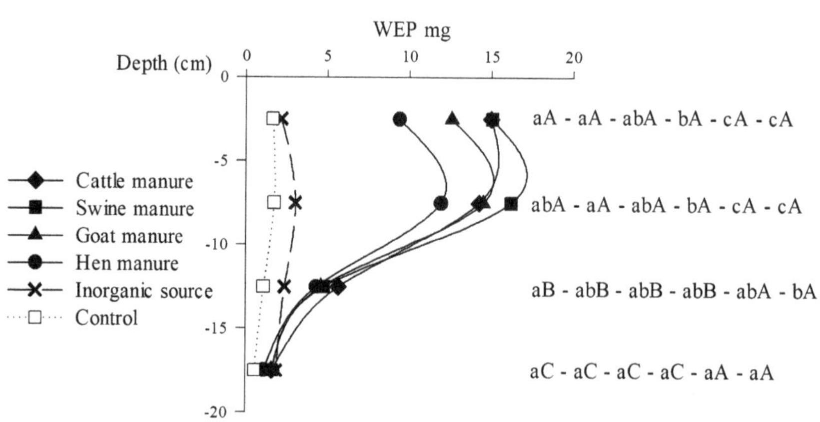

Fig. 3 Distribution of concentrations of P extracted with water from 20-cm columns filled with Dystric Xeropsamments after application of 10 PV. Treatments followed by the same letter do not differ significantly by Tukey test ($p < 0.1$); lower case letters compare sources within each depth and upper case letters compare depths with each source. The sequence of the letters follows the order of the legend of the treatments (cattle, swine, goat, hen manure, inorganic source, and control)

form of P in the column profile, which promoted lower accumulation of WEP in the soil resulting in concentrations similar to the control.

However, leached P for the inorganic source reached 34% of the added P, indicating that part of the added P was not transformed into water-soluble forms. In fact, the WEP balance in the soil before and after leaching, including leached P, indicates 33 mg less of WEP in the soil column treated with the inorganic source (Table 6). The fact that 33 mg of P have gone indicates that there was an interaction of P with the soil matrix, even though the soil had a sandy texture with the predominance of quartz and feldspars in the layers of 0–20 cm (Table 1). However, it is noteworthy that there was a small contribution of hematite (Table 1) to decrease leaching and promote adsorption of P in columns (Sims et al. 1998; Tarkalson and Leytem 2009).

Among the organic sources, hen WEP showed the lowest concentrations in layers from 0 to 5 and 5 to 10 cm (Fig. 3), which is consistent with low WEP added via this source (Table 6). WEP for goat manure, however, which was similar to that of hen, did not differ ($p < 0.1$) from cattle and swine. Regarding the accumulated leached P (Fig. 2), the treatment with goat manure was between the treatment with hen manure, which was smaller, and swine and cattle manure, which were higher (Table 6). This trend of WEP for goat manure, consistent in leached P and the remaining WEP in the soil, suggests an increase in solubility of P from this source during the course of the experiment. The nature and mineralization of manure are very heterogeneous and are dependent on the species of the animal, its age, diet, and the physical and chemical conditions of the soil. These conditions dictate the speed of decomposition and consequent solubilization of organic and inorganic forms of nutrients (Tedesco et al. 1999). Souto et al. (2005), in evaluating manure decomposition in the semiarid region of Paraíba, found that the highest rates of decomposition were goat and cattle manure, which were strongly influenced by rainfall events during the trial period.

The comparison between layers demonstrates that WEP for the organic sources was similar ($p < 0.10$) in the first two layers due to the mixing of the sources to the soil was performed in the layer 0–10 cm, but at the depth of 10–15 cm, its concentration was between the concentration of the 5–10- and 15–20-cm layers ($p < 0.10$), showing a vertical displacement of WEP for the organic source. It is worth investigating the reason why WEP did not relocate with the same intensity as the WEP from the inorganic source. One possibility is that this soluble form of P was formed in situ during 25 days of the experiment (15 days of incubation and 10 days of leaching), as a result of mineralization of the organic sources. The balance between inputs and outputs of P, which indicate greater WEP inputs that the outputs, suggests this possibility (Table 6). Therefore, the hen manure seems to be more resistant to mineralization than the other three sources.

It is noteworthy that the total applied PV was 15 L. Comparing to the average rainfall over the last 20 years, which is 917 mm (AESA 2017), or equivalent to 17.6 L, the organic sources still accumulated P in the columns in a form which is available to plants, unlike the inorganic source.

Adsorption and desorption of P

Although clay content in the soil in this study was low (Table 3), its characteristics may favor P adsorption in its matrix (Tables 1, 2).

The adsorption and precipitation of P in soil are directly linked to the texture and soil pH (Lee and Oa 2013) and especially the levels of Fe and Al (Pinto 2012). The results of amorphous Fe and Al, extracted with ammonium oxalate, increased with depth (Table 3) and are positively correlated with clay content ($r^2 = 0.90$ and 0.96, respectively). Kang et al. (2011) using a sandy soil ($Fe_{ox} = 0.20$; $Al_{ox} = 0.29$; and clay $= 21$ mg g^{-1}), similar to the one in our study, found adsorption of P in the soil, especially when observing the data from the treatment with the inorganic fertilizer. Tarkalson and Leytem (2009) also noted adsorption of P in

Table 6 WEP balance in the 20-cm columns with Dystric Xeropsamments treated or not with organic and inorganic sources with 62.5 mg of total P

Source	Cattle (mg)	Swine (mg)	Goat (mg)	Hen (mg)	Inorganic (mg)	Control (mg)
Soil WEPi	5.15	5.15	5.15	5.15	5.15	5.15
Source WEPi	17.4	16	4.75	6.18	62.5	0
Sum	22.6	21.2	9.9	11.3	67.6	5.15
Leached	2.46 ± 0.6	2.68 ± 0.6	1.68 ± 0.2	1.19 ± 0.2	22.3 ± 5.2	0.78 ± 0.1
Final observed WEP	48.2 ± 4.8	49.8 ± 1.7	44.0 ± 3.2	35.4 ± 1.2	12.3 ± 2.2	6.46 ± 1.7
Final calculated WEP	20.1	18.5	8.23	10.1	45.3	4.37
Difference	28.1	31.3	35.7	25.2	− 33	2.1

Source WEPi the amount of WEP in mg added via the source before leaching, *Sum* the sum of WEP in the original soil and added via source, *Final observed WEP* final real WEP extracted in water from soil in the columns, *Final calculated WEP* the difference between the sum[2] of WEP and leached WEP, *Difference* the difference between final observed WEP and final calculated WEP, ± standard error, mass of each layer equal to 2.65 kg

a soil with clay content of 62 mg g^{-1}. Therefore, the movement of P in soil profile may be reduced by small amounts of clay and/or Fe and Al oxides (Sims et al. 1998).

In the present study, the adsorption capacity of P adjusted for the Langmuir isotherm increased progressively with increasing depth (Fig. 4). Desorption values were consistent with the data adjusted by the Langmuir equation, because binding energy (KL) increased with the depth, and consequently reduced desorption kinetics of P (K_d) (Fig. 4 and Table 7). The relationship between Q_{max} and the amounts of Fe_{ox} and Al_{ox} is more evident when the experimental data regarding Fe_{ox} and Al_{ox} are used according to van der Zee and van Riemsdijk (1988). Based on studies by these authors, the ability of the poorly crystalline Fe and Al to adsorb P is approximately equal to half of the sum of their concentrations [(Fe_{ox} and Al_{ox})/2]. Table 7 shows the similarity between Q_{max} obtained by the Langmuir equation and according to van der Zee and van Riemsdijk (1988).

The binding energy and the degree of in saturation in 10–20 cm layers are higher than in the 0–10 cm layers, where P is believed to be transformed into forms which are not soluble in water (Table 8) (Gatiboni et al. 2008; Guardini et al. 2012; Conti et al. 2015).

The amounts of P desorbed by the Fe strips (Fe–P) showed an inverse relationship with P_{ads} estimated by Langmuir equation for the soil before leaching (Table 8). Analyzing this apparent discrepancy, one should remember that Fe–P was experimentally determined, with the P desorbing directly from soil onto the Fe strips. Furthermore, Fe–P maintained correlation with the values of water soluble P in a 1:10 (m/v) ratio

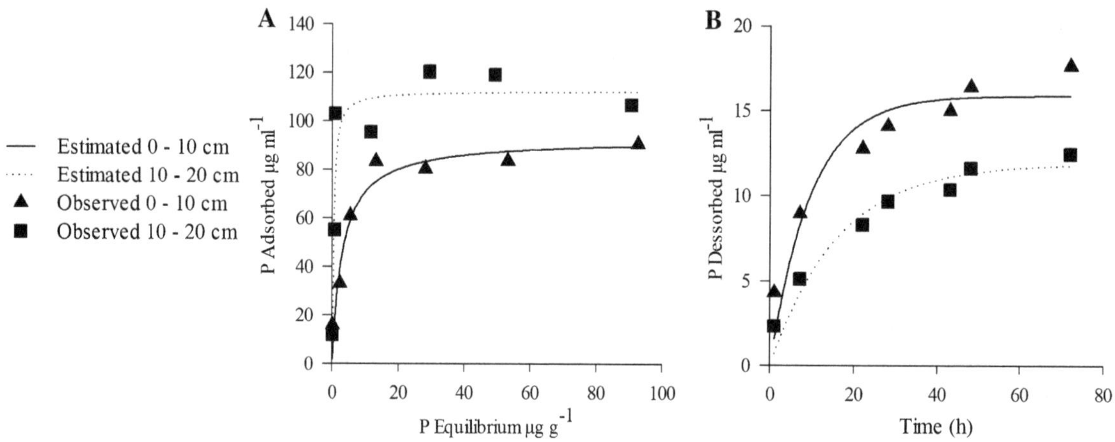

Fig. 4 **a** Langmuir isotherms for depths 0–10 and 10–20 cm. **b** Desorption of P at the depths 0–10 and 10–20 cm, using Fe strips

Table 7 Desorption parameters (Fe strips) and Langmuir isotherms adjusted for the depths of Dystric Xeropsamments in nonlinear regression

Depths cm	WEP	Desorption			Langmuir			Q_{max}
	Initial (µg ml^{-1})	Q_{in} (mg kg^{-1})	K_d (h^{-1})	r^2	K_L	Q_{max} (mg kg^{-1})	r^2	Fe and Al (mg kg^{-1})
0–10	0.12	16 ± 1.0	0.10 ± 0.03	0.94	0.32 ± 0.08	93 ± 5	0.98	103
10– 20	0.08	12 ± 0.8	0.06 ± 0.01	0.96	0.37 ± 0.07	128 ± 5	0.99	171

Table 8 Values of adsorbed P estimated by the Langmuir equation from the P content in equilibrium with water at the depths 0–10 and 10–20 cm of a Dystric Xeropsamments treated with organic and inorganic sources after leaching with 10 PV

Depth (cm)	Fe-strip desorption (mg)	Langmuir						
		Initial soil (mg)	Cattle (mg)	Swine (mg)	Goat (mg)	Hen (mg)	Inorganic source (mg)	Control (mg)
0–10	42.2	9.01	21.3	30.1	32.5	25.2	18.7	7.55
10–20	31.6	9.71	26.5	27.6	33.8	17.8	18.9	6.14
Total	73.8	18.7	47.8	57.7	66.3	43.0	37.6	13.7

[a]Determination extract 1:10 (m/v)

(Table 7, WEP initial and Q_{in}). On the other hand, P_{ads} estimated by Langmuir was calculated by adjusting the equation to the experimental data, and the parameters of this equation, especially K_L, showed considerable uncertainty (Table 7). Furthermore, P_{ads} estimated by Langmuir was derived from the difference between P added to an aqueous suspension and P found in the solution after 24 h of stirring. Thus, by inference, there is no reversibility in this phenomenon. It was found that, as the adsorption capacity of the solid phase increased, the resistance of P_{ads} to return to the solution also increased (Tiessen et al. 1993). Therefore, although the amounts of P_{ads} by Langmuir estimated the added P with reasonable approximation, it is not possible to draw conclusions as to the availability of this form of P for the plants.

The results of leaching and distribution of P in the columns show that there was a translocation of P from layers with lower binding energy to layers with higher binding energy, with the soil acting as a sink. Studies have shown that long-term fertilization with manure can promote changes in P adsorption influencing its dislocation into deeper layers of the soil profile (del Campillo et al. 1999; Silva and Menezes 2007; Galvão et al. 2008; Xavier et al. 2009; Tarkalson and Leytem 2009; Ojekami et al. 2011; Abdala et al. 2012; Bhattacharyya et al. 2015). In this regard, leaching of P may occur in soils saturated with P, even if there is no excessive input of P. Nelson et al. (2005) found higher values of leached P than P in the inputs for soils saturated with P when precipitation was above the average, indicating that desorption of P contributed to 50% of leached P.

The saturation of the adsorption sites in soil cannot be totally formed by phosphate ions. Although the phosphate ions are adsorbed preferentially, they have to compete with organic acids (Andrade 2001; Guppy et al. 2005a, b; Dotaniya et al. 2016). In this case, P leaching may occur even if the soil is not completely saturated with P.

According to del Campillo et al. (1999), 13–26 is an acceptable range for good crop development. The remaining concentrations of WEP in the 20-cm soil column for the cattle and swine manure were within this range, even after the application of the percolating solution equivalent to 851 mm of rainfall. This shows that these two source materials, the cattle and swine manure, provide high amounts of readily available nutrients for plants, because P leached the most from these sources, among all organic sources, and showed the highest amount of WEP remaining in the 20-cm soil columns.

Conclusions

The leaching of P in soil is dependent on the concentration of soluble P in the soil, whatever its source material from soil or added via organic or inorganic source. In this study occurred the leaching of P in columns of 20 cm, in which the greatest amount of P was leached from the inorganic source when compared with the organic sources with lower expression.

The availability of P by means of organic sources, manure from cattle, and pigs were the most made P, having been observed the greatest quantities in leachate (2.46 and 2.68 mg P) and recovered in the soil (48.2- and 49.8-mg WEP) after leaching, for the columns of 20 cm.

The reduction in leaching of P in the columns with 20 cm was related to the increase of the adsorption capacity of P from the soil in layers of 0–10 cm (93 mg kg^{-1}) in relation with the layers 10–20 cm (128 mg kg^{-1}). This indicated a significant reduction of the displacement of P. This suggests that the Fe and Al oxides present in a sandy soil, even in small quantities, decrease the mobility of P in the soil.

The model proposed by van der Zee and van Riemsdijk (1988) seems to be adequate to estimate the maximum P-adsorption capacity also in soil with sandy texture to the southern hemisphere; however, this confirmation needs further studies. Although the results of this study reveal adsorption of P in the layer of 10–20 cm in a Dystric Xeropsamments, new investigations about the saturation of P in these soils and about the influence of organic matter in the displacement of P in soil are still needed.

Acknowledgements The Coordenação de Aperfeiçoamento de Pessoal de Nível Superior (CAPES), by granting the Master's scholarship to the first author. The Conselho Nacional de Desenvolvimento Científico e Tecnológico (CNPq) through research funding and to Dr. Bruno O. Dias for the support.

References

Abdala DB, Ghosh AK, Silva IR, Novais RF, Venegas VHA (2012) Phosphorus saturation of a tropical soil and related P leaching caused by poultry litter addition. Agric Ecosyst Environ 162:15–23. https://doi.org/10.1016/j.agee.2012.08.004

Agência Executiva de gestão das águas do estado da Paraíba—AESA (2017) Monitoramento pluviométrico. http://site2.aesa.pb.gov.br/aesa/monitoramentoPluviometria.do?metodo=listarAnosChuvasAnuais. Accessed 25 Oct 2017

Alvarez VVH, Novais RF, Dias LE, Oliveira JA (2000) Determinação e uso do fósforo remanescente. Boletim informativo da SBCS 25:27–32

Andrade FV (2001) Ácidos orgânicos e adsorção de fosfato em Latossolos. Tese, Universidade Federal de Viçosa

Bartlett RJ, Ross DN (1988) Colorimetric determination of oxidizable carbon in acid soil solutions. Soil Sci Soc Am J 52:1191–1192. https://doi.org/10.2136/sssaj1988.03615995005200040055x

Bertol OJ, Fey E, Ravaretto N, Lavoranti OJ, Rizzi NE (2010) Mobilidade de P, Cu e Zn em colunas de solo sob sistema de semeadura direta submetido às adubações mineral e orgânica. Rev Bras Cienc Solo 34:1841–1850. https://doi.org/10.1590/S0100-06832010000600008

Bhattacharyya P, Nayak AK, Shahid M, Tripathi R, Mohanty S, Kumar A, Raja R, Panda BB, Lal B, Gautam P, Swain CK, Rou

KS, Dash PK (2015) Effects of 42-year long-term fertilizer management on soil phosphorus availability, fractionation, adsorption–desorption isotherm and plant uptake in flooded tropical rice. Crop J 3:387–395. https://doi.org/10.1016/j.cj.2015.03.009

Ceretta CA, Lorensini F, Brunetto G, Girotto E, Gatiboni LC, Lourenzi CR, Tiecher TL, Deconti L, Trentin G, Miotto A (2010) Frações de fósforo no solo após sucessivas aplicações de dejetos de suíno em plantio direto. Pesq Agropec Bras 45:593–602. https://doi.org/10.1590/S0100-204X2010000600009

Conti L, Ceretta CA, Ferreira PAA, Lorensini F, Lourenzi CR, Vidal RF, Tassinari A, Brunetto G (2015) Effects of pig slurry application and crops on phosphorus content in soil and the chemical species in solution. R Bras Ci Solo 39:774–787. https://doi.org/10.1590/01000683rbcs20140452

del Campillo MC, van der Zee SEATM, Torrent J (1999) Modelling long-term phosphorus leaching and changes in phosphorus fertility in excessively fertilized acid sandy soils. Eur J Soil Sci 50:391–399. https://doi.org/10.1046/j.1365-2389.1999.00244.x

Djodjic F, Bergström L, Ulén B, Shirmohammadi A (1999) Mode of transport of surface-applied phosphorus-33 through a clay and sandy soil. J Environ Qual 28:1273–1282. https://doi.org/10.2134/jeq1999.00472425002800040031x

Dotaniya ML, Datta SC, Biswas DR, Dotaniya CK, Meena BL, Rajendiran S, Regar KL, Lata M (2016) Use of sugarcane industrial by-products for improving sugarcane productivity and soil health. Int J Recycl Org Waste Agric 5:185–194. https://doi.org/10.1007/s40093-016-0132-8

Embrapa-Empresa Brasileira de Pesquisa Agropecuária (2011) Manual de Métodos de Análise de Solo. Embrapa Solos EMBRAPA—CNPS, Rio de Janeiro

Farias DR, Oliveira FHT, Santos D, Arruda JA, Hoffmann RB, Novais RF (2009) Fósforo em solos representativos do estado da Paraíba. I—Isotermas de adsorção e medidas do fator capacidade de fósforo. Rev Bras Cienc Solo 33:623–632. https://doi.org/10.1590/S0100-06832009000300015

Fink JR, Inda AV, Bavaresco J, Barrón V, Torrent J, Bayer C (2016) Adsorption and desorption of phosphorus in subtropical soils as affected by management system and mineralogy. Soil Till Res 155:62–68. https://doi.org/10.1016/j.still.2015.07.017

Galvão SRS, Salcedo IH (2009) Soil phosphorus fractions in sandy soils amended with cattle manure for long periods. Rev Bras Cienc Solo 33:613–622. https://doi.org/10.1590/S0100-06832009000300014

Galvão SRS, Salcedo IH, Oliveira FF (2008) Acumulação de nutrientes em solos arenosos adubados com esterco bovino. Pesq Agropec Bras 43:99–105. https://doi.org/10.1590/S0100-204X2008000100013

Gatiboni LC, Brunetto G, Kaminski J, Rheinheimer DS, Ceretta CA, Basso CJ (2008) Formas de fósforo no solo após sucessivas adições de dejeto líquido de suínos em pastagem natural. R Bras Ci Solo 32:1753–1761. https://doi.org/10.1590/s0100-06832008000400040

Guardini R, Comin JJ, Rheinheimer DS, Gatiboni LC, Tiecher T, Schmitt DE, Bender MA, Belli Filho P, Oliveira PAV, Brunetto G (2012) Phosphorus accumulation and pollution potential in a Hapludult fertilized with pig manure R Bras Ci Solo 36:1333–1342. https://doi.org/10.1590/s0100-06832012000400027

Guppy CN, Menzies NW, Blamey FPC, Moody PW (2005a) Do decomposing organic matter residues reduce phosphorus sorption in highly weathered soil? Soil Sci Soc Am J 69:1405–1411. https://doi.org/10.2136/sssaj2004.0266

Guppy CN, Menzies NW, Moody PW, Blamey FPC (2005b) Competitive sorption reactions between phosphorus and organic matter in soil: a review. Aust J Soil Res 43:189–202. https://doi.org/10.1071/SR04049

Kang J, Amoozegar A, Hesterberg D, Osmond DL (2011) Phosphorus leaching in a sandy soil as affected by organic and inorganic fertilizer sources. Geoderma 161:194–201. https://doi.org/10.1016/j.geoderma.2010.12.019

Lee Y, Oa SW (2013) Nutrient transport characteristics of livestock manure in a farmland. Int J Recycl Org Waste Agric 2:1. https://doi.org/10.1186/2251-7715-2-1

Mendonça ES, Matos ES (2005) Matéria orgânica do solo: Métodos de análise. Universidade Federal de Viçosa, Viçosa

Murphy J, Riley JP (1962) A modified single solution method for the determination of phosphate in natural water. Anal Chim Acta 27:31–36. https://doi.org/10.1016/S0003-2670(00)88444-5

Nascimento JT, Silva IF, Santiago RD, Silva Neto LF (2003) Efeito de leguminosas nas características químicas e matéria orgânica de um solo degradado. Rev Bras Eng Agr Amb 7:457–462. https://doi.org/10.1590/S1415-43662003000300008

Nelson NO, Parson JE, Mikkelsen RL (2005) Field-scale evaluation of phosphorus leaching in acid sandy soils receiving swine waste. J Environ Qual 34:2024–2035. https://doi.org/10.2134/jeq2004.0445

Novais RF, Smyth TJ, Nunes FN (2007) Fósforo. In: Novais RF, Alvarez VVH, Barros NF, Fontes RL, Cantaruti RB, Neves JCL (eds) Fertilidade do solo. SBCS, Viçosa, pp 471–550

Ojekami A, Ige D, Hao X, Akinremi O (2011) Phosphorus mobility in a soil with long term manure application. J Agric Sci 3:25–38. https://doi.org/10.5539/jas.v3n3p25

Oliveira FF, Salcedo IH, Galvão SR (2011) Adubação orgânica e inorgânica de batatinha em solos arenosos: produtividade, nutrientes na planta e lixiviação. Rev Bras Eng Agr Amb 15:1228–1234. https://doi.org/10.1590/S1415-43662011001200003

Pinto F A (2012) Sorção e dessorção de fósforo em solos de cerrado. Dissertation, Universidade Federal de Goiás

Piovesan RP, Favaretto N, Pauletti V, Motta ACV, Reissmann CB (2009) Perdas de nutrientes via subsuperfície em colunas de solo sob fertilização mineral e orgânica. Rev Bras Cienc Solo 33:757–766. https://doi.org/10.1590/S0100-06832009000400002

Roy B, Dzomback DA (1996) Colloid release and transport processes in natural and model porous media. Colloid Surface A 107:245–262. https://doi.org/10.1016/0927-7757(95)03367-X

Rubæk GH, Kristensen K, Olesen SE, Østergaard HS (2013) Phosphorus accumulation and spatial distribution in agricultural soils in Denmark. Geoderma 209–210:241–250. https://doi.org/10.1016/j.geoderma.2013.06.022

Sabourin E, Silveira LM, Tonneau JP, Sidersky P (2000) Fertilidade e agricultura familiar no Agreste Paraibano: Um estudo sobre o manejo da biomassa. CIRAD-TERRA/ASPTA, Esperança

Santos JS, Lima VLA, Júnior JCFB, Silva LVBD, Azevedo CAV (2010) Mobilidade de solutos em colunas de solo com água residuária doméstica de suinocultura. Rev Bras Eng Agr Amb 14:1226–1233. https://doi.org/10.1590/S1415-43662010001100013

Santos JCB, Souza Junior VSS, Corrêa MM, Ribeiro MR, Almeida MC, Borges LEP (2012) Caracterização de Neossolos Regolíticos da região semiárida do Estado de Pernambuco. Rev Bras Cienc Solo 36:683–695. https://doi.org/10.1590/S0100-06832012000300001

SAS. SAS/STAT 9.3 (2011) User's guide. SAS Institute, Cary

Schwertmann U (1964) Differenzierung der Eisenoxide des Bodens durch Extraktion mit saurer Ammoniumoxalat-Lösung. Z Pflanzenernaehr Bodenkd 105:194–202

Sharpley AN, Weld JL, Beegle DB, Kleinman PJA, Gburek WJ, Moore PA Jr, Mullins G (2003) Development of phosphorus indices for nutrient management planning strategies in the United States. J Soil Water Conserv 58:137–152

Sharpley AN, McDowell RW, Kleinnman JA (2004) Amounts, forms, and solubility of phosphorus in soils receiving manure. Soil Sci Soc Am J 68:2048–2057. https://doi.org/10.2136/sssaj2004.2048

Silva TO, Menezes RSC (2007) Adubação orgânica da batata com esterco e, ou, *Crotalaria juncea*. II—Disponibilidade de N, P e K no solo ao longo do ciclo de cultivo. Rev Bras Cienc Solo 31:51–61. https://doi.org/10.1590/S0100-06832007000100006

Sims JT, Simard RR, Joern BC (1998) Phosphorus loss in agricultural drainage: historical perspective and current research. J Environ Qual 27:227–293. https://doi.org/10.2134/jeq1998.0047242500 2700020006x

Snyder JD, Trofymow JA (1984) A rapid accurate wet oxidation diffusion procedure for determining organic and inorganic carbon in plant and soil sample. Commun Soil Sci Plant Anal 15:587–597. https://doi.org/10.1080/00103628409367499

Soares MR, Casagrande JC (2009) Adsorção e Modelos. In: Ribeiro MR et al (eds) Tópicos em Ciência do Solo. SBCS, Viçosa, pp 71–201

Soil Survey Staff (2014) Keys to soil taxonomy, 12th edn. USDA-Natural Resources Conservation Service, Washington, DC

Souto PC, Souto JS, Santos RV, Araújo GT, Souto LS (2005) Decomposição de estercos dispostos em diferentes profundidades em área degradada no semi-árido da Paraíba. Rev Bras Cienc Solo 29:125–130. https://doi.org/10.1590/S0100-06832005000100014

Systat Software (2006) SigmaPlot for windows. Version 10.0. Systat Software, San Jose

Tarkalson DD, Leytem AB (2009) Phosphorus mobility in soil columns treated with dairy manures and commercial fertilizer. Soil Sci 174:73–80. https://doi.org/10.1097/SS.0b013e31819750la

Tedesco MJ, Selbach PA, Gianello C, Camargo FA (1999) Resíduos orgânicos no solo e os impactos no ambiente. In: Santos GA, Camargo FA (eds) Fundamentos da matéria orgânica do solo: ecossistemas tropicais e subtropicais. Gênesis, Porto Alegre, pp 159–192

Thomas RL, Sheard RW, Moyer JR (1967) Comparison of conventional and automated procedures for nitrogen, phosphorus, and potassium analysis of plant material using a single digestion. Agron J 59:240–243. https://doi.org/10.2134/agronj1967.0002196200 5900030010x

Tiessen H, Abekoe MK, Salcedo IH, Owusu-Bennoah E (1993) Reversibility of phosphorus sorption by ferruginous nodules. Plant Soil 153:113–124

van der Zee SEATM, van Riemsdijk WH (1988) Model for long-term phosphate reaction kinetics in soil. J Environ Qual 17:35–41. https://doi.org/10.2134/jeq1988.00472425001700010005x

van der Zee SEATM, Fokkink LGJ, van Riemsdijk WH (1987) A new technique for assessment of reversibly adsorbed phosphate. Soil Sci Soc Am J 51:599–604. https://doi.org/10.2136/sssaj 1987.03615995005100030009x

Xavier FAS, Oliveira TS, Andrade FV, Mendonça ES (2009) Phosphorus fractionation in a sandy soil under organic agriculture in Northeastern Brazil. Geoderma 151:417–423. https://doi.org/10.1016/j.geoderma.2009.05.007

15

Effect of bagasse ash and filter cake amendments on wheat (*Triticum turgidum* L.var. *durum*) yield and yield components in nitisol

Assefa Gonfa[1] · Bobe Bedadi[2] · Anteneh Argaw[2]

Abstract

Purpose This experiment was initiated to evaluate the effect of filter cake and bagasse ash on the productivity of wheat under greenhouse conditions.

Methods Six levels of filter cake and bagasse ash each separately and control check were arranged in a completely randomized design with three replications.

Results At 100 tons/ha, both inputs resulted in the highest values of all the investigated traits. It was also found that yield and yield components that were obtained from bagasse ash overwhelmed those from filter cake treatments, except in tillers, dry biomass, and straw yield. Linear regression analysis revealed a significant and positive relationship between grain yield with that of total N, P, K, S, Ca, Mg, Cu, and Zn uptake. A linear relationship between the grain yield with that of N and Zn uptake was found, while the association between grain yield with total P, K, S, Ca, Mg and Cu uptake was quadratic.

Conclusion It can be concluded that filter cake and bagasse ash are good sources of nutrients to enhance wheat yield in acidic soil. To give a conclusive result, these inputs should be tested in field trials in different soil types.

Keywords Bagasse · Ethiopia · Filter cake · Soil acidity · Subsistence farmers

Introduction

Durum wheat (*Triticum turgidum* L var. durum) is the most important cereal crop grown in Ethiopia. Ethiopia is believed to be the center of genetic diversity for wheat (Tesfaye and Getachew 1991). It is primarily produced by the smallholder farmers under rainfed conditions at altitudes ranging from 1800 to 2800 meters above sea level (Tesfaye 1987). Prior to 1954, Ethiopia was an exporter of durum wheat grain and flour (Pinto 1971). However, consumer demand presently far exceeds domestic production and the country is even importing wheat products for the pasta industry.

Although many efforts have been made to develop high-yielding variety of wheat in Ethiopia, the productivity of wheat is still far below the potential yield reported elsewhere as a result of soil quality depletion and low soil fertility (Harden 2001; Lal 2001). The deficiency of some micronutrient besides widespread N and P (and localized K) deficiencies in major wheat-growing areas of Ethiopian has been recently reported (Hailu et al. 2015). These problems are aggravated by continuous nutrient depletion due to the complete harvest of whole plant biomass and losses in runoff and soil erosion (Vlek 1993; Sanchez et al. 1997), which have led to negative nutrient balances (Stoorvogel and Smaling 1990). Crop yield per hectare has eventually decreased (Vierich and Stoop 1990). As a result, most of the soil in eastern Africa is less responsive as well as non-responsive to macronutrients like N (Bationo et al. 1986).

Agro-ecology-based production systems have promoted sustainable agricultural production systems (Altieri 1995; Gliessman 1998). Agricultural practices based on organic amendments are strongly recommended for tropical agroecosystems. The application of organic inputs to the soil is increasing as both an environmentally favorable and as a means of improving soil organic matter (SOM) content in low-fertility soils. Organic leftovers, namely, plant biomass, manure, and waste from urban, industrial, and agricultural activity are identified as valuable natural resources that can

✉ Anteneh Argaw
antenehargaw@gmail.com

[1] Ethiopia Institute of Agricultural Research, Debre Zeit Agricultural Research Center, Debre Zeit, Ethiopia

[2] College of Agriculture and Environmental Sciences, School of Natural Resources Management and Environmental Sciences, Haramaya University, Haramaya, Ethiopia

be effectively used for sustainable production and ecosystem restoration. Such organic inputs are useful inputs for the potential restoration of soil organic matter and increase inherent soil nutrient supply capacity (Fernandez-Luqueno et al. 2009; Nurhidayati et al. 2018) since manures are often scarce like in Ethiopia (Lupwayi et al. 2000). Organic by-products originating from industrial processes represent an important source of nutrients for integrated soil fertility management.

Organic amendment could increase the soil P supplying capacity in acidic soils through increasing P mobility in the soil by chelating the soil Al and Fe, thereby enhancing P acquisition by stimulation of root growth achieved by organic amendment (Kretzschmar et al. 1991). The organic amendment may also increase the soil pH, the contents of exchangeable cations (K, Ca, Mg) and the base saturation (Kretzschmar et al. 1991; Geiger et al. 1992). Maize grain yields tended to increase with increasing contents of soil C, total N, extractable P, and exchangeable bases as result of organic input application (Tittonell et al. 2007).

After combustion, most of the inorganic nutrients in biomass ash could be a significant source of P, K, Mg and Ca (Bougnom and Insam 2009; Bougnom et al. 2009). Biomass ash application could increase soil pH and stimulated bacterial activities that in turn increase decomposition rate of the humus layer, N mineralization, and N availability to plants (Fritze et al. 1994; Kreutzer 1995; Haimi et al. 2000). A significant increase in growth and yield of different crops due to biomass ash application has been reported (Odiete et al. 2005; Adekayode and Olojugba 2010; Gupta et al. 2010). For instance, biomass ash application resulted in a better increase in grain yields of barley (*Hordeum vulgare* L.), canola (*Brassica rapa* L.), and pea (*Pisum sativum* L.) when compared to the effect of agricultural lime have been reported (Arshad et al. 2012). However, the comparative advantage of filter cake and bagasse ash application on the productivity of wheat in acidic nitisol soil is not well known. Hence, the aim of this work was to evaluate the effect of filter cake and bagasse ash application on yield of wheat and to evaluate the relationship between the nutrients uptake with grain yield of wheat.

Materials and methods

A greenhouse experiment was performed on a silty clay soil classified as Eutric Nitisols, collected from the Endibir area (Central Ethiopia). For soil collection, the fields of 20 farmers were purposely selected based on the extent of soil acidity. Soil samples were collected from 0 to 20 cm depth at random from the fields of these 20 farmers and mixed well. The dried soil sample was passed through a 4-mm sieve to eliminate coarse rock and plant material and thoroughly mixed to ensure uniformity. A subsoil sample of about 1 kg was taken, air dried, and passed through a 2-mm sieve and used for the determination of physical and chemical characteristics. A part of the homogenized soil sample was ground to pass through a 0.5-mm sieve for the analysis of N, organic matter (OM), Cu, Zn, Fe and Mn concentrations.

The soil was analyzed using standard procedure compiled by Sertsu and Bekele (2000). The soil had an average pH of 5.1 in a 1:2.5 soil-to-water ratio. The chemical composition of the soil before amendment contained 3.4% OM, 0.16% total N, 10 mg/kg available P, 35.18 mg/kg available sulfur; 0.54 meq/100 g soil of exchangeable acidity (Al + H); 44 mg/kg of Fe; 154 mg/kg of Mn; 0.56 mg/kg of Cu; 2.40 mg/kg of Zn, and 0.25, 1.89, 5.04 and 2.80 cmol (+)/kg exchangeable Na, K, Ca and Mg, respectively. The textural composition of the soil was 9% sand, 45.5% silt and 45.5% clay with the bulk density of 1.27 gm/cm^3. The volumetric soil water contents of the soil were 32 and 18% and 14% for field capacity, available water, and permanent wilting point, respectively.

Samples of both filter cake and bagasse ash were air dried and screened through a 2-mm sieve. Samples were sealed in polyethylene plastic containers until use for analysis or in the experiments. Triplicate samples from both filter cake and bagasse ash were analyzed for the selected chemical composition according to methods described for soil analysis. The chemical composition of both inputs is indicated in Table 1.

The pot experiments were conducted on wheat under greenhouse conditions using cleaned plastic pots of 24 cm height and 25 cm width. Each pot contained 4 kg of soil. There were 11 treatments comprising 5 levels of filter cake application (20, 40, 60, 80 and 100 tons/ha) and 5 levels of bagasse ash application (20, 40, 60, 80 and 100 tons/ha) separately, and the control check with 4 replications. Four replicates per treatment were established in a completely

Table 1 Chemical properties of filter cake and bagasse ash

Parameter	Filter cake	Bagasse ash
pH (H_2O)	8.0	10.2
EC (ds/m)	1.70	3.90
OM (%)	36.2	3.6
N (%)	1.77	0.17
Available P (mg/kg)	1.62	0.85
Available S (mg/kg)	1.67	1.29
K (%)	0.16	0.05
Ca (%)	1.3	1.45
Mg (%)	0.31	0.70
Fe (mg/kg)	3	3
Mn (mg/kg)	39	19
Cu (mg/kg)	1.24	0.40
Zn (mg/kg)	31.00	10.50

randomized design (CRD). A total of 44 experimental units were used during the experiment.

The pot plants were harvested when physiologically matured. Samples of grain and straw collected at harvesting were oven-dried in an oven at 65 °C till a constant weight was obtained and then ground and passed through a 0.5-mm screen for chemical analysis. The uptake of selected macronutrients and micronutrients was calculated by multiplying the concentration of each nutrient by the dry weight expressed in kg/plant.

Agronomic data were collected and assessed to evaluate the effect of filter cake and bagasse ash on durum wheat. The agronomic parameters and yield and yield components determined are number of tillers, grain yield (GY), aboveground biomass, harvest index (HI), spike length, number of kernels per spike, number of spikelet per spike, and 1000 kernel weights of durum wheat. Plant root length was also taken by flushing with tap water to assess the effect of filter cake and bagasse ash on root length as it was affected by soil acidity.

All analyses were carried out on the four replicates. Treatment effects were analyzed by one-way analysis of variance (ANOVA). Data analysis was performed with SAS version 9.0. Data on soil, yield and yield traits of the crop were compared at a significance level of 0.05. The means were separated by Tukey's test. Linear regression analysis of grain yields against nutrients uptake was performed to determine trends. The probability values of the slope indicated the level of significance of observed changes in yield.

Table 2 ANOVA data of filter cake treated agronomic parameters, yield and yield components

Parameters	*F* value	*P* value
PLH	159.27	<0.0001
TIL	20.93	<0.0001
SPL	18.93	<0.0001
SPLS	2.78	<0.0001
KPS	167.08	<0.0001
TKW	34.48	<0.0001
DBM	43.16	<0.0001
SY	9.85	0.0003
GY	266.57	<0.0001

PLH plant height; *TIL* tillers; *SPL* spike length; *SPLS* spikelet per spike; *KPS* kernels per spike; *TKW* thousand kernels weight; *DBM* dry biomass; *SY* straw yield; *GY* grain yield

Table 3 ANOVA data of bagasse ash treated durum wheat agronomic parameters and yield and yield components

Parameters	*F* value	*P* value
PLH	117.16	<0.0001
TIL	11.16	<0.0001
SPL	12.0	<0.0001
SPLS	56.78	<0.0001
KPS	5.79	0.0036
TKW	885.29	<0.0001
DBM	85.1	<0.0001
SY	10.01	0.0002
GY	1.58	0.0001

PLH plant height, *TIL* tillers, *SPL* spike length per spike, *SPLS* spikelet per spike, *KPS* kernels per spike, *TKW* thousand kernels per spike, *DBM* dry biomass, *SY* straw yield, *GY* grain yield

Results and discussion

Days of seedling emergence, heading, and physiological maturity

The ANOVA showed that filter cake and bagasse ash application did not significantly ($P > 0.05$) affect days to 50% seedling emergence but these amendments significantly influenced days to 50% heading and 90% physiological maturity (Tables 2 and 3). At early vegetative growth stage, when the rate of filter cake increased, the treatments which received high rates of filter cake (80 and 100 tons/ha) showed slowness in growth (Table 4). The delaying of wheat growth at early vegetative stage could be due to immobilization of nutrients (Butler et al. 2001). At high rates of filter cake application, filter cake enhances the microbial growth (Kaur et al. 2008) and favors immobilization (Tejada et al. 2006) though the C: N ratio < 15, as indicated in Table 1, should not occur immobilization (Powlson et al. 2001). In line with the present study, filter cake application enhanced N immobilization (Rasul et al. 2009). The rapid immobilization due to filter cake application might be due to the decomposition of an easily available carbohydrate fraction that may induce utilization of soil N by the soil microbial communities (Lentz and Ippolito 2012), thereby reducing the soil inorganic N (Khan et al. 2008). However, the plant that received high rates as well as low rates at 20 and 40 tons/ha of filter cake gradually showed good improvement over the control at a late stage of the plant.

Delaying in the days to 50% heading was also observed when the rates of filter cake increased. Likewise, days to physiological maturity was increased when filter cake application increased. The release of high N from the transiently immobilized N by the soil microbial biomass could be the cause of long days of 90% physiological maturity. Ellis and Cock (1992) found that excess N keeps vegetative growth, which results in delayed flowering and maturity. Bagasse ash treatments also recorded maximum days to 50% heading

Table 4 Effect of filter cake and bagasse ash application on durum wheat phenology

Treatment rates (t/ha)	Days to 50% emergence	Days to 50% heading	Days to 90% maturity
Filter cake			
0	4	60a	88e
20	4	54e	87e
40	4	55d	91d
60	4	56cd	95c
80	4	56cb	97b
100	4	57b	101a
LSD (0.05)	ns	1.02	1.15
CV (%)		1.20	0.82
Bagasse ash			
0	4	60a	88.0a
20	4	53b	85.5c
40	4	53b	86.0b
60	4	52d	86.5ba
80	4	51d	86.0b
100	4	48e	85.5c
LSD (0.05)	ns	0.98	0.54
CV (%)		1.23	0.44

Values along the column with the same letter are not significantly different at $\alpha=0.05$

(Table 3). When compared with filter cake, bagasse ash received plants showed lesser days to heading and physiological maturity. This difference may be due to more availability of plant nutrients at a reasonable time for bagasse ash application and having low nitrogen contents.

Root length of durum wheat

The data indicated that root length of experimental durum wheat was significantly and positively affected by filter cake and bagasse ash application (Table 5). Increasing rate of both inputs resulted in an increase in the root length of durum wheat. The highest value (31.2 and 28 cm) was obtained at 100 and 60 tons/ha of filter cake and bagasse ash application, respectively. The lowest values were recorded at the control. From this, it could be concluded that filter cake application increased root length more than bagasse ash. This may be due to high Ca and OM content of filter cake that suppresses the availability of toxic elements such as (Al^{3+}) which affect root growth through precipitation and complex formation (Munoz et al. 1990). Calcium displaces Al from exchange sites and makes it form precipitation in the form of $Al(OH)_3$ (Dee et al. 2002). Besides this, the filter cake could improve the soil physical properties and this may, in turn, facilitate the root growth.

Effect of filter cake and bagasse ash application rates on agronomic parameters and yield and yield components of durum wheat

Yield and yield components are soil fertility indicators and their positive response indicates the soil fertility improvement. Analysis of variance indicated that filter cake and bagasse ash rates application significantly affected the yield and yield components of wheat (Tables 2 and 3).

Plant height, dry biomass, and number of tillers

All treatments had shown positive response over the control in increasing plant height (Table 6). Dry biomass was significantly increased due to filter cake and bagasse ash application (Table 5). The maximum value was obtained with filter cake and bagasse ash application at 100 and 60 tons/ha, respectively, while the minimum value was recorded at the control treatment. The finding corresponds to that of Jamil et al. (2004) that filter cake and bagasse application positively affected durum wheat dry biomass. In general, increasing the application rates of filter cake increased dry biomass. However, increasing bagasse ash rates did not bring significant differences among treatments in dry biomass except that all treatments brought higher values as compared to the control. These different effects on biomass production between two inputs investigated might be because of the presence of readily available nutrients in bagasse ash.

The numbers of tillers per plant were significantly increased due to filter cake application but inconsistently (Table 6). The lowest value was recorded at control. Due to high N and P content in filter cake, it induced a better number of tillers and dry biomass when compared with bagasse ash application.

Table 5 Effect of filter cake and bagasse ash on experimental wheat root length

Treatment (t/ha)	0	20	40	60	80	100	Mean	LSD (5%)	CV%
R. length, FC	16.5e	30c	29d	31bc	31.1ba	31.2a	28.08	0.72	1.70
R. length, BA	16.5d	22.3c	25b	28a	25b	25b	23.63	0.61	1.70

Values with the same letter are not significantly different, t/ha, ton per hectare

FC filter cake, *R. length* root length (cm), *FC* filter cake, *BA* bagasse ash

Table 6 Effect of filter cake and bagasse ash application on wheat yield and yield components

Treatment rates (t/ha)	PLH (cm)	TIL (No)	SPL (cm)	SPLS (No)	KPS (No)	TKW (g)	DBM (g)	SY (g)	GY (g)	HI
Filter cake										
0	54.0f	1.2c	3.1c	7.4c	3.1e	42.2c	6.8d	5.9b	0.9e	0.15d
20	60.9e	2.0b	3.8b	8.8b	10.1d	49.6ba	9.6c	6.5b	3.1d	0.37c
40	64.5d	2.8a	4.0ba	8.9a	13.2c	50.8a	10.5c	6.0b	4.6c	0.47a
60	68.6c	2.8a	4.1ba	8.9a	14.4b	48.7b	13.6b	8.7a	4.9c	0.41bc
80	72.5b	2.7a	4.1ba	8.8a	15.3b	48.8b	13.2b	8.0a	5.2b	0.45ba
100	75.0a	2.7a	4.4a	8.8a	17.2a	49.4ba	14.7a	8.9a	5.8a	0.45ba
Mean	66.0	2.34	3.89	8.46	12.18	48.3	11.39	7.32	4.09	0.38
LSD (0.05)	1.85	0.43	0.30	0.30	1.18	1.48	1.36	1.33	0.33	0.05
CV (%)	1.86	12.15	5.14	2.38	6.41	2.04	7.91	12.02	5.38	8.6
Bagasse ash										
0	54.0	1.2c	3.1d	7.4c	3.1d	42.6e	6.8b	5.9e	0.9d	0.15d
20	64.9e	2.5a	3.9c	9.13b	12.2c	50.15d	10.9a	7.1c	3.9c	0.41c
40	67.8d	2.6a	4.1bc	8.88b	14.1b	51.4c	12.1a	7.6a	4.6b	0.43bc
60	71.9c	1.7b	3.9c	9.08b	15.1ba	58.5b	12.4a	7.3b	5.1a	0.47ba
80	75.4b	1.45cb	4.5ba	9.53a	15.8a	60.7a	10.6a	5.4f	5.2a	0.50ba
100	79.5a	1.65b	4.6a	9.50a	16.7a	58.7b	11.7a	6.7d	5.3a	0.56a
Mean	68.95	1.84	4.00	8.85	12.81	53.7	10.79	6.52	4.17	0.42
LSD (0.05)	2.52	0.44	0.46	0.37	1.61	0.71	1.96	0.13	0.42	0.09
CV (%)	2.43	15.81	7.65	2.79	8.46	0.86	12.08	1.31	6.67	14.14

Values with the same letter are not significantly different at $\alpha = 0.05$

PLH plant height; *TIL* tillers; *SPL* spike length; *SPLS* spikelet per spike; *No* number; *KPS* kernels per spike; *TKW* thousand kernels weight; *DBM* dry biomass; *SY* straw yield; *GY* grain yield; *No* number/counting; *HI* harvest index

Spike length, spikelet per spike and kernels per spike

The spike length, spikelet per spike and kernels per spike of durum wheat significantly increased as both inputs increased (Table 6). The maximum values were found at 100 tons/ha with both inputs whereas the minimum at the control treatment. Bagasse ash application resulted in a better mean value of spike length, spikelet per spike and kernels per spike over filter cake applications though higher tiller number was obtained with plants treated filter cake than bagasse ash. This might have been related to the high availability of nutrients in bagasse ash treated plants besides maintaining soil pH at an acceptable level (Rautaray et al. 2003). These authors showed that high K, P, Ca and Mg nutrients uptake due to high available nutrients delivered through fly ash integrated with chemical and organic fertilizer leads to higher yield in rice. Other authors found the increase in grain yield of rice by ash application as result of enhancing the nutrient uptake (Mittra et al. 2005).

Thousand kernels weight and straw yield

Thousand kernels weight is an important yield-determining component, which is reported to be a genetic characteristic of a plant and influenced sparsely by the environmental factors (Ayoub et al. 1994). It was found that both inputs significantly increased 1000 kernels weight (Table 6). At 40 and 80 tons/ha, filter cake and bagasse ash application resulted in the highest kernel weight, respectively. The lowest value was recorded at control. The mean values for 1000 kernels weight of durum wheat, after the wheat treated with filter cake and bagasse ash application, were 48.3 and 53.7 g, respectively. However, relatively higher 1000 kernels weight was obtained with filter cake when compared with bagasse ash treatment. This could be related to aphid and powdery mildew infection on the plants that received high rates of filter cake at maturity stage. High incidence of disease in filter cake amended plants might be due to the high N in the soil that may cause the plant to be more succulent and thus lead to susceptibility of the plants for disease.

The straw yield is another input response as well as an indicator of soil fertility status. The straw yield was significantly increased by filter cake and bagasse ash application (Table 6). The maximum value was recorded at 100 and 40 tons/ha for filter cake and bagasse ash application, respectively. The straw yield increased consistently with increasing application rate of filter cake but this trend along bagasse ash application rate was inconsistent.

Grain yield and harvest index

The results of statistically analyzed data revealed that the filter cake and bagasse ash application regardless of the rates significantly increased the grain yield when compared to the control treatment (Table 6). The highest value was found at highest rates of application. The minimum values were recorded at the control treatment. The yield increment due to applying such input could be attributed to the increase in available nutrients and suitable soil condition to beneficial microbial proliferation and activities around the plant root zone (Demeyer et al. 2001; Rasul et al. 2008). Other studies showed that 60 tons/ha of filter cake application increased the yield of cassava by 50% over the mineral fertilization (Ossom 2010). Ossom and Dlamini (2012) found the comparable amount of maize grain yield by filter cake and recommended chemical fertilizer application. In general, wheat grain yield consistently increased as the rate of applied filter cake increased to the highest level but showed the tendency of stopping increment after 60 tons/ha for bagasse ash application. In previous studies, Thind et al. (2012) found that 10 tons/ha increased the yield of wheat by up to 43.7% over the unfertilized plants in the alkaline loamy sand.

Harvest index (HI) was significantly increased by filter cake and bagasse ash application (Table 6). The highest values (0.45 and 0.56) were recorded at 100 tons/ha with filter cake and bagasse ash application, respectively, while the lowest values were obtained from control treatment. The value we have obtained in the present study after treatments, when compared with Hussians and Pan (1993) reported on average 0.34 for local cultivars and 0.44 for improved new varieties are high. Therefore, the highest values of harvest index (HI) in this study coincide with improved new varieties that were reported by Hussians and Pan (1993) as the experimental wheat was variety released from Debre Zeit Agricultural Research Center (DZARC) with the name "Yerer". In general, it was found that plants that were treated bagasse ash application had better HI when compared those treated filter cake applications.

Regression analysis between grain yield of wheat and nutrients uptake

The regression analysis between the grain yield and total uptake of selected macronutrients and micronutrients were significant at $P<0.05$ (Fig. 1a). Association between grain yield with N and Zn uptake was linear, implying that further increase in N and Zn uptake by applying inorganic fertilizer could improve the grain yield of wheat. This could be suggested that N and Zn the major limiting nutrients for wheat cultivation in the soil we used. This result indicates that organic fertilizer alone has not sufficiently supplied N to the plant (Herencia et al. 2007). With another nutrient uptake (P, S, Ca, Mg and Cu), grain yields exhibited quadratic relationship. This result could suggest that the rates below 100 tons/ha could be sufficient, although further investigations under field condition are needed.

A significant ($P<0.05$, $r^2=0.95$) association between grain yield and total N uptake with higher slope (2.783) for bagasse ash amended soils than the soil amended filter cake was demonstrated, implying the better effect of bagasse ash application on the total N uptake of wheat. The result might have been related with immobilization of native N by the high rate of filter cake application (Rautaray et al. 2003) and increase in N uptake by correcting the soil nutrients constraints through bagasse ash application (Aulakh and Malhi 2005). Ash application could also increase the decomposition of native soil organic matter through improving the microbial proliferation (Kaur et al. 2008) and activity (Lal et al. 1996), thereby improving the N availability (Wong and Wong 1986).

The association between grain yield and total P uptake were quadratic and significant ($P<0.05$, $r^2=0.96$) with both inputs (Fig. 1b). In case of total K uptake and grain yield was significant ($P<0.05$, $r^2=0.94$) and quadratic with a greater coefficient of the x-axis and y-intercept in plants that was amended bagasse ash than those amended filter cake (Fig. 1c). This result may indicate that bagasse ash could be a good source K for plants. Similar studies have shown that a strong association between K uptake and grain yield of wheat (Thind et al. 2012). Ash application increased the K availability in sandy loam acid lateritic soil (Rautaray et al. 2003). It was also found a significant ($P>0.05$, $r^2=0.96$) and the quadratic relationship between grain yield and total S uptake (Fig. 1d).

In bagasse ash and filter cake amended soils, a linear and quadratic relationship between grain yield and total Ca uptake were found, respectively (Fig. 1e). This figure also indicated that the higher y-intercept was recorded in filter cake amended soil than bagasse ash amended soils. This result could be related to the presence of high Ca in filter cake compared to bagasse ash. Quadratic association ($P>0.05$, $r^2=0.97$) between grain yield and total Mg uptake with a greater coefficient of x-axis and y-intercept was recorded when plants were treated with bagasse ash as compared with those treated with filter cake (Fig. 1f). This suggests that bagasse ash could be the good source of Mg when compared to the filter cake.

The relationship between grain yield and total Cu uptake was quadratic with a higher coefficient of the x-axis and y-intercept with plants treated bagasse ash than those with filter cake (Fig. 1g). Grain yield and total Zn uptake showed the linear association with higher y-intercept with bagasse ash amended plants than the filter cake amended plants (Fig. 1h), implying that bagasse ash is a good source of Zn and Cu when compared to the filter cake. Similar

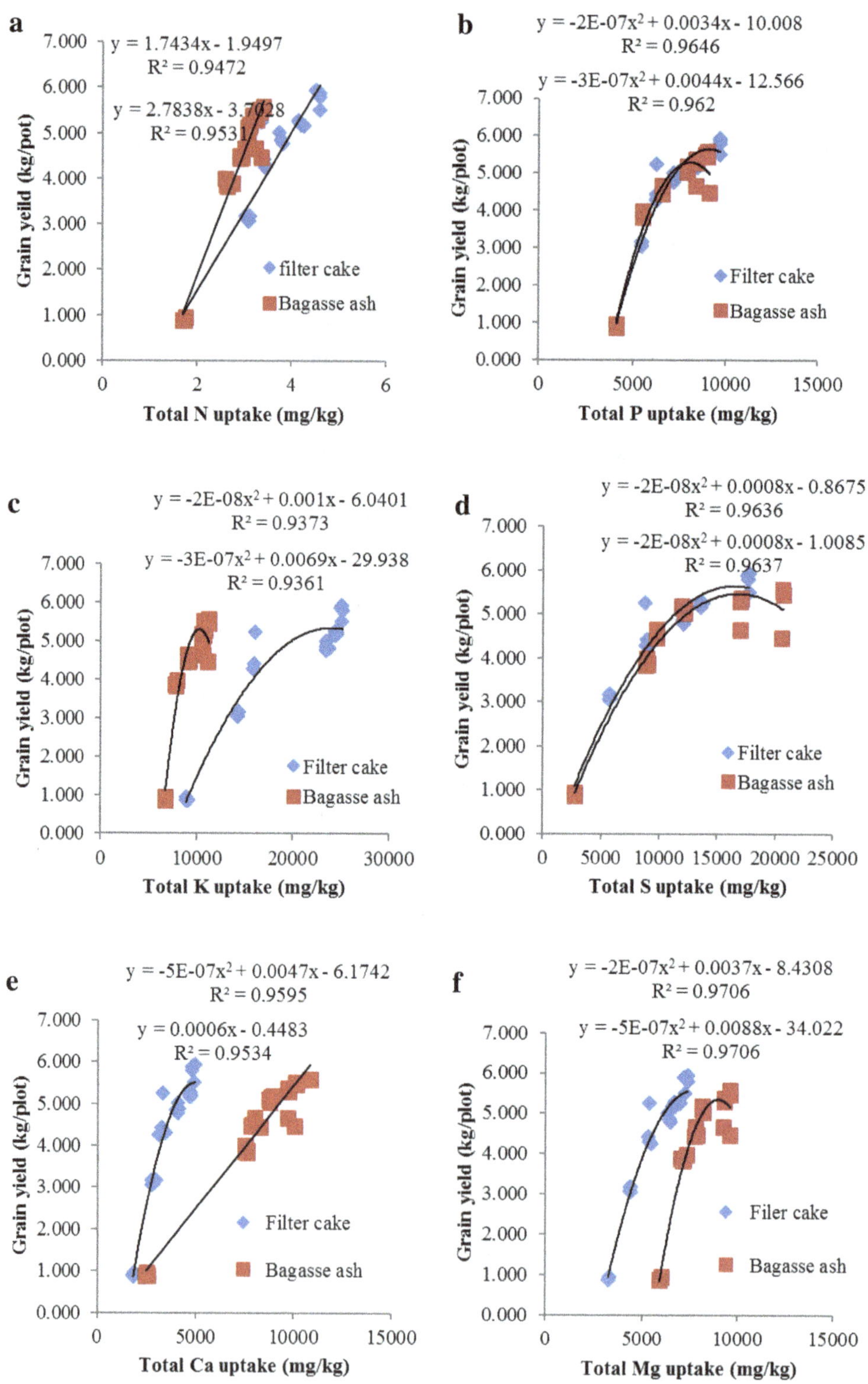

Fig. 1 Effect of filter cake and bagasse ash application rates on **a** Total N uptake, **b** Total P uptake, **c** Total K uptake, **d** Total S uptake, **e** Total Ca uptake, **f** Total Mg uptake, **g** Total Cu uptake and **h** Total Zn uptake of wheat under greenhouse condition

findings have been reported by Lentz and Ippolito (2012) who demonstrated that organic fertilizer with high nutrients content increased Cu, Zn, S, Mg, K, and N concentration of silage corn by 1.3 fold when compared to the control.

Conclusion

In general, due to immobilization of nutrient when the applied high rate of the filter cake, incorporating filter

Fig. 1 (continued)

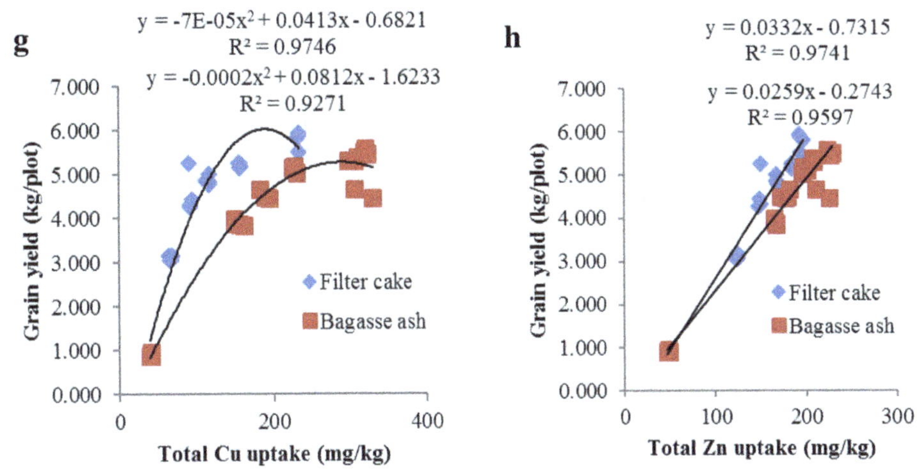

cake to the soil 1–2 months before sowing and undertaking decomposition before application to the soil is recommended. The result also found the need for relatively lower rate of bagasse ash when compared with filter cake to enhance wheat production in acidic and low fertile soils. To give a conclusive result for the end user and to fine-tune the rate for practical implementation, further research work under field conditions is needed. In general, the awareness of amending the acidic soil with filter cake and bagasse ash application, besides lime application, could correct the soil pH into acceptable range and improve the soil nutrients as well.

Acknowledgements The authors would like to acknowledge Ethiopian Institute of Agricultural Research (EIAR) for financial support, Matahara sugar Estate Farm for free support of organic inputs (filter cake and bagasse ash), and Haramaya University (HU) for facility support during my stay in the University Campus. Our acknowledgement also goes to Debre Zeit Agricultural Chemistry Laboratory for its full support in laboratory analytical service.

References

Adekayode FO, Olojugba MR (2010) The utilization of wood ash as manure to reduce the use of mineral fertilizer for improved performance of maize (*Zea mays* L.) as measured in the chlorophyll content and grain yield. J Soil Sci Environ Manag 1:40–45

Altieri MA (1995) Agroecology: the science of sustainable agriculture. Ann Arbor Press, Michigan

Arshad MA, Soon YK, Azooz RH, Lupwayi NZ, Chang SX (2012) Soil and crop response to wood ash and lime application in acidic soils. Agron J 104:715–721. https://doi.org/10.2134/agronj2011.0355

Aulakh MS, Malhi SS (2005) Interactions of nitrogen with other nutrients and water: effect on crop yield and quality, nutrient use efficiency, carbon sequestration, and environmental pollution. Adv Agron 86:342–409. https://doi.org/10.1016/S0065-2113(05)86007-9

Ayoub M, Guertin S, Lussier S, Smith DL (1994) Timing and levels of nitrogen fertility effects on spring wheat. Crop Sci 34:748–750. https://doi.org/10.2135/cropsci1994.0011183X003400030027x

Bationo A, Mughogho SK, Mokwunye AU (1986) Agronomic evaluation of phosphate fertilizer in tropical Africa. In: Mokwunye AU, Vlek PLG (eds) Management of nitrogen and phosphorus fertilizers in Sub-Saharan Africa developments in plant and soil sciences, 24th edn. Martinus Nijhoff Publishers, Dordrecht, pp 283–318

Bougnom BP, Insam H (2009) Ash additives to compost affect soil microbial communities and apple seedling growth. Bodenkultur 60:5–15

Bougnom BP, Mair J, Etoa FX, Insam H (2009) Composts with wood ash addition: a risk or a chance for ameliorating acid tropical soils? Geoderma 153:402–407. https://doi.org/10.1016/j.geoderma.2009.09.003

Butler TA, Sikora LJ, Steinhilber PM, Douglass LW (2001) Compost age and sample storage effects on maturity indicators of biosolids compost. J Environ Qual 30(6):2141–2148. https://doi.org/10.2134/jeq2001.2141

Dee BM, Haynes RJ, Meyer JH (2002) Sugar mill wastes can be important soil amendments. Proc S Afr Sugarcane Technol Assoc 76:51–60

Demeyer A, Nkana JCV, Verloo MG (2001) Characteristics of wood ash and influence on soil properties and nutrient uptake: an overview. Bioresour Technol 77:287–295. https://doi.org/10.1016/S0960-8524(00)00043-2

Ellis BG, Cock RL (1992) Soil management a world view of conservation. Krieger publish company, Malabar

Fernandez-Luqueno F, Reyes-Varela V, Martinez-Suare ZC, Reynoso-Keller RE, Mendez-Bautista J, Ruíz-Romero E, Lopez-Valdez F, Luna-Guido ML, Dendooven L (2009) Emission of CO_2 and N_2O from soil cultivated with common bean (*Phaseolus vulgaris* L.) fertilized with different N sources. Sci Total Environ 407:4289–4296. https://doi.org/10.1016/j.scitotenv.2009.04.016

Fritze H, Smolander A, Levula T, Kitunen V, Mälkönen E (1994) Wood ash fertilization and fire treatments in a Scots pine forest stand: effects on the organic layer, microbial biomass and microbial activity. Biol Fertil Soils 17:57–63

Geiger SC, Manu A, Bationo A (1992) Changes in a sandy Sahelian soil following crop residue and fertilizer additions. Soil Sci Soc

Am J 56:172–177. https://doi.org/10.2136/sssaj1992.0361599500 5600010027x

Gliessman SR (1998) Agroecology: ecological process in sustainable agriculture. Ann Arbor Press, Michigan

Gupta AK, Mishra RK, Sinha S, Lee BK (2010) Growth, metal accumulation and yield performance of *Brassica campestris* L. (cv. *Pusa Jaikisan*) grown on soil amended with tannery sludge/fly ash mixture. Ecol Eng 36:981–991. https://doi.org/10.1016/j.ecole ng.2010.04.003

Hailu H, Mamo T, Keskinen R, Karltun E, Gebrekidan H, Bekele T (2015) Soil fertility status and wheat nutrient content in Vertisol cropping systems of central highlands of Ethiopia. Agric Food Secur 4:19. https://doi.org/10.1186/s40066-015-0038-0

Haimi J, Fritze H, Moilanen P (2000) Responses of soil decomposer animals to wood-ash fertilization and burning in a coniferous forest stand. Forest Ecol Manag 129:53–61. https://doi.org/10.1016/s0378-1127(99)00158-9

Harden CP (2001) Soil erosion and sustainable mountain development: experiments, observations, and recommendations from the Ecuadorian Andes. Mt Res Dev 21:77–83. https://doi.org/10.1659/0276-4741(2001)021[0077:seasmd]2.0.co;2

Herencia JF, Ruiz-Porras JC, Melero S, Garcia-Galavis PA, Morillo E, Maqueda C (2007) Comparison between organic and mineral fertilization for soil fertility levels, crop macronutrient concentrations, and yield. Agron J 99:973–983. https://doi.org/10.2134/agronj2006.0168

Hussians DR, Pan WL (1993) Nitrogen efficiency components analysis: an evaluation of cropping systems difference in productivity. Agron J 85:898–905. https://doi.org/10.2134/agronj1993.00021 962008500040022x

Jamil M, Qasim M, Umar M, Subhan Abdus (2004) Impact of organic wastes (bagasse ash) on the yield of wheat (*Triticum aestivum* L.) in a calcareous soil. Int J Agri Biol 6(3):468–470

Kaur K, Goyal S, Kapoor KK (2008) Impact of organic fertilizers with and without chemical fertilizers on soil chemical properties and the establishment of nitrogen-fixing bacteria in the rhizosphere. Microbes Environ 23(4):313–316

Khan KS, Gattinger A, Buegger F, Schloter M, Joergensen RG (2008) Microbial use of organic amendments in saline soils monitored by changes in the $^{13}C/^{12}C$ ratio. Soil Biol Biochem 40:1217–1224. https://doi.org/10.1016/j.soilbio.2007.12.016

Kretzschmar RM, Hafner H, Bationo A, Marschner H (1991) Long- and short-term effects of crop residues on aluminum toxicity, phosphorus availability and growth of pearl millet in an acid sandy soil. Plant Soil 136:215–223

Kreutzer K (1995) Effects of liming on soil processes. Plant Soil 168–169:447–470

Lal R (2001) World crop land soils as source or sink for atmospheric carbon. Adv Agron 71:145–191. https://doi.org/10.1016/s0065 -2113(01)71014-0

Lal JK, Mishra B, Sarkar AK (1996) Effect of fly ash on soil microbial and enzymatic activity. J Indian Soc Soil Sci 44:77–80. https://doi.org/10.1016/0143-1471(86)90080-2

Lentz RD, Ippolito JA (2012) Biochar and manure affect calcareous soil and corn silage nutrient concentrations and uptake. J Environ Qual 41:1033–1043. https://doi.org/10.2134/jeq2011.0126

Lupwayi NZ, Girma M, Haque I (2000) Plant nutrient content of cattle manures from smallscale farms and experimental stations in the Ethiopian highlands. Agr Ecosyst Environ 78:57–63

Mittra BN, Karmakar S, Swain DK, Ghosh BC (2005) Fly ash a potential source of soil amendment and a component of integrated plant nutrient supply system. Fuel 84:1447–1451. https ://doi.org/10.1016/j.fuel.2004.10.019

Munoz MA, Colberg O, Dumas JA (1990) Chicken manure as an organic fertilizer. J Agric Univ P R 74:139–144

Nurhidayati N, Machfudz M, Murwani I (2018) Direct and residual effect of various vermicompost on soil nutrient and nutrient uptake dynamics and productivity of four mustard Pak^{-} Coi (*Brassica rapa* L.) sequences in organic farming system. Int J Recycl Org Waste Agric. https://doi.org/10.1007/s4009 3-018-0203-0

Odiete I, Chude VO, Ojeniyi SO, Okozi AA, Hussaini GM (2005) Response of maize to nitrogen and phosphorus sources in Guinea savanna zone of Nigeria. Niger J Soil Sci 15:90–101

Ossom EM (2010) Effects of filter cake fertilization on weed infestation, disease incidence and tuber yield of cassava (*Manihot esculenta*) in Swaziland. Int J Agric Biol 12(1):45–50

Ossom EM, Dlamini FT (2012) Effects of filter cake on soil mineral nutrients and maize (*Zea mays* L.) agronomy. J Trop Agric 89(3):141–150

Pinto FF (1971) Current wheat situation in Ethiopia. In: Proceedings of the first wheat workshop. El Batan, Mexico, D.F: CIMMYT, pp 21–24

Powlson DS, Hirsch PR, Brookes PC (2001) The role of soil microorganisms in soil organic matter conservation in the tropics. Nutr Cycl Agroecosyst 61:41–51

Rasul G, Khan KS, Müller T, Joergensen RG (2008) Soil-microbial response to sugarcane filter cake and biogenic waste compost. J Plant Nutr Soil Sci 171:355–360. https://doi.org/10.1002/jpln.200700094

Rasul G, Khan AA, Khan KS, Joergensen RG (2009) Immobilization and mineralization of nitrogen in a saline and alkaline soil during microbial use of sugarcane filter cake amended with glucose. Biol Fertil Soils 45:289–296. https://doi.org/10.1007/s0037 4-008-0333-z

Rautaray SK, Ghosh BC, Mittra BN (2003) Effect of fly ash, organic wastes and chemical fertilizers on yield, nutrient uptake, heavy metal content and residual fertility in a rice-mustard cropping sequence under acid lateritic soils. Bioresour Technol 90:275–283. https://doi.org/10.1016/S0960-8524(03)00132-9

Sanchez PA, Sheperd KD, Soule MJ, Place FM, Buresh RJ, Izac AM, Mokwunye AV, Kwesiga FR, Ndiritu CG, Woomer PL (1997) Soil fertility replenishment in Africa: an investment in natural resource capital. In: Buresh RJ et al (eds) Replenishing soil fertility in Africa, 51st edn. SSSA Spec. Publ., Madison, pp 1–46

Sertsu S, Bekele T (2000) Procedures for soil and plant analysis. National Soil Research Center. EARO, Addis Ababa, pp 15–23

Stoorvogel JJ, Smaling EMA (1990) Assessment of soil nutrient depletion in sub-Saharan Africa, 1983–2000. Rep. 28. Win and Staring Center for Integrated Land, Soil and Water Res., Wageningen

Tejada M, Garcia C, Gonzalez JL, Hernandez MT (2006) Organic amendment based on fresh and composted beet vinasse: influence on soil properties and wheat yield. Soil Sci Soc Am J 70:900–908. https://doi.org/10.2136/sssaj2005.0271

Tesfaye T (1987) Durum wheat breeding in Ethiopia. In: van Ginkel, M, Tanner DG (eds). In: Fifth regional wheat workshop for Eastern, Central and Southern Africa and the Indian Ocean. D.F. Mexico, CIMMYT, pp 18–22

Tesfaye T, Getachew B (1991) Aspects of Ethiopian tetraploid wheats with emphasis on durum genetics and breeding research. In: Mariam HG, Tanner DG, Hulluka M et al (eds) Wheat research in Ethiopia: a historical perspective. IAR/CIMMYT, Addis Ababa, pp 47–71

Thind HS, Singh Y, Singh B, Singh V, Sharma S, Vashistha M, Singh G (2012) Land application of rice husk ash, bagasse ash and coal fly ash: effects on crop productivity and nutrient uptake in rice–wheat system on an alkaline loamy sand. Field Crop Res 135:137–144. https://doi.org/10.1016/j.fcr.2012.07.012

Tittonell P, Vanlauwe B, De Ridder N, Giller KE (2007) Heterogeneity of crop productivity and resource use efficiency within smallholder Kenyan farms: soil fertility gradients or management intensity gradients? Agr Syst 94:376–390. https://doi.org/10.1016/j.agsy.2006.10.012

Vierich HID, Stoop WA (1990) Changes in West African savannah agriculture in response to growing population and continuing low rainfall. Agr Ecosyst Environ 31:115–132. https://doi.org/10.1016/0167-8809(90)90214-X

Vlek PLG (1993) Strategies for sustaining agriculture in sub-Saharan Africa. In: Rogland J, Lal R (eds) Technologies for sustaining agriculture in the tropics, 56th edn. ASA Spec. Publ., Madison, pp 265–277

Wong MH, Wong JWC (1986) Effects of fly ash on soil microbial activity. Environ Pollut Ser A Ecol Biol 40:127–144. https://doi.org/10.1016/0143-1471(86)90080-2

16

Influence of controllable variables on the composting process, kinetic, and maturity of *Stevia rebaudiana* residues

J. J. Castro-Fernández[1] · A. Serrato-González[1] · M. Ruiz-Montoya[1] · M. J. Díaz[1]

Abstract

Purpose *Stevia rebaudiana* residues composting is studied and optimized using a central composite experimental design

Methods The influence of controllable composting variables [moisture (40–70%), aeration (0.05–0.30 L_{air} min^{-1} kg^{-1}), and time (0–50 days)] on the temperature history and the properties of the compost (pH, organic matter) produced to determine suitable composting conditions. Mass balance and emitted gases of all the composting reactors have been done.

Results Compost with high degradation entails operating at 50 days of composting under high moisture content (70%) and aeration level with values close to 0.05 L_{air} min^{-1} kg^{-1}. Moreover, the composting process kinetic for *Stevia rebaudiana* residues has been studied. The magnitude of the kinetic parameters on the studied conditions varies among 1/42 and 1/46 for $1/K_1$ and 0.15 and 1.6 for K_2. Where $1/K_1$ is a value that measures the affinity between microorganisms and substrate and K_2 depends on the composting variable optimization.

Conclusions Both moisture and aeration affects positively and negatively the composting process. Moreover, low effect of aeration has been found. The values of $1/K_1$ and K_2 obtained showed higher values (higher degradation kinetic) under 55% moisture content and 0.30 L_{air} min^{-1} kg^{-1}.

Keywords *Stevia rebaudiana* residues · Composting controllable variables · Composting kinetic

Introduction

Among the most important, with worldwide medicinal and commercial uses, *Stevia rebaudiana* is found (Ahmad et al., 2016), because is increasingly being used as a sugar substitute (Magnuson et al. 2016). In that form, the steviol glycosides (active compounds in Stevia which is found mainly in leaves) are estimated to be 200–300 times sweeter than saccharose and have no influence on the glycemic index (Geuns 2010). In this sense, both, harvesting, to get its leaves, and pruning, which includes the elimination of all dead, dying, or diseased wood and branch to obtain a suitable regrowth of plants are necessary. This fact leads to *Stevia rebaudiana* residues (SRR), essentially, trunks and fine branches (0.5–5 cm thickness). The lack of markets for these wastes is a serious problem for the economic sustainability of this process. To enhance the added value of these wastes, composting could be an appropriate solution, because it is a natural way low cost of recycling organic matter and nutrients.

For extensive use of this technology, an adequate optimization is necessary. According to Haug (1993) and Peláez et al. (2004), composting process kinetic could be modelling using enzyme kinetic concepts and, in this sense, could be described through a first-order reaction (catalysed by enzymes).

Composting optimization entails finding the best values of the main involved variables to obtain a greater and faster degradation of the residues studied. The main factors controlling composting are: operating parameters (moisture content and aeration), pH, and temperature are not industrially altered, and nature of substrate (*C*/*N* ratio and particle size). Among them, temperature is a key factor to judge the process efficiency in composting (Ekinci et al. 2004; Lei and Vandergheynst 2000; Kianirad et al. 2009). Moreover, in accordance with Wang and Ai (2016) and Ponsá et al. (2009), a balance among moisture content, diffusion transport, and oxygen supply is necessary to obtain a suitable

✉ M. J. Díaz
dblanco@uhu.es

[1] Chemical Engineering Department, Facultad de Ciencias Experimentales, Chemical Process and Product Technology Research Center (Pro2TecS), Universidad de Huelva, Campus El Carmen, 21071 Huelva, Spain

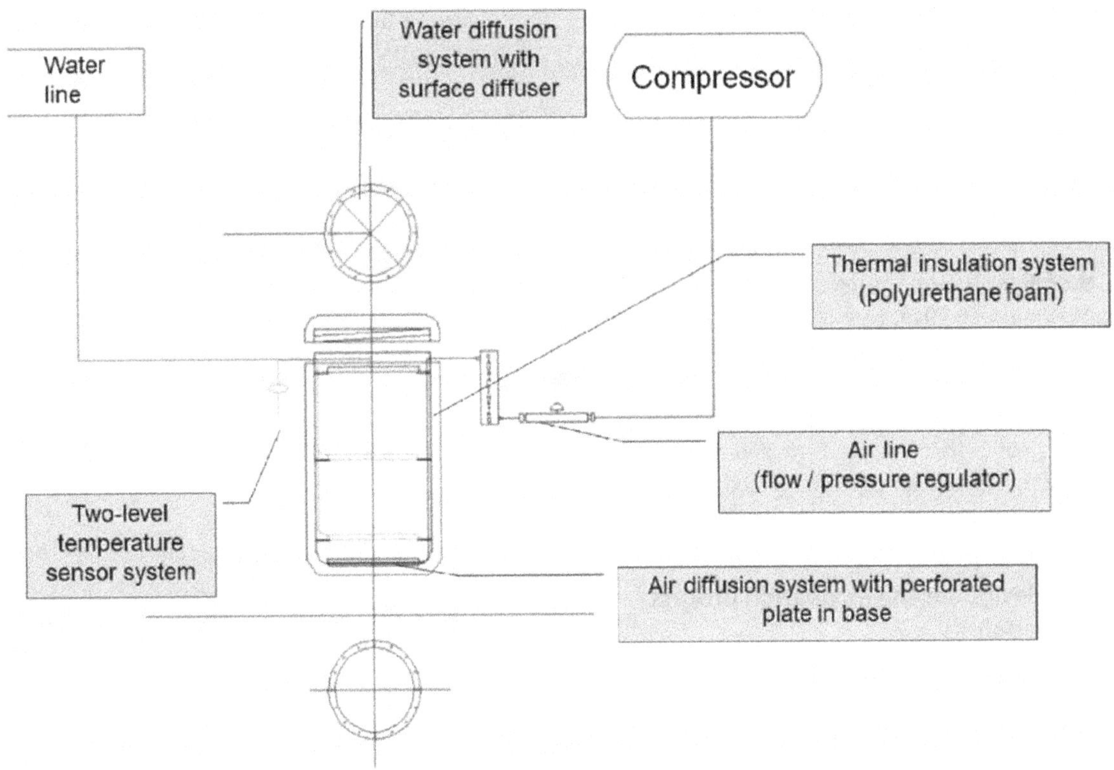

Fig. 1 Schematic reactor

composting process; hence, an adequate size, as physical pretreatment, is necessary in these residues.

On the other hand, knowledge of the degradation kinetics, to find the most suitable conditions for proper degradation, is necessary. Several kinetic data, by inductive and deductive models, have been obtained by Sánchez (2007), Bueno et al. (2008), and Wang and Ai (2016). In these models, references about the influence of several composting factors such (*C/N* ratio, moisture content, temperature, particle size, pH, mixing or turning frequency, and aeration rate) have been reported. However, to my knowledge, no previous studies on Stevia process composting have been found.

The main objective of this study is to measure the relative influence of the main controllable composting variables (moisture and aeration) and time on this process to obtain suitable industrial composting conditions for *S. rebaudiana* residues.

Materials and methods

Stevia residues (branches and twigs) with 0.5–5 cm in diameter were used. Plants were cut at ~ 5 cm above the soil. The material was collected, using closest individual sampling method, from mature plants growing at industrial scale plantation in Moguer (Huelva, Spain).

Table 1 Chemical composition of Stevia residues

Org matter (%)	85.13 ± 0.91[a]	Kjeldahl-N (%)	0.47 ± 0.20
Ash (%)	11.35 ± 0.33	P (%)	0.04 ± 0.01
Ca (%)	1.27 ± 0.02	Cu (mg kg^{-1})	8.9 ± 0.6
Mg (%)	0.33 ± 0.02	Fe (mg kg^{-1})	9.9 ± 1.3
Na (%)	0.12 ± 0.01	Mn (mg kg^{-1})	1.8 ± 1.1
K (%)	0.32 ± 0.01	Zn (mg kg^{-1})	1.6 ± 0.6

[a]Standard errors

Composting process

Stevia rebaudiana residues (SRR) were mixed, to get uniform initial material, and chipped, to ensure an adequate particle distribution during the process and an adequate initial porosity. About 20 kg mixture was placed in each reactor (Fig. 1). Relevant characteristics of the raw materials are reported in Table 1.

Cylindrical composting reactors were formed from an acrylic column (0.5 m in diameter and 1 m in depth). These reactors were insulated with polyurethane foam to minimize the conductive heat loss. Compressed air was introduced to the bottom of each reactor, through a perforated plate, and evenly distributed to the waste mixture. Two temperature sensors were fixed at the centre and on top of each reactor (K thermocouples, TMC6-HA). To obtain environmental temperature (Protimeter-MMS-Plus), an environmental temperature sensor was placed outside the reactors. Temperatures were recorded every 12 h in each reactor by two data loggers (HOBO U12-006).

Moisture (M), aeration (A), and time (t) have been used as independent and controllable composting variables. Each parameter has been established following a central composite experimental design (three levels for each variable). In that form, the values have been fixed in: 40%, 55 and 70% for moisture (Hong et al. 2013; Kim et al. 2015), 0.05, 0.175, and 0.3 $L_{air}\ kg^{-1}\ min^{-1}$ for aeration (Gao et al. 2010) and from 1 to 55 days for composting time with except to temperature (fixed in 90 days). During active composting phase, water losses were compensated to ensure and sustain initial conditions, due to they have been established as constants in the experimental design.

Experimental design for the incubation process and statistical analysis

A 2^n central composite factor design (Montgomery 1991) has been used to relate the dependent and independent (A, M, and t) variables. The central combination for the experimental design was: $A=0.175\ L_{air}\ kg^{-1}\ min^{-1}$, $M=55\%$, and $t=27$ days. The obtained results were implemented in SPSS system package to obtain multiple linear regressions (one for each dependent variable).

Independent variables were normalized using the following equation:

$$X_n = \frac{X - \bar{X}}{\left[\frac{(X_{max} - X_{min})}{2}\right]}, \quad (1)$$

where X is the absolute value of the independent variable concerned, $\bar{X}$ is the average value of the variable, and X_{max} and X_{min} are its maximum and minimum values, respectively.

The number of tests required was calculated as $N=2^n+2n+nc$. In this form, 2^n being the number of points constituting the factor design, $2n$ axial points, and nc central point. Under this study conditions, $N=10$.

The experimental results were fitted as second-order polynomial equation (Eq. 2):

$$Y = a_0 + \sum_{i=1}^{n} b_i X_{ni} + \sum_{i=1}^{n} c_i X_{ni}^2 + \sum_{i=1;j=1}^{n} d_i X_{ni} X_{nj} \qquad (i < j). \quad (2)$$

In this sense, only independent variables were those having a statistically significant coefficient (viz. those not exceeding a significance level of 0.05 in Student's t test and having a 95% confidence interval excluding zero) has been used in the obtained equations relating to both types of variables.

Mass balance and kinetic model

According to Haug (1993), composting process could be expressed as biodegradable solid oxidation (Eq. 3). In this balance, initial non-biodegradable solid mass is equal to the final non-biodegradable solid mass is assumed (Haug 1993):

$$C_aH_bO_cN_d + 0.5(ny + 2s + r - c)O_2 \rightarrow nC_wH_xO_yN_z + sCO_2 + rH_2O + (d - nz)NH_3, \quad (3)$$

$$\text{where}: r = 0.5[b - nx - 3(d - nz)] \quad (4)$$

$$\text{and } s = a - nw. \quad (5)$$

The stoichiometric oxygen required for decomposition can be determined from the organic feed decomposition. Carbon dioxide (CO_2), ammonium (NH_3), and water (H_2O) produced for organic decomposition were determined throughout Eqs. 3, 4, and 5. Nitrogen input is equal to nitrogen output is assumed and there have been no leachates. Moreover, the water added to maintain the moisture of each of the reactors has not been taken into account to obtain an adequate reactor comparison.

The kinetic model of Whang and Meenaghan (1980) and described by Haug (1993) was used. In essence, this model assumes the formation of an intermediate complex (CX^*) in which the hydrolytic enzyme (or free microorganism, X) is absorbed to an active site of the surface solid substrate (C, kg) yielding the product (P, kg):

$$C + X \underset{k_{-1}}{\overset{k_1}{\Leftrightarrow}} CX^* \overset{k_2}{\rightarrow} X + P. \quad (6)$$

Being k_1 the specific reaction (I) rate constant, k_{-1} the specific desorption rate constant, k_2 the specific reaction (II) rate constant

Assuming a quasi-equilibrium state for the intermediate complex:

$$K_1 = \frac{k_{-1} + k_2}{k_1}, \quad (7)$$

where K_1 is the dissociation constant of the microorganism–substrate complex. This constant is characteristic for the biological system considered. Its inverse, $1/K_1$ is a measure of the affinity between microorganisms and substrate (Whang and Meenaghan, 1980).

If R ($kg\ day^{-1}$) is the consumption rate of the substrate C (kg):

$$R = \frac{K_2(C)}{K_1 + C}, \quad (8)$$

where $K_2 = k_2\ (X_t)$ and X_t is the total microorganism concentration.

K_2 is a kinetic variable of the system and depends on the optimization parameters of the process such as temperature, moisture, *C/N*, pH, aeration, etc. It varies considerably depending on the experimental conditions. Its contribution to the reaction rate is directly proportional to its magnitude.

To determine the values of K_1 and K_2 from the experimental data, Eq. 8 is transformed as follows:

$$\frac{1}{R} = \frac{K_1}{K_2}\frac{1}{C} + \frac{1}{K_2}\text{(Lineweaver-Burke double reciprocal plot)}, \tag{9}$$

where C is the carbon content expressed on ash-free basis (kg) and R is the carbon loss per day (kg day^{-1}) and can be obtained each day from the tangent of the curve C (ash-free basis) vs. t (days).

The plot $1/R$ vs. $1/C$ permits the graphical estimation of the constants, K_1 and K_2.

Analytical methods

Samples (150 g approx.) were collected every 3 days in the early composting and every week in the mesophilic and maturation stages. All samples have been dried (60 °C) and ground to pass in a sieve of 0.5 mm diameter. On a significant sample of the original material, moisture was determined by drying at 105 °C to constant weight. The samples were analysed for: pH (1:5 *w*/*v*) using a pH electrode, total organic matter (OM) by loss on ignition (550 °C for 5 h) (Klute et al., 1986), and carbon was estimated as OM/1.8 (Haug 1993). Carbon was also expressed on ash-free basis, total P (HCl acid digest) using the ascorbic acid method (Jones 2001). Total K, Ca, Mg, Fe, Cu, Mn, and Zn are determined by total digestion of the compost in strong acid (concentrated HCl), with subsequent analysis by atomic absorption spectrometry (Variant SpectrA 220FS, Jones 2001) and the total Kjeldahl-N was determined by Kjeldahl digestion (Faithfull 2002).

Bulk density has been measured by following Schaub-Szabo and Leonard (1999) method. Particle density is calculated as an indirect estimation using Martinez (1992) method for high organic matter percentage materials. Under this method, density is calculated as an indirect estimation from organic matter and ash values obtained in calcination process. According to Boodt and Verdonck (1972), particle density of organic materials (OM, %) is 1.45 g cm^{-3} and that of ashes, as mineral matter (MM, %), is 2.65 g cm^{-3}:

$$\text{Particle density (g cm}^{-3}) = \frac{100}{\left(\frac{\text{OM}}{1.45} + \frac{\text{MM}}{2.65}\right)}. \tag{10}$$

The final compost characterizations are also shown in Table 1.

Results and discussion

To assess the relative influence of the selected independent variables (moisture, aeration, and time) on each dependent variable (temperature, pH, OM, and emitted gases) by substituting the values of the measured independent variables for each dependent variable, and applying a polynomial model analysis, the polynomial mathematical models have been obtained (Table 2). The average of three measurements has been used to obtain the equations. Suitable fits with values of R^2 greater than 0.94 in all cases have been obtained. The differences between the experimental and estimated (using the obtained equations) values have been lower than 5%.

To better envisage the influence of operational variables on measured parameters of compost, and to compare different conditions, the surface responses in for each dependent variable have been plotted.

Table 2 Polynomial mathematical models obtained for each dependent variable

	R^2	F-Snedecor
pH = 6.98 + 0.288 × Moist + 0.0564 × Aeration + 0.104 × Moist^2 + 0.075 × Moist × Time	0.97	52.36
OM = 82.746 − 3.1554 × Moist + 0.498 × Aeration − 3.27 × Time − 2.11 × Moist^2 + 1.299 × Aeration^2 + 1.254 × Time^2 + 1.223 × Moist × Aeration − 0.7 × Moist × Time	0.97	42.59
O_2 = 16.95 − 1.184 × Moist + 2.051 × Time + 2.565 × Moist^2 − 0.773 × Aeration^2 + 0.368 × Moist × Aeration + 0.978 × Moist × Time	0.99	135.61
CO2 = 4.66 + 0.85 × Moist − 1.088 × Time − 1.65 × Moist^2 − 0.722 × Time^2 − 0.766 × Moist × Time	0.98	74.15
VOC = 0.0259 − 0.034 × Moist − 0.059 × Aeration − 0.122 × Time − 0.052 × Moist^2 + 0.124 × Time^2 + 0.0435 × Moist × Time	0.96	23.88

OM organic matter percentage, *O_2* O_2 percentage, *CO_2* CO_2 percentage, *VOC* volatile organic compounds percentage, *moist* moisture normalized value, *time* time normalized value, and *aeration* aeration normalized value

Temperature and pH profiles

The temperature profiles (values expressed as mean of the two temperatures in each reactor) of the reactors are presented in Fig. 2. As expected, after the establishment of composting conditions, the temperature of the reactors began to rise to thermophilic temperatures in 4 days. A general tendency, during the first 15 days, of temperature elevation in all the reactors, has been observed. After that, as the transformation process takes place, a subsequent decrease. Because winter temperatures, in which the study was conducted, low-temperature peaks have been found. Moreover, the several ups and downs in the graphs of each reactor correspond to the water additions to compensate water losses during the process to maintain the initial experimental design moisture conditions. This fact is observed in all the reactors, a little drop during thermophilic stage is showed and high drop during mesophilic stage (with low-temperature values during 2 days) when it began to rise again.

For all experiments (with exception to reactors 8, 9, and 10), thermophilic temperatures (> 45 °C) were observed. The temperature profiles of reactors with high moisture content (70%) are, in general, higher than those found for medium (55%) and low (40%) moisture content. No significant aeration effect among reactor is observed. After these periods, composting temperatures began to drop to levels below 40 °C for all the reactors.

On the other hand, pH could be used as a suitable bio-oxidative phase evolution indicator (Nogueira et al. 1999). Furthermore, the pH levels will vary throughout the decomposition process and pH also significantly affects the composting process. In this sense, similar to that found for temperature, in Fig. 3, the evolution of pH with respect to time and aeration at three moisture contents is shown. In that figure, among the different studied aeration levels, a low pH variation during composting is observed. Nonetheless, under high moisture content (70%), pH increases gradually from 7.0 to 7.6. According to Nakasaki et al. (1993), it could be due to the metabolic degradation and proteolysis liberating ammonia compounds. However, under medium moisture content (55%), the pH maintains the same values during the whole composting process, understood to be between 6.9 and 7.1. As observed in Fig. 3, the pH, in reactors under low moisture content (40%), remains slightly below 7 during the composting process. Therefore, low sign of organic acid breakdown is found under these conditions.

Emitted gases profiles

The O_2 evolution with respect to time and aeration at three moisture contents is shown in Fig. 4. The mean O_2 concentration during the composting process was in the range 14–21% (Fig. 4), which indicated that the process for all the reactors was aerobic. The minimum suitable O_2

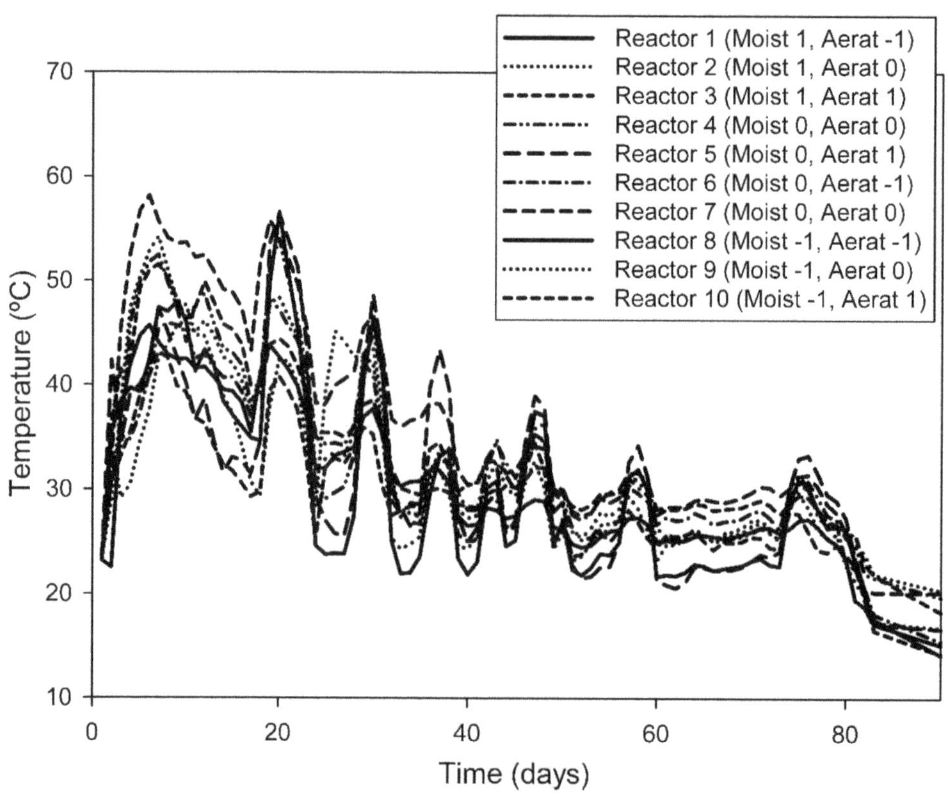

Fig. 2 Reactors temperature evolution

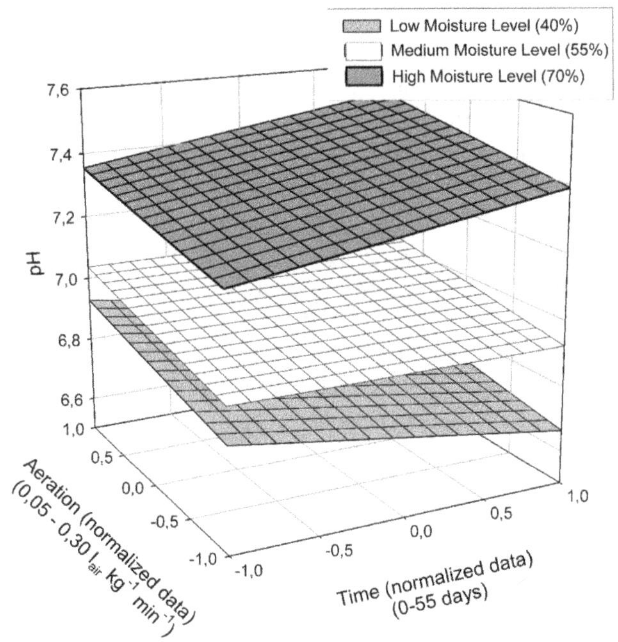

Fig. 3 pH evolution as a function of aeration and time at three moisture content

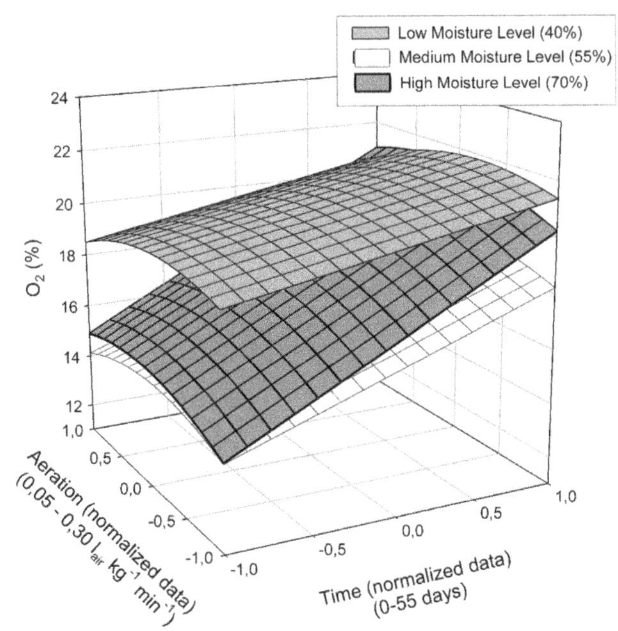

Fig. 4 O_2 evolution as a function of aeration and time at three moisture content

concentration in interstitial gas for optimal composting is 5% (Haug 1993). In general, a linear increase over time for all reactors is observed. In addition, although initially, differences in O_2 levels among the different aerations and moistures, these values have a high tendency to converge (near to 21%) as the process stabilizes (45–90 days).

The interstitial CO_2 evolution with respect to time and aeration at three moisture contents is shown in Fig. 5. Under lower operating times (0–15 days), a higher content of CO_2 is observed, mainly in higher moisture content (55–70%) reactors. This fact is also observed in the temperature evolution (Fig. 2) and pH (Fig. 3). Few statistical variations in the CO_2 concentration with respect to the aeration levels have been observed. Overall, the range of CO_2 concentrations within the reactor was higher during the initial monitoring period (thermophilic phase 1–15 days). After that, a progressive decrease to negligible values for all the reactors at 40 days is observed.

The partial depletion in O_2 and the increase in CO_2 were indicative of significant biological activity in the medium-to-high moisture (55–70%) reactors, and all studied aeration levels are considered suitable for Stevia residues composting process.

The volatile organic compounds' (VOCs') evolution with respect to time and aeration at three moisture contents is shown in Fig. 6. Emitted VOCs represent volatile breakdown products of aerobic and anaerobic processes (Maulini-Duran et al. 2013). A similar behaviour to that shown for CO_2 is observed in Fig. 6 for VOCs. On this occasion, high initial contents, within the thermophilic phase, are found. Moreover, low aeration influence and medium influence of moisture content are found. In this case, the lowest VOCs concentrations are observed in reactors with high moisture contents (70%). After that, a progressive decrease in the VOCs emission has been observed for all reactors and no significant statistical differences were found under the

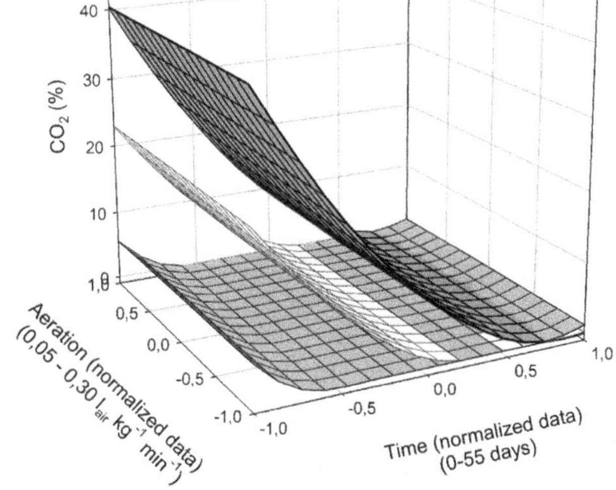

Fig. 5 CO_2 evolution as a function of aeration and time at three moisture content

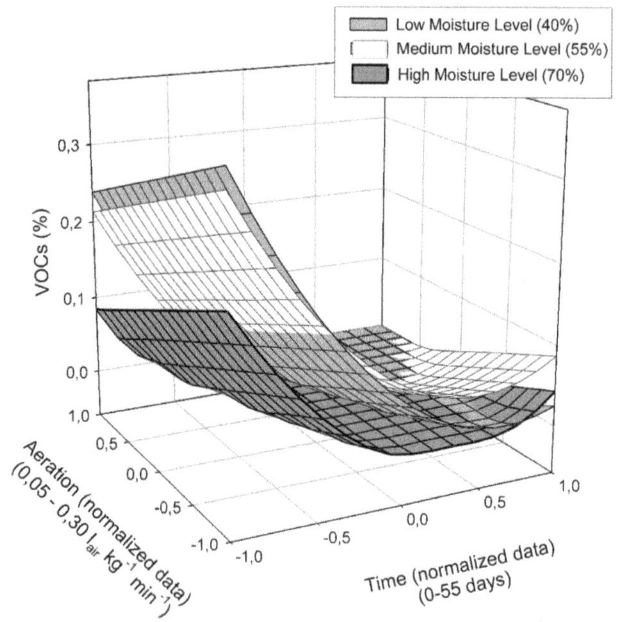

Fig. 6 Volatile organic compounds (VOCs) evolution as a function of aeration and time at three moisture content

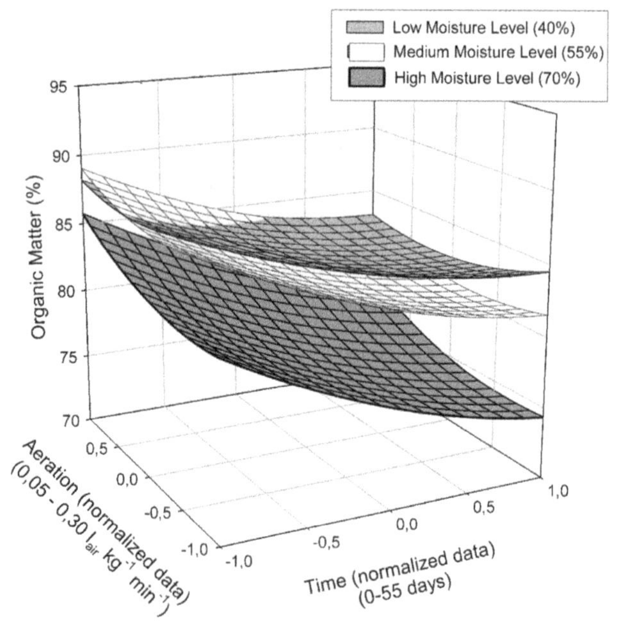

Fig. 7 Organic Matter evolution as a function of aeration and time at three moisture content

selected experimental composting conditions at medium-to-high (25–55 days).

Organic matter mineralization

The organic matter progression with respect to time and aeration at three moisture contents is shown in Fig. 7. As expected, during the composting process, organic matter content decreased for all reactors (Fig. 7). According to Kim et al. (2015), the composting process could be described by a rapid initial degradation followed by a longer slow degradation phase. However, a quasi-linear decrease has been observed for all reactors, and as described for previous parameters, low aeration influence and medium influence of moisture content is found. Loss of organic matter of 6.5, 6.8, and 7.2% for low (40%), medium (55%), and high (70%) moistures, respectively, have been recorded after 55 days of composting. In addition, low differences the organic matter evolution under low and medium moistures with respect to high moisture content have been calculated.

The organic matter balances for the studied reactors are shown in Table 3. The main gas losses, according to the proposed stoichiometry, consisted of carbon dioxide produced and water evaporated from the compost. Variations in organic matter losses between 3.1 and 13% have been found depending on the process conditions.

At the beginning and after the maturation phase, the reactors were weighed. The highest loss has been found for *R*2 (70.96% of the initial dry weight under 70% moisture and 0.05 L_{air} min^{-1} kg^{-1} aeration) followed by *R*7 (70.07% under, 70% moisture and 0.15 L_{air} min^{-1} kg^{-1} aeration).

Lower weight losses have been found for *R*10 (29.67% of the initial dry weight under 40% moisture and 0.30 L_{air} min^{-1} kg^{-1} aeration) followed by *R*9 (33.66% of the initial dry weight under 40% moisture and 0.15 L_{air} min^{-1} kg^{-1} aeration).

Calculated ammonium losses were higher for *R*7 (55% moisture and 0.30 L_{air} min^{-1} kg^{-1} aeration) followed by *R*3 (70% moisture and 0.30 L_{air} min^{-1} kg^{-1} aeration). Lower ammonium losses have been found for *R*8 (40% moisture and 0.05 L_{air} min^{-1} kg^{-1} aeration) followed by *R*9 (40% moisture and 0.15 L_{air} min^{-1} kg^{-1} aeration).

In that form, aeration is a determining factor in nitrogen losses as ammonia. Therefore, the use of high moisture and low aeration levels to reduce nitrogen losses is advisable to obtain suitable final compost.

In general, the total weight losses were higher under high moisture levels. This fact can be explained by the low water retention capacity of the Stevia residues, so it is necessary to maintain high levels of moisture for effective water absorption within the mass.

Particle and bulk density evolution

Particle and bulk density evolution of each reactor is also calculated. Similar values in the initial particle density have been calculated (1520–1550 kg m^{-3}). After 55 days of composting, *R*10 (40% moisture and 0.3 L_{air} min^{-1} kg^{-1} aeration) show the highest particle density value (1650 kg m^{-3}), followed by *R*2 (70% moisture and 0.175 L_{air} min^{-1} kg^{-1}

Table 3 Organic matter balances and $1/R$, $1/C$ data for Lineweaver–Burke correlations

Day	*R*1	*R*2	*R*3	*R*4	*R*5	*R*6	*R*7	*R*8	*R*9	*R*10
OM										
0	85.92	87.19	87.48	88.27	88.01	89.92	87.96	89.08	89.66	89.13
3	82.80	83.65	85.05	86.24	87.78	87.46	86.71	88.57	87.28	87.07
10	78.43	80.82	81.43	85.04	87.03	86.52	86.36	87.39	86.17	86.53
17	77.40	78.40	81.04	84.28	86.59	85.53	86.06	86.93	85.25	86.05
24	76.22	76.87	80.96	83.20	84.52	84.10	85.32	86.50	84.94	85.76
31	75.93	75.37	80.32	81.76	83.86	83.66	85.20	86.05	84.62	85.06
1/C										
0	0.0209	0.0206	0.0206	0.0204	0.0205	0.0200	0.0205	0.0202	0.0201	0.0202
3	0.0217	0.0215	0.0212	0.0209	0.0205	0.0206	0.0208	0.0203	0.0206	0.0207
10	0.0229	0.0223	0.0221	0.0212	0.0207	0.0208	0.0208	0.0206	0.0209	0.0208
17	0.0233	0.0230	0.0222	0.0214	0.0208	0.0210	0.0209	0.0207	0.0211	0.0209
24	0.0236	0.0234	0.0222	0.0216	0.0213	0.0214	0.0211	0.0208	0.0212	0.0210
31	0.0237	0.0239	0.0224	0.0220	0.0215	0.0215	0.0211	0.0209	0.0213	0.0212
1/R										
0	0.5776	0.6550	0.4512	0.3761	0.0430	0.4542	0.2327	0.0951	0.4410	0.3814
3	0.3466	0.2245	0.2869	0.0953	0.0595	0.0752	0.0277	0.0939	0.0881	0.0428
10	0.0820	0.1919	0.0308	0.0605	0.0346	0.0784	0.0235	0.0367	0.0732	0.0381
17	0.0934	0.1215	0.0064	0.0855	0.1649	0.1135	0.0588	0.0342	0.0247	0.0228
24	0.0229	0.1193	0.0510	0.1146	0.0522	0.0351	0.0095	0.0357	0.0251	0.0558
31	0.0299	0.1019	0.0013	0.1188	0.1617	0.0503	0.0341	0.0605	0.1589	0.0098
Slp	− 182.9	− 120.2	− 239.9	− 58.07	− 21.95	− 142.1	− 131.5	− 40.31	− 133.7	− 183.7
Int	4.357	2.953	5.373	1.390	0.533	3.129	2.859	0.894	3.018	3.942
R^2	0.955	0.735	0.977	0.923	0.8059	0.847	0.832	0.733	0.942	0.837
K_2	0.2295	0.3386	0.1861	0.7194	1.8762	0.3196	0.3498	1.1186	0.3313	0.2537
K_1	41.978	40.704	44.649	41.777	41.182	45.413	45.995	45.089	44.304	46.600

Slp slope, *Int* intercept

aeration, 1670 kg m^{-3}). The lowest value was recorded in *R*7 (55% moisture and 0.175 L_{air} min^{-1} kg^{-1} aeration, 1570 kg m^{-3}), followed by *R*5 (55% moisture and 0.3 L_{air} min^{-1} kg^{-1} aeration, 1570 kg m^{-3}). As expected, based on the results obtained by Van Ginkel et al. (1999) a direct relation between degradation and particle density is found, owing to increases in ash concentrations (with higher density) during the composting process.

On the other hand, a progressive increase in bulk density in all of the studied reactors is found, mainly due to a settlement effect. In accordance with the results showed by López-Real (1990), when the compost mix is constituted, a low bulk density is observed. However, during the process, the combined effects of biological degradation processes and change of water content due to transport processes, a marked influence on bulk density is noted. This effect is observed mostly in *R*2 (137.49 kg m^{-3} under 70% moisture and 0.175 L_{air} min^{-1} kg^{-1} aeration), followed by R4 (94.93 kg m^{-3} under 55% moisture and 0.175 L_{air} min^{-1} kg^{-1} aeration), have been observed. Low increase in *R*5 (3.38 kg m^{-3} under 55% moisture and 0.3 L_{air} min^{-1} kg^{-1} aeration), followed by *R*7 (5.08 kg m^{-3} under 55% moisture and 0.175 L_{air} min^{-1} kg^{-1} aeration), have been found. Similar data and evolution has been found by Mohee and Mudhoo (2005).

Kinetic constants' modelling

The first 30 day data (thermophilic temperatures and maximum degradation rate) have been used for the estimation of the kinetic constants.

Following the model proposed by Whang and Meenaghan (1980), to obtain the kinetic values, for each reactor, Lineweaver–Burke plots (1/*R* and 1/*R*) (figure not shown) were linearly correlated with high regression coefficients (Table 3). In this sense, due to the suitable regression coefficients obtained, the used kinetic model seems to be adequate to describe the thermophilic phase for Stevia residues composting process.

The polynomials' mathematical models for K_1 and K_2, similar to that used for the previous dependent parameters, have been obtained by substituting independent variables values of the for each dependent variable and applying a

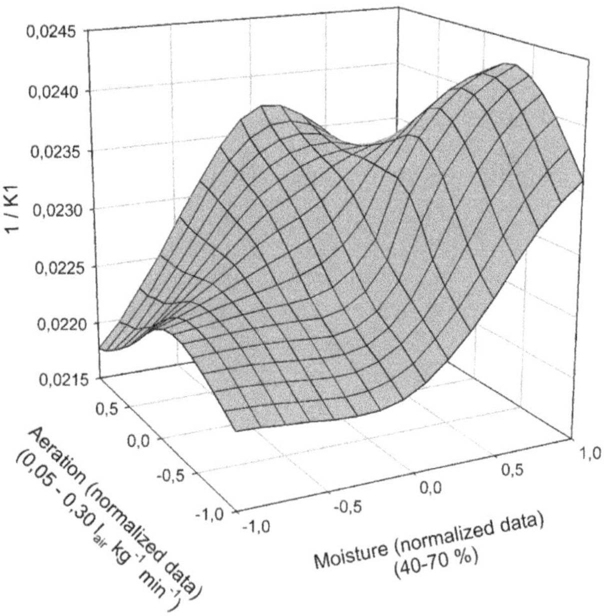

Fig. 8 $1/K_1$ evolution as a function of aeration and time

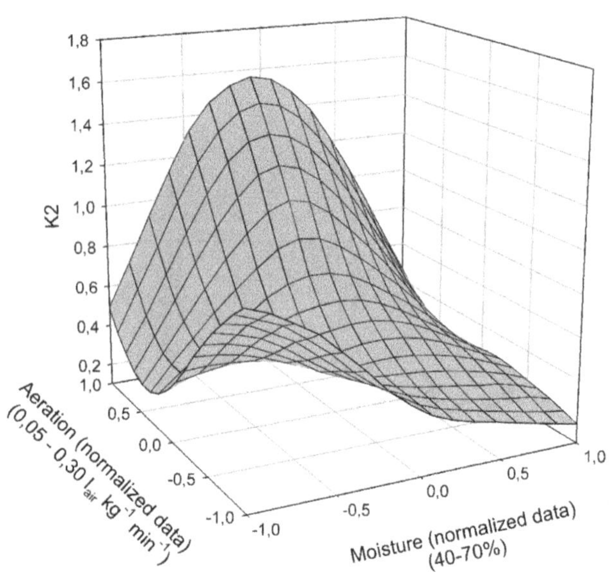

Fig. 9 K_2 evolution as a function of aeration and time

multiple linear regression analysis. Moreover, to identify the relative influence among independent variables on the dependent variables and to determine the values of the independent variables giving the K_1 and K_2 evolutions, the response surfaces for each variable were plotted three levels of the most strongly independent variable (Figs. 8 and 9).

Figure 8 show that the variable most strongly influencing the $1/K_1$ evolution is the moisture content, whereas aeration has the lowest effect on that evolution. In that form, higher $1/K_1$ values when the composting moisture is 70% are found. Furthermore, according to Whang and Meenaghan (1980), the stability of the substrate–microorganism complex is increased under optimum moisture (probably due to optimal balance among aeration–moisture) in that form medium aeration (0.175 L_{air} min^{-1} kg^{-1}) could be used to obtain maximum $1/K_1$ value.

According to that found for $1/K_1$ and as can be seen from Fig. 9, the K_2 values were less influenced by aeration than by the moisture. K_2 shows a high influence in aeration under medium moisture value (55%). Maximum K_2 value under medium moisture content and high aeration level was found. This value depends on the operational controllable variables of the process such as temperature, moisture, aeration and chemical conditions.

Among the studied variables, a $1/K_1$–K_2 balance shows that under 55–70% moisture, and 0.175 L_{air} min^{-1} kg^{-1} aeration could be a suitable option.

Conclusions

The composting process of *Stevia rebaudiana* residues is technically reachable and could be considered as a natural way to recycle these wastes. Among the studied composting conditions, [time (0–50 days), aeration (0.05–0.30 L_{air} min^{-1} kg^{-1}), and moisture (40–70%)] suitable temperature progression have been found for the reactors with high moisture content.

The kinetic model proposed is useful to describe the *Stevia rebaudiana* residues composting.

The values of $1/K_1$ (related to affinity between microorganisms and substrate) and K_2 (depends on the optimization parameters of the composting process) obtained using this model showed a optimal balance under 55% of moisture content, and 0.03 L_{air} min^{-1} kg^{-1} of aeration.

Acknowledgements The authors gratefully acknowledge the funding of this work from the Andalusian Regional Ministry of Economy, Innovation, Science, and Employment (Project number RNM 2323) and Ministry of Economy and Competitiveness (Spain National Programme for Research Amend at the Challenges of Society, CTQ2013-46804-C2-1-R and CTQ2017-85251-C2-1-R).

References

Ahmad N, Rab A, Ahmad N (2016) Light-induced biochemical variations in secondary metabolite production and antioxidant activity in callus cultures of *Stevia rebaudiana* (Bert). J Photochem Photobiol B Biol 154:51–56. https://doi.org/10.1016/j.jphotobiol.2015.11.015

Boodt MD, Rab A, Verdonck O (1972) The physical properties of the substrates in horticulture. Acta Hortic 3:37–44. https://doi.org/10.17660/actahortic.1972.26.5

Bueno P, Yáñez R, Ariza J, Díaz M (2008) Influence of environmental parameters on the composting kinetic of lignocellulosic residues.

Compost Sci Util 16:132–138. https://doi.org/10.1080/1065657x.2008.10702368
Ekinci K, Keener HM, Akbolat D (2004) Effect of thermocouple location on the optimum composting rate. Biosyst Eng 83(3):345–353. https://doi.org/10.1016/j.biosystemseng.2004.07.004
Faithfull NT (2002) Methods in agricultural chemical analysis: a practical handbook. Cabi Pub, Oxon
Gao M, Li B, Yu A, Liang F, Yang L, Sun Y (2010) The effect of aeration rate on forced-aeration composting of chicken manure and sawdust. Bioresour Technol 101:1899–1903. https://doi.org/10.1016/j.biortech.2009.10.027
Geuns JMC (2010) Stevia and steviol glycosides: properties, techniques, uses, exposure, toxicology, pharmacological effects. Euprint, Heverlee
Haug RT (1993) The practical handbook of compost engineering. Lewis Publishers, Boca Raton
Hong N, Chen J, Zhang XH, Chen TB (2013) Effect of turning on moisture content in sewage sludge composting. Adv Mat Res. 777:457–460. https://doi.org/10.4028/www.scientific.net/amr.777.457
Jones JB (2001) Laboratory guide for conducting soil tests and plant analysis. CRC Press, Boca Raton
Kianirad M, Muazardalan M, Savaghebi G, Farahbakhsh M, Mirdamadi S (2009) Effects of temperature treatment on corn cob composting and reducing of composting time: a comparative study. Waste Manag Res 28:882–887. https://doi.org/10.1177/0734242x09342359
Kim E, Lee DH, Won S, Ahn H (2015) Evaluation of optimum moisture content for composting of beef manure and bedding material mixtures using oxygen uptake measurement. Asian-Australas J Anim Sci 29:753–758. https://doi.org/10.5713/ajas.15.0875
Klute A, Weaver RW, Sparks DL, Dane JH, Topp GC (1986) Methods of soil analysis. Soil Science Society of America, Madison
Lei F, Vandergheynst J (2000) The effect of microbial inoculation and pH on microbial community structure changes during composting. Process Biochem 35:923–929. https://doi.org/10.1016/s0032-9592(99)00155-7
López-Real J (1990) Agro-industrial waste composting and its agricultural significance. Proc Fertil Soc 293:1–26
Magnuson BA, Carakostas MC, Moore NH, Poulos SP, Renwick AG (2016) Biological fate of low-calorie sweeteners. Nutr Rev 74:670–689. https://doi.org/10.1093/nutrit/nuw032
Martínez FX (1992) Propuesta de metodología para la determinación de las propiedades físicas de los sustratos. In: Proceedings I substrate workshop. Spanish Society for Horticultural Science (SECH) Sept 30–Oct 2. Villaviciosa (Asturias-Spain), pp 55–65
Maulini-Duran C, Artola A, Font X, Sánchez A (2013) A systematic study of the gaseous emissions from biosolids composting: raw sludge versus anaerobically digested sludge. Bioresour Technol 147:43–51. https://doi.org/10.1016/j.biortech.2013.07.118
Mohee R, Mudhoo A (2005) Analysis of the physical properties of an in-vessel composting matrix. Powder Technol 155(1):92–99. https://doi.org/10.1016/j.powtec.2005.05.051
Montgomery DC (1991) Design and analysis of experiments. Wiley, New York
Nakasaki K, Yaguchi H, Sasaki Y, Kubota H (1993) Effect of pH control composting of garbage. Waste Manag Res 11(2):117–125. https://doi.org/10.1006/wmre.1993.1013
Nogueira WA, Nogueira FN, Devens DC (1999) Temperature and pH control in composting of coffee and agricultural wastes. Water Sci Technol 40(1):113–119. https://doi.org/10.1016/S0273-1223(99)00371-6
Peláez C, Mejía A, Planas A (2004) Development of a solid phase kinetic assay for determination of enzyme activities during composting. Process Biochem 39:971–975. https://doi.org/10.1016/s0032-9592(03)00208-5
Ponsá C, Pagans E, Sánchez A (2009) Composting of dewatered wastewater sludge with various ratios of pruning waste used as a bulking agent and monitored by respirometer. Biosyst Eng 102(4):433–443. https://doi.org/10.1016/j.biosystemseng.2009.01.002
Sánchez A (2007) A kinetic analysis of solid waste composting at optimal conditions. Waste Manag 27:854–855. https://doi.org/10.1016/j.wasman.2006.07.003
Schaub-Szabo S, Leonard J (1999) Characterizing the bulk density of compost. Compost Sci Util 7:15–24. https://doi.org/10.1080/1065657x.1999.10701980
Van Ginkel JT, Raats PAC, Van Haneghem IA (1999) Bulk density and porosity distributions in a compost pile. Neth J Agrix Sci 47:105–121
Wang Y, Aai P (2016) Integrating particle physical geometry into composting degradation kinetics. Bioresour Technol 200:514–520. https://doi.org/10.1016/j.biortech.2015.10.073
Whang DS, Meenaghan GF (1980) Kinetic model of composting process. Compost Sci 21(3):44–46

17

Influence of compost supplemented with jatropha cake on soil fertility, growth, and yield of maize (*zea mays* L.) in a degraded soil of Ilorin, Nigeria

Adebayo Abayomi Olowoake[1] · **Oluremi Solomon Osunlola**[1] · **James Adebayo Ojo**[1]

Abstract

Purpose This study was conducted to determine the effect of compost supplemented with jatropha cake on maize (*Zea mays* L) yield in a degraded soil and their residual effects on soil fertility in Ilorin, Nigeria.

Methods Field trials were conducted at Kwara State University Teaching and Research Farm, Malete, in 2016/2017 cropping season. The treatments consisted of control, un-amended compost Grade B supplemented with Jatropha cake AJ (30% Grade B + 70% JC) at 1.5 t/ha, BJ (30% Grade B + 70% JC) at 2.0 t/ha, CJ (50% Grade B + 50% JC) at 2.5 t/ha including NPK at 60 kg N/ha. The treatments were arranged in a randomized complete block design (RCBD) and replicated three times.

Results Maize grain yield (3.1 t/ha) was obtained from CJ at 2.5 t/ha this was significantly ($p < 0.05$) greater than that of NPK treatment (2.2 t/ha) after the first cropping. At second trials, maize grain yield values gotten from CJ at 2.5 t/ha were also significantly ($p < 0.05$) greater than that of NPK values. Treatments CJ at 2.5 t/ha significantly ($p < 0.05$) improved soil pH, available P and exchangeable K.

Conclusion Fertilizer 50% Grade B + 50% Jatropha cake at 2.5 t/ha had a significant and positive effect on soil fertility after harvesting of maize when compared with NPK in both cropping.

Keywords Degraded soil · Fertilizer · Jatropha cake · Maize yield · Un-amended compost

Introduction

Reduction in tropical soil fertility and crop productivity has become a major concern, and indeed a great hindrance to achieving food sufficiency in the tropics. However, continuous and exhaustive use of highly priced mineral fertilizer materials, for improving crop productivity in the past 10 years, had been obviously connected to this problem as earlier reported to influence nutrient imbalances, leaching of nitrates and phosphates, marine eutrophication and groundwater contamination, microbial activities, soil acidity, and serious threats to man (Babajide and Olla 2014; Afe and Oluleye 2017; Crews and Peoples 2004). Unnecessary use of chemical fertilizers and pesticides adversely affects the soil environment, leading to decrease in crop productivity and production of possibly harmful food, unsafe for human feeding (Chaturvedi et al. 2013). Our traditional approach of use of synthetic fertilizer as a common fertilizer in most developing countries for crop productions can no longer be depended on especially as its use is being threatened with cost economics and absence when needed by farmers (Adeoluwa and Suleiman 2012). Thus, there is a need to consider another source of fertilizer.

Sivaprakashan (1991) reported that compost enhanced crop yield by improving nutrient status as well as growing of microbial activity in the soil. Similarly, nutrient supply by oil-based cakes which is a form of organic manure will not only decrease the reliance on chemical fertilizers, but also develop the soil structure, boost the growth and activity of useful organisms in the soil, ease the deficiency of secondary and micronutrients, and sustain higher yield due to better soil health (Tiwari 2002; Singh et al. 2006).

Jatropha cake produced during oil extraction in biodiesel production can be used as manure (Cano-Asseleih et al. 1989; Jayasingh 2003). Jatropha cake is opulent in nitrogen, phosphorous, and potassium and can be used as organic

✉ Adebayo Abayomi Olowoake
aolowoake@yahoo.com

[1] Department of Crop Production, Kwara State University, Malete, Nigeria

fertilizer. The percentages of nitrogen (N), phosphorous (P), and potassium (K) in jatropha cake were 3.2–4.5, 1.4–2.1, and 1.2–1.7%, respectively (Kumar and Sharma 2008). The occurrences of these elements in the cake were known as the organic nutrients sources that are even greater than that of chicken or cow farm yard manure (Olowoake 2017).

The large amount of de-oiled seed cakes produced by oil-extraction process is valuable as organic manure due to its high nitrogen content (up to 6.48%) and can be used to maintain fertility of soil. Supplementing compost with jatropha cake is not much familiar, but use of these two is of higher nutrient content as alternate of inorganic fertilizer. An organic fertilizer does more than supplying organic nutrients: it improves the water and nutrient holding capacity of soil. The organic matter over time combines the tiny clay particles into aggregates forming large opening space. This improves air infiltration, water infiltration, drainage and allows for deeper rooting depths. Build-up of toxic material in the soil is also unlikely. This work was carried out to know the possibility of using compost supplemented with jatropha cake on the yield of maize and its remaining effects on some soil physico-chemical properties on a degraded soil of Guinea savannah zone.

Materials and methods

Depiction of study area

The experimental site situated at the experimental plot of Kwara State University, Malete (08°42′48.5″N and 004°26′17.9″E), Ilorin, Nigeria, which lies in the southern guinea savanna zone of Nigeria. The yearly rainfall in the region is about 1200 mm and temperature varies among 33 and 34 °C during the period, with a dry spell from December to March. The Kwara State University land area forms part of the South-western region of Nigerian basement complex, a region of basement recurrence and plutonism during the Pan-African orogeny (Olowoake and Ojo 2014).

Soil analysis

Soil samples were randomly taken from 0 to 15 cm depth using soil auger before the study. The samples were bulked and air-dried and were taken to Laboratory for analysis. The particle size analysis was determined by hydrometer method (Bouycocous 1951). Organic carbon was determined by Walkley–Black dichromate digestion method as described by Nelson and Sommers (1982) and total nitrogen by Bremner and Mulvancy (1982). P was determined by Brays P1 method (Bray and Kurtz 1945). Exchangeable K, Na, Ca, and Mg were extracted by leaching the soil with neutral ammonium acetate (NH4OAC). K and Na concentrations were determined by flame photometer, while Mg and Ca were determined by atomic absorption spectrophotometer. Soil pH determination was carried out using a 1:2 (soil: water) ratio with a glass electrode pH meter and electrical conductivity in deionized water suspension of 1:5 (w/v).

Experimental layout

This experiment was set up under field condition with the objective of assessing the most effective compost supplemented with Jatropha de-oiled cake emanating from screen house. The field was cleared and ploughed. The experimental design used in this study was randomized complete block design (RCBD) replicated three times. The size of each plot was 3.0 × 3.0 (9 m^2), with a spacing of 0.5 m. The fertilizers were applied to the maize and the field trials had five major treatment combinations comprising the following: control, NPK 15–15–15 at 60 kg N/ha, AJ (30% Grade B + 70% JC) at 1.5 t/ha, BJ (30% Grade B + 70% JC) at 2.0 t/ha, and CJ (50% Grade B + 50% JC) at 2.5 t/ha. The chemical composition of Jatropha cake (JC) and Aleshinloye Grade B used in the study is given in Table 1. Aleshinloye Grade B (un-amended compost) is a commercial fertilizer from Aleshinloye fertilizer company, Ibadan, Oyo State Nigeria.

Planting of maize

Maize variety EVDT-W99 STR (*Zea mays* L.) was obtained from the International Institute of Tropical Agriculture (IITA), Ibadan, and Nigeria was used as a test crop. Maize was sown with a planting distance of 0.75 m × 0.25 mm at two plants per stand. However, maize plant was later thinned to one after seedling emergence. Compost supplemented with jatropha cake was applied using ring method 2 weeks

Table 1 Chemical composition of Aleshinloye Grade B and jatropha cake

Nutrient element (g/kg)	Concentration of Grade B fertilizer	Jatropha cake
Total N	11.7	34.1
Ca	23.4	0.3
Mg	2.4	8.39
K	20.9	2.2
Mn	106.67	0.01
Na	29.61	0.08
Fe	8195.39	2.1
Cu	16.98	0.02
P	7.6	0.7
Zn	19.9	0.08
EC (ds/m)	16.0	13.8
pH	9.3	8.5

before planting to improve mineralization, while NPK fertilizer was applied 2 weeks after planting. The plots weeds were cleared manually and free of weed throughout thee study. Maize was harvested at 12 weeks after planting. The second experiment was carried out to investigate the residual effects of fertilizer application.

Data collection

Five maize plants were randomly sampled and tagged per plot to determine growth parameters. The growth parameters collected include plant height, leaf area, stem girth, and number of leaves per plant while yield parameters; 100 seeds weight and grain yield were collected at 12 WAP.

Statistical analysis

Data collected from the study were analysed using statistical analysis system (SAS) for analysis of variance (ANOVA) and the treatments were compared using the Duncan's multiple range test (DMRT) ($p < 0.05$). Moreover, coefficient correlation was calculated to determine the relationship between maize yield and soil properties.

Table 2 Physico-chemical properties of experimental soil

Parameters	Soil test value
pH (H_2O)	5.7
EC (ds/m)	0.3
Org.C (gk/g)	6.84
Total N (g/kg)	3.1
P (mg/kg)	9.3
Exchangeable bases cmol/kg	
Mg	4.54
Ca	19.8
Na	0.78
K	1.52
Extractable micronutrients	
Cu	3.69
Fe	229.0
Mn	124.0
Zn	1.24
Textural class %	
Sand	75.8
Silt	13.4
Clay	10.8
Textural class	Sandy loam

Results

Table 1 shows the chemical analysis of fertilizer treatment used in the study. The total N in Aleshinloye Grade B is 11.7 g/kg and Jatropha cake is 43.1 g/kg. K in Aleshinloye Grade B is 2.09 g/kg and Jatropha cake is 2.2 g/kg. Ca in Aleshinloye Grade B is 23.4 g/kg and Jatropha cake is 0.3 g/kg. Mg in Aleshinloye Grade B is 2.4 g/kg and Jatropha cake is 8.39 g/kg. K in Grade B is 20.9 g/kg and Jatropha cake is 2.2 g/kg. P in Aleshinloye Grade B is 7.6 g/kg, while Jatropha cake is 0.7 g/kg. EC in Grade B is 16.0 ds/m and Jatropha cake is 13.8 ds/m. The physico–chemical properties of soil use are shown in Table 2. The soil textural class was sandy loam and slightly acidic with pH of 5.7.

Effects of fertilizers on growth and yield of maize at the first cropping

Table 3 shows the influence of fertilizers on growth and yield components of maize at the first and second cropping. Application of CJ at 2.5 t/ha performed better than the NPK fertilizer and control. Mean plant height of 152.6 cm was recorded for maize treated with 50% Grade B + 50% JC at 2.5 t/ha. The values obtained for this height was 13 and 33% higher than NPK and control, respectively. The highest stem girth (27.99 mm) recorded in maize was from treatment CJ (50% Grade B + 50% JC) at 2.5 t/ha. The lowest (13.71 mm) girth was recorded in the control and it was significantly ($p < 0.05$) lower than all other treatments. However, no significant difference ($p < 0.05$) among the values obtained for the stem girth of AJ and BJ at 1.5 and 2.0 t/ha, respectively. Maize under the CJ at 2.5 t/ha treatment performed best with 14.6 number of leaves. At 5% probability, there was no significantly difference between 2.0 t/ha BJ and 2.5 t/ha CJ at that gave 14.6 numbers of leaves. However, the treatment 2.0 t/ha BJ and 2.5 t/ha CJ were significantly ($p < 0.05$) higher than all other treatments as well as control. Application of CJ (50% Grade B + 50% JC) at 2.5 t/ha produced the highest leaves area with 501.6 cm^2. This values was not significantly different ($p < 0.05$) from BJ (30% Grade B + 70% JC) at 2.0 t/ha that had 493.9 cm^2, but was significantly ($p < 0.05$) different from control, NPK, and AJ (30% Aleshinloye Grade B + 70% Jatropha cake) at 1.5 t/ha. Maize grown with CJ at 2.5 t/ha consistently showed significantly ($p < 0.05$) higher grain yield (3.1 t/ha), and 100-seed weight (43.7 g) during the first planting. The grain yield from CJ at 2.5 t/ha was 58 and 29% higher than from control and NPK plot, respectively.

Influence of soil amendment on growth and yield of maize at second cropping

Among the fertilizer in the second cropping (residual) CJ at 2.5 t/ha produced the tallest plant of 112.6 cm, this values was significantly ($p < 0.05$) higher than values of 71.0, 76.6, 91.7, and 81.5 cm obtained from control, NPK, AJ at

Table 3 Effects of fertilizer types on growth and yield of maize at the first and second field cropping

Treatment	Rate (t/ha)	Plant height (cm)	Stem girth (mm)	Number of leaves	Leaves area cm^2	100 seed weight (g)	Grain yield (t/ha)
First cropping							
Control	0	102.8d	13.71d	11.3c	243.0d	18.0d	1.3c
NPK	0.06 kg N	123.1c	18.84c	12.0c	418.1c	31.9b	2.2b
AJ	1.5	141.5b	19.51b	13.1b	476.9b	24.7c	2.0b
BJ	2.0	149.9a	22.27b	14.3a	493.9a	22.3c	2.5b
CJ	2.5	152.6a	27.99a	14.6a	501.6a	43.7a	3.1a
Residual effect							
Control	0	71.0e	11.58c	11.9c	220.1d	10.0c	0.6c
NPK	0.06 kg N	76.6d	14.68b	12.2b	255.5c	11.9b	1.9b
AJ	1.5	91.7b	14.13b	12.9b	296.3b	13.7b	1.5b
BJ	2.0	81.5c	12.34c	12.9b	245.9c	11.0b	1.6b
CJ	2.5	112.6a	26.13a	13.6a	408.0a	18.7a	2.4a

Means having the same letter along the columns indicate no significant difference using Duncan's Multiple Range Test at 5% probability level

AJ—30% Aleshinloye Grade B + 70% Jatropha cake

BJ—30% Aleshinloye Grade B + 70% Jatropha cake

CJ—50% Aleshinloye Grade B + 50% Jatropha cake

1.5, and BJ at 2.0 t/ha, respectively. However, the plant height from CJ at 2.5 t/ha was 37 and 32% higher than from control and NPK plot, respectively. The stem girth ranged from 11.58 mm in control to 26.13 mm in CJ at 2.5 t/ha. Stem girth of maize plant with CJ at 2.5 t/ha was 56% greater than the value obtained from the control plants. The lowest stem girth recorded from control was 11.58 mm. The highest number of leaves of 13.6 was obtained from CJ at 2.5 t/ha which was significantly higher at 5% probability than other fertilizer treatments. The lowest mean number of leaves 11.9 was produced from control. The value of number of leaves obtained from CJ at 2.5 t/ha was 13% greater than the values produced from the control plant. The highest value of leaf area 408.0 cm^2 was produced with application of 2.5 t/ha of CJ. The least leaf area produced from the control plant has a value of 220.1 cm^2. The fertilizer 2.5 t/ha CJ at 5% probability level significantly influenced leaf area development in the residual experiment. The residual effects of application of compost resulted to higher yield components of maize than NPK and control. In maize, 100-seeds and grain yield obtained with the residual CJ at 2.5 t/ha compared favourable and significantly ($p < 0.05$) different from what was obtained with plants fertilized with NPK fertilizer. This rate gave about 8 and 75% more grain yield than NPK treatment and control, respectively.

Effects of fertilizer types on some soil chemical and physical properties at harvest during the first and second planting

Table 4 shows the soil analysis after maize harvest at the first and second planting. At the first harvesting stage of maize, pH value ranged from 5.5 in control plots to 6.1 in 2.5 t/ha CJ. Application of CJ at 2.5 t/ha increased soil pH value than other treatment including control and NPK. Furthermore, control soil had EC of 1.2 ds/m, while plot treated with compost supplemented with jatropha ranges from 4.8 to 7.3 ds/m. However, at residual harvest, soil treated with AJ at 1.5, BJ at 2.0 t/ha and CJ at 2.5 t/ha increased soil electrical conductivity content in the soil over control and NPK fertilizer plot. Soil available P content in soil ranged between 2.65 mg/kg from control plots and 44.29 mg/kg from NPK plots, which was higher than the plot treated with compost supplemented with Jatropha cake. However, there was a general increase in soil available P after comparing with the pre-cropping results (Table 2). The control soil had available K content of 2.14 cmol/kg, while that in plot treated with compost supplemented with Jatropha cake increased from 2.30 to 2.32 cmol/kg. Soil treated with AJ at 1.5, BJ at 2.0 t/ha,

Table 4 Effects of fertilizer types on some soil chemical and physical properties at harvest during the first and second planting

Treatment	Rate (t/ha)	EC (ds/m)	pH (H_2O)	Available P mg/kg	K cmol/kg	OC	N	Clay	Silt	Sand
First Harvest										
Control	0	1.2d	5.5e	2.65e	2.14e	15.2e	2.84b	80.0b	90.4c	820.6a
NPK	0.06 kg N	3.2c	6.0b	44.29a	2.28d	24.7b	2.19b	80.0b	130.4a	780.6c
AJ	1.5	4.8c	5.7d	6.32c	2.30c	17.9c	3.62a	80.0b	90.4c	820.6a
BJ	2.0	5.7b	5.9c	3.30d	2.31b	16.0d	2.24d	80.0b	110.4b	800.6b
CJ	2.5	7.3a	6.1a	6.75b	2.32a	31.2a	2.59c	120.0a	130.4a	740.6d
Second Harvest										
Control	0	0.7d	7.0d	7.7e	0.33e	7.5e	0.77e	18.2e	114.0d	867.8a
NPK	0.06 kg N	1.3c	7.2c	11.05c	0.41b	7.8d	0.81d	34.0a	126.0b	840.0e
AJ	1.5	6.1b	7.5b	13.93b	0.39c	14.6b	1.51b	24.6d	114.0d	861.4b
BJ	2.0	6.5b	7.2c	10.66d	0.36d	8.9c	0.92c	28.4b	116.2c	855.4c
CJ	2.5	7.5a	7.6a	20.44a	0.52a	15.9a	1.65a	27.0c	132.0a	841.0d

Means having the same letter along the columns indicate no significant difference using Duncan's multiple range test at 5% probability level

AJ—30% Aleshinloye Grade B + 70% Jatropha cake

BJ—30% Aleshinloye Grade B + 70% Jatropha cake

CJ—50% Aleshinloye Grade B + 50% Jatropha cake

and CJ at 2.5 t/ha increased soil available K content in the soil over NPK fertilizer plot.

Soil organic carbon ranged between 15.2 g/kg from control plot and 31.2 g/kg from plot with CJ at 2.5 t/ha. NPK plot had a total carbon content of 24.7 g/kg. There is a general increase in the values of organic C compared with pre-cropping (Table 2). Total N varied between 2.84 g/kg from control plot and 3.62 g/kg from 1.5 t/ha 30% Grade B + 70% JC plot. At residual harvesting stage of maize, pH value ranged from 7.0 in control plots to 7.6 in 2.5 t/ha CJ. Plot treated with CJ at 2.5 t/ha increased in pH value than other treatment including control and NPK. Soil available P content in maize plot extended between 7.7 mg/kg from control plots to 20.44 mg/kg from CJ at 2.5 t/ha plots. CJ fertilizer at 2.5 t/ha plots had available P value of 20.44 mg/kg which was greater than NPK plot. However, there was general increase in soil available P when compared with the first cropping values (Table 2). Control plots had available K content of 0.33 cmol/kg, while soil from plot treated with compost supplemented with Jatropha cake increased soil available K from 0.36 to 0.52 cmol/kg. Soil treated with CJ at 2.5 t/ha increased soil available K content in the soil over NPK fertilizer plot. Soil organic carbon ranged between 7.5 g/kg from control plot to 15.9 g/kg from CJ (50% Grade B + 50% JC) at 2.5 t/ha. NPK plots had nitrogen content of 0.81 g/kg N and varied between 0.77 g/kg from control plot and 1.65 g/kg from 50% Grade B + 50% JC at 2.5 t/ha plot.

Relationships between maize yield and soil properties

Table 5 presents Pearson correlation coefficients (r) between the soil parameters studied and maize yield at $p < 0.05$. At the first cropping, maize yield was positively correlated with the pH content (0.9202) and EC content (0.9361). However, at residual cropping, the yield was significantly and negatively correlated with the sand content (−0.8957) and positively correlated with K content (0.8876).

Table 5 Correlation coefficients between maize yields and soil chemical properties at the first and residual cropping

Soil properties	Correlation Coefficient (r)	
	First cropping	Residual cropping
EC	0.9361*	0.6073 NS
pH	0.9202*	0.7583 NS
Available P	0.0457 NS	0.8299 NS
K	0.8567 NS	0.8876*
OC	0.7498 NS	0.5787 NS
Total N	– 0.3559 NS	0.5855 NS
Clay	0.7442 NS	0.7334 NS
Silt	0.7564 NS	0.8375 NS
Sand	– 0.8498 NS	– 0.8957*

NS not significant

*Significant at five levels of probability ($p < 0.05$)

Discussion

The result of pH was in a range of 5.0–6.8 for tropical soil (Udo and Ogunwale 1977). Organic matter content 6.84 g/kg was below the critical level for maize crop production given as 30 g/kg (Agboola and Corey 1972a, b). The available P was 9.3 mg/kg, showing that it was lower than the critical level between 10 and 15 mg/kg (Adeoye and Agboola 1985). This indicates that the soil needs the addition of nutrient for crop optimum production. The total nitrogen was 3.1 g/kg, showing that it was greater than the critical level between 0.01 and 1.20 g/kg for the soils in the tropics (Bremner and Mulvancy 1982). A calcium value of 19.8 cmol/kg was higher than the critical level of 2.6 cmol/kg (Agboola and Corey 1972a, b). Exchangeable sodium value which was 0.78 cmol/kg was higher than the critical level for tropical soils between 0.48 and 0.94 cmol/kg. Exchangeable magnesium values of 4.54 cmol/kg were higher than the critical level of 0.15 cmol/kg. Exchangeable potassium values of 1.52 cmol/kg were greater than the critical level of 0.2 cmol/kg (Adeoye 1986).This indicates that potassium amendment is not needed for the optimum of maize crop production.

Maize growth and yield parameters significantly increased by highest values obtained from CJ at 2.5 t/ha. The fertilizer 2.5 t/ha CJ gave the higher maize grain yield of 3.1 t/ha. The high maize yield obtained from CJ at 2.5 t/ha could be due to the stability of the compost with Jatropha cake when applied to soil. The highest organic carbon and K after the harvest could be attributed to the relatively higher nutrient concentration in the compost supplemented with Jatropha cake.

The result from applied compost supplemented with Jatropha cake after harvesting of maize on soil fertility and its effect on some chemical properties such as soil pH, EC, soil available P, organic carbon, % N, exchangeable K, and clay content was presented. The increase in pH of the studied soil with the application of fertilizers was enhanced by contents of the various compost supplemented with Jatropha cake. The direction of the change in soil pH as a result of treatment application reflected the initial pH of the amendment material. The increase in pH could be due to the higher pH value of the jatropha cake (pH of 8.5) and Grade B fertilizer (9.3) in relation with soil (pH of 5.7). The rise in the soil pH level of the experimental soil after harvesting is an indication of the buffering capacity of compost with Jatropha cake applied. This enhanced the availability of the applied nutrients released from fertilizer resulting in increased crop growth and yield. Soil available P was assessed in the soil after maize had been harvested during the first cropping. Integration of composts into soil increases the salt content as well as soil electrical conductivity, especially if high rates of compost are applied because of the high salinity of composts (Angelova et al. 2013). The addition of compost and jatropha cake to the soils led to increase in EC values compared with control soil. The soils amended with compost had higher EC than the untreated soils. The soil EC increased with an increasing application rate of compost in soil as reported by Hossen et al. 2015. The EC of compost depends on the raw materials used for composting and their ion concentration (Atiyeh et al. 2001).

Furthermore, to ascertain that organic materials are known to release soil nutrients slowly in the soil (Agboola and Omueti 1982; Eghball et al. 2004; Tejada and Gonzalez 2007), the residual effect of compost supplemented with Jatropha cake on maize yield shows that 2.5 t/ha of 50% Grade B + 50% JC resulted to the maize grain yield of 2.4 t/ha. These results indicate that organic fertilizer has a high potential for building up residual nutrients in the soil. The nutrient added from NPK fertilizer was exhausted after the first cropping, leaving behind less nutrients that could sustain a good harvest during the second cropping.

The fertility study of the residual plots shows that residual soil organic carbon decreased in all the plots. The increase in the soil pH level of the plots after harvesting is an indication of the buffering capacity of the compost with Jatropha cake treatment applied. This enhanced the availability of the applied nutrients released from composts and jatropha cake, resulting in increased growth and yield. Adequate residual N was not observed in the plot fertilized with NPK, because much of the soil nitrogen in the inorganic fertilizer plots could have been lost to plant uptake and by other pathways (e.g., denitrification, ammonia volatilization, and leaching). In contrast, the soil N contents in the plots of maize the application of 50% Grade B + 50% JC at 2.5 t/ha had a slow but steady N supply capacity. The presence of compost with Jatropha cake contributed to the soil available P. The residual P increased in the second cropping of maize. Soil available P was assessed in the soil after maize had been harvested during second cropping. The available P contents increased more than in NPK fertilizer or control plots. This reflects the high level of P available in the compost supplemented with Jatropha cake used. Increase in the availability of P after organic fertilizer applications has been recorded by Eghball (2002) and Eghball et al. (2004). The amount of residual K in the soil is closely linked to the soil organic matter. By formation of organic complexes with the help of the organic substances existing in compost added to the soil potassium mobility in the soil is reduced (Miller and Donahue 1991). The residual K decreased in plot fertilized with NPK than other compost with jatropha cake.

Positive linear correlations observed between maize yield and pH, EC, K, and sand indicate that these properties were strongly correlated with maize yield. This was in agreement with the study carried out by Rodrigues et al. 2012 who found out that a positive correlation between maize yield and K contents in soils. Moreover, Montezano et al. 2006 observed that soil property such as sand content was negatively correlated to maize yield.

Conclusion

In summary, from this study, it can be deduced that jatropha cake used supplemented with compost can serve as an alternative to mineral fertilizer. Furthermore, application of fertilizer CJ at 2.5 t/ha increased soil productivity and led to higher maize yield than using NPK 15–15–15. Therefore, 50% Grade B + 50% JC at 2.5 t/ha is recommended for growing maize crop in the studied area.

References

Adeoluwa OO, Suleiman ON (2012) Effect of human urine on the growth of performances of *Jathropha curcas* seedlings and some soil health indices. Nig J Soil Sci 22(2):186–193

Adeoye GO (1986) Comparative studies of some extractants for sedimentary soil of South Western Nigeria. Ph.D. Thesis, Univ of Ibadan, Ibadan, pp 201

Adeoye GO, Agboola AA (1985) Critical levels for soil pH, available P, K, Zn and Mn and maize ear leaf content of P, Cu and Mn in sedimentary soils of south-western Nigeria. Fertil Res 6:65–71

Afe AI, Oluleye F (2017) Response of okra (*Abelmuschus esculenthus* L. Moench) to combined organic and inorganic foliar fertilizers. Int J Recycl Org Waste Agricult 6:281–287. https://doi.org/10.1007/s40093-017-0175-5

Agboola AA, Corey RB (1972a) Soil testing calibration for NPK for maize in the soils derived from metamorphic and igneous rocks of Western State of Nigeria. J W Afric Sci Assoc 19:93–100

Agboola AA, Corey RB (1972b) Soil testing calibration for NPK for maize in the soils derived from metamorphic and igneous rocks of Western State of Nigeria. J W Afric Sci Assoc 19(2):93–100

Agboola AA, Omueti JAI (1982) Soil fertility problems and its management in tropical Africa, In: Inter. Conf. On land clearing and develop. Proceedings Vol. 2, I.I.T.A., Ibadan, Nigeria

Angelova VR, Akova VI, Artinova NS, Ivanov KI (2013) The effect of organic amendments on soil chemical characteristics. Bulg J Agric Sci 19(5):958–971

Atiyeh RM, Lee Edward CA, Sulbar S, Metzger T (2001) Pig manure vermicompost as a component of a horticultural bedding plant medium. Effects on physiochemical properties and plant growth. Bioresour Technol 78:11–20

Babajide PA, Olla NO (2014) Performance of indigenous *Celosia argentea* variety and soil physico-chemical properties as affected by dual application of compost and single N-mineral fertilizer in Southern Guinea Savanna vegetation zone of Nigeria. J Bio Agric Healthcare 4(19):69–74

Bouycocous GJ (1951) Direction for making mechanical analysis by the hydrometer method. Soil Sci 42:225–229

Bray RH, Kurtz LT (1945) Determination of total and organic of phosphate in the soil. Soil Sci J 59:45–49

Bremner JM, Mulvancy CS (1982) Total Nitrogen. In: Page AL et al. (ed) Methods of soil analysis part 2, Agron Monogr. 9 (2nd edition)) ASA and SSSA, Madison, Wsc, pp 403–430

Cano-Asseleih LN, Plumbly RA, Hylands PJ (1989) Purification and partial characterization of the hemagglutination from seeds of *Jatropha curcas*. J Food Biochem 13:1–20

Chaturvedi S, Kumar A, Singh B, Nain L, Joshi M, Satya S (2013) Bioaugmented composting of *Jatropha* de-oiled cake and vegetable waste under aerobic and partial anaerobic conditions. J Basic Microb 53(4):327–335

Crews TE, Peoples MB (2004) Legume versus fertilizer sources of nitrogen: ecological trade-offs and human needs. Agric Eco Environ 102(3):279–297

Eghball B (2002) Soil properties as influenced by phosphorus- and nitrogen—based manure and compost applications. Agron J 94:128–135

Eghball B, Ginting D, Gilley JE (2004) Residual effects of manure and compost application on corn production and soil properties. Agron J 96:442–447

Hossen S, Islam N, Alam R, Baten A (2015) Effects of different rates of compost application on methane emission and crop yield in aman rice. J Agric Nat Res Sci 2(3):530–536

Jayasingh M (2003) The use of Bio-diesel by the Indian Railways. In: Proceedings of the National Workshop 5–8 August, Pune, India

Kumar A, Sharma S (2008) An evaluation of multipurpose oil seed crop for industrial uses (*Jatropha curcas* L.): a review. Ind Crops Prod 28:1–10. https://doi.org/10.1016/j.indcrop.2008.01.001

Miller RW, Donahue RL (1991) An introduction to soils and plant nutrient, 6th edn. Prentice Hall, USA, p 768

Montezano ZF, Corazza EJ, Muraoka T (2006) Variabilidade espacial da fertilidade do solo em área cultivada e manejada homogeneamente. Rev Bras Cienc Solo 30:839–847. https://doi.org/10.1590/s0100-06832006000500010

Nelson DW, Sommers LW (1982) Total carbon, organic carbon and organic matter. In: Page AL, Miller RH, Keeney DR (eds) Method of soil analysis, part 2, Chemical and microbiological properties, pp 539–579

Olowoake AA (2017) Nutrient Release Pattern from Compost Supplemented with Jatropha Cake on Alfisol of Ilorin, Nigeria. Int J P and Soil Sci 15((5):1–11. https://doi.org/10.9734/ijpss/2017/32523

Olowoake AA, Ojo JA (2014) Effect of fertilizer types on the growth and yield of *Amaranthus caudatus* in Ilorin, Southern Guinea Savanna Zone of Nigeria. Adv in Agric 3(4):121–125. https://doi.org/10.1155/2014/947062

Rodrigues MS, Corá JE, Fernandes C (2012) Spatial Relationships between soil attributes and Corn yield in no-tillage system. Rev Bras Cienc Solo 36:599–609. https://doi.org/10.1590/S0100-06832012000200029

Singh Y, Singh CS, Singh TK, Singh JP (2006) Effect of fortified and unfortified rice-straw compost with NPK fertilizers on productivity, nutrient uptake and economics of Rice (*Oryza sativa*). Indian J Agron 51:297–300

Sivaprakashan K (1991) Soil Amendment for crop disease management. In: Vidyasekharan P (ed) Basic research for crop disease management. Day Publishing House, New Delhi, pp 382–404

Tejada M, Gonzalez JL (2007) Application of different organic wastes on soil properties and wheat yield. Agron J 99:1597–1606. https://doi.org/10.2134/agronj2007.0019

Tiwari KN (2002) Nutrient management for sustainable agriculture. J Ind Soc Soil Sci 50:374–377

Udo EJ, Ogunwale JA (1977) Phosphorus fractions in selected Nigeria soils. SSSA J 46(6):1141–1146. https://doi.org/10.2136/sssaj1977.03615995004100060026x

18

Recycling of biogas digestates in plant production: NPK fertilizer value and risk of leaching

Trine A. Sogn[1] · Ivan Dragicevic[1] · Roar Linjordet[3] · Tore Krogstad[1] · Vincent G. H. Eijsink[2] · Susanne Eich-Greatorex[1]

Abstract

Purpose The main purposes of the study were to assess the NPK fertilizer value of biogas digestates in different soils and to evaluate the risk of unwanted nutrient leaching.

Methods The fertilizer value of digestates from anaerobic digesters was investigated in a greenhouse pot experiment with wheat in three different soils; silt, loam and sand. The digestates were based on different feedstock and had a low, dry matter content. The fertilizing effect of digestates was compared to mineral fertilizer and manure. To investigate the fate of excess nutrients in soil after the growing season, the pots were leached after harvest. A complementary soil column leaching experiment without plants was carried out in the laboratory.

Results The concentration of ammonium in digestates provided a good indicator of the nitrogen fertilizer value of the digestates. In the silt and loam, the ammonium N fraction in digestates had a fertilizer replacement value equal to that of mineral fertilizer N, whereas the replacement value was higher in the nutrient poor sandy soil. Digestates often have a ratio between nitrogen, phosphorus and potassium which is not favourable for plant growth. However, the suboptimal balance did not result in reduced plant growth or unwanted leaching from soil.

Conclusions The results show that digestates from biogas production based on fundamentally different feedstock are promising as NPK fertilizers. The N fertilization can simply be based on the digestate NH_4^+ concentration and, at least for wheat production, considerable variation in the concentrations of K and P can be tolerated.

Keywords Digestates · Fertilizer · Nitrogen · Phosphorus · Potassium

Introduction

The core of the bioeconomy is the change from use of fossil raw materials to sustainable production, refinement and utilization of renewable biomass. For instance, production of biogas using organic waste as feedstock represents a plausible activity within the bioeconomy. Organic waste becomes a valuable resource, and compared to fossil fuel utilization, use of biogas yields a reduced carbon footprint (Rehl and Müller 2013; Uusitalo et al. 2014). Biogas production through anaerobic digestion leaves organic residues, digestates, which are rich in nutrients. If these digestates are utilized in plant production, nutrients already within the nutrient cycle will be recycled and the sustainability of the biogas production process improves (Arthurson 2009; Ladanai and Vinterbäck 2009; Vaneeckhaute et al. 2013). In addition, use of organic fertilizers contributes to maintaining and/or improving soil quality (Hati et al. 2006).

Utilization of digestates may replace or at least reduce the use of mineral fertilizer in agronomic plant production. The liquid phase of digestates is usually rich in plant-available nutrients such as ammonium (NH_4^+), phosphate and potassium (K). Additionally, more nitrogen (N) will be plant available by microbial decomposition and mineralization of the digestates' solid phase. As anaerobic digestion produces methane (CH_4) gas from the organic feedstock, which is subsequently removed for use as biogas, the digestates have a

✉ Trine A. Sogn
trine.sogn@nmbu.no

1 Faculty of Environmental Sciences and Natural Resource Management, Norwegian University of Life Sciences, Ås, Norway

2 Faculty of Chemistry, Biotechnology and Food Science, Norwegian University of Life Sciences, Ås, Norway

3 Norwegian Institute of Bioeconomy Research, Ås, Norway

lower carbon to nitrogen ratio (C/N ratio) than the feedstock (Möller and Müller 2012). Although dependent on decomposition time, a C/N ratio lower than about 25 usually entails that the organic matter has a surplus of N relative to the N demand of the soil microbial community, which then can be mineralized to NH_4^+, i.e. plant-available mineral N (Ferris et al. 1998).

Since farmers have to base their N fertilizer strategy on farm level N budgets with relatively high precision, the mineral N fertilizer equivalents or replacement values of different digestates are of importance. Studies with manure have shown low plant availability and then little effect from the organic N on plant growth in the first year of application due to slow mineralization (Möller and Müller 2012; Webb et al. 2013). On the other hand, the NH_4^+ in the liquid phase is plant-available (Gutser et al. 2005) and may be directly comparable with mineral N fertilizer. Nevertheless, it is common practice for farmers to estimate the fertilization value of manure by considering a small part of organically bound N and a reduced effectiveness of the mineral N (Kirchmann 1985), the latter in order to allow for gaseous and/or leaching losses at application and during the growing season. A major challenge with all organic fertilizers is the predictability of the additional mineral N release by decomposition of the organic matter. Assessments of the fertilizer potential of digestates have been carried out (Alburquerque et al. 2012; Kuszel and Lorencowicz 2015; Cavalli et al. 2016; Riva et al. 2016), but due to the wide variety of feedstock used for biogas production, the efficiency and exact predictability of the fertilizer N value of digestates is difficult. Over time, organic N in the solid phase of digestates will be mineralized and NH_4^+ will be released. In a production system with a continuous plant cover, the residual effect due to mineralization is positive as it ensures a gradual supply of NH_4^+ to the plants. However, in cereal production, the timing of the mineralization is crucial. For example, release of mineral N late in the growing season, when plants have less requirement for N, or in seasons without plant cover, will increase the risk of nitrate (NO_3^-) leaching. In agricultural areas with only one growing season per year and then often months without plant cover, release of mineral N from organic matter decomposition may increase the risk of N leaching.

Next to N, K and phosphorus (P) are quantitatively the most important plant nutrients and essential to secure proper growth and development. Phosphorus and K in the feedstock are generally retained during the anaerobic digestion process (Zirkler et al. 2014). In addition to dissolved inorganic phosphate in the liquid phase, the solid phase of the digestates contains both inorganic and organic P (Kataki et al. 2017). As a surplus of phosphate in soils may eventually lead to eutrophication of freshwater resources and estuaries, it is important to avoid overdosing of P, by balancing the P and N input. In plants, K is not built into organic complexes but retained in cationic form. Consequently, the K in digestates derived from plant material is directly plant-available both in the solid and liquid phase. Some grass plant species are known to have a higher uptake of K than required in the metabolism at high K availability (Øgaard et al. 2001). An excessive K uptake may be problematic due to antagonisms with magnesium, resulting in nutrient imbalances with negative consequences for growth and quality (e.g. digestibility) (Masters and Thompson 2016). Thus, also the amount of K relative to N and P is important for optimized fertilization.

Organic material in soil is not only important as a source of plant nutrients, it is also important for soil porosity and structure, as well as the ability to store water and nutrients. Soils low in organic matter are often compact, have poor structure and low nutrient binding capacity, as well as a low ability to store water.

This study was aimed at evaluating the impact of widely different feedstocks for biogas production on the mineral N fertilizer replacement values of the resulting digestates, as well as the P and K supply to grain crops and potential losses through leaching. To assess these issues we have carried out a greenhouse experiment with wheat, using five different digestates derived from biogas reactors running on different combinations of manure, whey permeate, fish silage, food waste and sewage sludge, in three different soils. We have compared plant growth as well as nutrient leaching in treatments with digestates with data obtained from control treatments, including no fertilization, application of mineral fertilizer and fresh manure. A supplementary soil column leaching experiment without plants was carried out in the laboratory in order to study nutrient leaching from soils without plants.

Materials and methods

The digestates

The experiments in this study included the following treatments: 1. Control without fertilization (C), 2. Mineral fertilizer (MF), 3. Fresh manure (FM), 4. Digested manure (EDM), 5. Whey permeate co-digested with manure (EWM), 6. Whey permeate and fish silage co-digested with manure (EWMF), 7. Commercial digestates from source separated organic household waste (CF) and 8. Commercial digestates from source separated organic waste mixed with sewage sludge (CFS). The first three were controls (no fertilizer, mineral fertilizer and manure, respectively); the next three were experimental digestates produced in the biogas laboratory of the Norwegian University of Life Sciences and the Norwegian Institute of Bioeconomy Research, located 35 km south-east of the Norwegian capital Oslo, whereas the final two were digestates from commercial biogas plants.

The experimental digestates were produced on a laboratory scale in continuously stirred tank reactors (CSTR, Dolly, Belach Bioteknik, Stockholm, Sweden) operating at 37 °C with a working volume of 6 L, and a hydraulic retention time of 20 days.

Treatment 4 used an experimental digestate from a reactor fed with manure only (EDM). The experimental digestates used in treatments 5 and 6 were produced in reactors fed with a mixture of whey permeate and manure in a volume (mL) ratio of 70 to 230 (EWM) and 180 to 120 (EWMF), respectively. The latter digestate (EWMF) also included 6 mL fish silage in the feedstock mixture, in order to reduce the C/N ratio. Organic loading rates for EDM, EWM and EWMF were 3.4, 4.3 and 5.4 g volatile solids (VS) L^{-1} day^{-1}, respectively. The digestates used in treatments 7 and 8 were commercial digestates collected from municipal biogas plants. The digestate in treatment 7 (CF) was from a commercial biogas plant where biogas is produced on source separated organic household waste. The digestate in treatment 8 (CFS) was from a commercial biogas plant where biogas is produced based on source-separated organic household waste mixed with sewage sludge (50:50, based on fresh weight).

The pot experiment

A pot experiment was carried out in a greenhouse with wheat (*Triticum aestivum L.*var. Zebra) as a test crop. The pots had a soil volume of approximately 6.7 L. Assuming a soil depth of 20 cm, 300,000 pots correspond to one hectare (ha) of arable land. Three different soils: sand, silt and loam (Table 2), were used as growth media. The pots were irrigated regularly in order to keep the soil water content at 60% of the soils' water holding capacity (WHC) throughout the experiment. No water was leached from the pots during the experiment. The experiment consisted of the eight treatments as explained above. The mineral fertilizer treatment (MF) corresponded to a dose of approximately 120 kg N ha^{-1}, 20 kg P ha^{-1} and 180 kg K ha^{-1}. The NPK level was based on common fertilizer recommendation for wheat production in Norway, but adjusted in order to give optimal wheat growth in pots within a greenhouse. Because release of plant-available mineral N from the manure and the solid fraction of the digestates usually is low during the first growing season after application due to slow mineralization (Svensson et al. 2004), the amounts of manure and digestates were adjusted based on their inorganic N content, i.e. NH_4^+ concentration (Tables 1 and 3) in wet samples only. Since the treatments (1–8) varied in percent DM, the soil moisture was adjusted by application of deionised water according to weight after the initial addition of fertilizers in order to achieve a soil moisture at 60% WHC. To reduce plant growth effects due to differences in soil pH (Table 2), 20 and 10 g calcium carbonate ($CaCO_3$) per pot were added to the loam and sand, respectively, bringing the pH up to approximately 6.2.

Plants were harvested after 14 weeks of growth by cutting the wheat straws one cm above the soil surface, and divided into cereal grain and straw after drying at 60 °C to constant weight.

Leaching experiment after harvest

In order to evaluate the risk of nutrient leaching after the growing season, a simple leaching study was conducted after the harvest. For that purpose, the pots were first irrigated to 100% WHC and then leached with 1 L deionized water, added in doses of 250 mL each during the course of an hour. The leachate drained from the soil through a tube at the base of the pots.

Soil leaching experiment without plants

Another leaching experiment was carried out with bare soils in the laboratory. This study was intended to represent a worst-case scenario with extremely heavy rainfall just after

Table 1 Chemical characteristics of the manure and the digestates (treatments 3 to 8) used in the experiments

Treatments	DM %	pH	Tot C g kg^{-1}	Tot N[a] g kg^{-1}	NH_4–N + NO_3–N[b] g L^{-1}	NH_4–N + NO_3–N[c] g kg^{-1}	P g kg^{-1} DM	K g kg^{-1} DM
3. FM	5.5	7.4	414	21	11.4	25.4	6.4	62.3
4. EDM	4.7	7.7	351	27	0.9	20.0	11.5	82.0
5. EWM	4.6	7.6	382	25	1.4	29.3	8.4	76.0
6. EWMF	4.5	7.0	378	31	0.7	16.3	15.0	100
7. CF	2.9	7.6	397	57	1.1	39.5	9.1	46.3
8. CFS	1.7	6.9	318	42	0.9	53.3	16.0	8.6

[a]g per 100 g DM, i.e. measured in samples dried at 55 °C

[b]Measured in fresh samples

[c]Calculated based on the measured concentration in fresh samples and DM%

Table 2 Characteristics of the soils used in the pot and leaching experiments

	pH_N[a]	pH_L[b]	Tot. C g kg^{-1}	Tot. N g kg^{-1}	NO_3–N mg kg^{-1}	NH_4–N mg kg^{-1}	P-AL[c] mg kg^{-1}	K-AL[c] mg kg^{-1}	CEC cmol kg^{-1}	Sand (%)	Silt (%)	Clay (%)
Loam	4.8	6.0	24	2	18	3.0	73	100	12.9	45	38	17
Sand	5.1	6.0	4	0	0.2	1.3	15	17	4.2	94	3	3
Silt	6.2	6.2	18	1	28	1.4	73	180	12.1	2	93	5

[a]Prior to $CaCO_3$ addition

[b]After $CaCO_3$ addition

[c]P-AL and K-AL denote plant-available fractions and were measured in soil extracts with an ammonium acetate–lactate solution (0.1 M ammonium lactate and 0.4 M acetic acid, pH 3.75)

adding the different fertilizers. Plexiglass columns of 24 mm in diameter and 30 cm in length were filled with 15 cm of the same three soils as in the greenhouse pot experiment, i.e. sand, silt and loam (Table 2). In order to prevent soil loss during the leaching experiment, a plastic filter covered with PVC pellets (2 mm diameter) was placed at the bottom of the columns. The top of each column was also covered with PVC pellets in order to secure effective distribution of the added water. Each column was covered with tinfoil in order to prevent light exposure and thereby minimize algae growth within the columns. The treatments were the same as in the growth experiment (Table 3) and amounts of fertilizer added per volume of soil (mL per L soil) corresponded to those in the growth experiment. Prior to leaching, the soil columns were saturated with deionized water. The columns were then irrigated with deionized water at a flow rate of 1.3 mL h^{-1} corresponding to 500 mm of precipitation during a 7 days period. The leachates were collected at the base of the column every day during the 7 days and filtered prior to analysis.

Chemical analysis

Chemical characteristics of the different digestates, as well as of the manure used in the experiments, are given in Table 1. Total C and N in manure and digestates were measured in dried and crushed samples by dry combustion (Nelson and Sommers 1982) at 1050 °C, using a Leco CHN-1000 instrument (St. Joseph, Michigan, USA) according to the Dumas method (Bremner and Mulvaney 1982). NH_4^+ in fresh digestates and manures was measured by flow injection analysis (FIA) after extraction with 2 M potassium chloride (KCl) as described in Bremner (1965). For the determination of the total concentrations of P and K, the manure and digestates were digested in concentrated ultrapure nitric acid for 90 min by stepwise heating up to 250 °C using a Milestone Ultraclave and analysed using Inductively Coupled Plasma Mass Spectrometry (Agilent ICP-MS 8800 TripleQ).

The soil pH was measured in deionised water, with a soil to liquid ratio equal to 1:2.5. The total C and N in the soil was determined in dried and crushed samples by dry combustion (Nelson and Sommers 1982) at 1050 °C, using a Leco CHN-1000 instrument (St. Joseph, Michigan, USA) as described above for manure and digestates. The soil content of NH_4^+ and NO_3^- was measured by flow injection analysis (FIAstar 5000, based on ISO 13395-1996 and EN ISO 11732:1997) after extraction with 2 M KCl. Plant-available P (P-AL) and K (K-AL) were estimated by extraction with an ammonium acetate–lactate solution (0.1 M ammonium lactate and 0.4-M acetic acid, pH 3.75) (Egnér et al. 1960), followed by inductively coupled plasma optical emission spectrometry (ICP-OES, Perkin Elmer Optima 5300 DV).

Table 3 Treatments 1–8 expressed as element addition (mg) per pot

Treatments		Manure/digestate mL pot^{-1b}	N[a]:P:K ratio	N[a] mg pot^{-1b}	P	K
1.	C			–	–	–
2.	MF		4:1:6	396	75	600
3.	FM	286	4:1:10	400	101	980
4.	EDM	422	4:2:16	397	228	1626
5.	EWM	294	4:1:10	399	114	1028
6.	EWMF	540	4:4:24	399	365	2430
7.	CF	349	4:1:5	399	91	461
8.	CFS	441	4:1:1	399	119	64

[a]Inorganic N only

[b]The doses multiplied with 0.3 correspond to kg per hectare. The doses divided by 100 correspond to mg per soil column in the soil column experiment

The particle size distribution (i.e. percent of sand, silt and clay in the soil, Table 2) was determined by the pipette method (Elonen 1971).

The leachates from the pots after harvest and from the soil column experiment were analysed for ammonium (NH_4–N) and nitrate (NO_3–N) by flow injection analysis (FIAstar 5000, based on ISO 13395-1996 and EN ISO 11732:1997) and for other nutrients by ICP-OES (Perkin Elmer Optima 5300 DV).

Calculations and statistical analysis

The mineral N fertilizer replacement value is the ratio between the grain yield obtained in the digestates or manure treatments (treatments 3–8) and the grain yield obtained in the mineral N fertilizer treatment.

The total biomass and grain yield figures in t ha^{-1} are calculated by multiplying the amounts measured in g DM pot^{-1} with a factor 0.3. The factor is generated assuming a soil depth of 20 cm and thus 300,000 pots of 6.7 L correspond to one hectare (ha) arable land. The 300,000 ha are divided by 10^6 to convert from gram to tons of biomass or grain yield.

Each treatment, both in the column and pot experiments, was replicated three times for each soil. Treatment and soil effects were tested statistically by analysis of variance and multiple testing using the GLM and Student–Neuman–Keul's test (SNK) available in the SAS computer work package (SAS v.9.4). Differences were considered significant at $p < 0.05$.

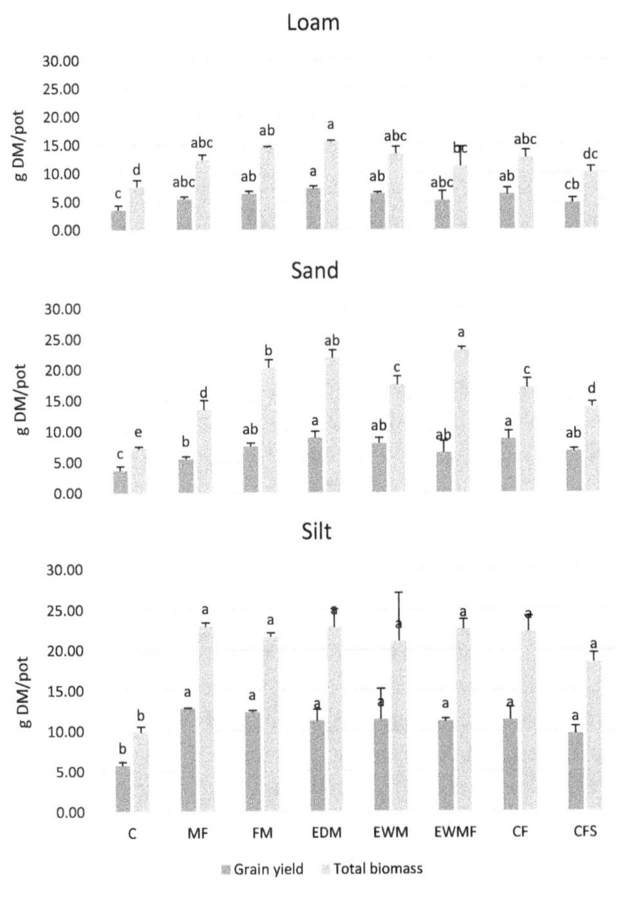

Fig. 1 Total biomass (g DM pot^{-1}) and grain yield (g DM pot^{-1}) in loam, sand and silt. The standard deviation within each treatment (n = 3) is indicated by the line extending the column. Within each soil, significant differences are indicated by letters

Results and discussion

Biomass and yield

Digestates generally had the same fertilizing effect on total biomass and grain yields as manure, with manure being similar to or better than mineral fertilizer (Fig. 1. Multiplying the figures (g DM pot^{-1}) with a factor 0.3 gives total biomass and grain yield in t ha^{-1}). Total biomass and grain yields were generally lower in sand and loam compared to silt, with an average grain yield across the treatments corresponding to 1.7 t ha^{-1} in the loam, 2.1 t ha^{-1} in the sand and 3.2 t ha^{-1} in the silt. The organic fertilizers showed a significantly positive biomass effect (not significant for the CFS treatment) relative to the mineral fertilizer in the sand only.

The variation in grain yield coincided generally with the variation in total biomass production. Figure 1 shows that in the silt and loam, the nitrogen fertilizer replacement value (NFRV) (Webb et al. 2013) for manure and digestates is approximately 1, when based on the NH_4^+ concentration measured in the fresh samples of the digestates (Table 1). In the sand, the NFRVs of manure and digestates seem to be higher than 1 (significant only for the EDM and CF treatments). The poor growth of wheat plants receiving mineral fertilizer may be connected to the high mobility of mineral N in sand. Due to their content of organic material, manure and digestates may contribute to keeping plant-available N in the top layer of the soil at the beginning of the growing season, thus ensuring a sufficient supply while the root systems is not yet well developed. Mineral fertilizer-based N will easily be washed down in the sandy soil due to lack of binding sites. The sand is a soil poor in nutrients and with a low content of organic matter (Table 2) and as such a growth medium in which positive effects of digestates are expected. In addition to the effect of the organic matter addition on improving nutrient availability, other plant growth conditions, such as water-holding capacity are likely to improve.

Since the manure and digestates varied in nutrient concentrations (Table 1), and the amount of manure and digestates applied to the pots in the different treatments was adjusted to give the same mineral N input, the input of P and K varied in the different treatments (Table 3). No additional fertilizer was applied in order to compensate for these differences. In the mineral fertilizer treatment, the ratio between N, P and K was 4:1:6. The commercial digestate based on food waste (CF) had almost the same balance between N, P and K as the mineral fertilizer, while the experimental digestate based on whey permeate, manure and fish silage (EWMF) represented the highest input of both P and K relative to the mineral fertilizer (4:4:24; Table 3). Although the mineral fertilizer treatment represents an adequate balance between the macronutrients N, P and K for wheat, it is still possible that growth in sand, which is poor in nutrients (Table 1), was stimulated by the extra supply of P and K in the manure and some of the digestates. However, also the digestates not particularly rich in P and K resulted in good yields on the sandy soil (Fig. 1) and the differences among the manure and digestates were not significant. There was no clear overall trend. The N to P ratio in the organic amendments was never lower than in the mineral fertilizer treatment, i.e. 4 to 1, but the supply of K was lower in treatments with commercial digestates (CF and CFS). No symptoms related to K deficiency were seen on the wheat plants in any of the soils and treatments. Thus, all in all it seems that all digestate treatments (Table 3) provided P and K in amounts and ratios which can be regarded adequate, i.e. within a range suitable for wheat production. This is in line with Tambone et al. (2010) who, based simply on chemical, spectroscopic and biological characteristics of several digestates appraised digestates to have good fertilizing properties just due to a high content of N, P and K in easily available form. However, no plant growth experiments were conducted in that study.

In the silt and loam, application of manure and digestates resulted in the same biomass production and grain yield as application of mineral fertilizer. Independent of treatment, the lowest biomass and grain yields were observed in the loam (Fig. 1). This was most likely due to limited air supply (Currie 1962) since the sieved loam provides a dense growth medium upon irrigation. Notably, in this study the dense loam was not improved by application of organic matter.

Leaching

Nitrogen

Nitrogen applied to soil as NH_4^+ will rapidly nitrify to NO_3^- if not taken up by plants. Both NH_4^+ and NO_3^- are plant-available, but cereals in general prefer NO_3^- (Cramer and Lewis 1993). A surplus of NO_3^- in soil relative to the plant requirement will be leached after harvest or denitrified. Generally, the NO_3^- leaching from the pots after harvest

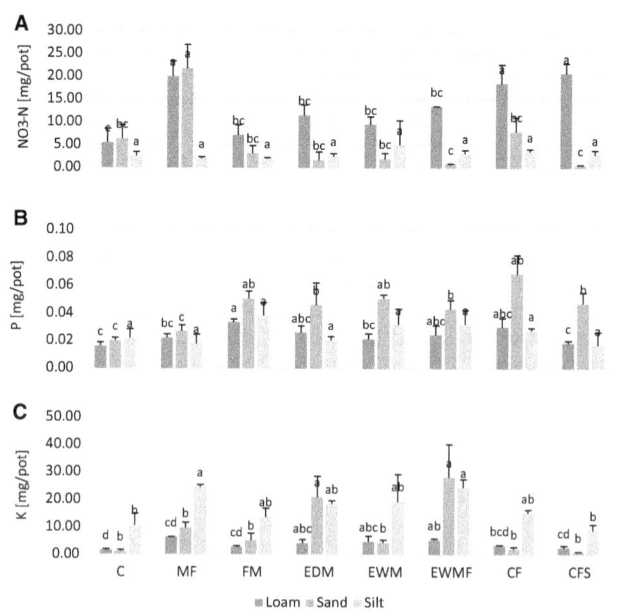

Fig. 2 Leaching of NO_3–N (**a**), total P (**b**) and total K (**c**) (mg pot^{-1}) measured after harvest in the pot experiment. The standard deviation within each treatment (n = 3) is indicated by the line extending the column. Within each soil, significant differences are indicated by letters

(Fig. 2a) was negatively related to the biomass growth, i.e. high biomass growth resulted in low NO_3^- leaching. Thus, across the fertilizer treatments, NO_3^- leaching was generally highest from the loam. Both in the sand and loam, high NO_3^- leaching was seen in the mineral fertilizer treatment (Fig. 2a). In these soils, the N given as NO_3^- (calcium nitrate) was not utilized by the plants. In the sand, the NO_3^- seemed to be leached towards the bottom of the pot before an effective root system had been developed. In the loam, root development might have been restricted by compactness, and thus, roots may physically not have been able to reach all the fertilizer NO_3^-. The compactness of the loam may restrict oxygen availability as well as root development, causing reduced uptake of mineral N during the period of vegetative growth when the plant N demand is at the highest. In the loam, NO_3^- leaching was reduced when manure and experimental digestates were applied (Fig. 2a), but the commercial digestate treatments (CF and CFS) showed the same high NO_3^- leaching as the MF treatment. Thus, the application of organic matter did not improve the N utilization in the compact loam in general. In the sand, rapid water flow towards the bottom of the pot may have limited the plants' ability to scavenge the nutrients. The NO_3^- leaching from pots treated with the manure or digestates was significantly lower compared to the MF treatment (Fig. 2a). This latter observation is in accordance with data from Svoboda et al. (2013) who, in field experiments with maize production in a sandy loam and a sand, found that NO_3^- leaching was lower upon treatment with digestates or manure, compared to treatment with mineral fertilizer. Interestingly in our experiments, NO_3^- leaching from sand receiving manure or digestates was mostly lower (significant for the EWMF and commercial digestates) than from sand not receiving any fertilizer. A similar inhibitory effect of organic fertilizer on NO_3^- leaching was found by Kramer et al. (2006) based on studies in an organic farming system. They suggested that the reduced NO_3^- leaching could be due to a positive effect of organic matter on the activity of the denitrifying microbial communities.

In the soil column leaching experiment, the amount of elements not retrieved in the leachate must be considered adsorbed in the soil or, in the case of N, possibly also lost as gaseous N species. Generally only 50% or less of the mineral N added was leached (Table 4) despite the very high irrigation rate equivalent to 500 mm of precipitation during 7 days. When applied on agricultural land, manure is usually mechanically incorporated into the top soil to avoid ammonia (NH_3) losses. In the column experiment,

Table 4 Total amounts (mg) of N, P and K applied (in) and leached (out) within 7 days in the soil column experiment

Soil	Treatments	N mg column^{-1}			P mg column^{-1}		K mg column^{-1}	
		N in	NO_3–N out	NH4–N out	P in	P out	K in	K out
Loam	1. C	–	1.1	0.0	–	0.0	–	1.0
	2. MF	3.6	1.5	0.1	0.8	0.0	6	1.4
	3. FM	4.1	1.2	0.0	1	0.0	9.9	1.3
	4. EDM	4	1.3	0.0	2.3	0.0	16.2	1.1
	5. EWM	4	1.0	0.0	1.1	0.0	10.1	1.1
	6. EWMF	4	2.1	0.0	3.7	0.0	24.3	1.1
	7. CF	4	1.0	0.0	0.9	0.0	4.6	1.1
	8. CFS	4	0.9	0.0	1.2	0.0	0.6	0.9
Sand	1. C	–	0.0	0.0	–	0.0	–	0.4
	2. MF	3.6	0.7	0.0	0.8	0.0	6	4.7
	3. FM	4.1	0.0	0.7	1	0.1	9.9	5.9
	4. EDM	4	0.0	0.7	2.3	0.1	16.2	7.5
	5. EWM	4	0.0	0.6	1.1	0.1	10.1	5.6
	6. EWMF	4	0.0	0.8	3.7	0.2	24.3	15.4
	7. CF	4	0.0	0.4	0.9	0.1	4.6	1.4
	8. CFS	4	0.0	0.3	1.2	0.0	0.6	0.8
Silt	1. C	–	1.1	0.0	–	0.0	–	3.1
	2. MF	3.6	1.3	0.0	0.8	0.1	6	5.7
	3. FM	4.1	0.5	0.1	1	0.1	9.9	7.4
	4. EDM	4	1.2	0.6	2.3	0.2	16.2	10.0
	5. EWM	4	0.9	0.3	1.1	0.1	10.1	5.6
	6. EWMF	4	0.9	0.3	3.7	0.1	24.3	7.2
	7. CF	4	0.7	0.1	0.9	0.1	4.6	5.0
	8. CFS	4	1.0	0.5	1.2	0.0	0.6	4.1

however, manure and digestates were applied on the top of the soil and washed into the soil with the waterfront. The air temperature in the laboratory was 20 °C. Although the soil surfaces were kept moist and also covered with pellets, the soil pH was 6 and slightly above (Table 2) after the liming and some N may have been volatilized as NH_3. In the loam, N was mainly leached as NO_3^-, in the sand mainly as NH_4^+, while in the silt, both forms occurred in the leachate with the major part being NO_3^- (Table 4). Although the loam is relatively compact and likely not rich in oxygen, the leaching data show that NH_4^+ must have been nitrified to NO_3^- before leaching, an observation that probably reflects a combination of high microbial activity and high retention. N leaching from the sand treated with organic fertilizers occurred mainly as NH_4^+ (Table 4), which is likely due to short retention times of non-adsorbed nitrogen in this porous soil. Still, N leaching in sand was generally less than 20% of the mineral N added. The cation exchange capacity of the sand is low, but might have been slightly improved by the organic matter addition. Thus, adsorption of NH_4^+ to negative surface charges may occur in the organic fertilizer treatments. The fact that NO_3^- leaching was found in the mineral fertilizer treatment, but not in sand receiving digestates or manure suggests also a positive effect of organic matter on the activity of denitrifying microbial communities (Kramer et al. 2006). In the silt, the soil with the highest natural content of NO_3^- (Table 2), N leaching was generally higher than in the sand, but at the same level as in the loam. In the silt, the main part of the N leached was as NO_3^-, but a noticeable amount also as NH_4^+. Thus, in this simple and well-controlled soil system, N species and amounts leached following addition of organic fertilizer depended on the soil type.

Phosphorus

The doses of total P in the digestates and manure treatments were equal to or higher than the amount of phosphate applied in the mineral fertilizer treatment (Table 3). The leaching of P in the pot experiment was generally very low (Fig. 2b). For all the three soils, P leaching from the mineral fertilizer treatment was as low as in the pots with no fertilization. Though very low, P leaching for the organic fertilizer treatments in the sand was significantly higher compared to the no fertilizer and mineral fertilizer treatments. It is interesting to note that the EWMF digestate had more than 4 times higher P content than the mineral fertilizer and that this was not reflected in the amounts of leached P in any of the soils. This result is in contrast to findings by Lu and O'Connor (2001) who found considerable leaching of P from organic wastes when applied to course texture soils with low P-adsorption capacity. However, the low leaching of P from digestates found in our experiment corresponds somewhat better to the findings by Haraldsen et al. (2011) who investigated the fate of P from different organic waste products in a similar greenhouse pot experiment with barley growing in sand. They found low P leaching in general, but a somewhat higher P leaching from the digestates compared to other organic waste products. In a review by Lu et al. (2012) it is described that since most of the P in biosolids is in recalcitrant form, biosolids P may generally be less subjected to runoff loss compared to other amendments when applied to agricultural lands.

Results from the column experiment also showed very low levels of leached P for all soils and treatments (Table 4). As the soil pH after liming was approximately 6 in all the three soils (pH_L, Table 2), neither sorption to Al- and Fe-oxides or precipitation with carbonates were processes expected to reduce the P leaching. At a soil pH of about 6, the P added in all treatments seems to be effectively chemisorbed by ligand exchange on the soil minerals even in the short time-span considered in this study. Independent of the natural plant-available P status of the soils (P-AL, Table 2), which was low in the sand and higher in the loam and silt, the added P was effectively retained within the soils. In a column leaching study including soils with different texture, P-status and pH ranging from 5.8 to 7.9, García-Albacete et al. (2014) found the potential P losses connected to digestate application not to be significantly dependent on soil type, but more related to digestate characteristics. The increase of the soils' P reserves may be positive for soils naturally low in P. In a long-term field study by Bedada et al. (2016), a build-up of the soil P status by addition of compost compared to mineral fertilizer was emphasised. However, in areas prone to erosion, P supplies exceeding plant requirements will increase the risk of eutrophication of water bodies through transport of particle-bound P (Sharpley et al. 2001).

Potassium

Among the three soils, the silt had the highest natural K content (Table 2) and, when no fertilizer was added, higher amounts of K leached from the silt after harvest compared to loam and sand (Fig. 2c). Among the fertilizer treatments, the commercial digestates (CF and CFS) represented the lowest K input (Table 3), and K leaching from the pots receiving these digestates did not differ from the control without any fertilizer addition. As plant growth did not appear to be K-limited in the CF and CFS treatments (Fig. 1) and the leaching of K was equal to the control with no fertilizer addition (Fig. 2c), the low K concentration in the CF and CFS treatments apparently still met the wheat plant requirements even in the K-poor sand.

The treatments with experimental digestates EWMF and EDM represented the highest K input (Table 3), i.e. 2.6 and

4 times, respectively, higher than for MF. The amount of K leached from loam and silt treated with EWMF and EDM was similar to the mineral fertilizer treatment, suggesting higher plant uptake and/or accumulation of K in the soil. For the sand, leached amounts of K were higher from the treatments with K-rich digestates and neither plant uptake nor accumulation in the soil seems to have played a major role. The addition of organic matter in the organic fertilizers has thus not improved K binding capacity in the sand. In the sand, which is naturally low in K (Table 2) and could thus have profited from an accumulation of K, most of the extra K was leached.

In accordance with the data from the pot experiments, the soil column leaching experiments showed generally a strong K adsorption in the loam (Table 4). For the other two soils, varying degrees of leaching were observed (Table 4). Also in the soil column experiment, the naturally higher K content in the silt compared to the sand was reflected in the amount of K leached. In the sand 45–100% of the K added in the different fertilizer treatments was leached, and leaching was highest for the treatments with the highest K input, i.e. the EWMF and the EDM treatments. In the silt, K leaching was high for EDM, but for the other treatments generally lower compared to sand.

Fertilizer K input in excess of plant requirements often has been found to persist in soil for years after application, improving the soil K status (Yu et al. 2009; Zhang et al. 2011). Considering both the results from the soil column experiment and the after harvest leaching study, the effect of a K input in excess of plant requirements apparently varies dependent of soil type. In the silt, which in general had the highest biomass production, the high amount of K added was to a large extent taken up by the plants. In the loam with lower biomass production, the extra K was primarily adsorbed by the clay minerals, while excess K in the sand was predominantly leached from the system.

Conclusion

The present study provides an overview of how a wide variety of digestates affects plant production and leaching of nutrients. The results show that the concentration of NH_4^+ in digestates (DM < 6%) provides a good indicator of the N fertilizer value of the digestates. Additionally, in the sand, digestate application reduced NO_3^- leaching. In longer-term applications of digestates in the field, additional N supply from mineralization of the organic material may benefit biomass production also in following growing seasons.

Particularly in the nutrient poor sand, the organic fertilizers had a positive effect on cereal yields. The positive effect may be related to an improved ability to keep plant nutrients within the root zone at an early stage of plant development.

Although most digestates had ratios between the plant macronutrients N, P and K that were not optimal for wheat growth, the variation in P and K content seemed to be within a range suitable for wheat production. The risk of P leaching was low, even in the sandy soil or at a P input that was up to 4 times higher compared to adequate mineral fertilizer additions. The variation in K input did not influence yield. In the sand, K from digestates in excess of plant requirements was leached from the soil, in the loam assimilated in the soil leading to a soil K-reserve build-up, while in the silt, the higher K availability was met by a higher biomass production and/or soil accumulation.

Overall, the data show that digestates from biogas production based on fundamentally different feedstocks are promising as fertilizers. The N fertilization can be based only on the NH_4^+ concentration of digestates and, at least for wheat production, considerable variation in the concentrations of K and P can be tolerated.

Acknowledgements We would like to thank Kurt R. Johansen for his help with the pot experiments, and Dr. Valentina Zivanovic, for her work with the soil and soil water analyses. This work was funded by the Norwegian Research Council (Project no. 203402/I10).

References

Alburquerque JA, de la Fuente C, Ferrer-Costa A, Carrasco L, Cegarra J, Abad M, Bernal MP (2012) Assessment of the fertiliser potential of digestates from farm and agroindustrial residues. Biomass Bioenerg 40:181–189. https://doi.org/10.1016/j.biombioe.2012.02.018

Arthurson V (2009) Closing the global energy and nutrient cycles through application of biogas residue to agricultural land—potential benefits and drawbacks. Energies 2:226–242. https://doi.org/10.3390/en20200226

Bedada W, Lemenih M, Karltun E (2016) Soil nutrient build-up, input interaction effects and plot level N and P balances under long-term addition of compost and NP fertilizer. Agric Ecosyst Environ 218:220–231. https://doi.org/10.1016/j.agee.2015.11.024

Bremner JM (1965) Inorganic forms of nitrogen. In: Black CA et al (eds) Methods of soil analysis, part 2. Agronomy, 1st edn. vol 9. American Society of Agronomy, Soil Science Society of America. pp 1179–1237

Bremner JM, Mulvaney CS (1982) Nitrogen-total. In: Page AL, Miller RH, Keeney DR (eds) Methods of soil analysis part 2. Chemical and microbial properties. vol 9.2. American Society of Agronomy, Soil Science Society of America. pp 595–624

Cavalli D, Cabassi G, Borrelli L, Geromel G, Bechini L, Degano L, Gallina PM (2016) Nitrogen fertilizer replacement value of undigested liquid cattle manure and digestates. Eur J Agron 73:34–41. https://doi.org/10.1016/j.eja.2015.10.007

Cramer MD, Lewis OAM (1993) The influence of nitrate and ammonium nutrition on the growth of wheat (*Triticum aestivum*) and maize (*Zea mays*) plants. Ann Bot 72:359–365. https://doi.org/10.1006/anbo.1993.1119

Currie JA (1962) The importance of aeration in providing the right conditions for plant growth. J Sci Food Agric 13:380–385. https://doi.org/10.1002/jsfa.2740130706

Egnér H, Riehm H, Domingo WR (1960) Untersuchungen über die

chemische Boden- Analyse als Grundlage des Nährstoffzustandes der Böden. II. Chemische Extraktionsmethoden zur Phosphor- und Kaliumbestimmung. Kungliga Lantbrukshögskolans Annaler 26:199–215

Elonen P (1971) Particle-size analysis of soil. Acta Agral Fenn 122:122p

Ferris H, Venette R, van der Meulen H, Lau SS (1998) Nitrogen mineralization by bacterial-feeding nematodes: verification and measurement. Plant Soil 203:159. https://doi.org/10.1023/A:1004318318307

García-Albacete M, Tarquis AM, Cartagena MC (2014) Risk of leaching in soils amended by compost and digestate from municipal solid waste. Sci World J 2014:1–9. https://doi.org/10.1155/2014/565174

Gutser R, Ebertseder T, Weber A, Schraml M, Schmidhalter U (2005) Short-term and residual availability of nitrogen after long-term application of organic fertilizers on arable land. J Plant Nutr Soil Sci 168:439–446. https://doi.org/10.1002/jpln.200520510

Haraldsen TK, Andersen U, Krogstad T, Sørheim R (2011) Liquid digestate from anaerobic treatment of source-separated household waste as fertilizer to barley. Waste Manag Res 29:1271–1276. https://doi.org/10.1177/0734242X11411975

Hati KM, Mandal KG, Misra AK, Ghosh PK, Bandyopadhyay KK (2006) Effect of inorganic fertilizer and farmyard manure on soil physical properties, root distribution, and water-use efficiency of soybean in Vertisols of central India. Bioresour Technol 97:2182–2188. https://doi.org/10.1016/j.biortech.2005.09.033

Kataki S, Hazarika S, Baruah DC (2017) Assessment of by-products of bioenergy systems (anaerobic digestion and gasification) as potential crop nutrient. Waste Manag 59:102–117. https://doi.org/10.1016/j.wasman.2016.10.018

Kirchmann H (1985) Losses, plant uptake and utilization of manure nitrogen during a production cycle. Acta Agr Scand Suppl 24:1–77

Kramer S, Reganold JP, Glover JD, Bohannan BJM, Mooney HA (2006) Reduced nitrate leaching and enhanced denitrifier activity and efficiency in organically fertilized soils. PNAS 103:4522–4527. https://doi.org/10.1016/B978-0-12-407247-3.00007-X

Kuszel M, Lorencowicz E (2015) Agricultural use of biogas digestate as a replacement fertilizers. Agric Agric Sci Procedia 7:119–124. https://doi.org/10.1016/j.aaspro.2015.12.004

Ladanai S, Vinterbäck J (2009) Global Potential of Sustainable Biomass for Energy. Report 013. SLU, Swedish University of Agricultural Sciences. Available at http://pub.epsilon.slu.se/4523/1/ladanai_et_al_100211.pdf

Lu P, O'Connor GA (2001) Biosolids effects on phosphorus retention and release in some sandy Florida soils. J Environ Qual 30:1059–1063. https://doi.org/10.2134/jeq2001.3031059x

Lu Q, He ZL, Stoffella PJ (2012) Land application of biosolids in the USA: a review. Appl Environ Soil Sci 2012:1–11. https://doi.org/10.1155/2012/201462

Masters DG, Thompson AN (2016) Grazing crops: implications for reproducing sheep. Anim Prod Sci 56:655–668. https://doi.org/10.1071/AN14517

Möller K, Müller T (2012) Effects of anaerobic digestion on digestate nutrient availability and crop growth: a review. Eng Life Sci 12:242–257. https://doi.org/10.1002/elsc.201100085

Nelson DW, Sommers LE (1982) Total carbon, organic carbon and organic matter. In: Page AL, Miller RH, Keeney DR (eds) Methods of soil analysis part 2. ASA, Madison, pp 539–579. https://doi.org/10.2134/agronmonogr9.2.2ed.c29

Øgaard AF, Krogstad T, Løes AK (2001) Potassium uptake by grass from a clay and a silt soil in relation to soil tests. Acta Agric Scand B-SP 51:97–105. https://doi.org/10.1080/09064710127617

Rehl T, Müller J (2013) CO2 abatement costs of greenhouse gas (GHG) mitigation by different biogas conversion pathways. J Environ Manag 114:13–25. https://doi.org/10.1016/j.jenvman.2012.10.049

Riva C, Orzi V, Carozzi M, Acutis M, Boccasile G, Lonati S, Tambone F, D'Imporzano G, Adani F (2016) Short-term experiments in using digestate products as substitutes for mineral (N) fertilizer: agronomic performance, odours, and ammonia emission impacts. Sci Total Environ 547:206–214. https://doi.org/10.1016/j.scitotenv.2015.12.156

Sharpley AN, McDowell RW, Kleinman PJA (2001) Phosphorus loss from land to water: integrating agriculture and environmental management. Plant Soil 237:287–307. https://doi.org/10.1023/A:1013335814593

Svensson K, Odlare M, Pell M (2004) The fertilizing effects of compost and biogas residues from source separated household waste. J Agric Sci 142:461–467. https://doi.org/10.1017/S0021859604004514

Svoboda N, Taube F, Wienforth B, Kluss C, Kage H, Herrmann A (2013) Nitrogen leaching losses after biogas residue application to maize. Soil Till Res 130:60–80. https://doi.org/10.1016/j.still.2013.02.006

Tambone F, Scaglia B, D'Imporzano G, Schievano A, Orzi V, Salati S, Adani F (2010) Assessing amendment and fertilizing properties of digestates through a comparative study with digested sludge and compost. Chemosphere 81:577–583. https://doi.org/10.1016/j.chemosphere.2010.08.034

Uusitalo V, Havukainen J, Manninen K, Höhn J, Lehtonen E, Rasi S, Soukka R, Horttanainen M (2014) Carbon footprint of selected biomass to biogas production chains and GHG reduction potential in transportation use. Renew Energy 66:90–98. https://doi.org/10.1016/j.renene.2013.12.004

Vaneeckhaute C, Meers E, Michels E, Ghekiere G, Accore F, Tack FMG (2013) Closing the nutrient cycle by using bio-digestion waste derivatives as synthetic fertilizer substitutes: a field experiment. Biomass Bioenerg 55:175–189. https://doi.org/10.1016/j.biombioe.2013.01.032

Webb J, Sörensen P, Velthof G, Amon B, Pinto M, Rodhe L, Salomon E, Hutchings N, Burczyk P, Reid J (2013) An assessment of the variation of manure nitrogen efficiency throughout Europe and an appraisal of means to increase manure-N efficiency. Adv Agron 119:371–442. https://doi.org/10.1016/B978-0-12-407247-3.00007-X

Yu W-T, Jiang Z, Zhou H, Qiang M (2009) Effects of nutrient cycling on grain yields and potassium balance. Nutr Cycl Agroecosys 84:203–213. https://doi.org/10.1007/s10705-008-9237-4

Zhang H-M, Yang X-Y, He X-H, Xu M-G, Huang S-M, Liu H, Wang B-R (2011) Effect of long-term potassium fertilization on crop yield and potassium efficiency and balance under wheat–maize rotation in China. Pedosphere 21:154–163. https://doi.org/10.1016/S1002-0160(11)60113-6

Zirkler D, Peters A, Kaupenjohann M (2014) Elemental composition of biogas residues: variability and alteration during anaerobic digestion. Biomass Bioenerg 67:89–98. https://doi.org/10.1016/j.biombioe.2014.04.021

19

Assessment and valorization of treated and non-treated olive mill wastewater (OMW) in the dry region

Haifa Rajhi[1] · Inès Mnif[2,3] · Mounir Abichou[1] · Ali Rhouma[1]

Abstract

Purpose The quantity of Olive by-products does not stop increasing and a water shortage that threatens the olive tree culture; require a serious valuation of these by-products. A comparative and valorization study of two kinds of OMW; Fresh OMW (FOMW), directly issue from three-phase continuous extraction factory and Disposal Evaporation Ponds OMW (DOMW) were done.

Methods Physico-chemical and biological parameters of OMWs and soil irrigated with OMWs, respectively, were determined. An antibacterial activity test of FOMW against Clinic Standard Bacteria was determined. A statistical analysis was performed for all defined parameters.

Results A significant increase of pH value of 6 and a lower failure of the EC in 8.94 (mS/cm^{-1}) were registered after OMW disposal in evaporation ponds. We registered a fall of BOD_5 and COD from 61.05–116.37 (g/L) to 55.67–103.82 (g/L), respectively. A significant increment of phenol compound removal was observed after OMW disposal. However, a switch of fatty acids distribution and content was observed, which several fermentation pathways could explain took place. This result suggested by a clear shift in biomass composition. An important soil fertility after DOMW soil irrigated was traduced by an important value of the germination index (170.55%) and efficient organic matter increment of 2.3%. A CMI rate of 32.76 ($\mu g\ mL^{-1}$) was determined by FOMW against different clinic standard bacteria.

Conclusion A spectacular soil fertility effect was obtained from DOMW soil spreading, that efficiently evaluate the OMW biological treatment. In addition, the FOMW was valorized as its powerful antibacterial.

Keywords Olive mill water valorization · Ponds evaporation · OMW compounds · Soil treatment · Soil fertility · OMW antibacterial activity

Introduction

Every year, olive oil production industry release large quantities of olive mill wastewater (OMW). They are estimated approximately 30 million m^3 in the world and 700 million m^3 in Tunisia (Mekki et al. 2013). They are obtained from olive tree after discontinuous press or from the continuous centrifugation (Niaounakis and Halvadakis 2006; Aggoun et al. 2016). The chemical characterization of OMW demonstrates that are highly rich of phenolic compounds, lipids and organic acids with a poor biodegradability that limits their treatment (Dermeche et al. 2013; Aggoun et al. 2016). Generally, different factors influence the characteristics of OMW such as the method of extraction, the technological process separation, the climatic conditions and the variety and the fruit maturity of olive tree (Dermeche et al. 2013; Aggoun et al. 2016). In fact, after oil extraction, OMW contained 98% of the total phenolic content of the olive fruits with concentration varying from 0.5 to 24 g/l (Paraskeva and Diamadopoulos 2006; Alaoui et al. 2016; Leouifoudi et al. 2014).

Different phenolic compounds were detected essentially for simple phenols and flavonoids and polyphenols resulting from polymerisation of the simple phenols (Comandini et al. 2014; Leouifoudi et al. 2014). They are phytotoxic and can

✉ Inès Mnif
inesmnif2011@gmail.com

1 Institute of Olive Trees, Zarzis 4170, Tunisia

2 Laboratoire d'amélioration des plantes et Valorization des Agroressources, University of Sfax, Road Soukra km 4, 3038 Sfax, Tunisia

3 National School of Engineers of Sfax, Sfax, Tunisia

inhibit plant growth limiting the use of OMW for irrigation. In addition, their direct release onto soil may affect soil's physical and chemical properties including soil porosity and pH (Niaounakis and Halvadakis 2006; Mekki et al. 2013). Moreover, release of OMW in the environment can discolour streams and rivers due to the high concentration of darkly coloured polyphenols. Indeed, discharge of the OMW in the environment has a serious environmental drawback (Paraskeva and Diamadopoulos 2006).

In addition to phenolic compounds, the highly reduced sugars content of OMW can stimulate microbial respiration, lowering therefore, dissolved oxygen concentrations (McNamara et al. 2008). Besides, the high phosphorus content can lead to eutrophication (McNamara et al. 2008). Likewise, the large-volume OMW and the seasonal nature of olive oil production limit their storage operation and their displacement (McNamara et al. 2008). Regardless, olive oil production remains a high polluting industry due to the utilization of large amounts of water and the production of large amounts of waste water and sludge.

To resolve these problems, pretreatment of OMW can improve the quality of the wastewater and remove some of its toxicity. Different strategies to know, physical–chemical, and biological process were applied to treat OMW. Dilution, evaporation, sedimentation, filtration and centrifugation were used largely as a physical process to treat OMW (Paraskeva and Diamadopoulos 2006; Villegas et al. 2016). Moreover, several biotechnological methods have been used to reduce the polluting load of OMW including decantation with lime and clay, coagulation–flocculation, electrocoagulation, natural evaporation and thermal concentration (Paraskeva and Diamadopoulos 2006; Villegas et al. 2016). Also, evaporation ponds and artificial ponds with very large surface areas that are designed to efficiently evaporate water by sunlight and exposure to the ambient temperatures can be used for OMW treatment (Barbera et al. 2013). Besides, different biological treatment technologies of OMW were reported as efficient strategies to clean OMW, including microbial degradation, aerobic and anaerobic process (codigestion with other effluents and composting) (Paraskeva and Diamadopoulos 2006; Villegas et al. 2016). It presents many advantages over other techniques including simplicity; less expensive for the reduction of pollutants that perfectly meets many sustainable social systems and ecological advantages (Barbera et al. 2013). Note in this regard that the arid southern Tunisia is a wide field of biochemical exploration and has many characteristics associated with the continuous presence of light and sunshine throughout the year.

Moreover, soil serves as a medium for organic waste disposal, both solid and liquid. OMW was known to have a fertilizing effect, assessing a possible valorization of these effluents in agronomy (Dakhli and Maalej 2017). Spreading represents an interesting solution for OMW treatment and disposal. In fact, the higher content of micronutrients, especially potassium and organic matter induce plant growth (Di Serio et al. 2008; Dakhli and Maalej 2017). However, the high organic load of OMW due mainly to the presence of polyphenols as well as short and long chain fatty acids may cause phytotoxic and antimicrobial effects of these effluents (Dakhli and Maalej 2017). The inhibition of the growth of microorganisms, especially bacteria, may reduce the mineralization process in the soil (Dakhli and Maalej 2017). For these reasons, the controlled spreading of OMW can increase the fertility of the soil and offer the opportunity to recycle various compounds.

Indeed, the goal of this work is constituted of two parts; the first one is to characterize different types of OMW. A comparison between Fresh OMW directly issue from three-phase continuous extraction factory and disposal evaporation ponds OMW in the south of Tunisia (Chammakh-Zarzis) was done. The second goal of this work was the investigation of the effects of different types of OMW spreading on the soil characteristics. The soil treated fertility was evaluated to evaluate the OMW spreading phytotoxicity potential. In addition, we tried to evaluate the antibacterial activities of the fresh OMW against different clinic bacteria, making the possibility of further production of antibiotic with high industrial and biotechnological values.

Materials and methods

Study zone

The study zone is located at Chammakh-Zarzis in the south-eastern Tunisia. The climate of this region is an arid Mediterranean climate with an annual average rainfall of 180 mm. The soil of this area is moderately deep with a slight texture, very filterable and relatively poor in organic matter (Abichou et al. 2009).

Meteorological parameters

Average monthly values of different meteorological factors of evaporation ponds are detailed in Fig. 1 including Month temperature (°C), Evaporation rate (mm per days), Relative humidity (%), Wind speed (m per second) and Dominant wind direction. The annual temperature and precipitation was 21 °C and 15.16 mm per year, respectively. The evaporation extended from 1 mm to 5.6 mm. In fact, the OMW evaporation rate can be considered around 4.03 mm per year. The maximal air humidity and wind speed was around 74% and 1.6 m per second, respectively, followed by (230°) as a maximal value of a wind direction Dig. The meteorological conditions described above can facilitate an efficient natural evaporation of OMW.

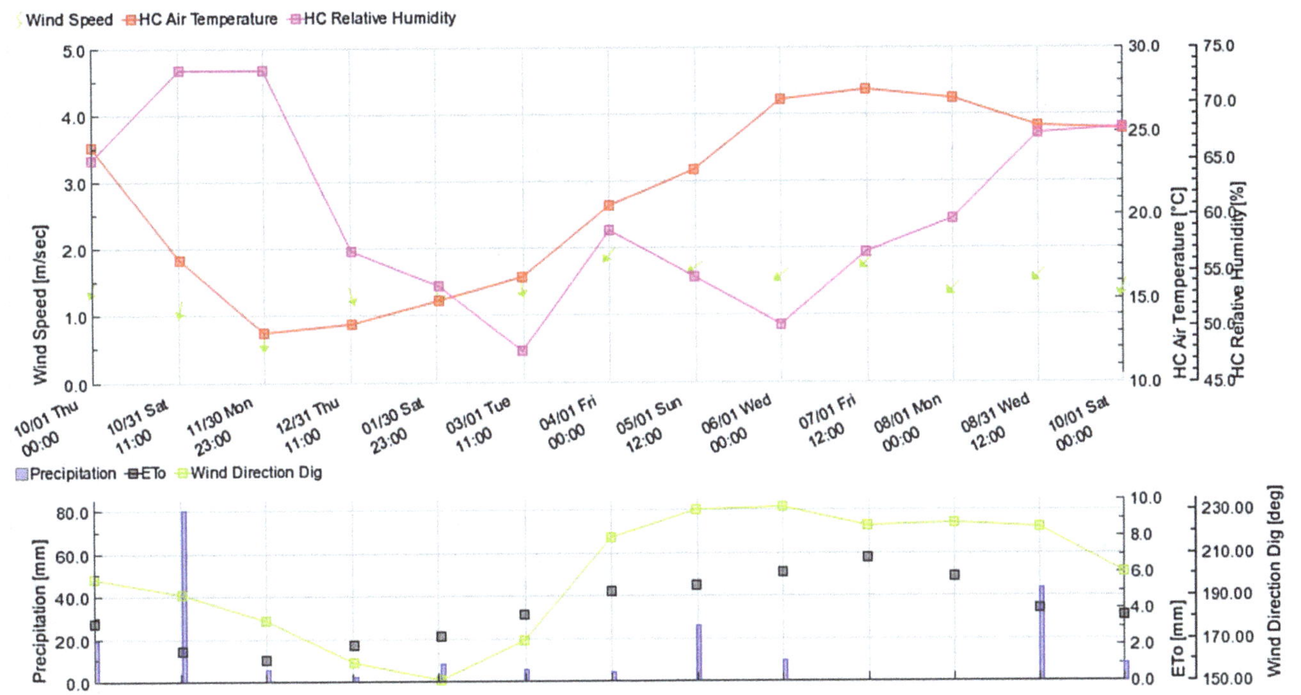

Fig. 1 The meteorological factors of study Zone; Month temperature (°C) Evaporation (mm per days); Relative humidity (%), Wind speed (m per second) and Dominant wind direction

Sampling

FOMW was taken from a three-phase continuous extraction factory located in the region of Chemmakh-Zarzis (Southern Tunisia, 33 36′N, 11°02′E). DOMW Samples were collected from the evaporation ponds after one year of fresh OMW disposal. The characteristics of ponds were in m: 70 of Length, 40 of Width, 4 of Height. The volume of ponds was 11200 m^3. The OMW was sprayed homogeneously on the sandy soil surface, previously tilled to a 20 cm of depth. DOMW and FOMW were sprayed in the soil olive tree culture plot during 5 years in the winter period (December–January) using at doses of 50 m^3/ha^{-1}. The dimension of soil plot was 1 ha. The soil sample was collected between two olive trees using the Hand auger tube *H-4268*. The soil sampling was taken in March. Each type of OMW was used in each different site. Non-OMW spreading soil (Untouched soil) was used as a control.

Physico-chemical analysis

Different physico-chemical parameters including organic matter, mineral content, biological oxygen demand (BOD), chemical oxygen demand (COD), electrical conductivity (EC) and pH were performed according to the Standard Methods for the Examination of Water and Wastewater, 20th Edition 1998. The pH value of soil was determined with pH meter XP with a pH 50 lab model. Electrical Conductivity (EC) of each site was determined Conductivity *inoLab WTW 7110 model*. Total Organic Carbon (TOC) was determined following the Walkley–Black method (Walkley 1947) and organic matter (OM) was calculated by multiplying the total carbon by 1.724.

Sugar and phenolic compounds analysis

High-performance liquid chromatography (HPLC 2010 PLUS, SHIMADZU) was used to quantify Sugar and Phenolic compounds. For sugar compounds identification, stationary phase consisted of NH_2 column (AQUASILC C18 with length and inner diameter of 150 and 30 mm, respectively) and mobile phase consisted of acetonitrile with a flow rate of (1 ml/min). For phenolic compound quantification, the mobile phase was an equal mixture of (A) 0.1% of formic acid in water and (B) 0.1% of formic acid in methanol with a flow of 0.4 ml/min. The column TSK (TSK gel with length and inner diameter of 30 cm and 7.8 mm, respectively) was used (Pérez et al. 1990; Kemal 1994). The extraction of soil polyphenols was realized according to FOLIN-DENIS method by the ethanol (Ranalli1997).

Fatty acid analysis

Free fatty acids were analyzed with a Shimadzu gas chromatograph system, (GC–MS GP2010ULTRA) adapted for capillary columns. For the analysis, a fused silica capillary column, 30 m × 0.25 mm × 0.25 µm film thickness, was used. The injector and detector temperatures were set at 200 °C with interface of 220. The column temperature was set at 50 °C, then raised to 250 °C with flow of 50 °C/minute, then increased from 200 to 230 °C at the rate of 50 °C/min. Peak heights were determined by integration software (D'Annibale et al. 1998).

Microbial enumeration

Microbial enumeration was done according to ISO 2718 (Technical Committee 1996). Results were expressed as the total number of colonies forming units (CFU). For the enumeration of total aerobic mesophilic bacteria, PCA plate were inoculated and incubated at 37 °C for 24 h. While, Sabouraud medium containing a bacterial inhibitor (chloramphenicol) was used to enumerate yeasts after incubation at 25 °C for 5 days. The microbial results were expressed by calculating the ratios of decimal logarithmic value of each flora (bacteria and yeasts). Triplicates plated were used in each test.

For soil Microbial estimation, a soil suspension was prepared as follows; 10 g of the soil sample were suspended in an Erlenmeyer flask containing 90 ml of a sterile solution (0.2% of sodium polyphosphate ($NaPO_3$) in distilled water, pH 7.0) and 10 g of sterile glass beads (1.5 mm diameter). The flask was shaken at 200 rpm for 2 h. Serial tenfold dilutions of the samples in a 0.85% NaCl solution were plated in triplicate on PCA at 37 °C and Sabouraud containing chloramphenicol at 25 °C for total bacterial counts and for yeasts counting, respectively.

Antibacterial activity study

Antibacterial activities of different OMWF concentration were studied in triplicate using glass tubes as described by (Aziz et al. 1998). The antimicrobial activity of the different test was tested against Gram-positive bacteria (*Staphylococcus aureus*) and Gram-negative bacteria (*Klebsiella pneumonia, Salmonella typhimurium* and *Enterobacter cloacae*) to determine the minimal inhibitor concentration (MIC). The tubes were incubated at 37 °C for 18 h. Results of bacterial growth inhibition were expressed by a visual examination: the inhibition was considered positive when there was no microbial growth in all the three tubes of the triplicate. The MIC was determined as the lowest concentration caused complete growth inhibition in the triplicate tubes of each treatment.

A Stock solution containing 10 mL of Mueller–Hinton medium and 1 mL of culture was prepared. A 1.8 mL of stock solution was mixed with 0.8 mL of the filtrate of FOMW extracted. A solution of 100, 50 and 25% dilution of OMW extracted range was prepared. Pure water was used as a control. The same protocol described above was repeated in triplicate for each strain.

Respirometric tests

Biological activity in the soil was achieved by measuring CO_2 evolution in the aerobic condition. Respirometric analysis was done according to the method of Ohlinger (1995). The soil sample was humidified to 50% of its water holding capacity, and then was incubated at 25 °C in the dark. Released CO_2 was trapped in NaOH solution and titrated with HCl.

Germination index evaluation

The phytotoxicity of the treated soils was evaluated by the measurement of the germination index according to the method proposed by (Zucconi et al. 1981). This technique consists of making germinate, in petri dish, 10 seeds of tomatoes with the different samples studied. Petri dish incubated during a 1 week in the darkness at 25 °C. The untouched soil was used as positive control. After incubation, the reading of the result is determined by counting the number of germinated seeds. The length of the roots of seeds having germinated measured in mm. The germination index "GI" is calculated according to the following formula:

$$\text{GI}(\%) = [\text{Number of seed germinated} \times \text{root length}] / [\text{Control Number of seed germinated} \times \text{Control root length}] \times 100$$

Statistical analysis

The Statistical Package for the Social Sciences (SPSS, Version 20.0) was used to perform the statistical analysis. The Phenol, fatty acids and sugar compounds of both OMW Samples (fresh and disposed in the evaporation ponds) values were compared using Student test at the "5% ($P = 0.05$)" significance level. Data are presented as mean ± standard deviation. Values were obtained from triplicate determinations and the differences were examined using one-way analysis of variance (ANOVA) followed by the Fischer's LSD (Least Significant Difference) post hoc test. Statistical significances of the correlations between datasets were calculated using Pearson's R values. At least three replicates were performed for each laboratory measurement.

Results and discussion

The physico-chemical properties

The physico-chemical properties of FOMW and DOMW are reported in Table 1. A significant increase of pH value of 6 and a lower failure of electrical conductivity of 8.94 (mS/cm^{-1}) were registered after OMW disposal in evaporation

Table 1 Physico-chemical properties of FOMW and DOMW

Samples	pH	EC (mS/cm[1])	BOD_5 (g/l)	COD (g/l)	OM (g/l)	MM (g/l)
FOMW	$5.045^a \pm 0.007$	$9.200^a \pm 0.28$	$61.050^a \pm 1.47$	$116.370^a \pm 2.81$	$89.590^a \pm 0.3$	$11.480^a \pm 0.2$
DOMW	$6.020^b \pm 0.02$	$8.940^b \pm 0.19$	$55.670^b \pm 0.89$	$103.820^b \pm 1.66$	$41.260^b \pm 0.02$	$5.300^b \pm 0.6$

Test Student was determined with FOMW and DOMW ($p = 0.01$, $*p < 0.05$). Mean scores in the same column with different letter ("a" and "b") are significantly different (for $p < 0.05$) (Results are compared for the same parameters and the two different OMW)

Table 2 Phenolic composition of FOMW and DOMW; reduction of single phenolic components was represented in removal percent

Phenolic compound	FOMW (mg/l^{-1})	DOMW (mg/l^{-1})	Removal (%)
Quinic acid	0.556^a	0.508^b	8.630
Protocatechuic acid	7.863^a	0.000^b	100
Caffeic acid	39.907^a	0.000^b	100
p-coumaric acid	38.617^a	0.000^b	100
Naringin	0.912^a	0.000^b	100
Hyperoside (quercetin-3-o-galactoside	1.600^a	0.313^b	80.430
Rutin	0.649^a	0.000^b	100
Salviolinic acid	5.064^a	0.000^b	100
Naringenin	0.309^a	0.122^b	60.510
Luteolin	14.924^a	1.708^b	88.550
Cirsiliol	5.658^a	2.570^b	54.570
Apegenin	0.551^a	0.244^b	55.710
Acacetin	14.447^a	14.774^b	–
Syringic acid	0.000^a	39.193^b	–

Test Student was determined with FOMW and DOMW ($p = 0.01$, $*p < 0.05$). Mean scores in the same line with different letter ("a" and "b") are significantly different (for $p < 0.05$) (Results are compared for the same parameters and the two different OMW)

ponds. The OMW are rich in organic matter expressed in terms of BOD_5 and COD. However, a fall of BOD_5 and COD from 61.05 and 116.37 (g/L) to 55.67 and 103.82 (g/L), respectively, after the disposal of OMW more than 1 year in evaporation ponds. These results suggest the occurrence of oxidative phenomena, which led to further removal of the acidic compounds in OMW (Paraskeva and Diamadopoulos 2006).

The phenol compounds

Table 2 summarizes the phenol compound concentration founded in OMWF and DOMW. Comparative study of phenol compounds between fresh and disposal OMW shows a significant difference in the concentration values ($p = 0.023$).

Regarding previous studies, it is not easy to compare phenol content in OMW in different experiments because phenol compounds are influenced by many factors, including the impact of geographical and climatic conditions.

OMW composition was studied in various studies (Bouknana et al. 2014; Leouifoudi et al. 2014; Alaoui et al. 2016). It was characterized by its complexity and it was found to be rich in hydroxytyrosol and secoiridoids derivatives (Peixoto et al. 2008; Leouifoudi et al. 2014). In this work, we tried to analyze phenolic compounds in OMW before and after ponds evaporation. An important failure of phenol compound concentration rate as well as the efficient removal of a large number of phenolic compounds was observed. Significant increment, ranging from 54 to 100% of phenol compound removal of the most compound after disposal of OMW in the evaporation ponds, was observed.

Phenol compounds such as protocatechuic acid, caffeic acid, *p*-coumaric acid, Naringin, Rutin and Salviolinic acid, were completely removed. The decrease and removal of phenol compounds may be due to their oxidation and their possible further transformations (Belaid et al. 2002). In fact, the complete oxidation of organic compounds to CO_2 and H_2O results from the split of the various aromatic compounds during the aerobic catabolism permitting their use in the cycle of the Krebs (Anderson and Dagley 1980). Regardless, a higher concentration rate of 39.193 of syringic acid compounds in DOMW was detected in this study. A previous study suggested that syringic acid is the least toxic phenol compound (Rahouti et al. 1997). These results showing the significant decrease of phenolic compounds along with their toxic effect are of great interest. However, previous studies showed that degradation of phenolic compounds is considered as an obstacle in the biological OMW treatment, including their natural breakdown is not easy (Tsioulpas et al. 2002; Martins et al. 2008).

The fatty acid content

The comparison of fatty acids between the fresh and disposal OMW shows a significant difference in the fatty acid concentration ($p = 0.01$). Figure 2 and Table 3 shows a significant change of the fatty acid distribution.

Several acids are completely removed such as ethylbenzene, styrene; ethyl 9-hexadecenoate, cyclobutane, 3-diphenyl-, trans-, cyclopropyl phenyl methane, benzene, 1,1′-(1,2-Cyclobutanedilyl) *bis*-, *trans*-, octadecanoic acid,

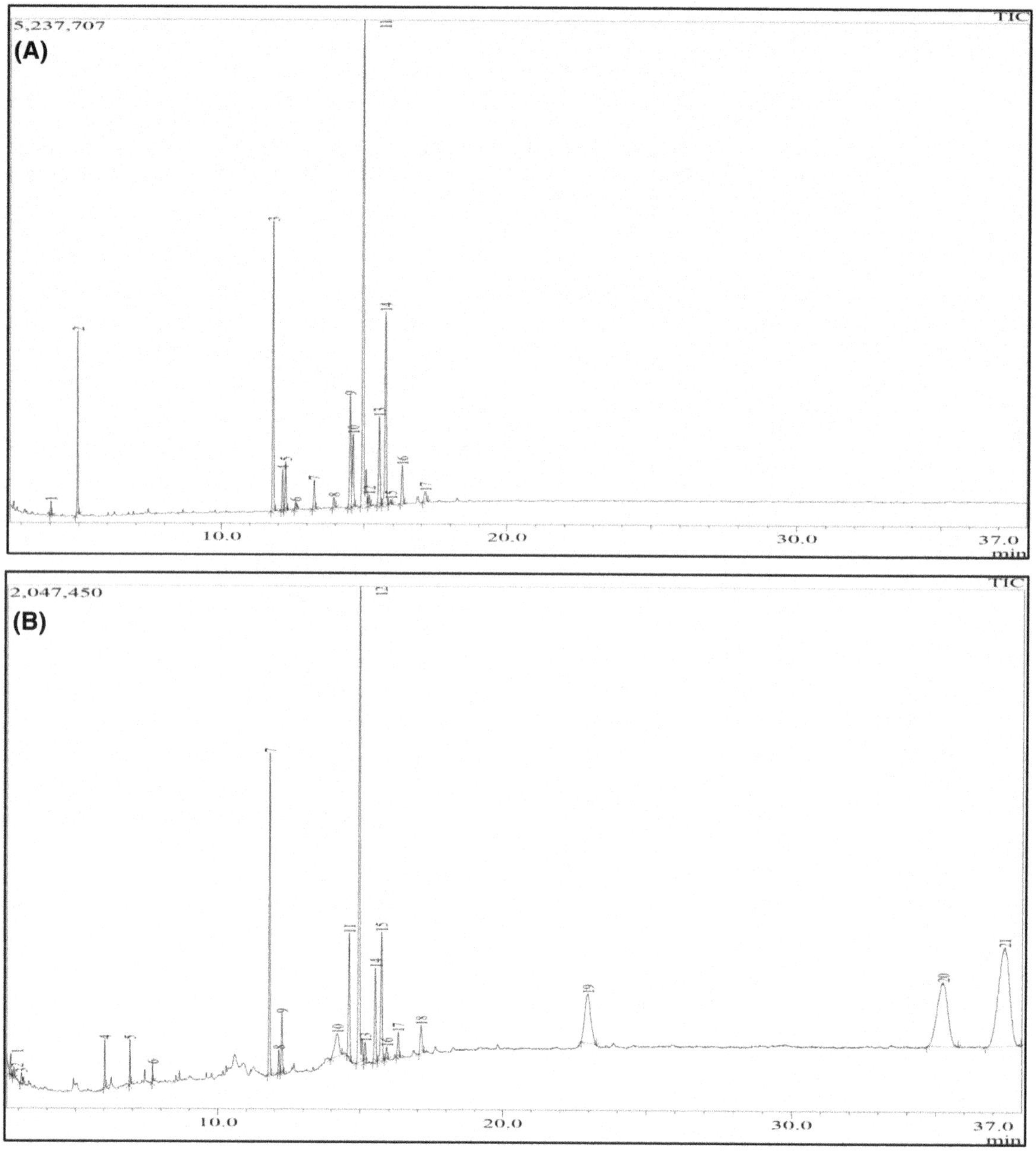

Fig. 2 Representative GC/MS chromatograms of the acid compounds of FOMW (**a**) and DOMW (**b**). The order and identification pic was detailed in Table 3

methyl ester, Octadecanoic acid, ethyl ester and 1-Eicosanol. Most of these acids have a little content that is quickly degraded. In addition, we note a remarkable failure for other acids such as, Hexadecanoic acid, methyl ester, 9-Hexadecenoic acid, methyl ester, (*Z*)-;hexadecanoic acid, ethyl ester; 9-Octadecenoic acid, methyl ester, (*E*)-; ethyl oleate 9,12-Octadecadienoic acid (*Z,Z*), methyl ester (*E*) and ethyl (9*Z*,12*Z*)-9,12-Octadecadienoate. However, an appearance of the other acids reduced to know Propionic acid, ethyl ester, propyl ester and acetic acid. Generally, these acids are a direct product of the main fermentation pathway, such as a propionic or acetic fermentation (Staples et al. 2001; Antonopoulou et al. 2010). OMW can produce glycerol and long chain fatty acids (LCFA). According to Schönfeld et al. (2004), long chain free fatty acids (C16–C48) are naturally occurring fats in olive fruits, which exhibit toxic effects towards microorganisms. Indeed, the fermentation of glycerol gives 1,2-propanediol and ethanol, which can be further oxidized to acetate and propionate (Rajhi et al. 2015). Moreover, LCFA can be degraded by β-oxidation releasing, therefore, one molecule of acetate per cycle (Rajhi et al. 2015). To know, a

Table 3 Fatty acids concentration rate in FOMW and DOMW

Fatty acids	FOMW (%)	DOMW (%)
Ethylbenzene	5.893[a]	0.000[b]
Styrene	5.893[a]	0.000[b]
Hexadecanoic acid, methyl ester	14.681[a]	8.972[b]
9-Hexadecanoic acid, methyl ester, (*Z*)-	2.113[a]	0.627[b]
Hexadecanoic acid, ethyl ester	2.651[a]	1.672[b]
Ethyl 9-hexadecenoate	0.433[a]	0.000[b]
Cyclobutane, 1,3-diphenyl-, trans-	1.800[a]	0.000[b]
Cyclopropylphenylmethane($C_{10}H_{12}$)	0.676[a]	0.000[b]
Benzene, 1,1′-(1,2-Cyclobutanedilyl) *Bis*-, *Trans*-	7.704[a]	0.000[b]
Octadecanoic acid, methyl ester	4.786[a]	16.789[b]
9-Octadecenoic acid, methyl ester, (*E*)-	34.252[a]	16.789[b]
Octadecanoic acid, ethyl ester	0.750[a]	0.722[b]
Ethyl Oleate	5.998[a]	3.286[b]
9,12-Octadecadienoic acid (*Z,Z*)-, Methyl ester	13.634[a]	4.805[b]
Acetic acid *n*-octadecyl ester	0.362[a]	0.474[b]
Ethyl (9Z,12*Z*)-9,12-Octadecadienoate #	2.729[a]	0.902[b]
1-Eicosanol	1.059[a]	0.000[b]
Propanoicacid, ethyl ester	0.000[a]	0.500[b]
Acetic acid, Propyl ester	0.000[a]	0.274[b]
3,4-Dihydrothienyl-[3,4,B]-5-carboxythiol	0.000[a]	0.324[b]
Cycloheptasiloxane, Tetradecamethyl-	0.000[a]	0.952[b]
Cyclooctasiloxane, Hexadecamethyl-	0.000[a]	0.629[b]
Cyclononasiloxane, Octadecamethyl-	0.000[a]	0.227[b]
1-Propen, 3-(2-Cyclopentenyl)-2-methyl-1,1-diphenyl-	0.000[a]	2.471[b]
Octacosanol	0.000[a]	1.391[b]
Benzene, 1,1′,1″,1‴-(1,3,4,6-hexanetetrayl)tetrakis-, (*R**,*R**)	0.000[a]	7.330[b]
(*E*)-3,3-Diphenyl-4-hexenoic acid	0.000[a]	16.831[b]
Methyl (*Z*)-3,3-diphenyl-4-hexenoate	0.000[a]	27.460[b]

Test Student was determined with FOMW and DOMW ($p=0.01$, $*p<0.05$). Mean scores in the same line with different letter ("a" and "b") are significantly different (for $p<0.05$) (Results are compared for the same parameters and the two different OMW)

possible triglyceride hydrolysis, β-oxidation presented in Eq. 1 can take place (Singh 1997).

$$\text{Fatty acids} + \text{ATP} + \text{CoA} - \text{SH} \rightarrow \text{acyl} - \text{CoA} + \text{AMP} + \text{PP}_i \quad (1)$$

Meanwhile, the increase of pH close to 6 led further to an eventual enhancement of the anaerobic digestion process of OMW, during their disposal in the evaporation ponds leading to the degradation of long chain fatty acids by methanogen communities (Speece 1996). Generally, long Chain Fatty Acids (LCFA), in anaerobic condition and at neutral pH can be ionized (Alves et al. 2009). However, studies have shown that LCFA inhibited anaerobic digestion (Neves et al. 2009; Palatsi et al. 2009). These inhibiting potential can be due to the complexity of lipids molecules in addition to their carbon chain length and their saturation (Kuang et al. 2006).

Sugar analysis

Carbohydrates are primarily represented by the parietal components; in particular cellulose and pectin; where they account for approximately 0.6% of the weight of the fresh pulp (Dermeche et al. 2013). Generally, they play an important role in texture olives. Previous studies have shown that the OMW is very rich in simple sugars (Dermeche et al. 2013). Analysis of these OMW shows the predominance of fructose, glucose, sucrose, palatinose and maltose. Sugars generated from different OMW samples are shown in Table 4. The comparison with Fresh and disposal OMW shows any significant difference ($p=0.14$) for the different sugar fractions. As can be seen, an accumulation of sucrose concentration is observed in DOMW. Despite, the sucrose rate was completely removed in DOMW.

Table 4 Sugar concentration rate in FOMW and DOMW; Test Student was determined with FOMW and DOMW ($p=0.14$, $p>0.05$)

Sugar compounds (g/l)	FOMW	DOMW
Fructose ($C_6H_{12}O_6$)	0.689^a	3.074^b
Glucose ($C_6H_{12}O_6$)	0.689^a	1.489^b
Sucrose ($C_{12}H_{22}O_{11}$)	0.053^a	0.000^b
Palatinose ($C_{12}H_{22}O_{11}$)	0.201^a	0.462^b
Maltose ($C_{12}H_{22}O_{11}$)	0.000^a	0.149^b

Test Student was determined with FOMW and DOMW ($p=0.14$, $p>0.05$). Mean scores in the same line with different letter ("a" and "b") are significantly different (for $p<0.05$) (Results are compared for the same parameters and the two different OMW)

With regard to DOMW, we noted an accumulation of fructose and glucose. They can be by-products of the oxidation of intermediates released by the complex chain carbon such as phenol or from long chain fatty acid degradation. Indeed, the OMW disposal in evaporation ponds led a further oxidation of the fermented product that enhances the COD degradation. The complexity of the results requires a more detailed analysis and a further treatment.

Microbial community

It is known that OMW contained all essential elements for microbial growth. However, they may contain several growth inhibitors, such as organic acids and phenolic compounds (Borja et al. 1992). As observed in Table 5, microbial community analysis before and after pond evaporation showed a clear shift in biomass composition. Bacterial aerobic mesophilic community present in the FOMW disappeared completely after disposal in pond evaporation, even though there was available substrate. Since, competing species or new species can be appeared (data not shown). Above all, the Archaea community defined as the act of anaerobic digestion. In fact, the pH increment has been previously reported (Section: The physico-chemical properties) would be enhancing the development of some Methanomicrobiales. It is well known that methanogens are more sensitive to low pH than other anaerobic microorganisms (Speece 1996). In addition, we showed a failure of yeast number. This can be explained by the reduction of the soluble oxygen in the pond evaporation.

Table 5 Microbial diversity in FOMW and DOMW

Samples	MB (CFU/mL)	Yeats (CFU/mL)
OMW F	1005×10^2	$38{,}100\times10^2$
OMW T	0	1632

Antibacterial activity

To valorize the FOMW as an antibacterial agent, MIC against different clinic bacterial strains (Gram-positive and Gram-negative bacteria) were determined. The antibacterial activity shows a MIC value of 32.76 (μg mL^{-1}) for the different studied strains (Table 6). A significant important MIC value was registered compared with others works. The MIC was established against bacterial isolates responsible for human intestine and respiratory tract infections such as *Staphylococcus aureus* (50.00 μg mL^{-1}) and *Klebsiella pneumoniae* (50.00 μg mL^{-1}) (Tafesh et al. 2011). The MIC defined in this study was the result of a combined effect of several phenolic compounds present in the FOMW extract. Indeed, previous studies suggested that bioactivity of the single phenolic component against clinic bacterial strains was found to be very low and required high concentration exceeding 1000 μg mL^{-1} to inhibit the growth of the these strains (Obied et al. 2007).

The observed antibacterial activities against Gram-negative and Gram-positive bacteria were similar. In contrast, previous study explains that there is a relationship between antibacterial activity and cell wall composition (Khoufi et al. 2008; Yangui et al. 2008). However, the comparison of antimicrobial data of plant extracts is very difficult, depends on the different factors, such bacterial strains, growth media, inoculum size, methods and the relative purity (Leouifoudi et al. 2015).

Soil chemical and biological and features, behaviour after FOMW and DOMW spreading

All OMW compounds are involved in several chemical and biological soil transformations, which influence their mobility and biodegradation. Indeed, just after spreading, the OMWs modify the soil. Table 7 summarizes the physico-chemical properties and biological quality of two soils treated with OMWF and OMWD, respectively. During the

Table 6 Antibacterial activity of different OMW F concentration (μg mL^{-1}) against the flowing bacteria: Gram-positive bacteria (*Staphylococcus aureus*) and Gram-negative bacteria (*Klebsiella pneumonias, Salmonella typhimurium* and *Enterobacter cloacae*)

Culture Studies	Different OMW F concentration (μg mL^{-1})			
	Control	32.76	65.52	131.04
Staphylococcus aureus	+	–	–	–
Klebsiella pneumoniae	+	–	–	–
Enterobacter cloacae	+	–	–	–
Salmonella typhimurium	+	–	–	–

(+) Bacterial growth observed and (–) No bacterial growth observed

Table 7 Physicochemical and microbial soil characteristics of two sites treated with OMWF and OMWD

Parameters	Non-treated soil	Soil treated FOMW	Soil treated DOMW
OM (%)	2.6±0.2	0.87±0.12	2.3±0.26
pH	8.7±0.16	8.64±0.01	7.66±0.01
EC (mS/cm^{-1})	0.86±0.05	1.76±0.15	0.48±0.02
Soil respiration (mg CO_2 g^{-1} dry soil)	0.22±0.02	0.66±0.02	0.3±0.01
GI (%)	100	61.65±3.03	170.55±1.88
Mesophilic bacterial (CFU g^{-1})	$18.93\times10^6\pm1$	$29\times10^6\pm1$	$50\times10^6\pm2$
Fungi (CFU g^{-1})	$51\times10^6\pm3.6$	$6\times10^6\pm1$	$12\times10^6\pm2$
Polyphenols contain (ppm)	–	1850±50	10± 800

Non-treated soil was used as a control
– not detected

experimented period, the soil treated with DOMW showed a significant failure of pH and EC values, compared to the soil FOMW spreading. As can be seen, the soil treated with DOMW showed an OM value close to the value found with control soil, which emphasizes the efficiency and the interest of the OMW treatment before use in soil irrigation. Results are similar to those published by Mahmoud et al. (2010) and Munir et al. (2016) showing an increase of soil organic matter after OMW treatments.

Moreover, this OM increase is accompanied by an important rate of soil fertility represented by GI and microflora biomass by 170.55% and 50 × 10^6 CFU g^{-1}, respectively. Indeed, OMW treatment entails the soil enrichment and the improvement of the yield on the culture. Mekki et al. (2013) showed that microbial growth could increase by soil fertility. The lowest polyphenols soil rate (10 ppm) was recorded when irrigated with DOMW. These results can be explained by the high decrease and removal of phenolic compounds for DOMW (see section *The Phenol Compounds*). In fact, phenolic compounds are non-degradable and can inhibit seed germination and decrease soil fertility. In addition, we note that low-molar-mass polyphenols remained in the DOMW did not affect the bacterial growth (50 × 10^6 CFU g^{-1}). This result corroborates with previous studies that showed a lower toxicity of Low-molar-mass polyphenols for bacteria (Fiorentino et al. 2003; Isidori et al. 2005).

However, higher concentrations of salts and polyphenols, found in the FOMW explained the highest rate of EC (1.76 mS/cm^{-1}) found in FOMW spreading soil. Results are similar to those presented by Rinaldi et al. (2003) explaining the increase of soil electrical conductivity. Generally, the increase of EC after direct OMW spreading was attributed to the highest contain of salt concentrations in the non-treated OMW (Di Serio et al. 2008).

Correlation analyses, among soil parameters of the studied sites indicated different significant trends (Table 8). A significant positive correlation between Polyphenols and EC ($r=0.732$) was established. In fact, the soil salinity increases with the content of polyphenols. These results agree with previous results showing a higher rate of soil EC due to the accumulation of polyphenols compounds (Di Serio et al. 2008). However, a positive correlation founded between Bacteria and the index of germination of ($r=0.775$). This result confirmed the increase of soil fertility by the bacterial growth. In addition, a positive correlation with soil organic matter and fungi biomass ($r=0.716$) confirmed that soil organic additives, (such as glucose or/and lipid OMW contain), can cause changes in soil microflora (Stenström

Table 8 Correlation matrix (Pearson's *r* values) between the different parameters determined in soils subjected to the two different treatments

	pH	EC	OM	Resp	IG	Poly	Bac	Fungi
pH								
EC	0.683*							
OM	– 0.310*	– 0.873**						
Resp	0.292	0.877**	– 0.957**					
GI	– 0.908**	– 0.908**	0.638	– 0.651				
Poly	0.29	0.732*	– 0.939**	0.953**	– 0.418			
Bac	– 0.948**	– 0.462	0.034	– 0.038	0.775*	0.242		
Fungi	0.418	– 0.340	0.716*	– 0.710*	– 0.045	– 0.883**	– 0.658	

pH, EC (Electrical Conductivity) OM (Organic Matter), Resp (Respiration), GI (Germination index), Poly (Polyphenol compounds), Bac (Bacterial mesophilic), Fungi (fungi)

$^*p<0.05$; $^{**}p<0.01$

et al. 2001). In fact, sugars and lipids can increase bacterial growth. Indeed, previous studies suggested that OMW enriched the soil in organic substrates affecting the nutritional status of the soil; influencing, therefore, the structure of the soil microbial communities (Barbera et al. 2013).

The significant negative correlation founded between polyphenols compound and fungi biomass ($r = -0.883$) could be explained by polyphenols phytopathogenic soil agents (Yangui et al. 2010; Debo et al. 2011). Of it made, the polyphenols content was negatively correlated with OM ($r = -0.939$), the result was in agreement with previous works (Cardinali et al. 2010). Having a phytotoxic effect and antimicrobial activities, polyphenols are the main limiting factor for spreading OMW. In fact, polyphenols are difficult to decompose. Indeed, polyphenols present a microbial toxicity suppressing soil microorganisms and inhibiting their growth (Obied et al. 2007). In addition, the high level of polyphenols in OMW can also pollute surface and groundwater resources (Obied et al. 2007).

However, the negative correlation between Respiration values and OM content ($r = -0.957$) seems to be strange. Indeed, this result was in contrast to previous works which showed a strongly positive effect with these between both soil parameters (Di Serio et al. 2008; Hoorman and Rafiq 2010). The negative correlation between the organic matter and soil respiration in our case could corroborate with previous studies which demonstrated that increasing soil OM is a strategy for sequestering carbon dioxide (CO_2) a greenhouse gas (Luo et al. 2010; Liu et al. 2014; Barton et al. 2016).

Conclusion

The natural biological treatment of OMW, which is obvious in the disposal of OMW in an evaporation pond promoted an important phenol compounds and long chain fatty acid degradation. Ponds disposal OMW also affected the microbial biomass composition. Both shift in microbial and LCFA composition was a consequence of an anaerobic digestion, which promoted further oxidation of organic by-products, thus generating more reduced product and led to disappearance of total aerobic mesophilic bacteria, which in turn prevented the growth of anaerobic bacteria. The profitable effect of DOMW was successfully presented after the DOMW soil spreading. Indeed, we note an important development of the soil microorganism communities accompanied by a very spectacular soil fertility without forgetting that FOMW can be also valorized as soon in the axis of powerful antibacterial activity against certain standard clinical species.

Acknowledgements The present work has been supported by grants to Central laboratory and Laboratory of Erémologie and Fight Against Desertification of Institute Arid Regions of Medenine.

References

Abichou M, Labiadh M, Cornelis W, Gabriels D, Ben Rouina B, Taamallah H, Khatteli H (2009) The Olive Mills Waste Water (OMW) as an organic amendment for controlling wind erosion in southern Tunisia by improving the soil surface structure. J Arid Land Studies 19–1:343–346

Aggoun M, Arhab R, Cornu A, Portelli J, Barkat M, Graulet B (2016) Olive mill wastewater microconstituents composition according to olive variety and extraction process. Food Chem 209:72–80. https://doi.org/10.1016/j.foodchem.2016.04.034

Alaoui NS, El Laghdach A, Stitou M, Bakkali A (2016) Treatment and valorization of olive mill wastewaters. Mediterr J Chem 5(3):458–464

Alves LC, Rubinger MM, Lindemann RH, Perpétuo GJ, Janczak J, Miranda DL, Zambolim L, Oliveira MRL (2009) Syntheses, crystal structure, spectroscopic characterization and antifungal activity of new N-R-sulfonyldithiocarbimate metal complexes. J Inorg Biochem 103:1045–1053. https://doi.org/10.1016/j.jinorgbio.2009.04.018

Anderson JJ, Dagley S (1980) Catabolism of aromatic acids in *Trichosporon cutaneum*. J Bacteriol 141:534–543

Antonopoulou G, Gaval N, Skiadas IV, Lyberatos G (2010) Influence of pH fermentative hydrogen production from sweet sorghum extract. Int J Hydrog Energy 35:1921–1928. https://doi.org/10.1016/j.ijhydene.2009.12.175

APHA, AWWA, WPCF (1998) Standard Methods for the Examination of Water and Wastewater American Public Health Association. Washington DC, USA. https://www.mwa.co.th/download/file_upload/SMWW_1000-3000.pdf

Aziz NH, Farag SE, Mousa LAA, Abo-Zaid MA (1998) Comparative antibacterial and antifungal effects of some phenolic compounds. Microbios 93:43–54

Barbera AC, Maucieri C, Cavallaro V, Ioppolo A, Spagna G (2013) Effects of spreading olive mill wastewater on soil properties and crops. Agric Water Manag 119:43–53. https://doi.org/10.1515/johh-2016-0017

Barton L, Hoyle FC, Stefanova KT, Murphy DV (2016) Incorporating organic matter alters soil greenhouse gas emissions and increases grain yield in a semi-arid climate. Agric Ecosyst Environ 231:320–330. https://doi.org/10.1016/j.agee.2016.07.004

Belaid C, Kallel M, Elleuch B (2002) Revue francophone d'écologie industrielle N°27-3ème trimestre 2002. Identification de nouveaux composés phénoliques présents dans les rejets liquides d'huileries d'olive (Margines). https://doi.org/10.4267/dechets-sciences-techniques.2389

Borja R, Martín A, Maestro-Durán R, Alba J, Fiestas JA (1992) Enhancement of the anaerobic digestion of olive mill waste water by the removal of phenolic inhibitors. Process Biochem 27:231–237. https://doi.org/10.1016/0960-8524(93)90097-U

Bouknana D, Hammouti B, Salghi R, Jodeh S, Zarrouk A, Warad I, Aouniti A, Sba M (2014) Physicochemical characterization of olive oil mill wastewaters in the eastern region of Morocco. J Mater Environ Sci 5:1039–1058

Cardinali A, Cicco N, Linsalata V, Minervini F, Pati S, Pieralice M, Tursi N, Lattanzio V (2010) Biological activity of high molecular weight phenolics from olive mill wastewater. J Agric Food Chem 58:8585–8590. https://doi.org/10.1021/jf101437c

Comandini P, Lerma-Garcia MJ, Massanova P, Simo-Alfonso EF, Toschi TG (2014) Phenolic profiles of olive mill wastewaters treated by membrane filtration systems. J Chem Technol Biotechnol. https://doi.org/10.1002/jctb.4416

D'Annibale A, Crestini C, Vinciguerra V, Sermanni GG (1998) The biodegradation of recalcitrant effluents from an olive mill by a white-rot fungus. J Biotechnol 61:209–218

Dakhli R, Maalej EM (2017) Olive-mill wastewater spreading

in Southern Tunisia: effects on a barley crop (*Hordeum vulgare* L.). J Agric Environ Int Dev 111:87–103. https://doi.org/10.12895/jaeid.20171.552

Debo A, Yangui T, Dhouib A, Ksantini M, Sayadi S (2011) Efficacy of a hydroxytyrosol-rich preparation from olive mill wastewater for control of olive psyllid, *Euphyllura olivina*, infestations. Crop Prot 30:1529–1534. https://doi.org/10.1016/j.cropro.2011.08.006

Dermeche S, Nadour M, Larroche C, Moulti-Mati F, Michaud F (2013) Olive mill wastes: biochemical characterization and valorization strategies. Process Biochem 48:1532–1552. https://doi.org/10.1016/j.procbio.2013.07.01

Di Serio MG, Lanza B, Mucciarella MR, Russi F, Iannucci E, Marfisi P, Madeo A (2008) Effects of olive mill wastewater spreading on the physico-chemical and microbiological characteristics of soil. Int Biodeterior Biodegrad 62:403–407. https://doi.org/10.1515/johh-2016-0017

Fiorentino A, Gentili A, Isidori M, Monaco P, Nardelli A, Parrella A, Temussi F (2003) Environmental effects caused by olive mill wastewaters: toxicity comparison of low-molecular-weight phenol components. J Agric Food Chem 51:1005–1009. https://doi.org/10.1021/jf020887d

Hoorman J, Rafiq I (2010) Understanding soil microbes and nutrient recycling. Agriculture and Natural Resources, The Ohio State University. https://ohioline.osu.edu/factsheet/SAG-16

Isidori M, Lavorgna M, Nardelli A, Parrella A (2005) Model study on the effect of 15 phenolic olive mill wastewater constituents on seed germination and *Vibrio fischeri* metabolism. J Agric Food Chem 53:8414–8417. https://doi.org/10.1021/jf0511695

Kemal UM (1994) Polyphénols, *O*-diphénols et acides phénoliques totaux dans les grignons d'olive et les margines. Oliv AE 51:34–35. https://doi.org/10.12895/jaeid.20171.552

Khoufi S, Aloui F, Sayadi S (2008) Extraction of antioxidants from olive mill wastewater and electro-coagulation of exhausted fraction to reduce its toxicity on anaerobic digestion. J Hazard Mater 151:531–539. https://doi.org/10.1016/j.jhazmat.2007.06.017

Kuang Y, Pullammanappallil P, Lepesteur M, Ho G-E (2006) Recovery of oleate inhibited anaerobic digestion by addition of simple substrates. J Chem Technol Biotechnol 81:1057–1063. https://doi.org/10.1002/jctb.1530

Leouifoudi I, Zyad A, Amechrouq A, Oukerrou MA, Mouse HA, Mbarki M (2014) Identification and characterisation of phenolic compounds extracted from Moroccan olive mill wastewater. Food Sci Technol 34:249–257. https://doi.org/10.1590/fst.2014.0051

Leouifoudi I, Harnafi H, Zyad A (2015) Olive mill waste extracts: polyphenols content, antioxidant, and antimicrobial activities. Ad Pharmacol Sci. https://doi.org/10.1155/2015/714138

Liu C, Meng LuM, Cui J, Li B, Fang C (2014) Effects of straw carbon input on carbon dynamics in agricultural soils: a meta-analysis. Global Change Biol 20:1366–1381. https://doi.org/10.1111/gcb.12517

Luo Z, Wang E, Sun OJ (2010) Soil carbon change and its responses to agricultural practices in Australian agro-ecosystems: a review and synthesis. Geoderma 155:211–223. https://doi.org/10.1016/j.geoderma.2009.12.012

Mahmoud M, Janssen M, Haboub N, Nassour A, Lennartz B (2010) The impact of olive mill wastewater application on flow and transport properties in soils. Soil Tillage Res 107:36–41

Martins F, Gomes-Laranjo J, Amaral C, Almeida J, Peixoto F (2008) Evaluation of olive oil mill wastewaters acute toxicity: a study on the mitochondrial bioenergetics. Ecotoxicol Environ Saf 69:480–487. https://doi.org/10.1016/j.ecoenv.2007.05.008

Mc Namara CJ, Anastasiou CC, O'Flaherty V, Mitchell R (2008) Bioremediation of olive mill wastewater. Int Biodeterior Biodegr 61:127–134

Mekki A, Dhouib A, Sayadi S (2013) Review: effects of olive mill wastewater application on soil properties and plants growth. Int J Recycl Org Waste Agricult 2:15

Neves L, Pereira MA, Mota M, Alves MM (2009) Detection and quantification of long chain fatty acids in liquid and solid samples and its relevance to understand anaerobic digestion of lipids. Biores Technol 100:91–96. https://doi.org/10.1016/j.biortech.2008.06.018

Niaounakis M, Halvadakis CP (2006) Olive processing waste management, vol 5, 2nd edn. Literature Review and Patent Survey. https://ia801204.us.archive.org/2/items/proposedplanamen01unit/proposedplanamen01unit.pdf. Accessed 1 Feb 2006

Obied HK, Bedgood DR Jr, Prenzler PD, Robards K (2007) Bioscreening of Australian olive mill waste extracts: biophenol content, antioxidant, antimicrobial and molluscicidal activities. Food Chem Toxicol 45:1238–1248. https://doi.org/10.1016/j.fct.2007.01.004

Öhlinger R (1995) Soil respiration by titration. In: Margesin R (ed) Methods in soil biology. Springer, Berlin, pp 95–98

Palatsi J, Laureni M, Andres MV, Flotats X, Nielsen HB, Angelidaki I (2009) Strategies for recovering inhibition caused by long chain fatty acids on anaerobic thermophilic biogas reactors. Biores Technol 100:4588–4596. https://doi.org/10.1016/j.biortech.2009.04.046

Paraskeva P, Diamadopoulos E (2006) Technologies for olive mill wastewater (OMW) treatment: a review. J Chem Technol Biotechnol 81:1475–1485. https://doi.org/10.1002/jctb.1553

Peixoto F, Martins F, Amaral C, Gomes-Laranjo J, Almeida J, Palmeira CM (2008) Evaluation of olive oil mill wastewater toxicity on the mitochondrial bioenergetics after treatment with *Candida oleophila*. Ecotoxicol Environ Saf 70:266–275. https://doi.org/10.1016/j.ecoenv.2007.11.003

Perez J, Cormenzana R, Martinez J (1990) Bacteria degrading phenolic acids isolated on a polymeric phenolic pigment. J Appl Bacteriol 69:38–42. https://doi.org/10.1111/j.1365-2672.1990.tb02909.x

Rahouti M, Seigle-Murandi F, Steiman R, Benoit-Guyod J-L (1997) Étude de la toxicité des composés phénoliques modèles de la.lignine vis-à-vis de 1040 souches de champignons. Actes Inst Agron Veto (Maroc) 17:165–172

Rajhi H, Puyol D, Martinez MC, Diaz EE, Sanz JL (2015) Vacuum promotes metabolic shifts and increases biogenic hydrogen production in dark fermentation systems. Front Environ Sci Eng 10:513–521. https://doi.org/10.1007/s11783-015-0777-y

Ranalli G (1997) Rheology of the lithosphere in space and time. In: Burg J-P, Ford M (eds) Orogeny through time. Geol Soc London Spec Publ, vol 121, pp 19–37

Rinaldi M, Rana G, Introna M (2003) Olive-mill wastewater spreading in southern Italy: effects on a durum wheat crop. Field Crops Res 84:319–326

Rusan MJM, Albalasmeh AA, Hanan I (2016) Treated olive mill wastewater effects on soil properties and plant growth. Water Air Soil Pollut 227:135

Schönfeld P, Kahlert S, Reiser G (2004) In brain mitochondria the branched-chain fatty acid phytanic acid impairs energy transduction and sensitizes for permeability transition. Biochem J 383:121–128. https://doi.org/10.1042/BJ20040583

Singh I (1997) Biochemistry of peroxisomes in health and disease. Mol Cell Biochem 167:1–29

Speece RE (1996) Anaerobic biotechnology (for industrial wastewater). Archae Press, Nashville. https://doi.org/10.1021/es00115a725

Staples CA, Williams JB, Craig GR, Roberts KM (2001) Fate, effects and potential environmental risks of ethylene glycol: a review. Chemosphere 43:377–383

Stenström J, Svensson K, Johansson M (2001) Reversible transition between active and dormant microbial states in soil. FEMS Microbiol Ecol 36:93–104

Tafesh A, Najami N, Jadoun J, Halahlih F, Riepl H, Azaizeh H (2011) Synergistic antibacterial effects of polyphenolic compounds from olive mill wastewater. Evid Complement Altern Med. https://doi.org/10.1155/2011/431021

Technical Committee, 1996 ISO/TC 34, Food products, Subcommittee SC 9, Microbiology, in collaboration with CEN Technical Committee CEN/TC 275, Food analysis—Horizontal methods

Tsioulpas A, Dimou D, Iconomou D, Aggelis G (2002) Phenolic removal in olive oil mill wastewater by strains of *Pleurotus* spp. in respect to their phenol oxidase (laccase) activity. Biores Technol 84:251–257

Villegas LGC, Mashhadi N, Chen M, Mukherjee D, Taylor KE, Biswas N (2016) A short review of techniques for phenol removal from wastewater. Curr Pollut Rep 2:157–167

Walkley A (1947) A critical examination of a rapid method for determining organic carbon in soils- effect of variations in digestion conditions and inorganic soil constituents. Soil Sci 63:251–264

Yangui T, Rhouma A, Triki MA, Gargouri K, Bouzid J (2008) Control of damping-off caused by *Rhizoctonia solani* and *Fusarium solani* using olive mill waste water and some of its. Crop Prot 27:189–197. https://doi.org/10.1016/j.cropro.2007.05.005

Yangui T, Sayadi S, Gargoubi A, Dhouib A (2010) Fungicidal effect of hydroxytyrosol-rich preparations from olive mill wastewater against *Verticillium dahliae*. Crop Prot 29:1208–1213. https://doi.org/10.1016/j.cropro.2010.04.016

Zucconi F, Pera A, Forte M, De Bertoldi M (1981) Evaluating toxicity of immature compost. BioCycle 22:54–57

20

Loss of phosphorus by runoff from soils after amendment with poultry litter co-composted with crop waste

Asma Saleem[1] · Muhammad Irshad[1,2] · An Ping[2] · Bushra Haroon[1]

Abstract

Purpose The study aimed to assess the influence of poultry litter (PL) after co-composting with sugarcane and cabbage waste on phosphorus (P) losses in runoff from soils under natural rainfall conditions.

Methods Co-composted PL was amended in silt loam and sandy clay soils. The soil applied with PL without agro-waste was considered as control treatment. Before the soil application, PL was co-composted with agro-wastes, i.e., sugarcane waste and cabbage waste at four levels (0, 25, 33, and 50%). Soils were packed in wooden trays layered with plastic. The surface soil was mixed with the co-composted PL at rates 200 and 400 kg P ha^{-1}. Runoff samples were collected from the sloped trays during two rainfall events and P concentration was determined.

Results Phosphorus concentration was found higher in the runoff in the PL treatment without agro-waste. Poultry litter application along with agro-waste profoundly lowered P losses in the runoff as compared to the control especially at higher application of agro-waste. Application of PL composted with higher level of agro-waste (%) reduced the P losses from soils. Phosphorus losses in the runoff enhanced with higher amount of PL application depending on the soil type and initial P content in the soil. Silt loam soil amended with co-composted poultry litter/agro-waste reduced P losses more significantly as compared to sandy clay soil.

Conclusion The application of manure amendments with agro-wastes decreased the losses of soluble P and would reduce detrimental environmental effects.

Keyword Phosphorus loss · Runoff · Rainfall events · Poultry litter · Crop waste

Introduction

Surface runoff is causing phosphorus (P) losses from soil which is responsible for eutrophication of water bodies (Pote et al. 1999). A well-established correlation between agricultural P losses to surface waters and eutrophication has been reported (Cassell et al. 1998; Correll 1998). Phosphorus application can result in increased risk of P delivery to surface waters. Application of excessive manure to soils for the longer period increases the P level in soil and/or surface water (Sharpley et al. 2004). Phosphorus is added to soil in the form of organic waste (Brogan et al. 2001). The use of poultry litter on the agricultural lands with elevated P levels needs special management. Fresh poultry litter may quickly release P in soil, whereas stable forms of organic matter may slowly release P sources. Composting is a biological process of decomposition in which organic matter is transformed to a stable product like humus. The process is carried out by microbes, which use organic waste for food and energy purposes (Chefetz et al. 1998). Composting of organic wastes is important due to the environmental and community constraints and enhanced agronomic value. McDowell and Sharpley (2003) reported the potential of soil P losses into runoff after application of P-containing materials. The release of P from soil to runoff has been well reported by the kinetics of P desorption from soil (Sharpley et al. 1981; Garcia-Rodeja and Gil-Sotres 1997). The salient P mobility controlling factors in surface runoff are transport and source such as manure or fertilizer (Pote et al. 1996; Sharpley et al. 1993). Runoff from cropland areas receiving poultry litter

✉ Muhammad Irshad
mirshad@ciit.net.pk

1 Department of Environmental Sciences, COMSATS Institute of Information Technology Abbottabad, Abbottabad, Pakistan

2 Arid Land Research Center Tottori University, Hamasaka cho, Tottori 680-0001, Japan

or compost may enhance P concentrations in water bodies (Eghball et al. 1996). The environmental issues related to poultry manure and agricultural residues application could be alleviated by composting before application to agricultural fields. Therefore, the present research was aimed to assess P losses from soils amended with poultry litter after co-composting with agro-wastes.

Materials and methods

Poultry litter (PL) was brought from poultry farms around Abbottabad, where these farms were in abundance. The wastes of sugarcane (*Saccharum officinarum*) and cabbage (*Brassica oleracea* var. capitata) collected from vegetable market of Abbottabad city (34.1558°N, 73.2194°E), Pakistan. These wastes were cut into smaller pieces < 5 cm. The PL was co-composted with sugarcane waste (SW) and cabbage waste (CW) in plastic bins (> 10 L) arranged in rows inside a tiled shed to protect from the rain. Poultry litter was mixed with SW and CW separately at three levels, i.e., 25, 33, and 50%. Poultry litter after composting without agro-waste was set as a control treatment. The material was co-composted for 120 days. The treatments were replicated thrice. The material was moistened periodically and aerated during the composting process. Additional water was added as required and moisture content in the poultry manure was kept at 30%. Samples were collected after mixing, and then air dried, crushed, and sieved (< 0.5 mm) to ensure homogeneity. The physico-chemical properties of the composted manure samples were determined (Table 1).

Table 1 Chemical properties of poultry litter and agro-wastes before composting

Parameters	Poultry litter	Sugarcane waste	Cabbage waste
pH (1:5)	7.8	6.8	2.6
EC (dS m^{-1})	9.4	2.7	0.2
Total C (g kg^{-1})	440.6	636	490
C:N	16	57	40
Total Ca (mg kg^{-1})	1234	307	500
Total Mg (mg kg^{-1})	409	187	120
Total Na (mg kg^{-1})	222	185	180
Total P (g kg^{-1})	13.6	5.3	0.26
Total K (mg kg^{-1})	721	8.1	1.7
Total N (g kg^{-1})	50.6	11.9	5.3
Total Fe (mg kg^{-1})	299	35.8	1.7
Total Cu (mg kg^{-1})	89	1.5	2.6
Total Mn (mg kg^{-1})	255	2.2	0.3
Total Zn (mg kg^{-1})	158	12.9	1.8

Total carbon was determined using dry-combustion method (Nelson and Sommers 1982). pH and EC were measured in manure and soil samples also. The pH was determined in the manure suspension with manure: water ratio of 1:5 by a pH meter (Model: HANNA HI 8520). Electrical conductivity (EC) in the manure suspension before and after composting was measured by an EC meter (Model: 4320 JENWAY) (in 1:5 water suspension). For total nitrogen, Kjeldahl N determination was performed according to APHA (1995). Total Kjeldahl nitrogen was estimated by digesting the compost sample in concentrated sulphuric acid and by distillation and trapping the ammonia in 0.1 N sulphuric acid titrated against standard 0.1 N sodium hydroxide solution (AOAC 1995). Carbon-to-nitrogen ratio was calculated based on the total C and total Kjeldahl N concentrations. Sample weighing 0.5 g was digested within a mixture (1:3) of per chloric ($HClO_4$) and nitric (HNO_3) acids for the determination of total concentrations of macro- and microelements [potassium (K), calcium (Ca), magnesium (Mg), sodium (Na), copper (Cu), iron (Fe), manganese (Mn), and zinc (Zn)], using an atomic absorption spectrophotometer (AAS) (Miller 1998). Total phosphorus (P) in the digest was determined by phospho-molybdate blue color method and absorbance was measured by a spectrophotometer at a wavelength of 710 nm (Olsen and Sommers 1982). Composted PL samples were fractionated into readily plant-available P, labile inorganic P (another plant-available fraction), sesquioxide-associated (Fe oxide and Al oxide) inorganic P- and Ca-associated P by sequential extraction with de-ionized water, 0.5 M $NaHCO_3$ (pH 8.5), 0.1 M NaOH, and 1 M HCl. The sequential extraction of inorganic P used in this study was as follows: 0.1 g sample of the compost (0.5 mm) was placed in a 50 mL centrifuge tube and was sequentially extracted with 30 mL each of de-ionized water, 0.5 M $NaHCO_3$, 0.1 M NaOH, and 1 M HCl. The extraction with each reagent was carried out in duplicate after 16 h of end-to-end shaking. After each shaking, tubes were shaken for 16 h at room temperature and centrifuged at 10000 rpm for 15 min. Supernatants were filtered and the P contents determined calorimetrically using the molybdenum blue method (Olsen and Sommers 1982).

Runoff experiment

A separate study was carried out to determine P losses from soils amended with composted manures. Soils were sampled from a depth of 0–20 cm from different sites of Abbottabad, where the annual rainfall occurred about 1160 mm. The physico-chemical properties of soils analyzed are given in Table 2. Soil texture was determined using pipette method (Gee and Bauder 1986). Soils were classified as silt loam and sandy clay based on textural analysis. Soil samples were sieved via a 2 mm sieve and packed with 10 kg soil in the

Table 2 Chemical composition of soils used for the study

Parameters	Unit	Silt loam	Sandy clay
Total C	g kg^{-1}	26.9	31.4
Total N	mg kg^{-1}	51.3	63.8
Total P	mg kg^{-1}	59.3	75.8
Total Ca	mg kg^{-1}	259.1	300.2
Total Mg	mg kg^{-1}	125.2	151.6
Total K	mg kg^{-1}	149.9	157.6
Total Na	mg kg^{-1}	29.1	37.3
Total Cu	mg kg^{-1}	48.3	61.7
Total Fe	mg kg^{-1}	69.6	61.6
Total Mn	mg kg^{-1}	61.5	75.5
Total Zn	mg kg^{-1}	71.3	103.8
Total Ni	mg kg^{-1}	2.3	2.7
Total Cd	mg kg^{-1}	0.7	1.3
EC (1:5)	dS m^{-1}	0.6	0.9
pH (1:5)		7.6	7.1

wooden trays layered at the bottom and sides with plastic sheets (tray size: 0.5 m × 0.5 m × 0.2 m). Soils in the trays were moistened to field capacity. The trays were installed in an open field. The outlet of each tray was connected to a plastic container. The trays were placed at a 5% slope. Triplicate trays were used for each treatment. Co-composted material was applied at the rate of 200 and 400 kg total P ha^{-1}. Surface runoff was collected in plastic bottles at the down slope end of the trays to evaluate runoff P losses. Time, duration, and precipitation level was determined by the automated weather station of CIIT Abbottabad. The data were collected after two consecutive rainfall events. Phosphorus contents in the runoff were determined to compare the loss of P from co-compost treated soil and untreated soil.

Statistical analysis

The data were statistically analyzed and analysis of variance (ANOVA) was done using Stat-view software. The least significant difference (LSD) was determined at $P \leq 0.05$ to calculate significant differences between treatments.

Results and discussion

Total P concentrations were found as 21.8, 19.6, 17.2, and 15.5 g kg^{-1} with the sugarcane waste (SW) at the application rate of 0, 25, 33, and 50%, respectively, after 120 days of composting. With the incorporation of cabbage waste (CW), the concentrations were 18.1, 16.7, and 15.1 g kg^{-1}, respectively (Table 3). The reduced organic P in the compost could be related to the lower P content in the plant waste when compared with the manure. Organic P in the control poultry litter was higher than that in the plant-waste treatments. However, the blending of poultry litter with the SW and CW reduced the P contents. The P fractions differed in the composted manure as HCl-P > water-P > NaHCO3-P > NaOH-P. Sugarcane waste amendment gave higher concentrations of P across all fractions than CW.

Phosphorus concentrations in the runoff were higher under the compost treatment without crop waste (Table 4). Poultry litter application along with agro-waste profoundly lowered P losses in the runoff as compared to the control. Phosphorus contents in the runoff decreased with the increasing ratio of agro-wastes in the compost treated soils. Phosphorus losses in the runoff were higher with the higher manure application rate. Silt loam soil amended with co-composted poultry litter/agro-waste reduced P losses more significantly as compared to sandy clay soil.

The differences in the P losses could be associated with the difference in the initial P content and soil texture. P contents in silt loam soil ranged from 31.2 to 24.5 mg L^{-1} after application 200 kg P ha^{-1} followed by 36.7 to 32.1 mg L^{-1} after application of 400 kg P ha^{-1} with poultry litter: sugarcane waste (SW) mixture, respectively, as compared to the sandy clay containing 35.5 to 31.2 mg L^{-1} in 200 kg P ha^{-1} followed by 39.7 to 35.4 mg L^{-1} at application rate of 400 kg P ha^{-1}. Compost without agro-waste contained 40.6 mg L^{-1} in silt loam and 44.1 mg L^{-1} P contents in runoff in sandy clay soil in SW-poultry litter. The study showed that co-compost containing agro-waste reduced P contents more effectively in the runoff.

Table 3 Phosphorus fractions (g kg^{-1}) in poultry litter after co-composting for 120 days with sugarcane and cabbage waste at different ratios

	Waste application (%)	Total	H_2O	$NaHCO_3$	NaOH	HCl
Sugarcane waste	0	21.8	8.6	3.5	1.9	12.7
	25	19.6	6.8	1.7	1.5	10.7
	33	17.2	4.3	0.9	1.3	8.7
	50	15.5	3.8	0.6	1.0	7.1
Cabbage waste	25	18.1	6.5	1.2	1.3	9.0
	33	16.7	4.0	0.5	1.7	8.1
	50	15.1	3.0	0.2	0.9	6.8
	LSD (0.05)	1.5	0.7	0.03	0.1	0.2

Table 4 Average phosphorus loss in runoff after two rainfall events from soils treated with co-composted poultry litter with agro-wastes at two application rates

Agro-waste	Soil	Agro-waste in PL (%)	200 kg P ha^{-1}	400 kg P ha^{-1}
			P in runoff (mg L^{-1})	
Sugarcane waste	Silt loam	0	40.6	43.4
		25	31.2	36.7
		33	27.4	34.4
		50	24.5	32.1
	Sandy clay	0	44.1	47.8
		25	34.5	39.7
		33	32.7	37.4
		50	30.2	35.4
		LSD (0.05)	2.4	2.3
Cabbage waste	Silt loam	25	26.3	31.2
		33	24.2	27.5
		50	21.1	25.1
	Sandy clay	25	30.9	35.3
		33	28.6	33.6
		50	26.1	31.7
		LSD (0.05)	2.6	3.1

Regardless of soil and application rates, P losses in the runoff were lowered with manure co-composted with cabbage waste in the order of (%) 50 > 33 > 25 > 0. Sandy clay soil decreased P contents as 30.9, 28.6, and 26.1 mg L^{-1} followed by 35.3, 33.6, and 31.7 mg L^{-1} after application of 25, 33, and 50% cabbage waste mixture in poultry manure amendment at 200 kg P ha^{-1} and 400 kg P ha^{-1}, respectively. Phosphorus losses increased with higher manure applications. Waste mixture at 50% in the poultry litter reduced maximum losses of P in the runoff. Spargo et al. (2006) reported higher infiltration and lower runoff from the soils after amendment with co-composted poultry litter and yard waste due to the improvement in soil physical properties and compared poultry litter to poultry litter and yard waste co-compost to reduce P contents in runoff. Gilley et al. (2002) reported that manure and compost application in disked soil did not increase dissolved and bioavailable fractions of P in runoff volume, but increased total P in runoff.

Tillage with cattle manure application significantly lowered dissolved P, particulate P, and total P, whilst disking with cattle manure amendment did not significantly reduce P constituents in the runoff volume (Gilley et al. 2007). Phosphorus losses were highly affected by the soil application rates of manure amendments mixed with sugarcane and cabbage waste. This substantial reduction of phosphorus losses in the bulk of runoff could be attributed to an increase in the water stable aggregates in soils after co-composts and compost application (McDowell and Sharpley 2001). Tarkalson and Mikkelsen (2007) had also supported an increased total P loss with increased application rates of broiler litter. Gilley et al. (2002) previously reported that, while manure and compost application to soil did not increase dissolved and bioavailable fractions of P in the runoff volume, they did increase the total amount of P in runoff. The observed substantial reduction of nutrient losses from soil to runoff was attributed to an increase in water stable aggregates in soils after co-composts and compost application (McDowell and Sharpley 2001). Tejada et al. (2009) also reported the importance of organic amendments in ameliorating soil composition by accommodating the stimulation and development of flocculation of clay minerals, which further promoted aggregation of soil particle.

Conclusions

It is concluded that compost application without agro-waste increased soil P losses. Co-compost of PL and agro-waste reduced P losses from soils. Soils amended with increasing ratios of sugarcane and cabbage waste in poultry litter apparently reduced P concentrations in the runoff. It is concluded that application of manure amendments with agro-wastes found to be a useful option in reducing P losses from soils. Further studies are needed to investigate the alterations in the soil physico-chemical properties after annual applications of composted manures in agricultural fields.

Acknowledgements The authors are thankful to the Higher Education Commission of Pakistan for funding this research through national research program for universities (NRPU-4831).

References

Brogan J, Crowe M, Carty G (2001) Developing a national phosphorus balance for agriculture in Ireland. Environmental Protection Agency, Wexford

Cassell EA, Dorioz JM, Kort RL, Hoffmann JP, Meals DW, Kirschtel D, Braun DC (1998) Modeling phosphorus dynamics in ecosystems: mass balance and dynamic simulation approaches. J Environ Qual 27:293–298

Chefetz BF, Adani P, Genevini F, Tambone Hadar Y, Chen Y (1998) Humic-acid transformation during composting of municipal solid waste. J Environ Qual 27:794–800. https://doi.org/10.2134/jeq1998.00472425002700040011x

Correll DL (1998) The role of phosphorus in the eutrophication of receiving waters: a review. J Environ Qual 27:261–266

Eghball B, Binford GD, Baltensperger DD (1996) Phosphorus movement and adsorption in a soil receiving long-term manure and fertilizer application. J Environ Qual 25:1339–1343. https://doi.org/10.2134/jeq1996.00472425002500060024x

Garcia-Rodeja I, Gil-Sotres F (1997) Prediction of parameters describing phosphorus-desorption kinetics in soils of Galicia (Northwest Spain). J Environ Qual 26:1363–1369. https://doi.org/10.2134/jeq1997.00472425002600050023x

Gee GW, Bauder JW (1986) Particle-size analysis. In: Klute A (ed), Method of Soil Analysis, Part 1, Agronomy series No. 9. Am. Soc. Agron. Soil Sci. Soc. Am. Inc Publ, Madison WI, USA, pp 383–441

Gilley JE, Risse LM, Eghball B (2002) Managing runoff following manure application. J Soil Water Conserv 57(6):530–533

Gilley JE, Eghball B, Marx DB (2007) Nutrient concentrations of runoff during the year following manure application. Trans ASABE 50(6):1987–1999

McDowell RW, Sharpley AN (2001) Approximating phosphorus release from soils to surface runoff and subsurface drainage. J Environ Qual 30:508–520. https://doi.org/10.2134/jeq2001.302508x

McDowell RW, Sharpley A (2003) The effects of soil carbon on phosphorus and sediment loss from soil trays by over land flow. J Environ Qual 32:207–214. https://doi.org/10.2134/jeq2003.2070

Miller RO (1998) Nitric-perchloric acid wet digestion in an open vessel. In: Kalra Y (ed) Handbook of reference methods for plant analysis. CRC Press, Washington, DC

Nelson DW, Sommers LE (1982) Total carbon, organic carbon and organic matter. In Methods of soil analysis. Part 2. Chemical and microbiological properties, ed. A.L. Page, R.H. Miller, and D.R. Keeney, 539–580. Madison, WI: ASA, SSSA

Olsen SR, Sommers LE (1982) Phosphorus. In: Page, A.L., Miller, R. H. and Keeney, D.R. (eds). Methods of Soil Analysis.Part 2. 2nd edn. Agronomy Series No. 9, American Society of Agronomy and Soil Science Society of America, Inc., Madison, WI, pp 403–430

Pote DH, Daniel TE, Sharpley AN, Moore PA, Edwards DR, Nichols DJ (1996) Relating extractable soil phosphorus to phosphorus losses in runoff. Soil Sci Soc Am J 60:855–859

Sharpley AN, Ahuja LR, Menzel RG (1981) The release of phosphorus to runoff in relation to the kinetics of desorption. J Environ Qual 10:386–391. https://doi.org/10.2134/jeq1981.00472425001000030029x

Sharpley AN, Daniel TE, Edwards DR (1993) Phosphorus movement in the landscape. J Prod Agric 6:492–500

Sharpley AN, McDowell RW, Kleinman PJA (2004) Amounts, forms, and solubility of phosphorus in soils receiving manure. Soil Sci Soc Am J 68:2048–2057

Spargo J, Gregory TKE, Marcus MA (2006) Repeated compost application effects on phosphorus runoff in the Virginia Piedmont. J Environ Qual 35(6):2342–2351. https://doi.org/10.2134/jeq2006.0105

Tarkalson DD, Mikkelsen RL (2007) Phosphorus losses in runoff after application of litter from broiler fed high available p corn feed. Commun Soil Sci Plant Anal 38:1805–1814. https://doi.org/10.1080/00103620701435571

Tejada M, Garcia-Martinez AM, Parrado J (2009) Effects of a vermicompost composted with beet vinasse on soil properties, soil losses and soil restoration. CATENA 77:238–247. https://doi.org/10.1016/j.catena.2009.01.004

21

Utilization of wheat straw for fungal phytase production

Zohre Shahryari[1,2] · **Mohammad H. Fazaelipoor**[3] · **Payam Setoodeh**[4] · **Ramkumar B. Nair**[2] · **Mohammad J. Taherzadeh**[2] · **Younes Ghasemi**[5,6]

Abstract

Purpose Wheat straw is an agricultural waste which can be used as a cost effective animal feed. However, high hemicellulose and phytic acid content in wheat straw prevents it as a primary feed choice. Utilization of wheat straw in solid-state fermentation may result in wheat straw valorization and enzyme production. In this study, phytase production in solid-state fermentation of wheat straw using *Aspergillus ficuum* and valorization of wheat straw were evaluated.

Methods A two-step experimental design procedure was employed for screening and optimization of influencing factors on phytase production. Effects of different nutritional and environmental factors were investigated by one factor at the time method (OFATM). To reach higher amounts of phytase, response surface methodology (RSM) was employed to optimize phytase production as a function of three of the most effective factors.

Results Optimization of the significant parameters resulted in an increase in the phytase activity from 0.74 ± 0.12 to a maximum of 16.46 ± 0.56 Units per gram dry substrate (U gds^{-1}). The high degree of the fungal phytase activity on wheat straw resulted in the decrease in phytic acid content by 57.4%, as compared to the untreated sample. Scanning electron microscopy (SEM) and FTIR structural analysis showed intensive fungal growth on wheat straw, and partial removal of hemicelluloses, lignin and phytic acid.

Conclusion The study demonstrated the feasibility of wheat straw utilization in solid-state fermentation using *Aspergillus ficuum* toward the production of phytase and valorization of wheat straw as an animal feed.

Keywords Wheat straw · Phytase · *Aspergillus ficuum* · Valorization · Solid state fermentation

✉ Mohammad H. Fazaelipoor
fazaelipoor@yazd.ac.ir

1 Department of Chemical Engineering, Faculty of Engineering, Shahid Bahonar University of Kerman, Kerman, Iran

2 Swedish Centre for Resource Recovery, University of Boras, 50190 Borås, Sweden

3 Department of Chemical and Polymer Engineering, Faculty of Engineering, Yazd University, Yazd, Iran

4 School of Chemical and Petroleum Engineering, Shiraz University, Shiraz, Iran

5 Department of Pharmaceutical Biotechnology, School of Pharmacy, Shiraz University of Medical Sciences, Shiraz, Iran

6 Pharmaceutical Sciences Research Center, Shiraz University of Medical Sciences, Box 71345-1583, Shiraz, Iran

Introduction

Wheat straw is the second most abundant lignocellulosic raw material in the world (Pensupa et al. 2013). The international grain council forecasted the annual world wheat production of 754 million tons in 2016 (ICG 2017) with a straw to grain ratio of 1.3 (for most wheat varieties). Surplus amount of straw has resulted in environmental and public health concerns attributing to the inefficiency of the conventional straw disposal or utilization methods. Currently, wheat straw is used as animal feed, as supporting materials (Panthapulakkal et al. 2006), as raw material for pulp and paper production (Nasser et al. 2015), and as a substrate for biogas, bioethanol, and mushroom production. Wheat straw is also burnt as a fuel and is added to soil for its maintenance (Ferreira et al. 2014; Huang et al. 2017; Mahboubi et al. 2017; Tomás-Pejó et al. 2017). Wheat straw contains 350–450 g kg^{-1} cellulose, making it an excellent potential source of energy for ruminants. However, intact straw is not an ideal feed for ruminants because of its high lignocellulose (200–300 g kg^{-1} hemicellulose and 80–150 g kg^{-1} lignin) and phytic acid content. Phytic acid, the main source of phosphorous in cereals, is not absorbed by monogastric animals, and due to its very strong chelating agent properties, it binds to metal ions (i.e., Ca^{2+}, Mg^{2+}, Zn^{2+}, Fe^{2+}) and proteins, making them unavailable for absorption. Furthermore, it inhibits the activity of a number of enzymes such as α-amylase, acid phosphatase and tyrosinase (Singh and Satyanarayana 2006a). Hence, phytic acid acts as an anti-nutritional factor in straw-derived feed components (Hurrell et al. 2003). On the other hand, phytic acid can find its way into waterways and consequently causes rapid algal growth and reduction in water oxygen content. This contributes to death of fish and hypoxia among other problems (Roopesh et al. 2006). These problems can be effectively addressed if the straw-derived feed diets are supplemented with phytase, an enzyme that degrades phytic acid. Phytase is produced naturally by several strains of bacteria, yeast, and filamentous fungi. Filamentous fungi have been widely used for the production of a wide range of extracellular enzymes and value-added supplements (Souza et al. 2015). Fungi-derived phytase has obtained considerable attention due to the ease of cultivation and high production yields (Roopesh et al. 2006). Various filamentous fungi such as *Aspergillus* sp. (Bhavsar et al. 2011; Moreira et al. 2014), *Mucor racemosus* (Bogar et al. 2003b), and *Sporotrichum thermophile* (Singh and Satyanarayana 2006b, Maurya et al. 2017), *Thermomyces lanuginosus* (Makolomakwa et al. 2017), and *Rhizopus oryzae* (Arora et al. 2017) have been reported to produce phytase. Several filamentous fungi and their enzyme products are used to break down the bond between the polysaccharides (cellulose and hemicellulose) and lignin in wheat straw to improve its digestibility through the process referred to as biological-pretreatment (Sindhu et al. 2016, Choudhary et al. 2016, Bagewadi et al. 2017). Many previous studies have also reported the ability of filamentous fungi to produce various commercial enzymes from wheat straw (Table 1).

However, only limited reports are available regarding the production of phytase from wheat straw using filamentous fungi (Singh 2014). The cultivation of filamentous fungi on wheat straw for phytase production could lead to the direct use of the crude product as a value-added supplement in feed mixtures (Bogar et al. 2003b). The aim of the present work is to evaluate phytase production through solid-state fermentation of wheat straw using filamentous fungi *Aspergillus ficuum*. Attempts were made to optimize the effective parameters on phytase production using response surface methodology. The effects of microbial growth on the physical and chemical characteristic of wheat straw were also investigated.

Table 1 Previous reports on wheat straw-derived fungal enzymes

Microorganism/s	Produced enzyme/s	Time (days)	Enzyme activity	Refs.
Aspergillus fumigatus	Endoglucanase	3	4.54 U mL^{-1}	Mehboob et al. (2014)
	Exoglucanase		3.81 U mL^{-1}	
	β-glucosidase		4.30 U mL^{-1}	
Schizophyllum commune	β-glucosidase	7	0.9 U mL^{-1}	Musatti et al. (2017)
	Xylanase	7	19 U mL^{-1}	
Aspergillus niger ADH-11	Cellulose	8	0.15 FPU[a] mL^{-1}	Patel et al. (2017)
	Xylanase	8	34.49 U mL^{-1}	
Penicillium janthinellum EMS-UV-8	Cellulase	8	1.2 FPU mL^{-1}	Sharma et al. (2015)
	CMCase	8	26.6 U mL^{-1}	
	β-Glucosidase	8	1.4 U mL^{-1}	
	Xylanase	8	76 U mL^{-1}	
Aspergillus niger	Cellulase	5	24 U g^{-1}	Pensupa et al. (2013)
Pleurotus ostreatus HP-1	Laccase	8	14.189 U g^{-1}	Thakur et al. (2013)
	Manganes peroxidase	8	562.80 U g^{-1}	
	Xylanase	8	10 U g^{-1}	
Fusarium concolor	Laccase	10	31.14U g^{-1}	Chang et al. (2012)
	Cellubiose	5	19.93U g^{-1}	
	Dehydrogenase	2	4.99 U g^{-1}	
	Xylanase	10	14 U g^{-1}	
	CMCase	5	15.64 U g^{-1}	
	FPase			

[a]Filter paper unit

Materials and methods

Materials

Wheat straw was obtained from a local farm in Shiraz, Iran. It was dried in a forced air oven at 60 °C for 24 h, milled and sieved (0.21–0.84 mm). All the chemicals used in this research work were of analytical grade and purchased from Sigma-Aldrich, Germany.

Microorganism

The fungus *Aspergillus ficuum* PTCC 5288 was maintained on Sabouraud Dextrose Agar (SDA) slants at 30 °C for 4 days and stored at 4 °C. Inoculum was prepared by adding 10 mL of sterile distilled water with Tween 80 (1 mL L^{-1}), to fully sporulated slant using a sterile pipette. The spores were gently agitated using a needle under strict aseptic conditions. The total spore count was determined by the Neubauer counting method.

Solid state fermentation of wheat straw

Solid-state fermentation of wheat straw (5 g dry weight) was carried out in 250 mL Erlenmeyer flasks. Flasks were sterilized at 120 ± 1 °C and 15 psi for 20 min. After cooling, the substrate was moistened (under aseptic conditions) with a sterilized solution containing the additives (carbon source, nitrogen source, mineral salt) mentioned below, inoculated with the spore suspension of *A. ficuum* (10^6 spores gds^{-1}) and incubated for 86 h (the time needed for the fully growth of the fungus on straw). The moisture level of 600 g kg^{-1}, the substrate particle size of 0.70 ± 0.14 mm, and the temperature of 30 °C were used for all the tests, except otherwise specified. After fermentation, the fermented straw was mixed with 120 mL of 0.1 M sodium acetate buffer (pH 5.5) at room temperature and agitated for 120 min at 180 rpm to extract the crude enzyme. Finally, the enzyme extract was centrifuged at 13000*g* and 4 °C for 15 min and the supernatant was collected for phytase activity determination. All the experiments described below have been performed in duplicate (except otherwise specified), and all the data points presented are the mean of duplicate runs.

Screening of the significant factors affecting phytase production

Fermentation experiments were carried out using one factor at the time (OFAT) experimental approach to screen among the effective factors on phytase production. Various carbon and nitrogen sources and mineral salts were examined to find out the required additives to the straw. The effects of variation in temperature, moisture, and substrate particle size on phytase production were also investigated.

Comparison of different carbon sources as inducers for the fungal growth and phytase production

An experiment was conducted to select the best carbon source to supplement wheat straw for rapid growth of the fungi. Mannitol, glucose, lactose, maltose, fructose, and sucrose were tested for this purpose. Each of the materials was added to the straw to the concentration of 0.1 g gds^{-1}. The fermentation was conducted as described in Sec.2.3. A control run (without carbon source) was also included for this experiment.

Comparison of different nitrogen sources as supplements for the fungal growth and phytase production

Malt, ammonium sulfate, yeast extract, tryptone, sodium nitrate, ammonium nitrate, peptone and ammonium chloride were compared to find out the best nitrogen sources among them. Each of the materials was added to the wheat straw to the concentration of 0.05 g gds^{-1}. Glucose (as the best carbon source) was added to the straw (0.1 g gds^{-1}). The fermentation was conducted as described in "Solid state fermentation of wheat straw". A control run (without nitrogen source) was also included for this experiment.

Comparison of different mineral salts as supplements for the fungal growth and phytase production

KCl, KH_2PO_4, K_2HPO_4, $MgSO_4$, $ZnCl_2$, $MnSO_4$, $NaCl$, and $FeSO_4$ were added to the straw to the concentration of 0.005 g gds^{-1}. Glucose (0.1 g gds^{-1}) and ammonium sulfate (0.05 g gds^{-1}) were also added to the straw. The fermentation was conducted as described in "Solid state fermentation of wheat straw". A control run (without any mineral salts) was also included for this experiment.

The effects of temperature, moisture, and particle size on the fungal growth and phytase production

Three sets of experiments were conducted to evaluate the effects of temperature (in the range of 25–35 °C), moisture (in the range of 400–750 g kg^{-1}), and particle size (in the range of 0.25–0.70 mm) on the fungal growth and phytase production. The fermentation process was carried out as described in "Solid-state fermentation of wheat straw". All the samples contained 0.1 g gds^{-1} glucose, 0.05 g gds^{-1} ammonium sulfate, and 0.005 g gds^{-1} manganese sulfate.

Response surface approach for optimizing phytase production

Response surface method (RSM) was applied to optimize phytase production as a function of moisture level, and glucose and ammonium sulfate concentrations, which were identified as the most effective factors on phytase production by the OFAT screening process. Each factor was studied at three different levels following the Box–Behnken design. The fermentation process was conducted as described in "Solid-state fermentation of wheat straw". The experimental design for this part is depicted in Table 2.

A total of three replicates were carried out for each experimental set, and the average values are presented with standard deviation.

Analytical methods

Spectrophotometric (PG Instrument Ltd., UK) enzyme assay method was performed as described by Coban et al. (2015). One unit of phytase (U) was defined as the amount of enzyme that released 1 μmol/min of inorganic phosphorus from 1.5 mM sodium phytate under the assay conditions. The enzyme activity was expressed in units per gram of dry substrate (U gds^{-1}), unless otherwise specified.

Structural analysis

Scanning electron microscope (SEM) (Tescanvega 3, Czech Republic) was used for the structural analysis of wheat straw. Fermented wheat straw was oven dried at 50 °C for 72 h. Dry fermented wheat straw was fixed on the aluminum stub and SEM image was captured. The investigation of the structural changes in the fermented wheat straw was also carried out using a Fourier Transform Infrared (FTIR) spectrometer (Impact 410 iS10, Nicolet Instrument Corp., Madison, WI, USA). The spectra of the fermented and untreated wheat straw were obtained with an average of 64 scans and resolution of 4 cm^{-1} from 600 to 4000 cm^{-1}. The spectrum data were developed by Nicolet OMNIC 4.1 software (Nair et al. 2017).

Table 2 Experimental design matrix and the results from Box–Behnken design for phytase production

Run	Glucose (g gds^{-1})	Ammonium sulfate (g gds^{-1})	Moisture (g kg^{-1})	Phytase activity (U gds^{-1})	
				Observed	Predicted
1	0.00	0.00	600	0.75 ± 0.11	1.58
2	0.30	0.00	600	3.70 ± 0.21	3.73
3	0.00	0.10	600	2.55 ± 0.13	2.52
4	0.30	0.10	600	14.37 ± 0.57	13.54
5	0.00	0.05	400	0.15 ± 0.07	-0.34
6	0.30	0.05	400	3.27 ± 0.24	3.58
7	0.00	0.05	800	1.05 ± 0.10	0.74
8	0.30	0.05	800	9.50 ± 0.32	9.99
9	0.15	0.00	400	0.29 ± 0.05	-0.05
10	0.15	0.10	400	2.77 ± 0.18	3.29
11	0.15	0.00	800	2.17 ± 0.12	1.65
12	0.15	0.10	800	8.73 ± 0.36	9.07
13	0.15	0.05	600	13.22 ± 0.69	13.16

Determination of phytic acid

A modified protocol of Harland and Oberleas (1977) was followed for the extraction of phytate from the untreated and fungal fermented wheat straw. The phytate content was measured based on the reaction between ferric ion and sulfosalicylic acid (Wade reagent) following the rapid colorimetric method described by Latta and Eskin (1980).

Results and discussion

Solid-state fermentation (SSF) of wheat straw using *Aspergillus ficuum* was carried out to produce phytase. Phytase production from wheat straw was found to improve its digestibility as observed with its structural analysis. To reach the maximum phytase production, efforts were made in this study to optimize the physical and nutritional parameters affecting the fungal phytase production, which is explained below.

Fungal phytase from wheat straw: screening for culture parameters

The effects of nutrient parameters such as carbon source inducers, nitrogen sources and mineral salts together with the physical parameters such as temperature, moisture content and wheat straw particle size, on phytase production were evaluated. Among the different carbon sources, glucose was found to yield the maximum phytase production (Table 3). Glucose has previously been reported as the most effective carbon source for phytase production from various microorganisms such as *Myceliophthora thermophile* (Hassouni et al. 2006), *Pseudomonas aeruginosa* (Sasirekha et al. 2012), and *Aspergillus flavus* (Gaind and Singh 2015). The optimum nitrogen source was selected by supplementation of the media with 0.1 g gds^{-1} of glucose together with 0.05 g gds^{-1} of different nitrogen sources. Compared to other nitrogen sources, ammonium sulfate showed considerable improvement in the phytase production (Table 3). Similar observations were made by Bogar et al. (2003a), Singh and Satyanarayana (2006b), and Bala et al. (2014), demonstrating the significance of ammonium sulfate as

Table 3 Effect of different nutritional parameters on phytase production

Carbon source (0.1 g gds^{-1})	Phytase activity (U gds^{-1})	Nitrogen source (0.05 g gds^{-1})	Phytase activity (U gds^{-1})	Mineral salts (0.005 g gds^{-1})	Phytase activity (U gds^{-1})
Manitol	1.602 ± 0.11	Malt extract	4.044 ± 0.18	KCl	7.252 ± 0.34
Glucose	**3.242 ± 0.24**	**$(NH_4)_2SO_4$**	**8.912 ± 0.05**	KH_2PO_4	2.285 ± 0.19
Lactose	0.886 ± 0.16	Yeast extract	2.936 ± 0.13	K_2HPO_4	3.057 ± 0.27
Maltose	1.795 ± 0.21	Tryptone	1.690 ± 0.06	$MgSO_4$	8.679 ± 0.23
Fructose	2.948 ± 0.19	$NaNO_3$	4.042 ± 0.20	$ZnCl_2$	5.930 ± 0.48
Sucrose	2.255 ± 0.30	NH_4NO_3	7.805 ± 0.28	**$MnSO_4$**	**10.200 ± 0.24**
Control	0.739 ± 0.12	Peptone	4.672 ± 0.28	NaCl	9.041 ± 0.11
		NH_4Cl	6.923 ± 0.10	$Fe_2(SO_4)_3$	9.024 ± 0.43
		Control	3.210 ± 0.07	Control	8.900 ± 0.25

the most favorable nitrogen source for phytase production. Among various mineral salts added to the cultivation media, $MnSO_4$ exhibited maximum increase in phytase production.

Both the phosphate salts (KH_2PO_4 and K_2HPO_4) had an inhibitory effect on phytase production, which could be attributed to the phosphate existence in media and lack of need for phytate hydrolyzing and phytase production. Similar observations on the inhibitory effect of phosphate on phytase production have been reported previously. Exclusion of phosphate from the medium enhanced phytase production by *A. niger* (Żyła and Gogol 2002) and *Sporotrichum thermophile* (Singh and Satyanarayana 2006a).

Concerning the physical culture conditions, it was observed that the temperature variation in the range of 27–35 °C had no significant effect ($p \geq 0.05$) on phytase production (Table 4).

Similarly, variation in wheat straw particle sizes in the range of 0.297 to 0.840 mm showed no significant effect on phytase production. However, reduction in particle sizes lower than 0.297 mm led to a decrease in the phytase production (Table 4). This drop might be due to compaction and consequent formation of agglomerates that contribute to a lower oxygen transfer (Schmidt and Furlong 2012). Table 4 shows that moisture variation significantly affects phytase production by the fungus. In the investigated range of 400–750 g kg^{-1}, phytase production increased with the moisture content of the substrate and reached an optimum value at the moisture content of 650 g kg^{-1}.

Response surface optimization of the effective process parameters

The significant factors influencing phytase production were optimized using response surface method. The predicted and observed responses are in Table 2.

The experimental data were fitted to a quadratic equation with phytase activity (U gds^{-1}) as a function of the concentration of glucose (A), ammonium sulfate (B), and moisture content (C). The model is:

$$\begin{aligned}\text{Phytase activity} = &-48.83 + 32.64A + 104.70B \\ &+ 0.170C + 295.66AB + 0.044AC \\ &+ 0.102BC - 173.75A^2 \\ &- 1564.80B^2 - 0.0001C^2\end{aligned} \tag{1}$$

Table 5 shows the analysis of variance for the model. P value for the model was less than 0.0001, indicating that phytase activity is affected significantly by the model parameters. The test of lack of fit was insignificant (P

Table 4 Effect of different physical parameters on phytase production

Temperature (°C)	Phytase activity (U gds^{-1})	Moisture content (g kg^{-1})	Phytase activity (U gds^{-1})	Substrate particle size (mm)	Phytase activity (U gds^{-1})
25	9.400 ± 0.11	400	2.086 ± 0.15	0.707–0.840	10.192 ± 0.29
27	9.871 ± 0.16	450	5.534 ± 0.38	0.595–0.707	10.127 ± 0.15
30	10.220 ± 0.39	500	8.028 ± 0.53	0.420–0.595	10.024 ± 0.12
32	10.418 ± 0.13	550	9.859 ± 0.44	0.297–0.420	10.159 ± 0.34
35	8.533 ± 0.42	600	10.213 ± 0.36	0.250–0.297	9.184 ± 0.27
		650	10.318 ± 0.47	0.210–0.250	8.327 ± 0.19
		700	9.215 ± 0.17		
		750	8.517 ± 0.73		

Table 5 Analysis of variance (ANOVA) for the quadratic model

Source	Sum of squares	*df*	Mean square	*F* value	Prob > *F*
Model	502.07	9	55.79	84.44	< 0.0001
A	86.72	1	86.72	131.27	< 0.0001
B	57.84	1	57.84	87.54	< 0.0001
C	28.01	1	28.01	42.40	0.0003
AB	19.67	1	19.67	29.77	0.0009
AC	7.10	1	7.10	10.75	0.0135
BC	4.16	1	4.16	6.30	0.0404
A^2	64.35	1	64.35	97.41	< 0.0001
B^2	64.44	1	64.44	97.54	< 0.0001
C^2	139.79	1	139.79	211.60	< 0.0001
Residual	4.62	7	0.66		
Lack of fit	2.81	3	0.94	2.07	0.2467
Pure error	1.81	4	0.45		
Total	506.70	16			
$R^2 = 0.9909$	Adj $R^2 = 0.9791$		Pred$R^2 = 0.9056$		

value = 0.2467) indicating good fitness of the model with the experimental data. The values of coefficients of determination (R^2 = 0.9909, Pred R^2 = 0.9056, Adj R^2 = 0.9791) indicated the suitability of the model to predict phytase activity as the function of the model parameters (Ghoshoon et al. 2015; Żyła and Gogol 2002).

Using the model surface plots were generated showing interactions between factor pairs (Fig. 1). All the plots show a maximum value of phytase activity as the function of model parameters. The plots indicate that there is a unique optimum value of phytase activity as the function of moisture, and glucose and ammonium sulfate concentrations.

Response surface optimization predicted the maximum phytase activity of 14.51 U gds^{-1} in the medium containing 0.17 g gds^{-1} glucose, 0.068 g gds^{-1} ammonium sulfate and 655 g kg^{-1} moisture content. To verify the accuracy of the model for the prediction of the optimum value, an additional experiment was conducted with the mentioned values of moisture, and glucose and ammonium sulfate concentrations. The phytase activity of 14 U gds^{-1} was obtained which is reasonably close to the value predicted by the model (3.64% deviation).

To find if extension of the incubation time could improve the maximum phytase production, solid-state fermentation was performed for varying incubation time, with the culture parameters being constant at their optimum value.

The results (shown in Fig. 2) suggested that phytase production increased with the increase in incubation time, with the maximum enzyme production (16.4 U gds^{-1}) at 144 h.

The enzyme activity, however, declined with further incubation, which could be attributed to the reduced nutrient level and enzyme deactivation. The straw-derived fungal phytase obtained in the present study was higher than phytase produced from other crop residues such as sesame oil cake (0.34 U gds^{-1}) (Singh and Satyanarayana 2008), Groundnut oilcake (24.3 U gds^{-1}) (Roopesh et al. 2006), soybean meal (16 U gds^{-1}), wheat bran (4.5 U gds^{-1}) (Chantasartrasamee et al. 2005) and cotton seed cake (1.10 U gds^{-1}) (Singh 2014). Furthermore, *Aspergillus ficuum* strain (in the present study) showed higher phytase production in the solid-state fermentation of wheat straw, as compared to other filamentous fungal strains such as *Aspergillus oryzae* (Singh 2014).

To determine the effect of fungal fermentation on wheat straw, phytic acid was extracted from untreated and fermented wheat straw using hydrochloric acid. Analysis of the crude extract indicated a sharp decrease in phytic acid content from 23.5 ± 0.67 mg kg^{-1} in the untreated straw to 10 ± 0.83 mg kg^{-1} in fungal fermented straw, which indicates that the treatment reduces the phytic acid content of the straw effectively.

Structural analysis of fermented wheat straw

Scanning electron microscopy (SEM) analysis was used to characterize the structural and surface morphology of the fermented wheat straw (Fig. 3).

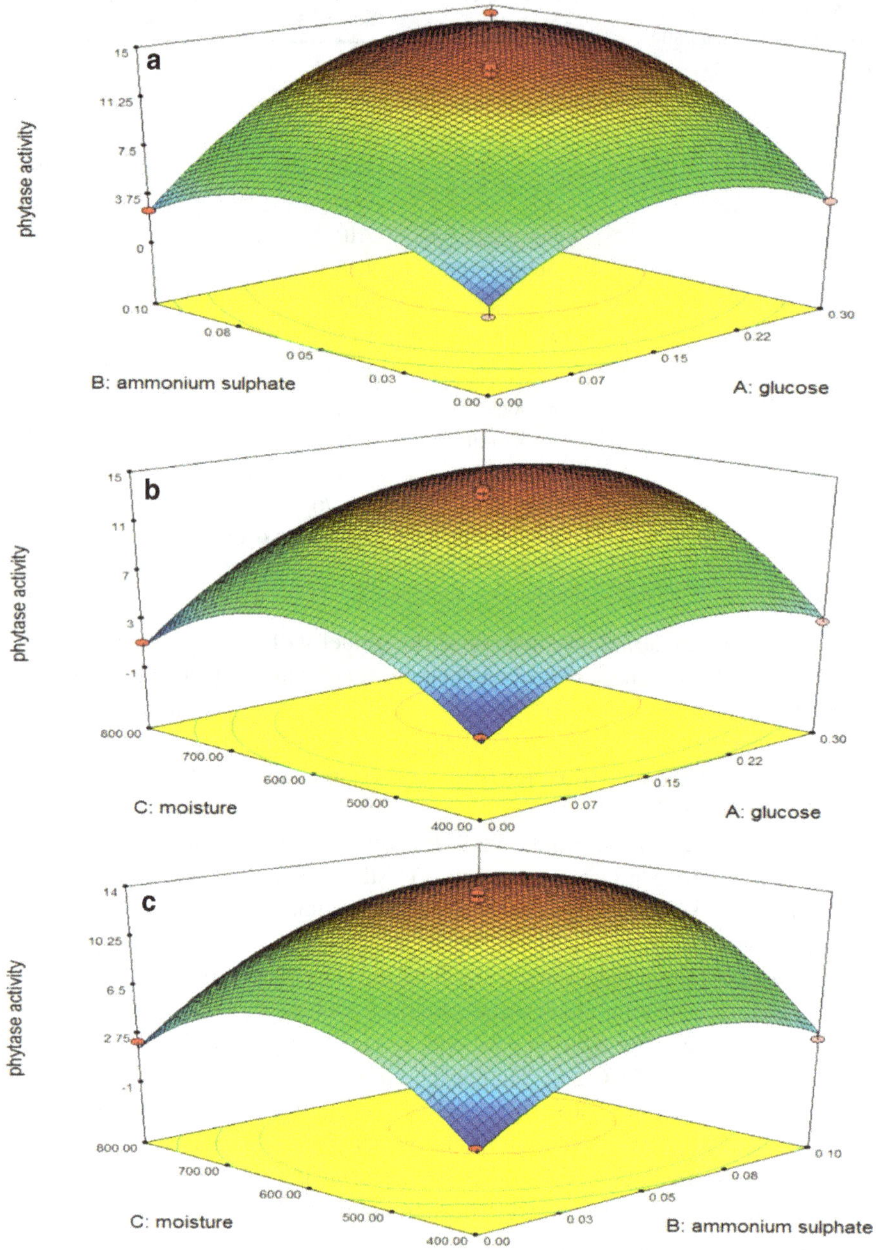

Fig. 1 Factorial interaction of **a** ammonium sulfate (g gds^{-1}) and glucose (g gds^{-1}), **b** moisture (g kg^{-1}) and glucose (g gds^{-1}) and **c** moisture (g kg^{-1}) and ammonium sulfate (g gds^{-1}) on phytase production (U gds^{-1})

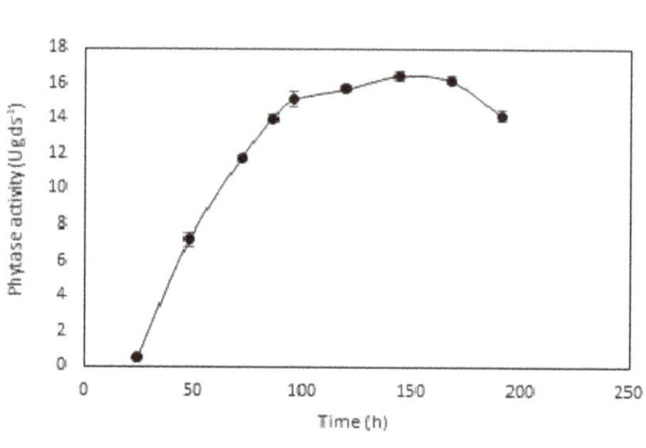

Fig. 2 Phytase production during incubation time

The SEM image showed intense fungal growth on wheat straw and hyphae penetration in straw pores and channels (Fig. 3, bottom). The untreated wheat straw (control) image, however, showed the packed and inaccessible surface structures (Fig. 3, top). The fungal growth on wheat straw resulted in the destruction and modification of the fibrous structure and increase in the porosity. Fermented wheat straw also showed more accessible surfaces compared to the unfermented wheat straw (Fig. 3). The changes in the functional groups of wheat straw after fermentation were analyzed based on the FTIR spectra of the untreated and fermented wheat straw (Fig. 4). Peak assignments are summarized in

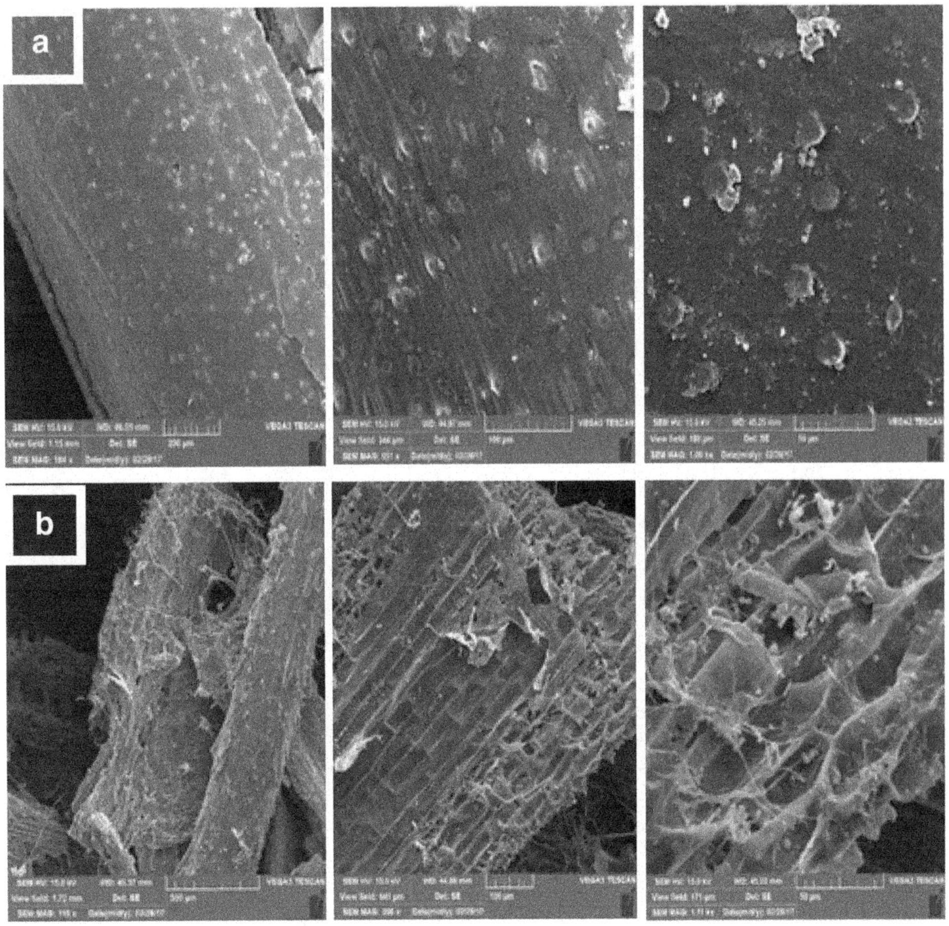

Fig. 3 Scanning electron microscopy (SEM) image indicating the inaccessible and packed structure of untreated wheat straw (top), and fungal hyphae penetration together with germinated spores in the channels and pores on fermented wheat straw (bottom)

Table 6, based on the previous literature reports (Kaushik et al. 2010; Shen et al. 2011).

The analysis of the FTIR spectra depicts significant changes in the structure of the fermented wheat straw compared to untreated wheat straw. Decrease in the polymerization of the cellulose chains was clearly indicated in the fermented wheat straw by the decrease in the absorbance of hydrogen bonds (O–H), represented at peak 3300 cm^{-1} (Oh et al. 2005) (Fig. 4). The absorbance intensity of the peak at 1730 cm^{-1} (that corresponds to the acetyl group) was reduced from 0.036 for untreated wheat straw to 0.032 for fermented wheat straw. Hemicelluloses bind to lignin via acetyl group. Therefore, this can be an indication of cleavage between hemicelluloses and lignin. A sharp peak at 1030 cm^{-1} corresponds to the phytate content (He et al. 2007; Ishiguro et al. 2003) and has significantly decreased in fermented wheat straw, indicating the partial removal of phytic acid (Fig. 4).

Conclusion

Solid-state fermentation of wheat straw using *Aspergillus ficcum* is a feasible way to valorize the straw as an animal feed. Addition of glucose as an inducer and ammonium sulfate enhance phytase production considerably. Maximum phytase activity of 16.4 U gds^{-1} was achieved at the optimum medium conditions using 0.17 g gds^{-1} glucose, 0.068 g gds^{-1} ammonium sulfate and 655 g kg^{-1} moisture after 144 h of incubation at 30 °C. Scanning electron microscopy (SEM) analysis of the fermented wheat straw showed high intensity of fungal growth resulting in the structural destruction and larger accessible surface area in the wheat straw. Chemical analysis and FTIR measurements of the wheat straw revealed the partial removal of phytic acid (57.4% reduction), hemicelluloses and lignin.

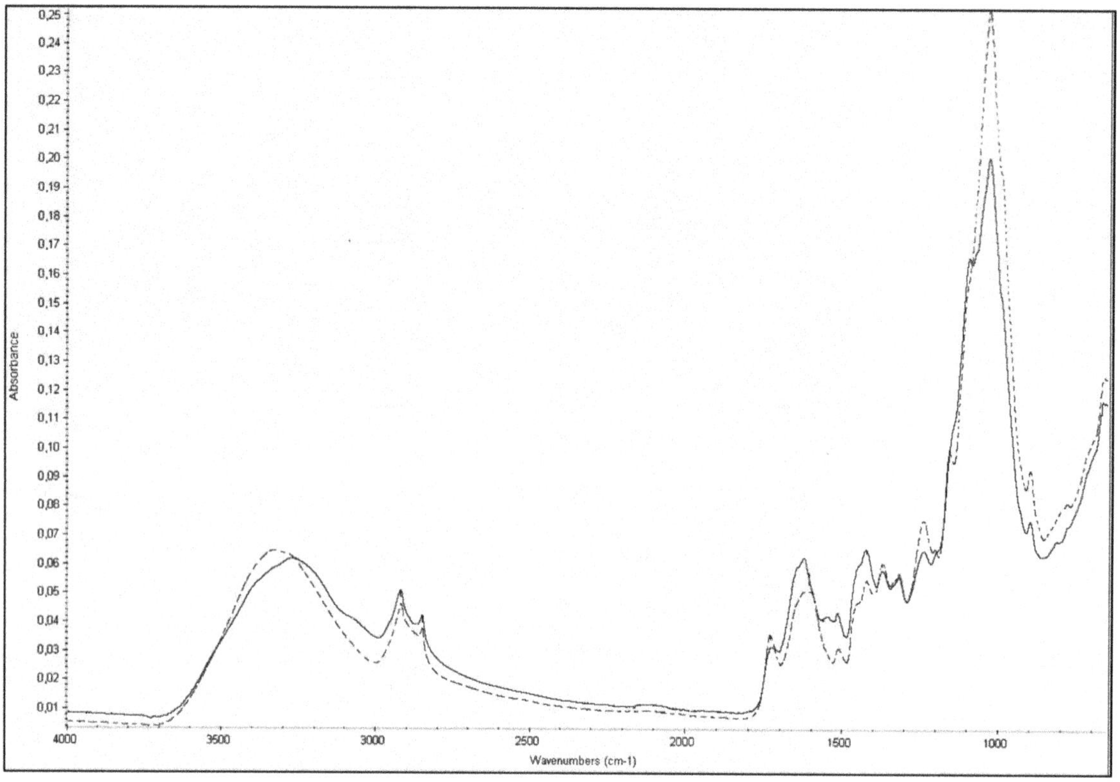

Fig. 4 FTIR spectra of untreated (dashed line), and fermented wheat straw (continuous line)

Table 6 Characteristics and variations in FTIR spectra of untreated and fermented wheat straw

Wavenumber (cm^{-1})	Functional group	Untreated wheat straw	Fermented wheat straw
3300	–OH stretching intramolecular hydrogen band	0.065	0.063
2919	C–H stretching band	0.047	0.052
1730	C=O stretching of acetyl or carboxylic acid	0.036	0.032
1620	C=C stretching of the aromatic ring	0.052	0.063
1510	C=C stretching of the aromatic ring	0.033	0.044
1422	C–H2 symmetric bending	0.056	0.066
1370	C–H asymmetric bending	0.061	0.059
1317	C–H2 wagging	0.058	0.056
1200	C–O–C asymmetric stretching bond	0.076	0.065
1030	C–O stretching band	0.252	0.2
897	C–O–C stretching at β-glucosidic linkage	0.094	0.077

Acknowledgements This study was financially supported by Grant no.: 950508 of the Biotechnology Development Council of the Islamic Republic of Iran. Zohre Shahryari is grateful to the Pharmaceutical Sciences, Research Center, Shiraz University of Medical Science for providing the facility for experiments and Vahid Niknezhad for his technical assistance and valuable discussions.

References

Arora S, Dubey M, Singh P, Rani R, Ghosh S (2017) Effect of mixing events on the production of a thermo- tolerant and acid-stable phytase in a novel solid-state fermentation bioreactor. Process Biochem 61:12–23. https://doi.org/10.1016/j.procbio.2017.06.009

Bagewadi ZK, Mulla SI, Ninnekar HZ (2017) Optimization of laccase production and its application in delignification of biomass. Int J Recycl Org Waste Agric 6:351–365. https://doi.org/10.1007/s40093-017-0184-4

Bala A, Jain J, Kumari A, Singh B (2014) Production of an extracellular phytase from a thermophilic mould Humicola nigrescens in solid state fermentation and its application in dephytinization. Biocatal Agric Biotechnol 3:259–264. https://doi.org/10.1016/j.bcab.2014.07.002

Bhavsar K, Kumar VR, Khire J (2011) High level phytase production by Aspergillus niger NCIM 563 in solid state culture: response surface optimization, up-scaling, and its partial characterization. J Ind Microbiol Biotechnol 38:1407–1417. https://doi.org/10.1007/s10295-010-0926-z

Bogar B, Szakacs G, Linden J, Pandey A, Tengerdy R (2003a) Optimization of phytase production by solid substrate fermentation. J Ind Microbiol Biotechnol 30:183–189. https://doi.org/10.1007/s10295-003-0027-3

Bogar B, Szakacs G, Pandey A, Abdulhameed S, Linden JC, Tengerdy RP (2003b) Production of phytase by Mucorracemosus in solid-state fermentation. Biotechnol Prog 19:312–319. https://doi.org/10.1021/bp020126v

Chang AJ, Fan J, Wen X (2012) Screening of fungi capable of highly selective degradation of lignin in rice straw. Int Biodeterior Biodegradation 72:26–30. https://doi.org/10.1016/j.ibiod.2012.04.013

Chantasartrasamee K, Ayuthaya DIN, Intarareugsorn S, Dharmsthiti S (2005) Phytase activity from *Aspergillus oryzae* AK9 cultivated on solid state soybean meal medium. Process Biochem 40:2285–2289. https://doi.org/10.1016/j.procbio.2004.03.019

Choudhary M, Sharma PC, Jat HS, Nehra V, McDonald AJ, Garg N (2016) Crop residue degradation by fungi isolated from conservation agriculture fields under rice-wheat system of North-West India. Int J Recycl Org Waste Agric 5:349–360. https://doi.org/10.1007/s40093-016-0145-3

Coban HB, Demirci A, Turhan I (2015) Microparticle-enhanced *Aspergillus ficuum* phytase production and evaluation of fungal morphology in submerged fermentation. Bioprocess Biosyst Eng 38:1075–1080. https://doi.org/10.1007/s00449-014-1349-4

Ferreira L, Nilsen P, Fdz-Polanco F, Pérez-Elvira S (2014) Biomethane potential of wheat straw: influence of particle size, water impregnation and thermal hydrolysis. Chem Eng J 242:254–259. https://doi.org/10.1016/j.cej.2013.08.041

Gaind S, Singh S (2015) Production, purification and characterization of neutral phytase from thermotolerant *Aspergillus flavus* ITCC 6720. Int Biodeter Biodegr 99:15–22. https://doi.org/10.1016/j.ibiod.2014.12.013

Ghoshoon MB, Berenjian A, Hemmati S, Dabbagh F, Karimi Z, Negahdaripour M, Ghasemi Y (2015) Extracellular production of recombinant L-Asparaginase II in *Escherichia coli*: medium optimization using response surface methodology. Int J Pept Res Ther 21:487–495. https://doi.org/10.1007/s10989-015-9476-6

Harland B, Oberleas D (1977) A modified method for phytate analysis using an ion-exchange procedure: application to textured vegetable proteins [Soybeans]. Cereal Chem 54:827–832

Hassouni H, Ismaili-Alaoui M, Gaime-Perraud I, Augur C, Roussos S (2006) Effect of culture media and fermentation parameters on phytase production by the thermophilic fungus *Myceliophthora thermophila* in solid state fermentation. Micol Appl Int 18:29–36

He Z, Honeycutt CW, Xing B, McDowell RW, Pellechia PJ, Zhang T (2007) Solid-state Fourier transform infrared and 31P nuclear magnetic resonance spectral features of phosphate compounds. Soil Sci 172:501–515. https://doi.org/10.1097/SS.0b013e318053dba0

Huang C et al (2017) An integrated process to produce bio-ethanol and xylooligo saccharides rich in xylobiose and xylotriose from high ash content waste wheat straw. Bioresour Technol. https://doi.org/10.1016/j.biortech.2017.05.109

Hurrell RF, Reddy MB, Juillerat M-A, Cook JD (2003) Degradation of phytic acid in cereal porridges improves iron absorption by human subjects. Am J Clin Nutr 77:1213–1219. https://doi.org/10.1093/ajcn/77.5.1213

ICG International Grains Council (2017) Market Report 2017. http://www.igcint/downloads/gmrsummary/gmrsumme.pdf. Accessed June 2017

Ishiguro T, Ono T, Nakasato K, Tsukamoto C, Shimada S (2003) Rapid measurement of phytate in raw soymilk by mid-infrared spectroscopy. Biosci Biotechnol Biochem 67:752–757. https://doi.org/10.1271/bbb.67.752

J-h Shen, Z-m Liu, Li J, Niu J (2011) Wettability changes of wheat straw treated with chemicals and enzymes. J Forest Res 22:107–110. https://doi.org/10.1007/s11676-011-0134-3

Kaushik A, Singh M, Verma G (2010) Green nanocomposites based on thermoplastic starch and steam exploded cellulose nanofibrils from wheat straw. Carbohydr Polym 82:337–345. https://doi.org/10.1016/j.carbpol.2010.04.063

Latta M, Eskin M (1980) A simple and rapid colorimetric method for phytate determination. J Agric Food Chem 28:1313–1315. https://doi.org/10.1021/jf60232a049

Mahboubi A, Ylitervo P, Doyen W, De Wever H, Molenberghs B, Taherzadeh MJ (2017) Continuous bioethanol fermentation from wheat straw hydrolysate with high suspended solid content using an immersed flat sheet membrane bioreactor. Bioresour Technol 241:296–308. https://doi.org/10.1016/j.biortech.2017.05.125

Makolomakwa M, Puri AK, Kugen P, Singh S (2017) Thermo-acid-stable phytase-mediated enhancement of bioethanol production using *Colocasia esculenta*. Bioresour Technol 235:396–404. https://doi.org/10.1016/j.biortech.2017.03.157

Maurya AK, Parashar D, Satyanarayana T (2017) Bioprocess for the production of recombinant HAP phytase of the thermophilic mold Sporotrichum thermophile and its structural and biochemical characteristics. Int J Biol Macromol 94:36–44. https://doi.org/10.1016/j.ijbiomac.2016.09.102

Mehboob N, Asad MJ, Asgher M, Gulfraz M, Mukhtar T, Mahmood RT (2014) Exploring thermophilic cellulolytic enzyme production potential of *Aspergillus fumigatus* by the solid-state fermentation of wheat straw. Appl Biochem Biotechnol 172:3646–3655. https://doi.org/10.1007/s12010-014-0796-3

Moreira KA et al (2014) Optimization of phytase production by *Aspergillus japonicus* Saito URM 5633 using cassava bast as substrate in solid state fermentation. Afr J Microbiol Res 8:929–938. https://doi.org/10.5897/AJMR2013.6565

Musatti A, Ficara E, Mapelli C, Sambusiti C, Rollini M (2017) Use of solid digestate for lignocellulolytic enzymes production through submerged fungal fermentation. J Environ Manage 199:1–6. https://doi.org/10.1016/j.jenvman.2017.05.022

Nair RB, Lundin M, Lennartsson PR, Taherzadeh MJ (2017) Optimizing dilute phosphoric acid pretreatment of wheat straw in the laboratory and in a demonstration plant for ethanol and edible fungal biomass production using Neurospora intermedia. J Chem Technol Biotechnol 92:1256–1265. https://doi.org/10.1002/jctb.5119

Nasser RA, Hiziroglu S, Abdel-Aal MA, Al-Mefarrej HA, Shetta ND, Aref IM (2015) Measurement of some properties of pulp and paper made from date palm midribs and wheat straw by soda-AQ pulping process. Measurement 62:179–186. https://doi.org/10.1016/j.measurement.2014.10.051

Oh SY, Yoo DI, Shin Y, Seo G (2005) FTIR analysis of cellulose treated with sodium hydroxide and carbon dioxide. Carbohydr Res 340:417–428. https://doi.org/10.1016/j.carres.2004.11.027

Panthapulakkal S, Zereshkian A, Sain M (2006) Preparation and characterization of wheat straw fibers for reinforcing application in injection molded thermoplastic composites. Bioresour Technol 97:265–272. https://doi.org/10.1016/j.biortech.2005.02.043

Patel H, Divecha J, Shah A (2017) Microwave assisted alkali treated wheat straw as a substrate for co-production of (hemi) cellulolytic enzymes and development of balanced enzyme cocktail for its enhanced saccharification. J Taiwan Inst Chem Eng 71:298–306. https://doi.org/10.1016/j.jtice.2016.12.032

Pensupa N, Jin M, Kokolski M, Archer DB, Du C (2013) A solid state fungal fermentation-based strategy for the hydrolysis of wheat straw. Bioresour Technol 149:261–267. https://doi.org/10.1016/j.biortech.2013.09.061

Roopesh K, Ramachandran S, Nampoothiri KM, Szakacs G, Pandey A (2006) Comparison of phytase production on wheat bran and oilcakes in solid-state fermentation by *Mucor racemosus*. Bioresour Technol 97:506–511. https://doi.org/10.1016/j.biortech.2005.02.046

Sasirekha B, Bedashree T, Champa K (2012) Optimization and partial purification of extracellular phytase from *Pseudomonas aeruginosa* p6. Eur J Exp Biol 2:95–104

Schmidt CG, Furlong EB (2012) Effect of particle size and ammonium sulfate concentration on rice bran fermentation with the fungus *Rhizopus oryzae*. Bioresour Technol 123:36–41. https://doi.org/10.1016/j.biortech.2012.07.081

Sharma B, Agrawal R, Singhania RR, Satlewal A, Mathur A, Tuli D, Adsul M (2015) Untreated wheat straw: potential source for diverse cellulolytic enzyme secretion by *Penicillium janthinellum* EMS-UV-8 mutant. Bioresour Technol 196:518–524. https://doi.org/10.1016/j.biortech.2015.08.012

Sindhu R, Binod P, Pandey A (2016) Biological pretreatment of lignocellulosic biomass—an overview. Bioresour Technol 199:76–82. https://doi.org/10.1016/j.biortech.2015.08.030

Singh B (2014) Phytase production by Aspergillus oryzae in solid-state fermentation and its applicability in dephytinization of wheat ban. Appl Biochem Biotechnol 173:1885–1895. https://doi.org/10.1007/s12010-014-0974-3

Singh B, Satyanarayana T (2006a) A marked enhancement in phytase production by a thermophilic mould *Sporotrichum thermophile* using statistical designs in a cost-effective cane molasses medium. J Appl Microbiol 101:344–352. https://doi.org/10.1111/j.1365-2672.2006.02921.x

Singh B, Satyanarayana T (2006b) Phytase production by thermophilic mold *Sporotrichum thermophile* in solid-state fermentation and its application in dephytinization of sesame oil cake. Appl Biochem Biotechnol 133:239–250. https://doi.org/10.1385/ABAB:133:3:239

Singh B, Satyanarayana T (2008) Phytase production by a thermophilic mould Sporotrichum thermophile in solid state fermentation and its potential applications. Bioresour Technol 99:2824–2830. https://doi.org/10.1016/j.biortech.2007.06.010

Souza PMD et al (2015) A biotechnology perspective of fungal proteases. Braz J Microbiol 46:337–346. https://doi.org/10.1590/s1517-838246220140359

Thakur S, Shrivastava B, Ingale S, Kuhad RC, Gupte A (2013) Degradation and selective ligninolysis of wheat straw and banana stem for an efficient bioethanol production using fungal and chemical pretreatment. 3 Biotech 3:365–372. https://doi.org/10.1007/s13205-012-0102-4

Tomás-Pejó E et al (2017) Valorization of steam-exploded wheat straw through a biorefinery approach: bioethanol and bio-oil co-production. Fuel 199:403–412. https://doi.org/10.1016/j.fuel.2017.03.006

Żyła K, Gogol D (2002) In vitro efficacies of phosphorolytic enzymes synthesized in mycelial cells of *Aspergillus niger* AbZ4 grown by a liquid surface fermentation. J Agric Food Chem 50:899–905. https://doi.org/10.1021/jf010551o

22

Availability and dynamics of organic carbon and nitrogen indices in some soils amended with animal manures and ashes

Toyin B. Olowoboko[1] · Jamiu O. Azeez[1] · Olanrewaju O. Olujimi[2] · Oluwatoyin A. Babalola[1]

Abstract

Purpose The study evaluated dried manures and manure ashes as alternative products from animal manures, and determined the dynamics of soil nitrogen availability indices and organic carbon release in some soils amended with dried manures and ashes under laboratory incubation, screen-house and field experiments.

Methods Cattle, goat and poultry manures were collected and air dried to produce dried manures after which part of the dried manures were burnt in open space at a temperature range of 320–450 °C to produce manure ashes. Treatments which were no amendment (control), dried manures of cattle, goat and poultry manure ashes of cattle, goat and poultry and NPK 15-15-15 at 120 kg N ha^{-1} were applied to soil. Samples were taken fortnightly after amendment incorporation in the incubation experiment; in screen-house experiment and field experiments, plant seeds were established after amendment incorporation; thereafter, soil samples were taken fortnightly. Soil samples were analyzed for soil organic carbon (SOC), NH_4^+–N and NO_3^-–N using a spectrophotometer.

Results Incineration of manures increased pH, exchangeable cations and carbon, while nitrogen in manures ashes was comparable to nitrogen in dried manures. The application of manure ash increased SOC, NH_4^+–N and NO_3^-–N by 182, 102, 128% in incubation experiment while 64 and 628% increase in SOC and NH_4^+–N was recorded in screen-house experiment, respectively, relative to the control. In the field experiment, a significant increase in NO_3^-–N was recorded at the onset of incorporation; nitrification was more pronounced than ammonification under incubation and field conditions.

Conclusion Dried manures are not superior to manure ashes and incorporation of manures ashes increased the SOC, NH_4^+–N and NO_3^-–N though changes with increasing weeks were erratic.

Keywords Nitrification · Nitrogen availability indices · Nutrient dynamics · Soil organic carbon

✉ Toyin B. Olowoboko
rachy_blare26@yahoo.com

Jamiu O. Azeez
azeezjo@funaab.edu.ng

Olanrewaju O. Olujimi
olujimio@funaab.edu.ng

Oluwatoyin A. Babalola
babalolaoa@funaab.edu.ng

1 Department of Soil Science and Land Management, Federal University of Agriculture Abeokuta, Abeokuta, Nigeria

2 Department of Environmental Management and Toxicology, Federal University of Agriculture Abeokuta, Abeokuta, Ogun state, Nigeria

Introduction

There is recent interest in the use of agricultural wastes by dominant resource poor farmers in sub-Saharan Africa. Farm wastes such as crop residues and animal excreta are used directly or composted (Ayeni and Adeleye 2008) for cop production. Incorporation of organic matter through farmyard manure has been shown to improve soil structure, water retention capacity, infiltration rate, hydraulic conductivity and decrease bulk density (Edmeades 2003). Organic manures have also been reported to enhance nutrient use efficiency and reduce the need for chemical fertilizers, to improve the soil fertility and soil health (Hao et al. 2008). More recently, Law-Ogbomo et al. (2011) recorded improved nitrogen, phosphorus, potassium, calcium, and organic carbon status of the soil treated with poultry manure.

Soil organic carbon (SOC) refers only to the carbon component of organic compounds; it includes the microbial biomass and slow and passive humus pools, but not residue or litter (Griffin et al. 2013). Soil organic matter is closely related to soil carbon and the product of its decomposition is humus. Humus is a major determinant of soil cation exchange capacity, particularly in weathered soils of sub-Saharan Africa. Nitrogen is a major soil nutrient and closely related to soil carbon; it is a determinant nutrient for plant growth (Hossain et al. 2007). Soil nitrogen forms depend primarily on the processes of immobilization and organic matter mineralization and its supply through animal manures is transformed by microbial activity from organic nitrogen to ammonium and further on to nitrate by nitrification during mineralization. The turnover of manure N in the soil, mineralization–immobilization process and subsequent N availability for plant growth are all influenced by manure composition (Chadwick et al. 2000).

Hence, in order to predict the net mineralization of nitrogen in animal manures, transformation of nitrogen from organic into inorganic form and immobilization processes need to be understood (Saka et al. 2017) since these processes are mediated through microbial activities and could be biochemical (Vel and Swarnam 2013).

Animal wastes are abundant all over the world with Nigeria producing about 227,500 tons of fresh waste each day (Oyeleke et al. 2003). Due to the large amount of the wastes being produced, they are increasingly being regarded as an industrial waste that needs to be managed in an environmentally sound manner since they cannot be continuously applied to land because the volume of its generation exceeds usage (Ksawery et al. 2012). The incineration of animal manure is an option to effectively reduce the volume, concentrate fertilizer nutrients and remove offensive odor, though some nitrogen might also be lost to volatilization (Yusiharni et al. 2007). The difficulty in the prediction of manure nutrients to plants is due to turnover processes and microbial mediated changes in nutrient dynamics of the manures (Azeez and Van Averbeke 2010). Hence, a more predictable trend could be gotten when the manure is ashed and its contents changed to a product more inorganic. Therefore, incorporation of animal manures and their ashes as timely to supply carbon and nitrogen without posing any injury to soil and plants is very important in soil nutrients replenishment to support crop production; this necessitated the research to

1. Characterize dried manures and their ashes.
2. Evaluate the effect of dried manures and ashes on organic carbon, ammonium nitrogen and nitrate nitrogen in some soils of Abeokuta under incubation (laboratory), screen-house and field conditions.
3. Determine the dynamics of soil nitrate, ammonium and organic carbon from different animal manures and their ashes in incubation (laboratory), screen-house and field conditions.

Materials and methods

Soil collection and analysis

The top soil samples (0–20 cm) were collected from four locations in Ogun state in a forest-savanna transitional ecological zone. Samples from Alabata and Osiele are classified as Arenic Paleudalf (Adetunji 1991) from basement complex while samples from Itori and Papalanto are Oxic Tropodaulf from sedimentary parent materials. The soils were air dried to constant weight at room temperature (23–27 °C) and sieved with 2 mm mesh sieve. Sub-samples from the soils were collected and analyzed for the following properties: Soil pH was estimated in 1:2 (soil:water) using glass electrode pH meter (Orion research model 201) and electrical conductivity was estimated in 1:5 (soil:water) with an EC 215 conductivity meter (Hanna instrument) according to Kaira and Maynard (1991). Particle size was determined by hydrometer method (Bouyoucos 1965). Total nitrogen was analyzed using Kjeldahl method. Organic carbon was determined colorimetrically according to Baker (1976). Exchangeable cations were extracted with 1 N ammonium acetate; Na and K in the extract were determined by flame photometry, while Ca and Mg were determined by atomic absorption spectrophotometer. Ammonium nitrogen and nitrate nitrogen were determined spectrophotometrically (UV-3200 PCS spectrophotometer) according to Cataldo et al. (1975).

Manure collection and analysis

Cattle, goat and poultry manures were collected from the Federal University of Agriculture, Alabata (FUNAAB) farm and air dried to produce dried manures, while part of the dried manures was burnt in open space at a temperature range of 320–450 °C to produce manure ash. Dried manures were homogenized with a hammer mill after which pH, exchangeable bases, total nitrogen and total organic carbon were determined according to Cataldo et al. (1975).

Experimental design

Incubation and screen-house experiments were laid out in completely randomized design while the field experiment was laid out in randomized complete block design. The experiments were replicated thrice with eight treatments which included no amendment (control), dried cattle manure (DCM), dried goat manure (DGM), dried poultry manure (DPM), cattle manure ash (CMA), goat manure ash (GMA), poultry manure ash (PMA) and NPK 15-15-15 applied at120 kg N ha^{-1}.

Laboratory incubation

The photograph of the laboratory incubation is given in Fig. 1. One hundred grams of air-dried soil (collected from the four locations—Alabata, Osiele, Itori and Papalanto) which had passed through 2 mm sieve was dispensed into 200 g capacity plastic and tightly covered. Dried manure, their ashes and NPK were applied at a rate of 120 kg N ha^{-1} separately into each plastic and mixed. The rate applied is equivalent to 0.04 g NPK, 2.06 g CMA, 1.99 g GMA, 2.23 g PMA, 1.68 g DCM, 1.58 g DGM and 1.49 g DPM. The soil and amendment mixture were watered to field capacity and kept in a dark cupboard for 8 weeks. Occasionally, the moisture content was adjusted to field capacity by adding distilled water. Temperature of the incubation cupboard was determined daily with a thermometer and average temperature of 27.5 °C was recorded. Soil samples were taken at 0, 2, 4, 6, 8 weeks of incubation and analyzed for ammonium nitrogen, nitrate nitrogen and organic carbon content according to method described earlier.

Screen-house experiment

This was a two cycle experiment in a screen-house (made of mesh). In the first cycle, five kilograms of soil from two locations (Alabata and Papalanto) was dispensed into pots with treatment applied as mentioned earlier. The rate applied is equivalent to 1.79 g NPK, 103.00 g CMA, 99.25 g GMA, 116.50 g PMA, 83.75 g DCM, 78.80 g DGM and 74.40 g DPM. The soil samples in the pots were watered to field capacity and three seeds of maize (BR-9928-DMR-SR-Y) were sown per pot. The plants were thinned to one plant per pot after 2 weeks. The plants were watered when necessary and monitored in the screen-house for 8 weeks. Soil samples were also taken at 0, 2, 4, 6 and 7 weeks after planting. Soil samples were taken from the pots by collecting samples round the plant root at 0–15 cm depth with the aid of a calibrated spoon, after which samples were bulked to become a representative sample from each pot. Collected soil samples were analyzed for ammonium nitrogen, nitrate nitrogen and organic carbon. This process was repeated in the second maize cycle but soil samples were only collected at 0, 4 and 6 weeks after planting.

Fig. 1 Photograph of laboratory incubation

Field experiment

This experiment was carried out at the Teaching and Research Farms, Federal University of Agriculture (FUNAAB), Abeokuta Ogun state Nigeria. Dried manure treatments were applied manually 2 weeks before planting. Manure ashes treatments and NPK fertilizer were applied 2 weeks after planting. The rate applied is equivalent to 0.93 kg NPK, 55.38 kg CMA, 53.33 kg GMA, 60.00 kg PMA, 45.00 kg DCM, 42.35 kg DGM and 40.00 kg DPM to plots of 3 m by 4 m dimension. Maize seeds were sown at a spacing of 25 cm by 75 cm. Soil samples were taken at 0, 2, 4, 6, 8, 10 weeks after planting. Soil samples were collected randomly within plot at a depth of 0–15 cm after which samples were bulked as a representative sample of each plot. Collected soil samples were analyzed for ammonium nitrogen, nitrate nitrogen and organic carbon.

Statistical analysis

Data collected were analyzed for their variance by using the software package SAS (1999). Significant treatments were separated using LSD at 5% level of probability.

Results

Soil characteristics

Characterizations of experimental soils are shown in Table 1. The pH of soils used for the experiment ranged from 7.63 in Itori to 6.07 in FUNAAB. Electrical conductivity values were 0.38, 0.43, 0.73, 0.26 and 0.40 dS m^{-1} in soils from Alabata, Osiele, Itori, Papalanto and FUNAAB, respectively. Calcium and sodium were highest in soil from Itori while the lowest values were observed in Papalanto. Magnesium ranged from 2.21 cmol kg^{-1} in soil from Osiele to 1.05 cmol kg^{-1} in soil from Alabata while potassium ranged from 0.39 cmol kg^{-1} in FUNAAB to 0.18 cmol kg^{-1} in soil from Papalanto. Total N in FUNAAB soil was 0.25%; this is followed by 0.23% in Itori, 0.21% in Alabata, 0.19% in Osiele and 0.18% Papalanto. Value of total organic carbon was highest in Osiele (55.9 g kg^{-1}), and the lowest in Itori (43.7 g kg^{-1}). All the experimental soils are loamy sandy except soil from Papalanto which is sandy in texture.

Table 1 Characterization of experimental soils, dried manures and manure ashes

Soil	pH	EC dS m^{-1}	Ca (cmol kg^{-1})	Mg (cmol kg^{-1})	K (cmol kg^{-1})	Na (cmol kg^{-1})	Total N %	Total org C g kg^{-1}	Texture
Alabata	7.16	0.38	5.13	1.05	0.27	0.47	0.21	52.5	Loamy sand
Osiele	6.75	0.43	4.06	2.21	0.25	0.47	0.19	55.9	Loamy sand
Itori	7.63	0.73	6.71	1.34	0.34	0.60	0.23	43.7	Loamy sand
Papalanto	6.84	0.26	2.56	1.71	0.18	0.31	0.18	51.3	Sand
FUNAAB	6.07	0.4	5.63	1.13	0.39	0.58	0.25	53.2	Loamy sand

Treatments	pH	Ca (cmol kg^{-1})	Mg (cmol kg^{-1})	K (cmol kg^{-1})	Na (cmol kg^{-1})	Total N (%)	Total Org C g kg^{-1}
Dried cattle manure	8.7e	30.1d	15.0d	10.2f	10.4f	3.2abc	51.3c
Cattle manure ash	10.15c	33.7c	43.0b	21.9d	22.4d	2.6abc	52.5bc
Dried goat manure	9.6d	17.8f	20.0d	38.7c	38.4c	3.4ab	53.2ab
Goat manure ash	10.9a	25.3e	42.0b	75.2a	76.8a	2.7abc	55.9a
Dried poultry manure	8.2f	44.1b	30.7c	15.3e	15.2e	3.6a	51.8bc
Poultry manure ash	10.45b	75.5a	63.0a	55.5b	58.4b	2.4c	43.7d

Means with the same letter(s) within the same column are not significantly different at $p < 0.05$

Characterization of dried manures and manure ash

Characteristics of dried manure and manure ashes are shown in Table 1. The pH values ranged from 10.90 in GMA to 8.20 in DPM. pH values of manure ashes were in the order of GMA > PMA > CMA while the pH of dried manures were in the order of DGM > DCM > DPM.

The equivalent amount of calcium, magnesium, potassium and sodium content of dried manures is 30.1, 15.0, 10.20, 10.4 cmol kg^{-1} for DCM, 17.8, 20.0, 38.7, 38.4 cmol kg^{-1} for DGM and 44.1, 30.7, 15.3, 15.2 cmol kg^{-1} for DPM while the values for manures ashes were 33.7, 43.0, 21.9, 22.4 cmol kg^{-1} for CMA, 25.3, 42.0, 75.2, 76.8 cmol kg^{-1} for GMA and 75.5, 63.0, 55.5, 58.4 cmol kg^{-1} for PMA. There was no significant difference in values of total organic carbon (TOC) of dried manures and manures ashes except in poultry manure ash TOC was 33% lower compared with its dried manure. Dried poultry manure had the highest total nitrogen followed by dried goat manure and dried cattle manure. Total organic carbon was in the order of GMA > D GM > CMA > DPM > DCM > PMA.

Effect of amendments on soil organic carbon (SOC)

Table 2 shows that during incubation experiment, no significant differences were observed among dried manures and manure ashes in soil from Alabata at 0 weeks after incorporation (0 WAI), though soil organic carbon (SOC) increased with the application of manure ashes and dried manures except in dried cattle manure relative to the control. At 2 WAI, application of NPK produced significantly higher SOC (182%) than the control and manure amendments. However, the effect of the amendments was not significant at 0 and 4 WAI in soil collected from Osiele. Effect of manure ashes on the SOC did not differ significantly from their dried manures during the period of observation, although SOC was numerically higher with the application of manure ashes in comparison to their dried manures at 2 WAI in soil collected from Osiele. Soil from Itori amended with DCM had the highest SOC of 112.5 g kg^{-1}, which was significantly higher than other amendments and the control (Table 3).

In the first cycle of the screen-house experiment (Table 4), soil from Alabata released significantly higher SOC when applied with DGM. However, control, GMA and DPM had similar effect while DCM, PMA and NPK did not differ from one another. Manure ashes with the exception of PMA differed from their dried manures and a significant increase of 23% and decrease of 29% were recorded in CMA and GMA, respectively, relative to their dried manures at 0 WAI. However, the effect of the amendment was not noticeable in the second cycle of the experiment because no significant difference was observed in amended soil relative to control soil. In Papalanto, the value of SOC in control soil did not differ from amended soils at 0 and 4 WAI in the 1st cycle of screen-house experiment. More organic carbon was released from manure ashes when compared with their dried manures at 0 WAI, while dried manure increased SOC than manure ashes at 4 WAI. Application of PMA significantly increased SOC at 2 and 8 WAI.

Dynamics of soil organic carbon

The amount of SOC released had an inconsistent pattern with increasing weeks for amendment and control (Tables 2, 3, 4 and 5); when high organic carbon is released at a particular week, there is a reduction in release at the succeeding

Table 2 Effect of amendments and incubation time on soil organic carbon (SOC) of soils from Alabata and Osiele

Treatments	0 WAI	2 WAI	SOC 4 WAI	($g\ kg^{-1}$) 6 WAI	8 WAI	Lsd
Alabata						
Control	63.3a	44.0b	65.6bcd	40.4bc	82.4a	13.3
Dried cattle manure	57.2a	70.0b	81.4ab	47.4a	75.7ab	28.0
Cattle manure ash	101.3a	62.1b	74.7abc	36.4c	65.3ab	74.3
Dried goat manure	63.7a	56.7b	87.8a	40.8bc	66.7ab	24.0
Goat manure ash	64.7a	56.7b	63.3bcd	39.3bc	77.7ab	24.0
Dried poultry manure	104.6a	62.6b	61.7cd	43.9abc	79.3ab	11.2
Poultry manure ash	65.3a	52.8b	61.2cd	41.6bc	61.2b	18.9
NPK (15-15-15)	80.2a	124.2a	53.0d	41.6bc	59.8b	13.3
Osiele						
Control	71.3a	57.7ab	67.2a	61.3b	69.4ab	11.7
Dried cattle manure	72.6a	54.6b	70.6a	49.5cd	81.2ab	23.5
Cattle manure ash	101.4a	70.9ab	69.1a	48.1d	74.8ab	48.9
Dried goat manure	80.0a	60.7ab	86.9a	50.0cd	98.0a	31.7
Goat manure ash	87.5a	75.3a	91.0a	55.1bcd	81.4ab	28.0
Dried poultry manure	65.6a	59.9ab	83.3a	57.5bc	86.4ab	34.9
Poultry manure ash	73.6a	65.9ab	74.3a	49.0cd	68.2b	11.5
NPK (15-15-15)	67.2a	65.1ab	91.9a	72.8a	72.5ab	38.4

Means with the same letter(s) within the same column are significantly different at $p<0.05$

WAI weeks after incorporation, *Lsd* differences across the weeks for each treatment

Table 3 Effect of amendments and incubation time on organic carbon of soils from Itori and Papalanto

Treatment	0 WAI	2 WAI	SOC 4 WAI	($g\ kg^{-1}$) 6 WAI	8 WAI	Lsd
Itori						
Control	90.1ab	56.0bc	55.8b	60.8ab	67.7d	26.0
Dried cattle manure	112.5a	52.1c	92.6ab	52.4b	96.3abc	30.0
Cattle manure ash	61.9c	64.4bc	157.5a	62.1ab	78.6bcd	29.0
Dried goat manure	70.4bc	67.2bc	115.9ab	77.9a	107.4a	81.6
Goat manure ash	73.3bc	124.6a	104.3ab	47.1b	75.2cd	32.1
Dried poultry manure	71.0bc	70.1bc	72.5b	53.4b	80.0bcd	15.7
Poultry manure ash	66.3bc	132.2a	91.0ab	59.1ab	76.6cd	50.1
NPK (15-15-15)	85.1bc	78.8b	90.0ab	55.5ab	101.0ab	26.4
Papalanto						
Control	68.3abc	52.1b	71.8a	37.5c	77.5ab	27.6
Dried cattle manure	62.7abc	54.1b	68.1a	47.2bc	74.4ab	16.6
Cattle manure ash	62.3abc	54.4b	81.4a	37.6c	56.8b	27.3
Dried goat manure	58.9bc	62.2b	63.5a	51.5ab	77.6ab	18.1
Goat manure ash	53.4c	64.5b	91.0a	64.2a	80.0ab	27.0
Dried poultry manure	84.1a	52.1b	83.7a	64.5a	60.4b	20.5
Poultry manure ash	80.2ab	98.6a	77.0a	61.6a	101.5a	62.9
NPK (15-15-15)	69.3abc	85.7a	90.5a	48.9abc	57.1b	19.0

Means with the same letter(s) within the same column are significantly different at $p<0.05$

WAI weeks after incorporation, *Lsd* differences across the weeks for each treatment

week. With the application of DCM, the highest and the lowest amount of SOC were released at 4 and 6 WAI, respectively, in the soil from Alabata during incubation experiment (Table 2). SOC decreased significantly by 42% at 4 WAI and increased again at 60% at 8 WAI. Similar trend was observed in soil from Osiele. In soil from Itori (Table 3), the

Table 4 Effect of amendments and time on organic carbon of soils from Alabata and Papalanto in screen-house experiment

Treatments	SOC (g kg^{-1}) Cycle 1					Lsd*	SOC (g kg^{-1}) Cycle 2			Lsd**
	0 WAI	2 WAI	4 WAI	6 WAI	7 WAI		0 WAI	4 WAI	6 WAI	
Alabata										
Control	89.8b	63.0cd	64.5b	82.1ab	70.0ab	17.1	67.7a	52.2a	58.0a	1.75
Dried cattle manure	73.1c	84.1ab	68.6b	67.2ab	67.7ab	17.2	68.6a	68.6a	55.2a	1.70
Cattle manure ash	89.7b	66.3cd	69.1b	99.2a	68.6ab	47.5	66.7a	54.9a	54.7a	0.37
Dried goat manure	105.2a	82.3ab	93.8b	76.0ab	66.7ab	35.2	69.5a	62.1a	57.6a	0.96
Goat manure ash	75.2bc	71.6bc	67.7b	56.2b	82.1a	37.1	67.7a	58.5a	55.2a	1.62
Dried poultry manure	75.7bc	66.7c	71.5b	57.0b	75.1ab	25.3	75.1a	57.6a	62.5a	2.70
Poultry manure ash	67.7c	93.8a	148.8a	59.8ab	72.4ab	21.2	72.4a	59.5a	64.4a	3.21
NPK (15-15-15)	68.2c	51.0d	90.6b	67.3ab	58.0b	39.5	58.0a	67.4a	51.3a	2.99
Papalanto										
Control	123.8a	75.6bcd	78.0a	82.5ab	67.4bc	65.8	67.4ab	67.2ab	66.4a	2.97
Dried cattle manure	80.8a	80.1bc	83.6a	76.2abc	68.6ab	15.7	68.6ab	63.0ab	66.9a	1.99
Cattle manure ash	119.6a	66.5d	59.8a	78.5abc	74.3abc	19.8	74.3ab	63.3ab	61.8a	1.44
Dried goat manure	89.6a	86.0b	76.2a	86.8a	81.8abc	23.0	81.8a	72.1a	59.1a	2.10
Goat manure ash	89.4a	83.8bc	74.4a	71.2abc	66.0bc	14.6	66.0ab	62.3ab	64.7a	0.93
Dried poultry manure	84.6a	82.9bc	80.5a	64.9bc	89.2ab	17.5	76.3ab	68.1ab	72.5a	1.28
Poultry manure ash	88.2a	131.2a	75.4a	63.3c	92.0a	19.6	76.1ab	62.8ab	71.0a	3.27
NPK (15-15-15)	80.3a	72.8cd	76.8a	66.0bc	64.6c	26.2	64.6b	54.3b	61.6a	1.33

Means with the same letter(s) within the same column are significantly different at $p<0.05$

WAI weeks after incorporation

*Differences across the weeks for each treatment in cycle 1

**Differences across the weeks for each treatment in cycle 2

Table 5 Effect of amendments and time on organic carbon of soil in field experiment

Treatments	0 WAI	2 WAI	SOC (g kg^{-1}) 4 WAI	6 WAI	8 WAI	10 WAI	Lsd**
Control	59.9b	127.0a	60.3b	49.5c	63.5a	76.2a	20.9
Dried cattle manure	73.7b	77.5b	104.0a	71.7ab	75.4a	83.7a	46.7
Cattle manure ash	57.4b	65.9bc	58.1b	61.9abc	67.5a	75.6a	15.1
Dried goat manure	84.2b	68.4bc	59.4b	70.0abc	69.8a	89.4a	19.7
Goat manure ash	82.3b	68.1bc	52.5b	48.8c	63.5a	65.4a	22.3
Dried poultry manure	115.7a	66.0bc	80.5ab	74.2a	67.8a	74.9a	21.8
Poultry manure ash	63.5b	67.0bc	69.0ab	58.9abc	64.1a	78.9a	17.5
NPK (15-15-15)	79.3b	59.8c	56.2b	51.3bc	59.4a	67.8a	19.4

Means with the same letter(s) within the same column are significantly different at $p<0.05$

WAI weeks after incorporation

**Differences across the weeks for each treatment

amount of SOC released fortnightly differed significantly. Table 2 shows that with the application of CMA, soil from Alabata did not differ in the amount of organic carbon content released during the period of observation. The highest and the lowest values were recorded at 0 and 6 WAI, respectively, during incubation. Similar response was observed in soil from Osiele only that a significant reduction of 52% was observed at 6 WAI relative to 0 WAI.

During incubation, responses observed with the application of DGM in soil from Alabata and Osiele were similar to that observed when DCM was applied (Table 2). On the other hand, application of DGM to soil from Itori resulted in optimal release of SOC at 4 WAI, though changes in SOC with increasing weeks were not significant in soil from Itori and Papalanto (Table 3). In soil from Papalanto (Table 4), changes in the amount of SOC released with increasing weeks were not significant in the first cycle of the screen-house experiment but it reduced significantly with increasing weeks in both soils during the second cycle. The application of GMA to soil from Alabata (Table 2) significantly increased SOC by 98% at 8 WAI in comparison to 6 WAI. In soil from Osiele, SOC decreased significantly after 4 WAI by 39% and it increased again by 48% after 6 WAI. Organic carbon of soil from Itori (Table 3) significantly increased after 0 WAI by 70% and, thereafter, there was continuous reduction between 2 WAI and 6 WAI, leading to a significant reduction was observed at 6 WAI. However, a decrease in SOC with increasing weeks was observed in soil from Papalanto in the first cycle while SOC decreased with increasing weeks in soil from Alabata and vice versa in soil with Papalanto in the second cycle. A decrease with increasing weeks was observed until 6 WAI after which SOC increased during the field experiment (Table 5).

With the application of DPM, SOC decreased with increasing weeks up to the 6 WAI after which a significant increase was observed in soil from Alabata (Table 2). However, in the second cycle of the screen-house experiment, changes in SOC were erratic but significant. Significantly, higher SOC was recorded at 0 WAI when compared to succeeding weeks in field experiment. Values of SOC after 0 WAI did not change with increasing weeks.

In Table 2, the amount of organic carbon released in soil from Alabata and Osiele at weeks prior 6 WAI did not differ with the incorporation of PMA. SOC decreased significantly at 6 WAI and increased at 8 WAI by 32 and 32%, respectively, in the soil from Alabata. During screen-house experiment, SOC significantly increased with increasing weeks up to 4 WAI after which a significant decrease was noted at 4 WAI in both soils. In the second cycle of the experiment, SOC changes in soil were significant but not predictable. Changes observed during field experiment (Table 5) were similar to that observed in screen-house experiment.

Effect of amendment on soil ammonium nitrogen

This study confirms that in the incubation study, NH_4^+–N did not significantly differ as a result of dried manures and manure ashes addition to soil, though increases were noted in Figs. 2 and 3. In Fig. 2, at 0 WAI, NH_4^+–N content of soil from Alabata was highest with the application of PMA (40% increase) when compared with DCM. The effect of the manure and ash amendment was not significant; however, at 2–8 WAI, ammonium values in the amended soils did not significantly differ from the control soil. Ammonification process in soils amended with dried manures was lower relative to soils amended with manure ashes at 6 and 8 WAI. The conversion of organic nitrogen to ammonium nitrogen was greatly enhanced with application of PMA by 24% above the control soil at 0 WAI in soil from Osiele. During the first cycle of screen-house experiment, amendment addition led to lower but similar ammonium nitrogen value, then in the control soil at 0 WAI (Fig. 4) values in all amended soil did not differ significantly in soil from Alabata. At the second cycle (Table 6), amendment effect was significant at 4 WAI and 6 WAI. At 4 WAI, significantly lower NH_4^+–N was released from the control soil when compared with CMA, DGM, DPM, and GMA. A significant increase and decrease of 107 and 74% were observed in soil amended with CMA and PMA compared with their respective dried manures.

Dynamics of soil ammonium nitrogen

The concentration of ammonium nitrogen in soil changed over the weeks although a non-predictable or erratic pattern of release was recorded with increasing weeks since control, dried manures, manure ash and NPK had different pattern of release (Figs. 2, 3, 4). The highest NH_4^+–N values in soil from Alabata was recorded at 0 WAI for all the amendment except DCM (Fig. 2). NH_4^+–N increased to its peak at 2 WAI and, thereafter, it decreased across the weeks with DCM. As a result of CMA application, NH_4–N was highest at 0 WAI, it reduced at 2 WAI and increased sharply at 4 WAI after which it decreases across the remaining weeks with the decrease being significant at 8 WAI. Similar response was observed with DGM. GMA released significantly higher NH_4^+–N at 0 WAI compared with the other weeks where NH_4^+–N decreases with increasing weeks. This was also similar to the sequence observed with DPM. Poultry manure ash amended soils had the highest NH_4^+–N at 0 WAI; it decreased at 2 WAI and increased across the weeks until a decrease occurred at 8 WAI although all the changes motivated by PMA were similar.

Figure 3 shows the ammonium values of soil collected from Itori across the weeks. Generally, for control and amendments except PMA, the highest NH_4^+–N values were released at 0WAI. NPK released lower NH_4^+–N at 2, 4 and 6 WAI, with significantly lower NH_4^+–N at 6 WAI in comparison with 8 WAI. The incorporation of DPM produced the highest NH_4^+–N at 0 WAI with significant decrease at the following weeks. Dried manure of goat reduced NH_4^+–N at 2 WAI but increased it at 4 WAI after which NH_4^+–N reduced across the following weeks. Similar response was observed with GMA. NH_4^+–N reduced significantly across the weeks with the application of DCM and similar NH_4^+–N sequence was observed in the control. NPK incorporation to soil from Papalanto reduced NH_4^+–N across the weeks though decrease was only significant at 8 WAI relative to 0 and 2 WAI. DCM released significantly higher NH_4^+–N at 0 WAI when compared with other weeks. Similar response was observed with CMA. GMA increased NH_4^+–N at 6 WAI though erratic pattern was observed at preceding weeks; a significant decrease occurred at 8 WAI relative to 6 WAI. An erratic pattern of NH_4–N release was also observed with the application of PMA.

Fig. 2 Changes in ammonium nitrogen of soils from Alabata and Osiele with incubation time

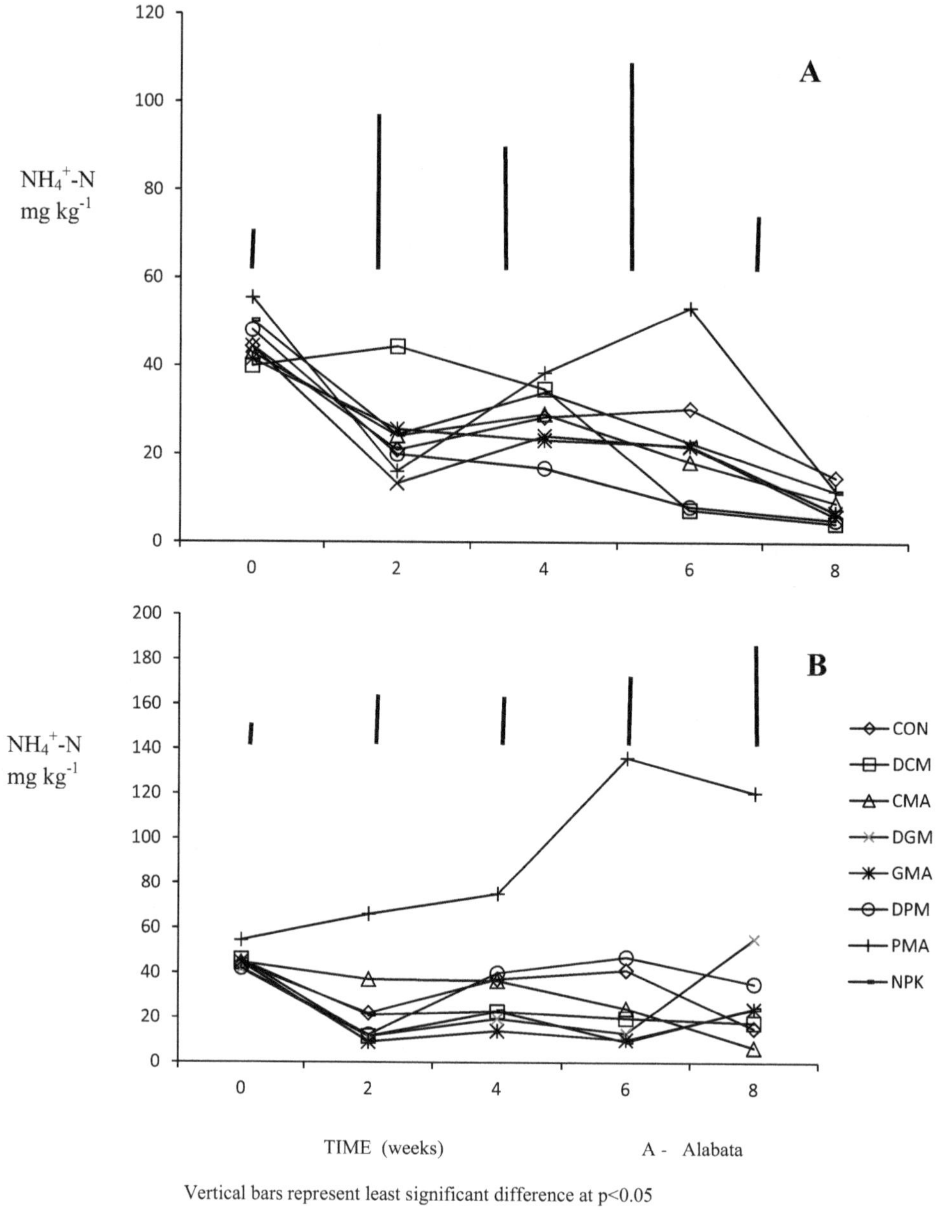

Table 6 shows the changes in the ammonium nitrogen of soil from Alabata and Papalanto in the second cycle of the screen-house; DCM increased soil NH_4^+–N across the weeks though significant increase was only recorded at 6 WAI. There was reduction in NH_4^+–N released across the weeks with the application of CMA; meanwhile, NH_4^+–N at 0 WAI was significantly higher than NH_4^+–N at 4 and 6 WAI in soil from Alabata while significant decrease was observed at 6 WAI in Papalanto. Similar response was observed with DGM. The lowest NH_4^+–N was released by PMA at 0 WAI, a similar but lower value was obtained at 4 WAI after which a sharp significant increase was observed at 6 WAI. NH_4^+–N increased across the weeks with the application of NPK though the values did not differ significantly.

An erratic pattern of release was observed with the application of PMA and the highest NH_4^+–N was recorded at 0 WAI in the field experiment even though changes across the weeks were not significant (Fig. 8). DPM released significantly higher NH_4^+–N at 8WAI than NH_4^+–N at 2 and 6 WAI. A non-predictable pattern of release was observed across the weeks with GMA however; a significant increase only occurred at 4 WAI relative to 2 WAI.

Effect of amendment on soil nitrate nitrogen

The study showed that application of manure ashes increased the concentration of nitrate nitrogen in comparison with the control in incubation experiment. Generally,

Fig. 3 Changes in ammonium nitrogen of soils from Itori and Papalanto with incubation time

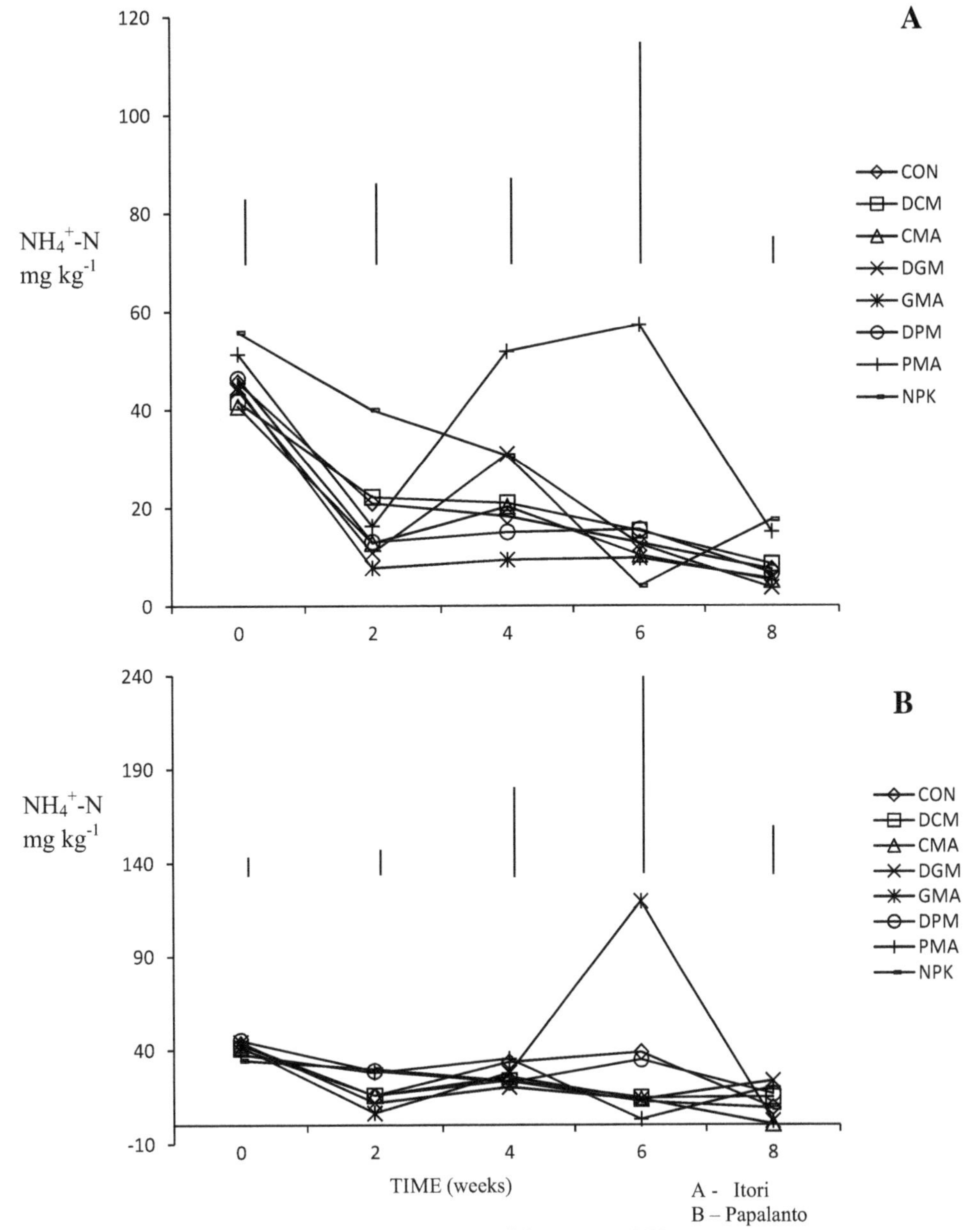

manure ash incorporation led to increases in nitrate nitrogen when compared with their dried manures although some of the increase did not significantly differ (Figs. 5 and 6). In soil collected from Alabata (Fig. 5), application of amendments increased NO_3–N at 0WAI, a significant increase of 254% was observed with PMA relative to control. Amendment effect was significant with the application of GMA at 2, 6 and 8 WAI; however, PMA and CMA did not differ from their dried manures but a significant increase in the amount of NO_3^-–N released was observed with GMA relative to DGM. The application of NPK to soil from Papalanto enhanced the release of NO_3^-–N significantly above other amendments. A significant increase of 198% was observed with NPK when compared to control. At 2 WAI, the highest NO_3^-–N observed in the control did not significantly differ from other amended soil except GMA amended soil. Manure ashes and dried manures except PMA had similar effect on the conversion of ammonium to nitrate at 4 WAI despite that the highest NO_3^-–N observed with PMA was 153% more than that released in control. The response of soil to amendment at 6 WAI was similar to that observed at 0WAI. At 8 WAI, amendment did not differ except with NPK and DCM. In field experiment, application of poultry manure ash encouraged significantly higher NO_3^-–N at 335% over the control soil at 0 WAI as shown in Fig. 8. Manure ash increased NO_3^-–N

Fig. 4 Changes in soil ammonium nitrogen in screen-house (cycle 1)

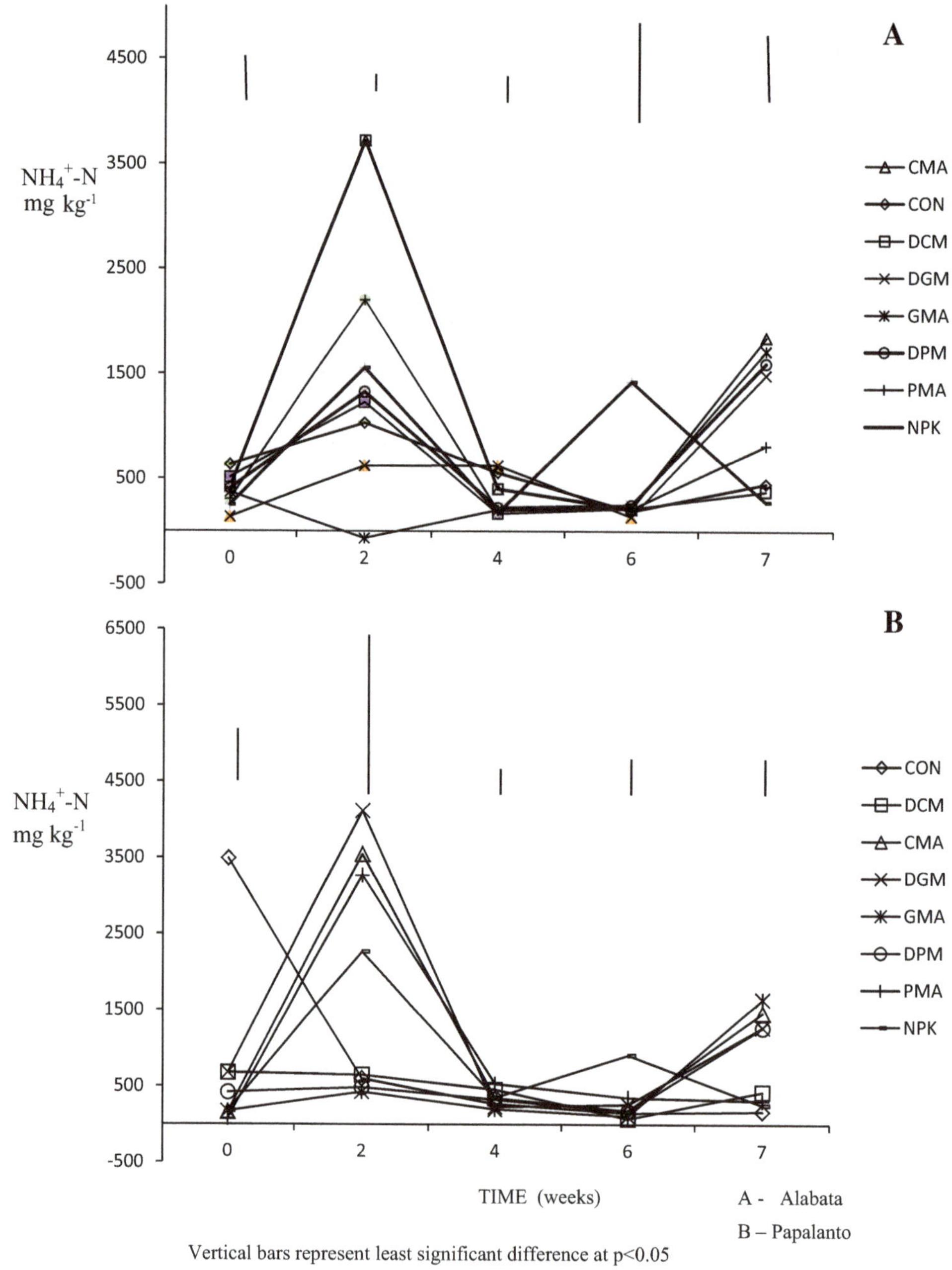

than their dried manures but a significant increase was observed with PMA with value of NO_3^-–N of 224% higher than its dried manure. However, at 6 WAI, CMA recorded the highest NO_3^-–N at 59% higher than the control soil. It also increased NO_3^-–N by 750% in comparison to its dried manure. At 8 WAI, control soil had the highest NO_3^-–N that did not significantly differ with the application of amendment except DCM and GMA where the reduction of 365 and 379% was recorded. CMA increased NO_3^-–N at 209% of its dried manure. However, at 10 WAI, application of CMA increased NO_3^-–N than their dried manures. It also recorded the highest and lowest increase of 154 and 64% above DPM and control, respectively.

Dynamics of soil nitrate nitrogen

Figure 5 shows that in soil from Alabata and Osiele, there was no significant difference in nitrate nitrogen released across the weeks with DCM. For the control, nitrate increased at 2 WAI and later decreased across the weeks with a significant reduction at 8 WAI. In the control soil from Osiele, an increase was recorded at 2 WAI relative 0 WAI but NO_3^-–N decreased after 2 WAI to 6 WAI and a significant increase was observed at 8 WAI. Application of CMA increased NO_3^-–N at 2 WAI while a reduction was observed at 4 and 6 WAI but a significant increase of 199% in NO_3^-–N was observed at 8 WAI. DGM released

Table 6 Effect of amendments and time on Soil ammonium nitrogen and nitrate nitrogen of Alabata and Papalanto in screen-house (cycle 2)

Treatment	NH_4^+–N 0 WAI	($mg\ kg^{-1}$) 4 WAI	6 WAI	Lsd*	NO_3^-–N 0 WAI	4 WAI	($mg\ kg^{-1}$) 6 WAI	Lsd**
Alabata								
Control	450.2b	315.8c	460.2b	97.99	55.44c	209.56a	182.34a	97.9
Dried cattle manure	376.2b	592.0bc	3223.7a	218.94	133.89a	269.86a	249.46a	218.9
Cattle manure ash	1843.8a	1227.2a	653.7b	54.13	73.94bc	151.83a	218.02a	54.13
Dried goat manure	1486.2a	721.5a	691.5b	397.38	114.50ab	232.23a	230.44a	397.3
Goat manure ash	1714.3a	1110.3abc	1147.0b	165.44	126.71ab	199.93a	228.64a	165.4
Dried poultry manure	1597.2a	1275.7a	783.2b	270.76	53.16c	280.69a	141.78a	270.7
Poultry manure ash	481.2b	320.7c	789.3b	72.08	98.54abc	171.22a	101.58a	72.08
NPK (15-15-15)	277.5b	992.8ab	1159.3b	87.16	79.33bc	175.16a	75.52a	87.16
Papalanto								
Control	167.2b	363.3c	953.7bcd	362.46	90.41a	136.04bc	232.95b	91.28
Dried cattle manure	274.5b	2336.7a	388.0d	665.11	55.93a	502.15a	181.10bc	100.1
Cattle manure ash	1461.5a	1319.5b	666.8cd	459.62	56.75a	190.96b	226.85b	114.2
Dried goat manure	1270.3a	777.0bc	1523.0b	447.46	56.60a	199.93b	234.66b	179.8
Goat manure ash	1640.3a	949.7bc	3213.8a	1494.1	84.35a	83.28c	382.68a	142.2
Dried poultry manure	1264.2a	458.3bc	1288.0bc	241.82	61.38a	119.17bc	192.03bc	95.3
Poultry manure ash	308.3b	2226.8a	548.8d	1890.8	51.73a	62.46c	150.76bc	71.45
NPK (15-15-15)	252.8b	289.8c	2646.3a	740.5	41.64a	118.85bc	119.21c	57.00

Means with the same letter(s) within the same column are significantly different at $p < 0.05$

WAI Weeks after incorporation

*Lsd across the weeks for ammonium nitrogen across the weeks

**Lsd across the weeks for nitrate nitrogen across the weeks

the highest NO_3–N at 0 WAI after which NO_3^-–N decreased across the weeks in soil from Alabata. Application of GMA significantly increased NO_3^-–N across the weeks with the highest significant increase at 8 WAI in both soils. For PMA, NO_3^-–N at 0 and 8 WAI did not differ until after 6 WAI, where significantly higher NO_3^-–N was released in both soils despite that there was a decrease after 0 WAI. NO_3^-–N was similar across the weeks in control soil from Itori although the highest NO_3^-–N was observed at 8WAI as presented in Fig. 6. For all the amendments, the highest NO_3^-–N was released at 8WAI. NO_3^-–N was lowest at 0 WAI and increased with increasing weeks for GMA and DPM. NO_3^-–N reduced at 2 WAI for PMA and NPK after which subsequent increase was recorded up to 8 WAI. The lowest NO_3^-–N was at 0WAI for DCM, CMA, and DGM, only that DCM decreased NO_3^-–N at 2 WAI with increases in subsequent weeks. For CMA, NO_3^-–N decreased at 2 WAI, increased at 4 WAI, decreased again at 6 WAI and finally increased significantly at 8 WAI. A non-predictable pattern was observed with DGM across the weeks with a final considerable increase at 8 WAI.

In the screen-house experiment (Fig. 7), Application of DCM led to increase in NO_3^-–N up to 2 WAI after this NO_3^-–N decreased till 6 WAI. The lowest NO_3^-–N was recorded at 0 WAI with CMA, a significant increase occurred at 2 WAI and after 4 WAI, NO_3^-–N decreased across the weeks. Similar response was observed with DGM. Significantly higher NO_3^-–N was recorded at 4 WAI when compared with NO_3^-–N at preceeding and succeeding weeks with the application of GMA. A decrease in NO_3^-–N released across weeks was observed with the application of DPM as the highest NO_3^-–N was produced at 0 WAI. In the second cycle of screen-house experiment (Table 6), soil from Alabata had the highest release of nitrate nitrogen at 4WAI; however, the incorporation of CMA allowed significant increase in NO_3^-–N across the weeks. The amount of NO_3^-–N released with PMA increased significantly at 4 WAI in comparison with 0 WAI while a considerable decrease occurred at 6 WAI. In soil collected from Papalanto, NO_3–N was lowest at 0 WAI, with the application of DCM it increased significantly at 4 WAI after which an abrupt reduction was observed at 6 WAI. A decline was recorded in NO_3^-–N from 0 to 4 WAI, an abrupt significant increase occurred at 6 WAI with GMA. DPM encouraged significant increase in NO_3^-–N with increasing weeks. Nitrate nitrogen increased across the weeks as a result of PMA addition though increase was significant only at 6 WAI.

Fig. 5 Changes in nitrate nitrogen of soils from Alabata and Osiele with incubation time

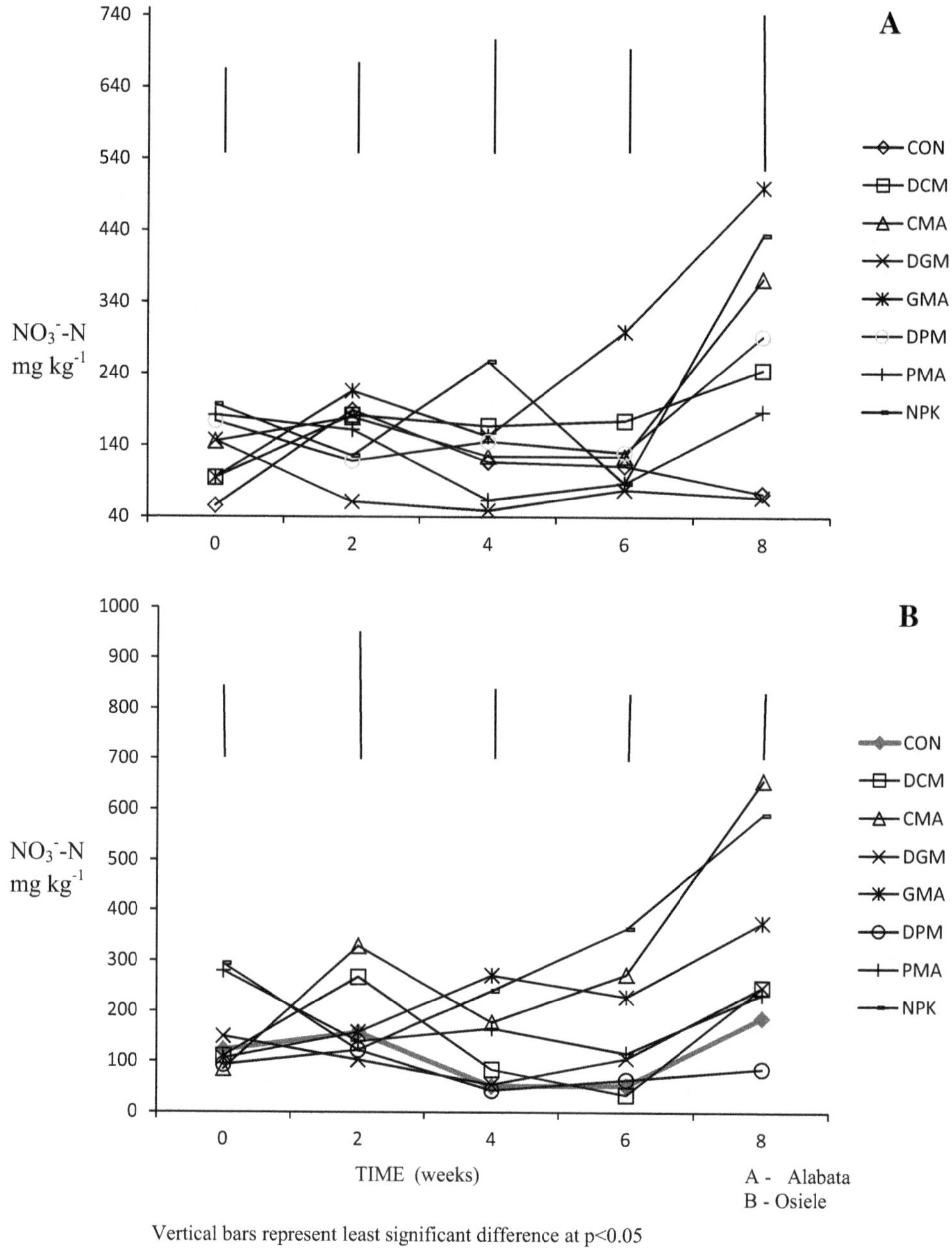

The pattern of soil nitrate nitrogen during field experiment is shown in Fig. 8. In the control soil, NO_3^-–N was significantly lower at 0 and 4 WAI, NO_3^-–N increased significantly across the weeks with the highest at 8 WAI. DCM application produced a considerable increase in NO_3^-–N at 2 WAI after which it decreased across the weeks. There was no significant difference in NO_3^-–N released by CMA across the weeks but the highest value was at 8 WAI. DGM significantly increased NO_3^-–N at 2 WAI in comparison to other weeks. Similar response was observed with NPK. There was no significant difference in NO_3^-–N at 0 to 8 WAI with PMA; the highest and lowest NO_3^-–N was observed at 8 and 10 WAI, respectively.

Discussion

The experimental soils were slightly acidic to slightly alkaline in pH and the highest pH value in soil from Itori could be due to its proximity to Ewekoro cement company and the consequent pollution with calcium dust (Azeez 2004). Incineration of manure increased the pH,

Fig. 6 Changes in nitrate nitrogen of soils from Itori and Papalanto with incubation time

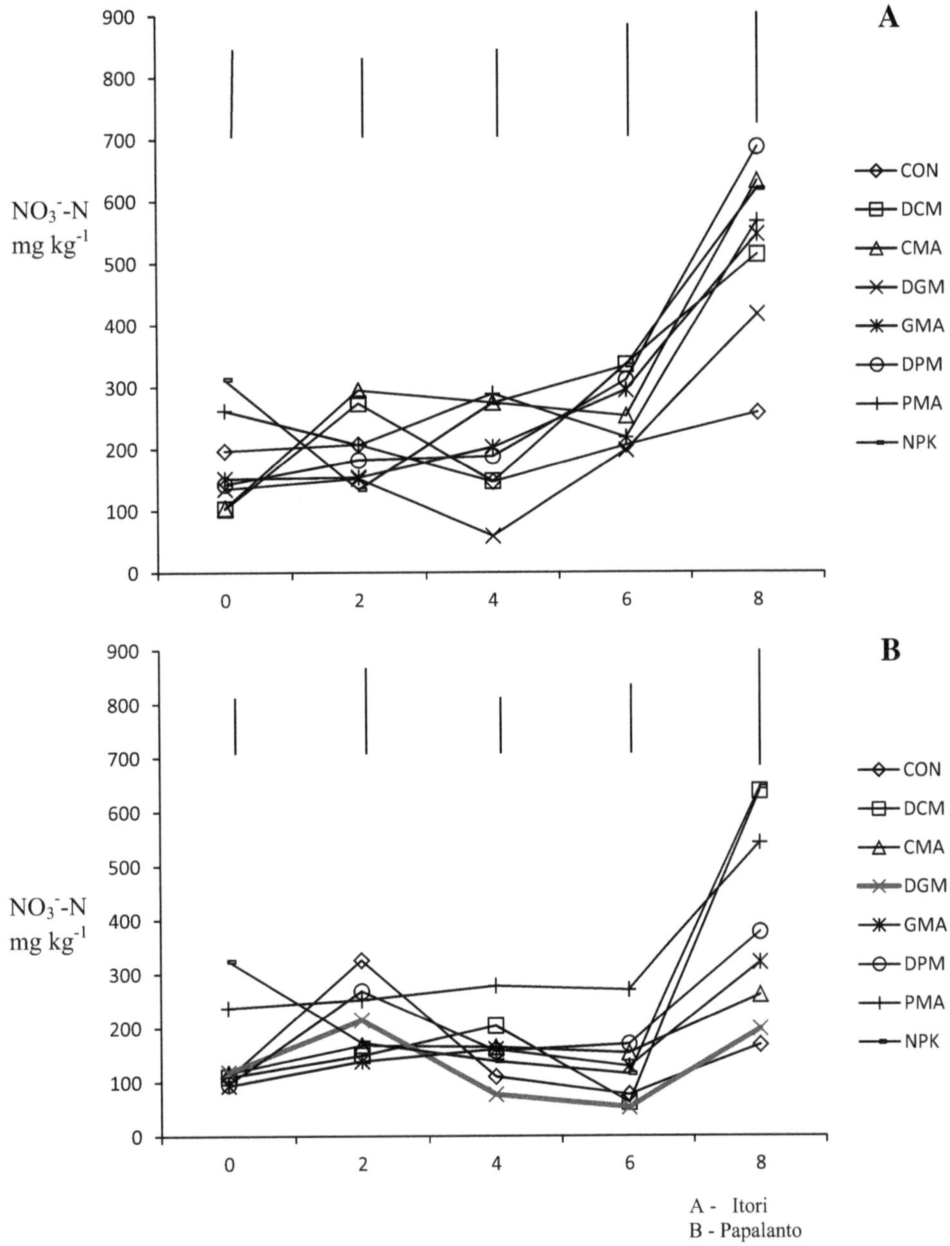

Vertical bars represent least significant difference at $p<0.05$

exchangeable cations, and carbon–nitrogen ratio. This is due to the burning off the non-nutrient part of the manure and the concentration of the nutrients in the manure when ashed. This supports the findings of Lal and Ghuman (1989) that the conversion of biomass to ash increased the levels of cations and other nutrient element. The pH of dried manures was strongly alkaline, while that of manure ashes was considered very strongly alkaline. This would enable mineralization and release of nutrients when they are applied to soil. Significantly, higher pH was observed in manure ashes in comparison to their dried manures as this could be attributed to the concentration of basic cations that contributed to increasing soil pH.

Exchangeable cations were medium to high in the soils. The total organic carbon of the soils was observed to be very high. Even though the total nitrogen content of the soils were very high, this may not necessarily signify a high content of available nitrogen for plant uptake, especially if the organic N content of such soils is high. Soils from Alabata, Osi-ele, Itori and FUNAAB were loamy sand, while soil from Papalanto was sand (USDA 2010). There was a significant increase in calcium, magnesium, potassium and sodium content of manures after ashing. GMA is richer in sodium

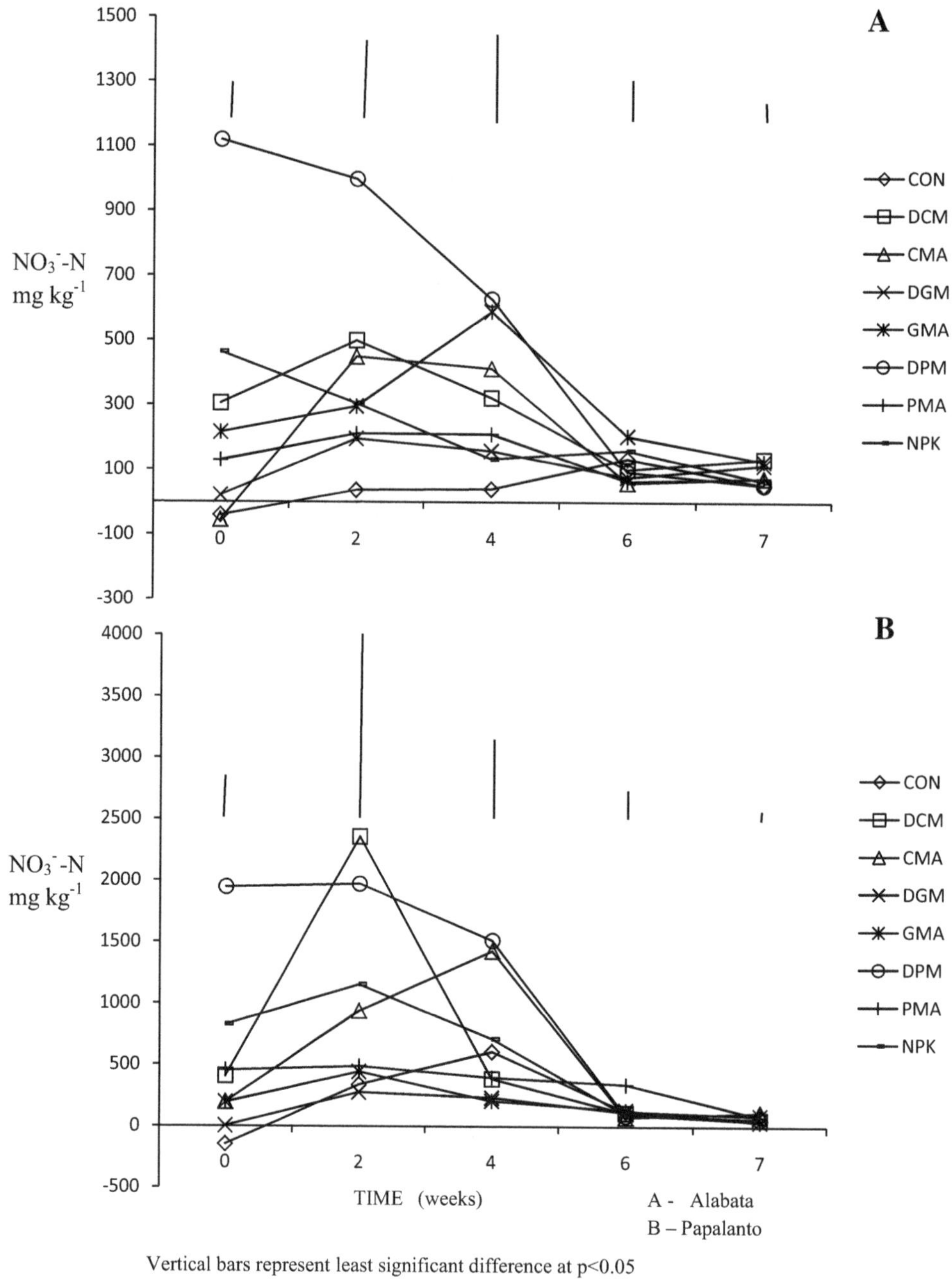

Fig. 7 Changes in soil nitrate nitrogen in screen-house (cycle 1)

and potassium when compared to cattle and poultry manure ash. This supports the findings of Azeez and Van Averbeke (2010) that abundance of manure K is in the order of goat manure > poultry manure > cattle manure. However, poultry manure ash is richer than goat manure ash and cattle manure in terms of calcium and magnesium contents. This is as a result of calcium and magnesium salts that are specifically added to poultry diets for egg production (layers).

No significant reduction in total organic carbon and nitrogen of manures after burning at 320–450 °C. The total nitrogen of manures did not significantly differ from that of manure ash; this shows that incineration of manure at this temperature did not lead to significant loss of nitrogen. This is in disagreement with the findings of Teppei et al. (2012) that C and N contained as organic compounds were incinerated and gasified as nitrogen and carbon dioxide. The non-conformity of this result to the results of Teppei et al. (2012) could be attributed to the difference in temperature of ashing since the dried or composted manures were incinerated at 700 °C in Teppei et al. (2012). The highest total nitrogen of dried poultry manure reported in this study is in conformity with Azeez and Van Averbeke (2010). The carbon–nitrogen ratio follows the order of goat manure ash > cattle manure ash > dried goat manure > dried cattle manure > dried poultry manure = poultry manure ash (data not shown). This implies that dried poultry manure and its ash is expected to

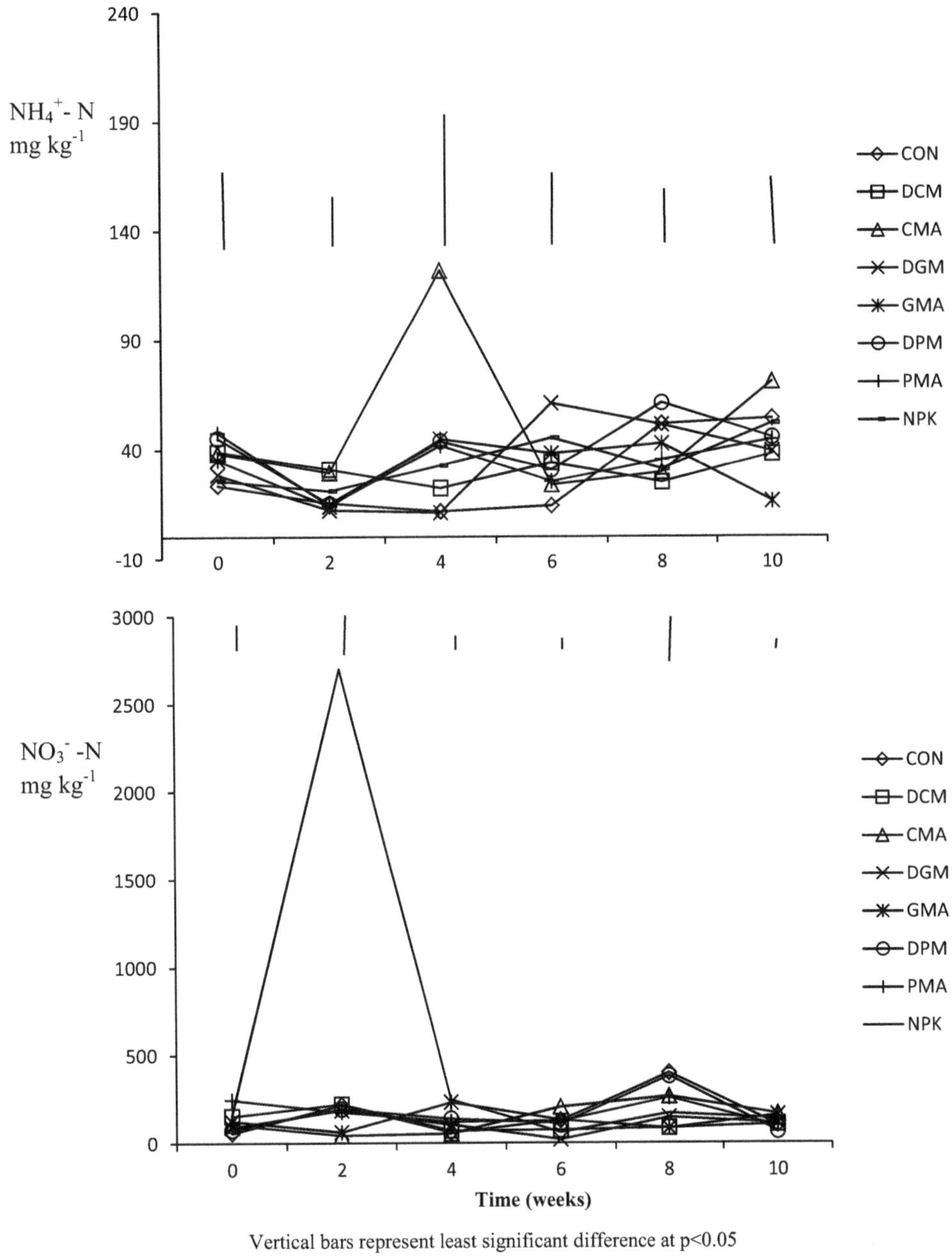

Fig. 8 Changes in soil ammonium nitrogen and nitrate nitrogen in field evaluation

mineralize faster than other ashes; nutrients will be released rapidly from dried poultry manure when compared with dried manures of goat and cattle. This supports the findings of Qiu et al. (2008) that poultry manure is expected to mineralize faster than cattle manure and least in goat manure. However, immobilization of nutrients is likely to occur in goat and cattle manures.

Incorporation of amendments to soil increases the soil organic matter with a major component as organic carbon. Generally, amended soils recorded higher organic carbon when compared with the control soil. This conforms to the result of Ojeniyi (2000) who reported that increase in the level of soil organic carbon and organic matter was expected since organic manures have the ability of increasing soil organic matter content. It has also been reported that the contents of some major nutrients in the soil were slightly dependable on the level of organic matter (Adeniyan and Ojeniyi 2005). Amendment effect on soil organic carbon was environment and soil type dependent according to the findings of this study. This is because the effect of manures and manure ashes was different relative to the type of environment; controlled environment (incubation and screen-house) and non-controlled environment (field). In the field experiment, only manure ash of poultry differed from its dried manure, the similarities between DGM, DCM and their ashes could be attributed to similar organic carbon content in these amendments allowing them to behave the same way in the field experiment. Dried manures and manure ash did not have residual effect since amendment effect was not significant at the second cycle of the experiment. The abrupt increase in the SOC after incorporation of NPK could be attributed to mineralization processes in the soils due to their native fertility and the presence of microbes.

Organic carbon was least released from the amendments at 6 WAI while at 4 WAI more of it was released in most of the amendment. The probable reason for the high release of organic carbon at 4 WAI may be attributed to increase mineralization of organic matter at 4 WAI, thereby increasing SOC, which is a major component of organic matter. The inconsistency in dynamics of SOC may be due to the usage of organic carbon by microbial community as energy and food sources at the first week, thereby decreasing it at the succeeding weeks. In a similar study, it was reported that the proliferation of microbial population leads to competition and death of microbes; this was adduced as the major reason as for inconsistency in organic carbon changes with time (Azeez and Van Averbeke 2010). The higher content of organic carbon release at 0 WAI may be because of enhanced organic manure and manure ash mineralization at the onset of incorporation. This is similar to the findings of (Sajal and Abul Kashem 2014) who reported that the highest amount of soil organic carbon at the beginning of incubation experiment was indicative of a larger pool of the less resistant fractions that were available to be broken down and recycled, thus, resulting in lower contents remaining at the end incubation. Mannivannan et al. (2009) and Gulser et al. (2010) also found that the addition of organic residues increases the soil organic carbon level initially and with the course of time organic carbon content decreases in soil up to a certain period.

This study confirms that in the incubation study, NH_4^+–N did not significantly differ as a result of dried manures and ashes addition to soil despite that increases were observed; this signifies that ammonification process is controlled not only by the characteristics of added material but also by the soil environmental condition including the microbial community. The rapid conversion of ammonium to nitrate through nitrification process could also be the rationale for the similar ammonium content of dried manure amended soil and ash amended soil. Additionally, it could also be because nutrients in the manure had not been equilibrated with the soil. Azeez and Van Averbeke (2010) reported that ammonium mineralized by the manures could be trapped by soil constituents or temporarily immobilized by soil microorganism, or further transformed into nitrate or lost as ammonia under favorable soil conditions. The increase in NH_4^+–N of amended soils is in line with the findings of Sajal and Abul Kashem (2014) that the application of organic manures produced higher amount of available NH_4^+–N in amended soils compared to control soil. Poultry manure ash had the smallest carbon:nitrogen ratio which enhanced mineralization process when compared to other amendments; this could have been the rationale for its ability to release higher NH_4^+–N than other amendments.

The concentration of ammonium nitrogen in soil changed over the weeks although a non-predictable or erratic pattern of release was recorded with increasing weeks since control, dried manures, manure ash and NPK had different pattern of releasing ammonium. The increase at 0 WAI and subsequent decrease was similar to the findings of Duffera et al. (1999) who observed that the processed swine lagoon solids increased the concentrations of ammonium nitrogen and during the first week and by the second to fourth week after application, the ammonium nitrogen concentrations dropped. The increase in soil ammonium with the application of PMA after 2 WAI could be because poultry manure ash had most of its nitrogen in ammonium form and its lower C:N could have enhanced ammonification process after 2 WAI. David et al. (2009) reported the increase in ammonium content during the onset of composting period in which it is biologically transformed into NO_3–N/NO_2–N via nitrification. This study confirms that in screen-house experiment, ammonification process was higher than nitrification process since more ammonium nitrogen than nitrate nitrogen was recorded across the weeks. Uptake of available nitrogen by plant or immobilization of nitrate could be the reason for the lower nitrate nitrogen content in amended soils during screen-house experiment.

The findings in this study conform to Abassi et al. (2007) who reported that soil amended with organic manures accumulated significantly more nitrate than soil without organic manures. Biological process like immobilization in which microbial community use part of the available nitrate as energy source during the mineralization could be the reason for the non-significance in the nitrate nitrogen of soil amended with ash relative to soil amended with dried manures. The significant increase in NO_3^-–N under field condition may be attributed to the fact that NPK contained the available form of nitrogen as NO_3^-–N and the sharp increase at 2 WAI was because it was incorporated at 2 WAI. The low NO_3^-–N stimulated by NPK at subsequent weeks after could be attributed to leaching or uptake of available nitrogen. This is supported with the findings of Adeniyan et al. (2011) who reported that the nutrients released from NPK fertilizer were for a short period because leaching of nutrients may be higher in the soil treated with NPK fertilizer than the soil treated with organic manures.

In incubation study, nitrate nitrogen increased across the weeks for most of the amendments. Nitrification process was optimum at 8 WAI for all amendment and control while nitrification was lower at 0 WAI. Although in soil from Alabata and Itori, immobilization occurred at 4 WAI with the application of PMA and DGM while in soil from Osiele and Papalanto, immobilization occurred at 6WAI with the application of DCM and PMA. Changes in nitrate observed across the weeks were erratic in screen-house and field experiment; however, during screen-house experiment, the least amount of nitrate was released at the end of the experiment in both soils due to nitrogen losses either through immobilization or uptake.

Conclusion

Dried manures and manure ashes differ in their characteristics; incineration of dried manures led to increase in pH and concentration of nutrients especially exchangeable cations. Dried manures and manure ashes are comparable in terms of total organic carbon and total nitrogen content. Generally, the addition of amendment increased soil organic carbon, the effect of dried manures and manure ash on soil organic carbon is environment dependent, and the addition of amendment did not have residual effect on soil organic carbon. Ammonification and nitrification process was increased with the application of manure ash, especially PMA and the conversion of organic nitrogen to ammonium nitrogen is not only controlled by the characteristics of added material but also by the soil environmental condition. The release of nitrogen through nitrification process depends on the type of soil and amendment. Soil organic carbon and nitrogen availability release indices were not consistent with increasing time hence erratic. Nitrification is more pronounced under incubation and field conditions than ammonification process and vice versa was observed in screen-house experiment.

Acknowledgements The authors wish to acknowledge soils of Forest Island in Africa (SOFIIA-FUNAAB) for its support in laboratory analyses. SOFIIA is a collaborative research project funded through the UK Government's Royal Society-DFID Africa Capacity Building Initiative. Opinions expressed and conclusions arrived at are those of the authors and are not necessarily to be attributed to the project.

References

Abassi MK, Hina M, Khalique A, Khan SR (2007) Mineralization of three organic manures used as nitrogen source in a soil incubated under laboratory conditions. Commun Soil Sci Plant Anal 38:1691–1711. https://doi.org/10.1080/00103620701435464

Adeniyan ON, Ojeniyi SO (2005) Effect of poultry manure and NPK 15-15-15 and combination of their reduced levels on maize growth and soil chemical properties. Niger J Soil Sci 15:34–41

Adeniyan ON, Ojo AO, Akinbode OA, Adediran JA (2011) Comparative study of different organic manures and NPK fertilizer for improvement of soil chemical properties and dry matter yield of maize in two different soils. J Soil Sci Environ Manag 2(1):9–13

Adetunji MT (1991) An evaluation of the soil nutrient status for maize production in South Western Nigeria. J Agric Res 8:101–113

Ayeni LS, Adetunji MT, Ojeniyi SO, Ewulo BS, Adeyemo AJ (2008) Comparative and cumulative effect of cocoa pod husk ash and poultry manure on soil and maize nutrient contents and yield. Am Eurasian J Sustain Agric 2(1):92–97

Azeez JO, Adetunji MT (2004) Comparative effects of organic and inorganic fertilizer on soil chemical properties: an incubation experiment. Proceedings of the 29th annual conference of the Soil Science Society of Nigeria. Dec 6–10 pp 198–205

Azeez JO, Van Averbeke W (2010) Nitrogen mineralization potential of three animal manures applied on a sandy clay loam soil. Biores Technol 101:5645–5651. https://doi.org/10.1016/j.biortech.2010.01.119

Baker KF (1976) The determination of organic carbon in soil using a probe colorimeter. Lab Pract 25:82–83

Bouyoucos GH (1965) A calibration of the hydrometer method for testing mechanical analysis of soils. Agric J 43:434–438

Cataldo DA, Haroon M, Schrander LE, Youngs VL (1975) Rapid colorimeteric determination of nitrate in plant tissue by nitration of salicylic acid. Commun Soil Sci Plant Anal 6:71–80. https://doi.org/10.1080/00103627509366547

Chadwick DR, John F, Pain BF, Chambers BJ, Williams JC (2000) Plant uptake of nitrogen from the organic nitrogen fraction of animal manures: a laboratory experiment. J Agric Sci Camb 134:159–168. https://doi.org/10.1017/s0021859699007510

David FW, Steven KM, Thomas LR, Heekwon A (2009) Effects of a livestock manure windrow composting site with a fly ash pad surface and vegetative filter strip buffers on sediment, nitrate, and phosphorus losses with runoff. Agric Eng publ. pp 1–10 http://lib.dr.iastate.edu/cgi/viewcontent.cgi?article=1229&context=abe_eng_pubs

Duffera M, Robarge WP, Mikkelsen RL (1999) Estimating the Availability of nutrients from processed swine lagoon solids through incubation studies. Bio Technol 70:261–268. https://doi.org/10.1016/s0960-8524(99)00039-5

Edmeades DC (2003) The long-term effects of manures and fertilizers on soil productivity and quality. Nutr Cycl Agroecosyst 66:165–180

Griffin E, Hoyle FC, Murphy DV (2013) Soil organic carbon. In report card on sustainable natural resource use in agriculture. Department of Agriculture and Food, South Perth, Western Australia. Accessed online 20Oct 2017

Gulser C, Demir Z, Ic S (2010) Changes in some soil properties at different incubation periods after tobacco waste application. J Environ Bio 31:671–674

Hao X, Godlinski F, Chang C (2008) Distribution of phosphorus forms in soil following long-term continuous and discontinuous cattle manure applications. Soil Sci Soc Am J 72:90–97. https://doi.org/10.2136/sssaj2006.0344

Hossain MA, Hamid A, Nasreen S (2007) Effect of nitrogen and phosphorus fertilizer on N/P uptake and yield performance of groundnut (*Arachis hypogaea* L.). J Agric Res 45:119–127

Kaira YP, Maynard DG (1991) Method manual for forest soil and plant analysis. Information Report NOR-X-319. Forestry Canada, Northwest Region, Northern Forestry Centre, Edmonton, Alberta. http://web.unbc.ca/~arocenaj/lab/kalra_maynard.pdf

Ksawery K, Robert JB, Tjalfe GP, Baiq EY (2012) Ash from the thermal gasification of pig manure—effects on ryegrass yield, element uptake, and soil properties. Soil Res 50:406–415. https://doi.org/10.1071/SR12075

Lal R, Ghuman BS (1989) Biomass burning in windows after clearing a tropical rain forest: effect on soil properties, evaporation and crop yields. Field Crop Res 22(4):247–255. https://doi.org/10.1016/0378-4290(89)90025-7

Law-Ogbomo KE, Egharevba RKA, Osaigbovo AU (2011) The effect of poultry manure on growth and yield of *Amaranthus cruentus* and properties. Benson Idahosa University Nigeria, book of abstract of the 29th annual national conference of Horticultural Society of Nigeria 24th–29th July, University of Makurdi Nigeria. p 59

Mannivannan S, Balamurugan M, Parthasarathi K, Gunasekaran G, Ranganathan LS (2009) Effect of vermi-compost on soil fertility and crop productivity-beans (*Cycleolus vulgaris*). J Environ Bio 30:275–281

Ojeniyi SO (2000) Effect of goat manure on soil nutrients content and okra yield in rainforest area of Nigeria. Appl Trop Agric 5:20–23

Oyeleke SB, Onibagjo HO, Ibrahim K (2003) Degradation of animal wastes (cattle dung) to produce methane (cooking gas). Proceedings of the 5th annual conference of Animal Science of Nigeria (SAN), pp 168–169

Qiu S, McComb AJ, Bell RW (2008) Ratios of C, N, P in soil water direct microbial immobilization- mineralization and N availability in nutrient amended sandy soil in south western Australia. Agric Eco Environ 127:93–99. https://doi.org/10.1016/j.agee.2008.03.002

Sajal R, Abul Kashem MD (2014) Effects of organic manures in changes of some soil properties at different incubation periods. Open J Soil Sci 4:81–86. https://doi.org/10.4236/ojss.2014.43011

Saka HB, Azeez JO, Odedina JN, Akinsete SJ (2017) Dynamics of soil nitrogen availability indices in a sandy clay loam soil amended with animal manures. Int J Recycl Org Waste Agric 6:167–178. https://doi.org/10.1007/s40093-017-0165-7

Teppei K, Arata K, Manabu Y (2012) The chemical characteristics of ashes from cattle, swine and poultry manure. J Mater Cycl Waste Manag 15(1):106–110. https://doi.org/10.1007/s10163-012-0089-2

USDA (2010) Soil texture calculator. United State development of agriculture natural resources conservation service soils. http://www.nrcs.usda.gov/wps

Vel AM, Swarnam TP (2013) Nitrogen release pattern from organic manures applied to an acid soil. J Agric Sci 5(6):174–184. https://doi.org/10.5539/jas.v5n6p174

Yusiharni BE, Ziadi H, Gilkes RJ (2007) A laboratory and glasshouse evaluation of chicken litter ash, wood ash and iron smelting slag asliming agents and P fertilizers. Aust J Soil Res 45:374–389. https://doi.org/10.1071/sr06136

23

Malting barley response to integrated organic and mineral nutrient sources in Nitisol

Kassu Tadesse[1] · Asrat Mekonnen[1] · Almaz Admasu[1] · Wubengda Admasu[1] · Dawit Habte[1] · Amare Tadesse[1] · Bahiru Tilahun[1]

Abstract

Purpose Depletion of soil fertility accounts for the low yield and quality of malting barley (*Hordeum vulgare*) in Ethiopia. Mineral fertilizers can counteract nutrient depletion but are unaffordable by low-input farmers. Organic amendments can contribute to correct soil degradation but are often unavailable in the required amounts. Therefore, this experiment was conducted to investigate the effects of combined use of organic and mineral amendments on the yield and quality of malting barley, and on selected soil chemical properties at Lemu-Bilbilo district in the southeastern highlands of Ethiopia.

Methods Eight treatments were arranged in randomized complete block design with three replications to compare sole and integrated applications of organic and mineral nutrient sources on crop performance attributes and selected soil characteristics.

Results Application of 2.82 t ha^{-1} compost or 1.07 t ha^{-1} farmyard manure (FYM) along with 18–10 kg ha^{-1} N–P gave 4234–4443 kg ha^{-1} grain yield with 15–17 and 32–39% increase over sole organic amendments and control, respectively, and 757% economic benefit. Combining organic and mineral fertilizers also gave 7–17 and 1–6% increase in available soil P and total N content of the soil.

Conclusion Combined application of 50% recommended compost (RC) or 50% recommended FYM (RFYM) + 50% recommended NP (RNP) or 33% RC + 33% RFYM + 33% RNP enhanced yield, grain quality, soil attributes and economic benefits. Therefore, integration of organic and mineral amendments is recommended as best agronomic and economic optimum soil fertility management options for sustainable malting barley production in the southeastern highlands of Ethiopia.

Keywords Compost · Ethiopia · Farmyard manure · *Hordeum vulgare* · Mineral fertilizers

Introduction

Barley (*Hordeum vulgare L.*) ranks fifth next to teff (*Eragrostis tef*), maize (*Zea mays*), sorghum (*Sorghun bicolor*) and wheat (*Triticum aestivum*) in Ethiopia, and third after wheat and teff in the southeastern highlands of Ethiopia. It is grown annually on 1 million hectares of land in a wide range of environments in the country (CSA 2014). Malting barley is one of the most important crops for food, feed, malt and income generation for many smallholder farmers in the highlands of Ethiopia (FAO 2013; Mulatu and Lakew 2011).

Malting barley accounts for about 90% of the raw materials cost in beer production in Ethiopia (USDA GAIN 2014). It has become a very important industrial crop in the country since the establishment of Saint George Brewery in 1974. The demand for malting has increased further since the establishment of Assela malt factory in 1984. The inauguration of Gonder malt factory in 2015 and the blooming of many more breweries created greater demands for malting barley (Berhane et al. 2016).

Ethiopia has favorable environment and considerable market opportunities for increased production of high-quality malting barley; however, its production has not been expanded enough to benefit most barley growers, malt factories and breweries (Berhane et al. 2016). Regardless of the huge potential, domestic production could satisfy only 55% of the raw malt barley demand of the country; the remainder has been imported from different countries. The share of malting barley production is about 10–15% as compared to food barley (Berhane et al. 2016).

✉ Kassu Tadesse
kasstad96@yahoo.com

1 Kulumsa Agricultural Research Center, Ethiopian Institute of Agricultural Research, P. O. Box 489, Kulumsa, Ethiopia

Low soil fertility and poor agronomic practices are among the major constraints responsible for the low productivity of malting barley in Ethiopia (Gete et al. 2010). Low soil fertility in turn resulted from farming without replenishing nutrients over time, continuous cropping of cereals, removal of crop residues, nutrient leaching, low level of fertilizer usage and unbalanced application of nutrients. The problem of soil fertility degradation is most pronounced in the highlands of Ethiopia, where most of human and livestock population concentrated, and malting barley is produced (Getachew et al. 2014b). The average annual top soil loss from agricultural lands is estimated to be 137 tons ha^{-1} year^{-1} for the Ethiopian highlands (Gete et al. 2010). Regardless of the sever soil degradation, the use of external inputs to maintain soil fertility for increasing malting barley production is still low. In 2014, for example, only 5973 ton of fertilizer was used to nourish barley sown on 91,617 hectare of land in the study area indicating the rate used (65 kg ha^{-1}) was lower than the existing recommendation, which was 100 kg ha^{-1} diammonium phosphate (18–20 kg ha^{-1} N–P) (CSA 2014).

Proper fertilizer management is required to counteract soil fertility depletion, increase production at required quality and satisfy the increasing demand of malting barley. However, experiences showed that focusing on separate interventions will no longer work. There has been a five-fold increase in fertilizer application since the 1980s, but the increase in the national cereal yield was only 10% (Gete et al. 2010). Besides, continuous applications of inorganic fertilizers alone resulted in deterioration of soil health in terms of soil physical, chemical and biological properties (Getachew et al. 2014b). The results of previous studies indicated that application of inorganic or organic fertilizers alone did not bring a sustainable increase in yields (Getachew et al. 2014a).

While mineral fertilizers such as diammonium phosphate and urea are widely used for cereals production in Ethiopia, such amendments are no longer affordable for resource-poor farmers at recommended application rates. On the other hand, organic amendments are often not available on farm at the amounts required (Agegnehu and Amede 2017). The best alternative is integrating organic and mineral fertilizers for tackling soil fertility depletion and sustainably increasing crop yields (Agegnehu and Amede 2017; Getachew et al. 2014b; Gete et al. 2010).

However, information on the integrated uses of organic and mineral amendments and their effects on yield and grain quality of malting barley, and soil chemical properties in the southeastern highland of Ethiopia, where this study was conducted, is scarce. This field experiment was, therefore, conducted to contribute for the development of appropriate nutrient management strategies that can sustain malting barley production, improve the livelihood of farmers and satisfy the increasing demands from the brewing industry in Ethiopia. The specific objectives were to investigate the effects of combined uses of organic and mineral amendments on the yield and quality of malting barley and selected soil chemical properties. The specific objectives also included determination of the economically optimum application rates of integrated organic and inorganic fertilizers for improving the yield and attributes of malting barley.

Materials and methods

Description of study sites

A 2-year study was conducted on five farmers' fields at Lemu-Bilbilo district in the southeastern highlands of Ethiopia between 2013 and 2014. Lemu-Bilbilo district has a high potential for both food and malting barley production. Its climate is characterized as warm temperate sub-humid to humid (Atlas of Arsi Zone 2002). The dominant soil at the experimental sites is classified as Nitisol (IUSS Working Group 2014). The study sites were located between 7°10′14″ to 7°40′20″N latitude and 39°4′59″ to 39°38′56″E longitude at altitudes ranging from 2633 to 2755 m a.s.l. (Fig. 1).

The long-term mean annual rainfall was 1053 mm. The rainy season starts in March and continues to October with the highest precipitation occurring in July and August (Fig. 2). The mean annual minimum and maximum temperatures of the study area are 7.07 and 19.42 °C, respectively. November and March are the coldest and hottest months with temperatures of 6.09 and 21.34 °C, respectively.

Soil sampling procedures and chemical analysis

A total of 21 composite soil samples, three for each location, were collected at a depth of 0–40 cm before planting for soil characterization. Composite soil samples were also collected immediately after harvesting from each site for investigating the changes in soil chemical properties due to treatments application. The soil samples were air dried, grounded to pass to 2 mm sieve, and analyzed for pH using 1:2.5 soil: water ratio (Mclean 1982); total nitrogen (N) using Kjeldahl digestion with sulfuric acid catalyzed by potassium sulfate (He et al. 1990) and available phosphorous (P) using Mehlich III extraction method (Mehlich 2008) at the soil and plant nutrition laboratory of Kulumsa Agricultural Research Center.

Preparation of compost and farmyard manure (FYM)

Compost was prepared by pit method from 40% maize and wheat crop residues, 25% animal manure, 20% green alfalfa leaves, 10% ash and 5% forest soil. FYM was made of fresh cow dung, urine and organic bedding materials for dairy

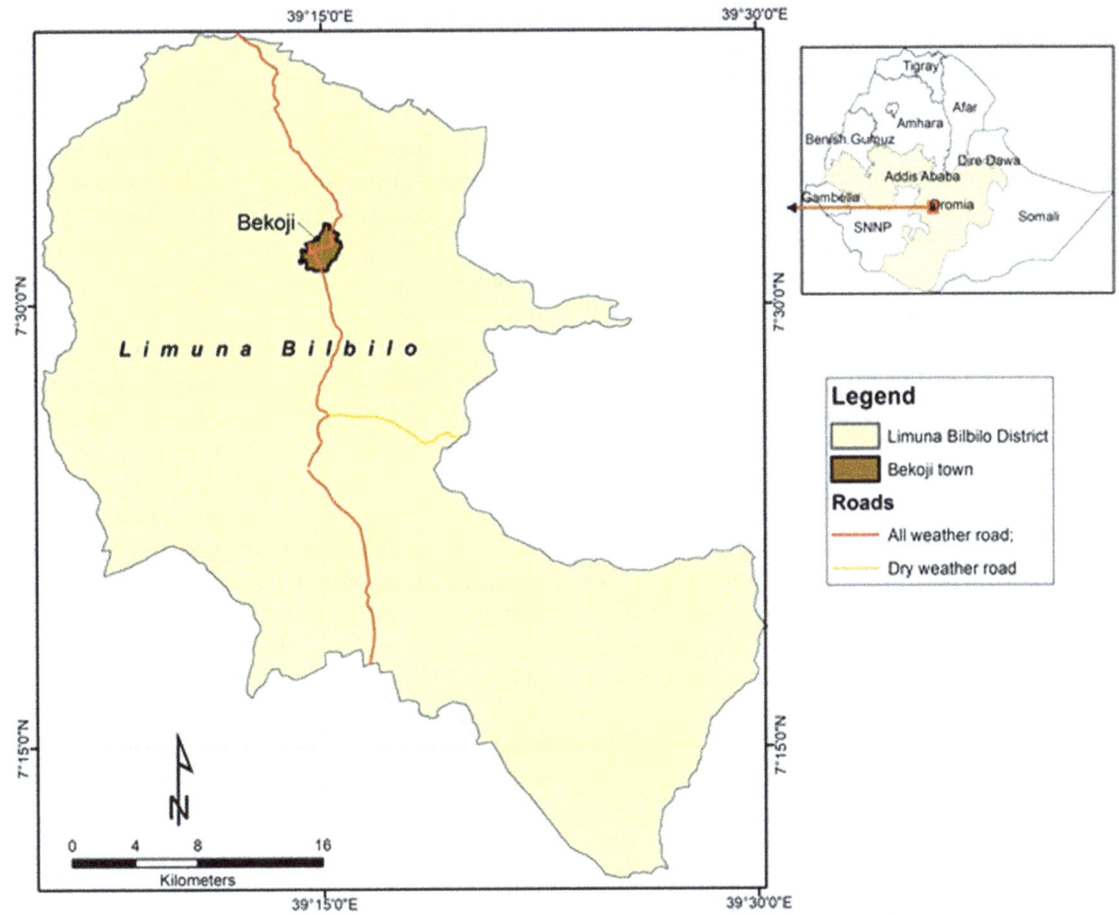

Fig. 1 Map of the experimental sites at Lemu-Bilbilo district in southeastern Ethiopian highlands

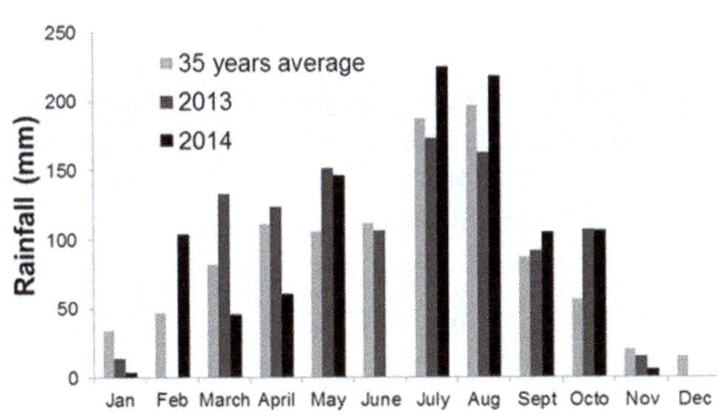

Fig. 2 Monthly total for 2013 and 2014 cropping seasons, and 35 years average rainfall at Lemu-Bilbilo district in southeastern Ethiopian highlands

cows, and dried for 1 month in the shade. The N and moisture contents of the organic amendments were determined just before application. The amounts of compost and FYM used for this experiment were adjusted to the same rate of 36 kg ha^{-1} N (N equivalence). Accordingly, the amounts of compost used in 2013 and 2014 were 5.9 and 5.37 t ha^{-1}, respectively, on dry weight basis. The corresponding amounts for FYM were 2.05 and 2.2 t ha^{-1}, respectively (Table 1).

Experimental set-up and procedure

The experiment had eight treatments, which consisted of sole and various combinations of mineral fertilizers (DAP

Table 1 Properties and amounts of compost and farmyard manure used in the field experiments

Sample type	Total nitrogen (%)		Moisture (%)		Fresh matter (t ha^{-1})		Dry matter (t ha^{-1})	
	2013	2014	2013	2014	2013	2014	2013	2014
Compost	0.61	0.67	28.2	32.6	8.22	7.97	5.90	5.37
FYM	1.76	1.64	82.7	63.1	11.82	5.95	2.05	2.20

and urea) and organic amendments (compost and FYM). Details of the treatments are shown in Table 2. The treatments were arranged in randomized complete block design with three replications.

The seedbed was tilled by traditional ox plough. Compost and FYM were applied as dry matter, uniformly spread and incorporated in the top 15 cm soil layer by hand hoe 2 weeks prior to planting. All P and half of N were uniformly applied to all plots, except for the control treatment, as DAP and Urea, respectively; while the remaining half of N was top-dressed at tillering stage. All experimental plots at each site were seeded with two-row malting barley (cv. *Holker*) at the recommended seeding rate of 125 kg ha^{-1}. Sowing was conducted in the 3rd to 4th weeks of June. Seeds were drilled by hand at 0.20 m spacing between rows for all sites in the plot size of 2.6 m × 4 m. The spacing between plots and replications was 0.5 and 1 m, respectively. Broad-leaved weeds were controlled using a herbicide, known as 2,4-D while grass weeds were manually removed. Insecticides, "Fenitrothion" and "tilt" were applied to control shoot fly and scald, respectively.

Data collection

Grain yield and above-ground biomass were harvested from the 6 m^2 (2 m × 3 m) net plot areas. Above-ground biomass weight was determined after air-drying. The dried biomass was threshed manually; the grains were cleaned from straws and weighed. The moisture contents of the weighed grain samples were determined using a moisture tester device, and adjusted to the standard moisture content of 12.5% for conversion to hectare basis.

Grain protein concentrations were determined using near infrared reflectance spectroscopy. The number of tillers per plant was counted from samples of ten plants per plot. Number of spikes per plant was counted from ten sample points of 0.5 m length each.

Economic analysis

The procedure of partial budget analysis as recommended by CIMMYT (1988) was followed to estimate the economic benefits of integrated use of organic and mineral fertilizers. The values of total variable costs (TVC) were calculated based on farm gate prices of inputs. The variable costs estimated were the prices of fertilizers and labor for application of organic amendments before planting. Hence, TVC was calculated as the sum of the costs of compost, FYM, DAP, urea and labor. Grain and straw yields of malting barley were adjusted downwards by 10% prior to calculation. All variable costs and gross field benefits were estimated based on the average values over locations and seasons. All treatments were listed based on increasing sequence of TVC. One dominated treatment with higher variable cost, but lower net benefit was discarded from further consideration through dominance analysis. For the remaining treatments, marginal rate of return (MRR) was calculated. The minimum acceptable rate of return considered to declare economical profitability in this study was greater than or equal to 100%. Based on this criterion, one treatment was eliminated from further analysis. Farmers in the study area used 65 and 75 kg ha^{-1} DAP in 2013 and 2014, respectively. 70 kg ha^{-1} DAP, mean value, was used to estimate the variable cost for the farmers' practice. Though the mean yield of malting barley in 2013 and 2014 in the study area was 2.71 t ha^{-1}, the yield from the control treatment of this study, 3.2 t ha^{-1}, was used as the farmers' practice due to the difficulty to get estimate of the straw yield from their fields. Sensitivity analysis was conducted employing the same data used for the calculation of MRR with values greater than the minimum rate of return, i.e., 100%. Hence, four treatments were considered for this study (Table 7). Sensitivity analysis was made based on the assumption that all variable costs increase by 10% annually with in the coming 3 years, which make 30%, while keeping the costs of produce constant.

Table 2 Treatment setup for 2013 and 2014 cropping seasons

Treatment no.	Description of treatments
1	Control (no input)
2	Recommended nitrogen and phosphorous (RNP) from mineral fertilizers (36–20 kg ha^{-1} N–P)
3	Recommended farmyard manure (RFYM) based on nitrogen (N) equivalence (2.05 and 2.2 t ha^{-1} in 2013 and 2014, respectively)
4	Recommended compost (RC) based on N equivalence (5.9 and 5.37 t ha^{-1} in 2013 and 2014, respectively)
5	50% RFYM + 50% RNP from mineral fertilizer
6	50% RC + 50% RNP from mineral fertilizer
7	50% RFYM + 50% RC
8	33% RFYM + 33% RC + 33% RNP from mineral fertilizer

Statistical analysis

All yield, yield components, grain quality and soil chemical properties data were subjected to analysis of variance using the general linear model procedure (Proc GLM) of SAS statistical package version 9.0 (SAS institute 2002). Mean separation was done using least significant difference test at the 5% level of probability.

Results

Soil chemical properties

Soil analysis results for the samples collected prior to planting showed that the pH, available P and total N contents of the study sites ranged between 4.5–5.6, 11–34 mg kg^{-1} and 0.22–0.32%, respectively. Mean values for pH, available P and total N of the study sites were 5.02, 20.71 mg kg^{-1} and 0.28%, respectively (Table 3).

Analysis of the soil samples collected after harvesting malting barley showed that application of either organic amendments or mineral fertilizers or their combinations significantly ($p < 0.001$) increased the available P and total N contents in the soil (Table 4). Spatial variabilities also significantly affected the soil chemical properties.

Yield and yield components of malting barley

Grain yield, total above-ground biomass, number of tillers per plant, plant population and grain protein content of malting barley were significantly ($p < 0.001$) increased by combined application of organic and mineral nutrients. The result further indicated that season and sites significantly affected most of the measured variables (Table 5).

Yields, yield-related components and grain protein concentration of malting barley were greater with all combined organic and mineral amendments than sole applications of either compost or FYM or integration of both (Table 6). The highest grain (4527 kg ha^{-1}) and biomass (11,514 kg ha^{-1}) yields of malting barley were obtained from the application of recommended rate of N and P (RNP) (36–20 kg ha^{-1} N–P) from mineral sources (DAP and Urea). Compared to application of full dose of N and P from mineral sources, statistically similar grain (4443 kg ha^{-1}) and biomass (11,241 kg ha^{-1}) yields were obtained from the application of 50% RNP (18–10 kg ha^{-1} N–P) from mineral sources + 50% from compost (2.95 and 2.69 t ha^{-1} in 2013 and 2014, respectively). Similarly, statistically equivalent grain (4234 kg ha^{-1}) and biomass (10,960 kg ha^{-1}) yields were obtained from the application of 50% RNP from mineral sources + 50% from FYM (1.03 and 1.1 t ha^{-1} in 2013 and 2014, respectively). Statistically equivalent grain yield (4037 kg ha^{-1}) was also obtained from the application of

Table 3 Soil chemical analysis results for the experimental sites prior to application of organic and mineral amendments

Chemical properties	Values	
	Ranges	Mean
pH	4.5–5.6	5.02
Available phosphorous	11.0–34.0 mg kg^{-1} soil	20.71 mg kg^{-1} soil
Total nitrogen	0.22–0.32%	0.28%

Table 4 Mean soil chemical properties as influenced by combined application of organic and mineral amendments after harvesting of malting barley at Lemu-Bilbilo district in southeastern Ethiopian highlands

Factor	Soil quality parameters	
	Available P (mg kg^{-1})	Total N (%)
Control (no input)	22^{c}	0.28^{d}
RNP	25^{a}	0.29^{abc}
RFYM	25^{ab}	0.29^{ab}
RC	25^{ab}	0.28^{bcd}
50% RFYM + 50% RNP	25^{ab}	0.29^{a}
50% RC + 50% RNP	24^{ab}	0.29^{abc}
50% RFYM + 50% RC	23^{bc}	0.28^{cd}
33% RFYM + 33% RC + 33% RNP	25^{ab}	0.29^{abc}
Mean	24.13	0.29
CV	9.5	3.35
LSD	2.18	0.001

Table 5 Analysis of variance of treatment effects on yield, yield components and grain quality of malting barley at Lemu-Bilbilo district in southeastern Ethiopian highlands

Effect	Yield, yield components and quality parameters				
	No. of tillers/plant	Spike m^{-2} (no)	Grain yield (kg ha^{-1})	Biomass yield (kg ha^{-1})	Grain protein (%)
Rep	ns	*	ns	ns	ns
Treatment	***	***	***	***	***
Year	***	***	*	***	ns
Location	***	***	***	***	ns
Mean	3.98	497.52	3959.09	9866.39	9.89
CV	13.93	8.15	18.27	18.55	5.27
LSD	0.45	28.42	525.95	1330.7	0.62

ns not significant at 0.05,* significant at 0.05 and *** highly significant at 0.001 probability levels

Means within each column having no common superscripts are significantly different ($p < 0.05$)

Table 6 Table of means for the effect of treatments on yield, yield components and grain quality of malting barley at Lemu-Bilbilo district in southeastern Ethiopian highlands

Factors	Yield, yield components and quality parameters				
	No. of tillers plant^{-1}	Spike m^{-2} (no)	Grain yield (kg ha^{-1})	Biomass yield (kg ha^{-1})	Grain protein (%)
Year					
2013	6.50^{a}	523.0^{a}	4034.2^{a}	10,182.2^{a}	10.02^{a}
2014	3.14^{b}	305.8^{b}	3658.8^{b}	8602.9^{b}	9.76^{a}
Integrated nutrient					
Control (no input)	3.58^{c}	452.7^{d}	3200.2^{d}	7471.6^{e}	9.04^{e}
RNP	4.17^{ab}	514.5^{a}	4527.3^{a}	11,514.8^{a}	10.65^{a}
RFYM	3.58^{c}	483.1^{bc}	3668.0^{cd}	8578.2^{de}	9.57^{cde}
RC	3.92^{bc}	475.9^{bcd}	3793.2^{bc}	9627.9^{cd}	9.66^{cde}
50% RFYM + 50% RNP	4.17^{ab}	489.1^{ab}	4234.4^{ab}	10,960.2^{ab}	10.20^{ab}
50% RC + 50% RNP	4.08^{ab}	459.8 ^{cd}	4443.0^{a}	11,241.3^{ab}	10.37^{a}
50% RFYM + 50% RC	3.92^{bc}	470.6b^{cd}	3769.9^{bc}	9596.9^{cd}	9.45^{de}
33% RFYM + 33% RC + 33% RNP	4.42^{a}	490.5^{ab}	4036.6^{abc}	9940.2^{bc}	10.18^{abc}

RNP is recommended N and P from mineral fertilizer sources, RFYM is recommended FYM and RC is recommended compost. Results with the same letters are statistically indifferent

33% RNP (12–7 ha^{-1} N–P) from mineral sources + 33% from compost (1.97 and 1.79 t ha^{-1} in 2013 and 2014, respectively) + 33% from FYM (0.68 and 0.73 t ha^{-1} in 2013 and 2014, respectively) (Table 6). The lowest yield and yield components were obtained from the control treatment.

Results depicted that large number of productive tillers per plant (4.42) was recorded from the application of 33% RNP from FYM + 33% from compost + 33% from mineral fertilizers (Table 6). However, this result was found statistically similar to the application of RNP from mineral fertilizers (4.17), 50% RNP from mineral fertilizers + 50% from FYM (4.17), and 50% RNP from mineral fertilizers + 50% from compost (4.08). The smallest numbers of productive tillers per plant were obtained from the control treatment, and application of RNP from either compost or FYM.

The maximum number of spikes m^{-2} (515 plants m^{-2}) was obtained from the application of RNP from mineral sources (Table 6). However, it was found statistically similar to the application of 33% RNP from mineral fertilizers + 33% from compost + 33% from FYM (491 plants m^{-2}); and the application of 50% RNP from mineral fertilizers + 50% from FYM (489 plants m^{-2}). The smallest number of spikes m^{-2} (453 plants m^{-2}) was obtained from the application of no input (Table 6).

Yield and yield components were also significantly ($p < 0.001$) affected by seasons (Tables 6) and locations (Fig. 3). Statistically, higher grain yield (4034 kg ha^{-1}), biomass yield (10,182 kg ha^{-1}), seedling density (6.5) and spikes m^{-2} (523 plants m^{-2}) were recorded in 2013 than 2014. Due to locational differences, gain yield varied from 3352 to 4858 kg ha^{-1} (Fig. 3).

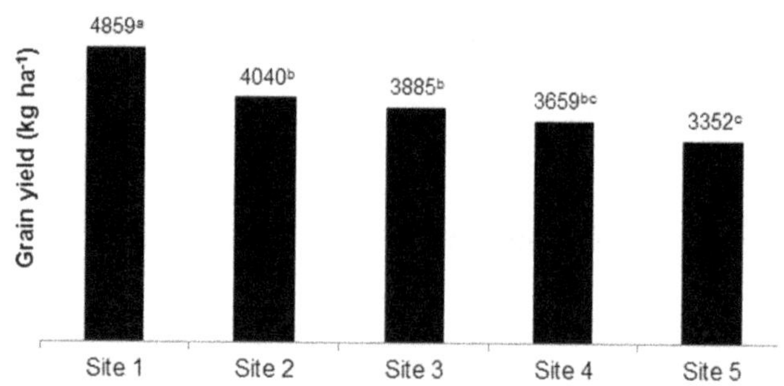

Fig. 3 Grain yield of malting barley as influenced by location at Lemu-Bilbilo in southeastern Ethiopian highlands

Grain protein contents of malting barley

Analysis of variance further revealed that the grain protein concentrations of malting barley was very significantly ($p < 0.001$) affected by the integrated use of organic and mineral fertilizers (Table 6). The highest grain protein content (10.7%) was recorded from the application of RNP from mineral fertilizers. However, statistically similar grain protein contents were obtained from the applications of 50% RNP from mineral fertilizers + 50% from compost (10.4%); 50% RNP from mineral fertilizers + 50% from FYM (10.2%); and 33% RNP from mineral fertilizers + 33% from compost + 33% from FYM (10.2%) (Table 6). The lowest grain protein content (9%) was obtained from the control plot.

Economic analysis

Combined applications of organic and mineral amendments were superior to applications of either organic amendments or mineral fertilizers and economically profitable since they gave acceptable rate of return over 100%. The marginal rate of return (MRR) generally varied from 0.85 to 26.20 (Table 7). The highest MRR was obtained from the application of 50% RNP from mineral fertilizers + 50% from FYM (26.24) followed by 33% RNP from mineral fertilizers + 33% from FYM + 33% from compost (8.94), and full dose of RNP from compost (3.91). The net benefits improved further with combined application of 50% RNP from mineral sources + 50% from compost (7.57). All of the four treatments used for sensitivity analysis gave MRR values in excess of 100%.

Discussions

Soil chemical properties

Application of either organic amendments or mineral fertilizers or their combinations gave 7–17% (equivalent to 1.4–3.8 mg kg^{-1}) increase in available soil P and 1–6% (equivalent to 0.003–0.02% N) increase in total N content in the soil compared to the control treatment (Table 4). The improvements in available soil P and total N contents attributed to the additions of the aforementioned nutrients from both organic and mineral sources. Previous works also reported that integrated use of organic and mineral fertilizers resulted in the highest available soil P (Getachew et al. 2015) and total N contents (Getachew et al. 2014a) in the soil. The application of organic fertilizers directly influences the availability of inherent or applied P (Sharma et al. 1990). The increase in total N content of the soil could be attributed to the direct additions of N through decompositions of compost and/or FYM.

Yield and yield components of malting barley

In this study, the combined application organic and mineral fertilizers resulted in similar yields to the RNP from mineral fertilizers. The yield of malting barley and its components were substantially improved with all integrated uses of mineral and organic amendments. Combined applications of mineral and organic fertilizers from either compost or FYM at a ratio of 1:1 or 1:1:1 gave superior grain and biomass yield advantages as compared to sole application of organic amendments, either from compost or FYM or their combinations (Tables 5 and 6). Application of 50% RNP from mineral fertilizers + 50% from compost resulted in grain yield advantages of 17 (650), 18 (673) and 39% (1243 kg ha^{-1}) as compared to RNP from compost, 1:1 ratio of compost and FYM, and the control treatment, respectively. The corresponding increments in biomass yields were 17 (1613), 17 (1645) and 51% (3770 kg ha^{-1}), respectively. Similarly, application of 50% RNP from mineral fertilizers + 50% from FYM resulted in grain yield advantages of 15 (566), 12 (465) and 32% (1034 kg ha^{-1}) as compared to application of RNP from FYM, 1:1 ratio of compost and FYM, and the control treatment, respectively. The corresponding increments in biomass yields were 28 (2382), 14 (1364) and 47% (3489 kg ha^{-1}), respectively. This implied that the integrated use of organic and mineral fertilizers resulted in yield benefits greater than using them alone. Similar results were also reported on barley (Agegnehu et al. 2016a; Assefa 2015), wheat (Demelash et al. 2014; Getachew et al. 2014b), teff (Getachew et al. 2014b), maize (Mahmood et al. 2017; Wakene et al. 2005), Okra (Afe and Oluleye 2017) and tomato (Rajaie and Tavakoly 2016). The improved yields of malting barley due to combined application of organic and mineral amendments resulted from positive changes to the soil, including increased soil pH, available P and total N, and possibly other macro- and micronutrients.

The grain (3200 kg ha^{-1}) and biomass (7472 kg ha^{-1}) yields obtained from the control treatment were inferior.

Table 7 Economic analysis for the use of integrated organic and mineral amendments on malting barley at Lemu-Bilbilo district in southeastern Ethiopian highlands

Treatments	Adjusted grain yield (kg ha^{-1})	Adjusted straw yield (kg ha^{-1})	Gross field benefits (US$ ha^{-1})	Total cost that vary (US$ ha^{-1})	Net benefit (US$ ha^{-1})	Dominance	Marginal rate of return
RFYM	3301.20	4419.18	1819.22	24.23	1794.99		
RC	3413.88	5251.23	1934.20	47.62	1886.58		3.91
Farmers practice	2880.18	3844.26	1586.32	50.43	1535.90	D	
33% RFYM + 33% RC + 33% RNP	3632.94	5313.24	2036.96	57.96	1979.01		8.94
50% RFYM + 50% RNP	3810.96	6053.22	2174.01	62.99	2111.02		26.20
50% RC + 50% RNP	3998.70	6118.47	2263.03	73.38	2189.65		7.57
RNP	4074.57	6288.75	2310.17	98.87	2211.30		0.85

Applications of RNP from FYM or compost and their combinations could not also result in higher grain and biomass yields of malting barley due, possibly, to the fact that the nutrients especially N and P were not immediately available during the season to fully nourish the crop. Hence, under low soil fertility condition, integrated application of organic and mineral fertilizers is highly required because of their additive effects (Agegnehu et al. 2016a; Liu et al. 2012). Zahir and Mian (2006) reported that plots, which received N solely from FYM produced significantly lowest grain yield compared to other fertilizer treatments.

The number of productive tillers of barley m^{-2} was improved with integrated application of organic and mineral fertilizers. The current result is in agreement with Mitiku et al. (2014). The smallest numbers of tillers per plant could be attributed to the limited supply of plant nutrition from compost and/or FYM at the initial stages of growth.

Yield and yield components were also significantly affected by seasonal variations, particularly rainfall. Rainfall was very intense in 2014 cropping season, which was above the long-term average of the area for all of the months in the cropping season (June to November) (Fig. 2). The highest amount of rainfall in 2014 could have induced excess accumulation of soil moisture in the root zone, which could attribute to water logging and consequent reduction in yields compared to 2013 cropping season.

Variations in grain yields among the five studied sites were attributed to the differences in the initial nutrients contents and pH values of the soils. There were differences in the total N and available P contents among the soils prior to planting (Table 3). The site where the highest grain yield was recorded had relatively higher available P and total N contents (data not shown) compared to the other four sites. The sites with the highest (4859 kg ha^{-1}) and lowest (3352 kg ha^{-1}) grain yields (Table 4) had pH value of 5.6 and 4.5, respectively. The available P could have been fixed more at the low pH soils and become unavailable for malting barley.

Grain protein contents of malting barley

Similar to yield and yield components, integrated application of organic and mineral fertilizers resulted in optimum concentrations of grain protein. Agegnehu et al. (2016b), Yolcu (2015) and Miller et al. (2004) also reported significant improvements in the grain protein contents of malting barley owing to integration of mineral fertilizers with organic amendments such as biochar, cattle manure and composted manure. Applications of full dose of RNP from either compost or FYM or 1:1 compost: FYM resulted in statistically lower grain protein contents of 9.7, 9.6 and 9.5%, respectively (Table 6). This could be attributed to the slow release of N from the organic amendments.

The grain protein concentrations of malting barley recorded from all treated plots were within the threshold range of grain protein content set by Assela Malt factory for malting purpose, 9.5–11%. Despite the fact that integrated applications of organic and mineral fertilizers gave optimum values in this study, care need to be given to timing of N application as it significantly influences the grain protein contents of malting barley. Previous studies including Kara (2010) showed that grain protein content of malting barley increased with late-season N application. Therefore, it should be taken into account that N top-dressing of malting barley at its late stage will increase protein content in grain beyond the upper limit for malting use. The N requirement of the crop at its latest stage is expected to be fulfilled by the mineralization of organic amendments so that growers need to be cautious about top-dressing at booting stage. Though grain protein content of malting barley did not surpass the upper threshold limit during the 2-year study, there may be a chance of increased mineralization of N at the late stage of the crop development depending on environmental conditions. Therefore, seeding malting barley with low protein content, such as the variety used in this study, lowers the risk of the grain protein content exceeding beyond the threshold limit (O'Donovan et al. 2015). Besides, monitoring of grain quality on long-term application of organic amendments is required as the buildup and mineralization of N may increase grain protein concentrations.

Economic analysis

Combined application of 1:1 compost: mineral fertilizer could enable farmers to earn a return of US$7.57 for every US$1.0 investment, which implied high profitability due to integration of the locally available organic materials with mineral fertilizers. This economic benefit is on top of soil quality improvements, which we could not quantify and include their equivalent economic values in this study. Integration of FYM with mineral fertilizers also resulted in an attractive increase in farmers' income. Demelash et al. (2014) reported economical profitability with 844% MRR owing to integration of compost with mineral fertilizers in wheat production. Otinga et al. (2013) also reported an increased net benefit of over 33% in response to combined application of FYM and mineral fertilizer in maize production. Sole application of mineral fertilizers brought relatively lower revenue to farmers compared to integration with organic amendments. However, in the absence of organic amendments, investment on mineral fertilizers resulted in a very high net income to farmers with a MRR of 13.94 (Table 7) compared to the existing farmers' practice in the study area. The economic profitability agrees with the agronomic results.

Recalculating the partial budget with the possible higher prices of total variable costs confirmed that the MRR were still above 100% minimum rate of return implying that the same recommendation could hold true. If the prices of all variable costs rise by 30% within the coming 3 years, farmers who can make the decision to apply 2.82 t ha^{-1} compost along with 18–10 kg ha^{-1} N–P will earn additional US$5.59 for every US$1.0 investment. The amount of compost and/or FYM that provided optimum agronomic and economic yields could further be reduced and hence their MRR increased by vermicomposting. Shah et al. (2015) reported that farm manure inoculated with local species of earthworms produced best quality manure with enriched nutritional status including 3% N. This result was found almost two fold of the values obtained in the current study (1.76 and 1.64% N in 2013 and 2014, respectively) implying that the amount of organic fertilizers and their associated costs could further be reduced by enhancing their quality.

Conclusions and recommendation

The present study revealed that combined application of 2.82 t ha^{-1} compost or 1.07 t ha^{-1} FYM with 18–10 kg ha^{-1} N–P or 1.88 t ha^{-1} compost + 0.71 t ha^{-1} FYM + 12–7 kg ha^{-1} N–P boosted yield, grain protein concentration, soil attributes and economic benefits. Results further showed that productivity of malting barley and soil quality could be more sustainable with integration of organic amendments and mineral fertilizers than with the use of either organic amendments or mineral fertilizers alone. To counteract the decline in soil fertility on the one hand and to cope with the increasing price of the imported mineral fertilizers on the other hand, the integrated use of organic and mineral fertilizers is found to be very important. Therefore, farmers of the study area and other similar agroecologies are advised to apply 50% recommended compost (RC) + 50% recommended NP (RNP) or 50% recommended farm yard manure (RFYM) + 50% RNP or 33% RFYM + 33% RC + 33% RNP for optimum ecological and economic benefits. Viable soil fertility strategies have to be implemented to make use of the available manure and crop residue resources for improving soil fertility and crop productivity by minimizing their competitive demands for fuel and feed.

Acknowledgements The authors would like to thank Ethiopian Institute of Agricultural Research for the provision of funding and logistics for the execution of this research activity. For their unreserved efforts from planting to harvesting and proper data management, all members of Land and Water Resources Research team of Kulumsa Agricultural Research Center are greatly acknowledged. The management and laboratory technician of Sinana Agricultural Research Center are also greatly acknowledged for the provision of analytical services. We would also like to acknowledge the anonymous reviewers, whose comments and suggestions significantly improved this manuscript.

References

Afe AI, Oluleye F (2017) Responses of Okra (*Abelmuschus esculenthus* L. Moench) to combined organic and inorganic foliar fertilizers. Int J Recycl Org Waste Agric 6(3):189–193. https://doi.org/10.1007/s40093-017-0166-6

Agegnehu G, Amede T (2017) Integrated soil fertility and plant nutrient management in tropical agro-ecosystems: a review. Pedosphere 27(4):662–680. https://doi.org/10.1016/S1002-0160(17)60382-5

Agegnehu G, Nelson PN, Bird MI (2016a) Crop yield, plant nutrient uptake and soil physicochemical properties under organic soil amendments and nitrogen fertilization on Nitisols. Soil Tillage Res 160:1–13. https://doi.org/10.1016/j.still.2016.02.003

Agegnehu G, Nelson PN, Bird MI (2016b) The effects of biochar, compost and their mixture and nitrogen fertilizer on yield and nitrogen use efficiency of barley grown on a Nitisol in the highlands of Ethiopia. Sci Total Environ 569–570:869–879. https://doi.org/10.1016/j.scitotenv.2016.05.033

Assefa W (2015) Response of barley (*Hordium vulgare L.*) to integrated cattle manure and mineral fertilizer application in the vertisol areas of south Tigray, Ethiopia. J Plant Sci 3(2):71–76. https://doi.org/10.11648/j.jps.20150302.15

Atlas of Arsi Zone (2002) Regional state of Oromia: Arsi zone planning and economic development office and rural development project for Arsi and Bale. Arsi, Addis Ababa

Berhane L, Chilot Y, Wondimu F (2016) Malt barley research and development in Ethiopia: opportunities and challenges. In: Dawit A, Eshetu D, Getnet A, Abebe K (eds) Proceedings of the National Conference on agricultural research for Ethiopian renaissance. Ethiopian Institute of Agricultural Research, Addis Ababa, pp 11–20

CIMMYT (International Maize and Wheat Improvement Centre) (1988) From agronomic data to farmer recommendation: an economics training manual completely revised edition. CIMMYT, Mexico. http://repository.cimmyt.org:8080/xmlui/bitstream/handle/10883/859/25152.pdf

CSA (Central Statistical Agency of Ethiopia) (2014) Report on area and production of major crops for 2013/2014 (private peasant holdings, Meher Season). Statistical Bulletin No. 532. Addis Ababa, Ethiopia

Demelash N, Bayu W, Tesfaye S, Zidat F, Sommer R (2014) Current and residual effects of compost and inorganic fertilizer on wheat and soil chemical properties. Nutr Cycling Agroecosyst 100(3):357–367. https://doi.org/10.1007/s10705-014-9654-5

FAO (2013) FAOSTAT 2013: Food and Agriculture Organization (FAO), Rome, Italy. http://faostat.fao.org

Getachew A, Berhane L, Paul NN (2014a) Cropping sequence and nitrogen fertilizer effects on the productivity and quality of malting barley and soil fertility in the Ethiopian highlands. Arch Agron Soil Sci 60(9):1261–1275. https://doi.org/10.1080/03650340.2014.881474

Getachew A, vanBeek C, Michael IB (2014b) Influence of integrated soil fertility management in wheat and tef productivity and soil chemical properties in the highland tropical environment. J Soil Sci Plant Nutr 14:532–545. https://doi.org/10.4067/S0718-95162014005000042

Getachew A, Michael IB, Paul NN, Adrian MB (2015) The ameliorating effects of biochar and compost on soil quality and plant growth on a Ferralsol. Soil Res 53(1):1–12. https://doi.org/10.1071/SR14118

Gete Z, Getachew A, Dejene A, Shahidur R (2010) Fertilizer and soil fertility potential in Ethiopia: Constraints and opportunities for enhancing the system. IFPRI, Addis Ababa, Ethiopia. http://www.ifpri.org/publication/fertilizer-and-soil-fertility-potential-ethiopia

He XT, Mulvaney RL, Banwart WL (1990) A rapid method for total nitrogen analysis using microwave digestion. Soil Sci Soc Am J 54(6):1625–1629. https://doi.org/10.2136/sssaj1990.03615995005400060019x

IUSS Working Group WRB (2014) World reference base for soil resources 2014. International soil classification system for naming soils and creating legends for soil maps. World Soil Resources Reports No. 106, FAO, Rome. http://www.fao.org/3/a-i3794e.pdf

Kara B (2010) Influence of late-season nitrogen application on grain yield, nitrogen use efficiency and protein content of wheat under Isparta ecological conditions. Turk J Field Crop 15:1–6. http://www.field-crops.org/assets/pdf/product5132156434e57.pdf

Liu J, Schulz H, Brandl S, Miehtke H, HuweB Glaser B (2012) Short-term effect of biochar and compost on soil fertility and water status of a Dystric Cambisol in NE Germany under field conditions. J Plant Nutr Soil Sci 175:698–707. https://doi.org/10.1002/jpln.201100172

Mahmood F, Khan I, Ashraf U, Shahzad T, Hussain S, Shahid M, Abid M, Ullah S (2017) Effects of organic and inorganic manures on maize and their residual impact on soil physico-chemical properties. J Soil Sci Plant Nutr 17(1):22–32. https://doi.org/10.4067/S0718-95162017005000002

Mclean EO (1982) Soil PH and lime requirement. In: Page AL (ed) Methods of soil analysis. Part 2. Chemical and microbiological properties. American Society of Agronomy, Madison, pp 199–224. https://doi.org/10.2134/agronmonogr9.2.2ed.c12

Mehlich A (2008) Mehlich 3 soil test extractant: a modification of Mehlich 2 extractant. Commun Soil Sci Plant Anal 15:1409–1416. https://doi.org/10.1080/00103628409367568

Miller JJ, Beasley BW, Larney FJ, Olson BM (2004) Barley dry matter yield, crop uptake, and soil nutrients under fresh and composted manure containing straw or wood-chip bedding. Can J Plant Sci 84:987–999. https://doi.org/10.4141/P03-208

Mitiku W, Tamado T, Singh TN, Teferi M (2014) Effect of integrated nutrient management on yield and yield components of food barley (*Hordeum vulgare L.*) in Kaffa Zone, Southwestern Ethiopia. Sci Technol Arts Res J 3(2):34–42. https://doi.org/10.4314/star.v3i2.5

Mulatu B, Lakew B (2011) Barley research and development in Ethiopia—an overview. In: Mulatu B, Grando S (eds) Barley research and development in Ethiopia. Holeta, Ethiopia, pp 1–16

O'Donovan JT, Anbessa Y, Grant CA, Macleod AL, Edney MJ, Izydorczyk MS, Turkington TK, Juskiw PE, Lafond GP, May WE, Harker KN, Johnson EN, Beres BL, McAllister TA, Smith EG, Chapman W (2015) Relative responses of new malting barley cultivars to increasing nitrogen rates in western Canada. Can J Plant Sci 95:831839. https://doi.org/10.4141/cjps-2014-415

Otinga AN, Okalebo JR, Njoroge R, Emong'ole M, Six L, Vanlauwe B, Merckx R (2013) Partial substitution of phosphorus fertilizer by farmyard manure and its localized application increases agronomic efficiency and profitability of maize production. Field Crop Res 140:32–43. https://doi.org/10.1016/j.fcr.2012.10.003

Rajaie M, Tavakoly AR (2016) Effects of municipal waste compost and nitrogen fertilizer on growth and mineral composition of tomato. Int J Recycl Org Waste Agric 5(4):339–347. https://doi.org/10.1007/s40093-016-0144-4

SAS Institute (2002) SAS Institute Inc. Cary

Shah RU, Abid M, Qayyum MF, Ullan R (2015) Dynamics of chemical changes through production of various composts/vermicompost such as farm manure and sugar industry wastes. Recycl Org Waste Agric 4(1):39–51. https://doi.org/10.1007/s40093-015-0083-5

Sharma PK, Verma TS, Gupta JP (1990) Ameliorating effects of phosphorus, lime and animal manure on wheat yield and root cation exchange capacity in degraded Alfisols of North–West Himalayas. Fert Res 23:7–13. https://doi.org/10.1007/BF02656127.pdf

USDA GAIN (USDA'S Global Agriculture Information Network) (2014) USDA GAIN: Ethiopia Grain and Feed Annual Report. GAIN Report Number: ET1401. Available at http://www.thefarmsite.com/reports/contents/EthiopiaGrains9April2014.pdf

Wakene N, Heluf G, Friesen DK (2005) Integrated use of farmyard manure and NP fertilizers for maize on farmers' fields. J Agr Rural Dev Tropic 106(2):131–141. https://www.jarts.info/index.php/jarts/article/view/89

Yolcu H (2015) Effect of applications of organic solid cattle manure on Hungarian vetch and barley intercropping mixtures grown on soils of different depths. Grass Forage Sci 70:428–438. https://doi.org/10.1111/gfs.12120

Zahir S, Mian IA (2006) Effect of integrated use of farmyards manure and urea on yield and nitrogen uptake of wheat. J Agric Biol Sci 1(1):60–65. https://www.arpnjournals.com/jabs/research_papers/jabs_0706_10.pdf

24

An environmental evaluation of food waste downstream management options: a hybrid LCA approach

Ramy Salemdeeb[1] · Mohammad Bin Daina[2] · Christian Reynolds[3] · Abir Al-Tabbaa[1]

Abstract

Purpose Food waste treatment methods have been typically analysed using current energy generation conditions. To correctly evaluate treatment methods, they must be compared under existing and potential decarbonisation scenarios. This paper holistically quantifies the environmental impacts of three food waste downstream management options—incineration, composting, and anaerobic digestion (AD).

Methods The assessment was carried out using a novel hybrid input–output-based life cycle assessment method (LCA), for 2014, and in a future decarbonised economy. The method introduces expanded system boundaries which reduced the level of incompleteness, a previous limitation of process-based LCA.

Results Using the 2014 UK energy mix, composting achieved the best score for seven of 14 environmental impacts, while AD scored second best for ten. Incineration had the highest environmental burdens in six impacts. Uncertainties in the LCA data made it difficult determine best treatment option. There was significant environmental impact from capital goods, meaning current treatment facilities should be used for their full lifespan. Hybrid IO LCA's included additional processes and reduced truncation error increasing overall captured environmental impacts of composting, AD, and incineration by 26, 10 and 26%, respectively. Sensitivity and Monte Carlo analysis evaluate the methods robustness and illustrate the uncertainty of current LCA methods. Major implication: hybrid IO-LCA approaches must become the new norm for LCA.

Conclusion This study provided a deeper insight of the overall environmental performance of downstream food waste treatment options including ecological burdens associated with capital goods.

Keywords Anaerobic digestion · Incineration · Composting · Food waste · Hybrid life cycle assessment · Capital goods

Introduction

In November 2016, 194 countries signed the world's first comprehensive climate agreement in Paris. This agreement has been adopted overwhelmingly to tackle climate change by introducing binding emissions targets (United Nations 2015; European Commission 2016). Food waste has been identified as a major contributor to climate change with 8% of global anthropogenic greenhouse gas emissions (GHGE) which is attributed to improper food waste management (FAO 2013).

In spite of various efforts to reduce food waste arisings per capita, quantities of food waste continue to increase due to population growth. In spite of the UK success in introducing intervention programmes that have reduced food waste arisings per capita by 24%, the overall household food waste arisings in the UK have flat lined at around 7 million tonnes per year, with further large-scale reductions not yet possible (Parry 2014; Quested and Andrew 2017). As food waste is unlikely to be significantly reduced in the short term, greater efforts must be made to reduce its environmental impacts. This could be achieved by first quantifying environmental burdens associated with food waste treatment, identifying processes with the highest environmental impacts (i.e.,

✉ Ramy Salemdeeb
rsalem@zerowastemena.org

[1] Department of Engineering, University of Cambridge, Trumpington Street, Cambridge CB2 1PZ, UK

[2] Supreme Council for the Environment, Manama, Bahrain

[3] Department of Geography, Faculty of Social Sciences, The University of Sheffield, Sheffield, UK

hotspot analysis) and finally introducing measures to mitigate these impacts.

Life Cycle Assessment (LCA) is a systematic approach that is widely used to evaluate the environmental impacts of food waste treatment options (Morris et al. 2013; Laurent et al. 2014a, b). A standardised method prescribed by the International Organisation for Standardization (BSI 2006), LCA, adopts a process-based modelling approach where a system is modelled using an inventory of processes representing inputs, outputs and potential environmental burdens of the system. The literature is abundant with studies that have used LCA to assess different food waste treatment options. However, a comparison of the results is difficult due to difference in methodological approaches, functional units, system boundaries, and variation in waste composition and geographical coverage (Khoo et al. 2010; Fruergaard and Astrup 2011; Boldrin et al. 2011; Kim et al. 2011; Grosso et al. 2012; Evangelisti et al. 2014; Laurent et al. 2014a; Eriksson et al. 2015; Buratti et al. 2015; Ahamed et al. 2016; Jeswani and Azapagic 2016; Bernstad Saraiva Schott et al. 2016).

In addition, reviewed studies suffer from inconsistency and significant variations in results (Cleary 2009; Bernstad and la Cour 2012). The inconsistency observed in food waste management LCA studies can be attributed to several LCA methodological and data-related limitations (Heijungs and Guinée 2007; Cleary 2009; Curran 2012). One of the main drawbacks of conventional process-based LCA is the difficulty of drawing a comprehensive systematic boundary around the studied system which ensures the full inclusion of all components of the system, including both direct and indirect processes. Therefore, the cutoff criteria in system boundaries could potentially lead to system incompleteness, alternatively referred to as truncation error (Suh et al. 2006). The degree of incompleteness varies with the type of system, but it can reach up to 50% in some cases (Lenzen 2001; Suh et al. 2006). Moreover, the results of studies by Lenzen (2001), and Miller and Blair (2009) suggest that even extensive process-based inventories for complex systems fail to include all direct and indirect environmental impacts.

Another factor that is often overlooked and which leads to large variations and substantial underestimations in results is the exclusion of environmental impacts associated with capital goods, a term commonly used to refer to goods, services and energy inputs required at the construction stage of a food waste treatment facility such as composting and anaerobic digestion plants and associated infrastructure. According to a review study, 88% of waste management-LCA studies exclude capital goods. Even more concerning is that 26% of these studies claim that capital goods have insignificant environmental impacts (Laurent et al. 2014a, b).The few studies which did investigate the environmental impacts of capital goods concluded that they could constitute 20% of the total environmental life cycle impacts and contribute substantially to various environmental impact categories such as abiotic resource depletion, climate change, and toxicity (Finnveden et al. 2005; Frischknecht et al. 2007; Brogaard and Christensen 2016).

To counter limitations of LCA, this study adopts a state-of-the-art approach to providing a gate-to-grave assessment of the processing of 1 tonne of household food waste (the functional unit of the study) in the UK using three well-established downstream management options: incineration; composting; and anaerobic digestion (AD). A hybrid input–output-based life cycle assessment, developed by Salemdeeb et al. (2017), was adopted in order to expand system boundaries to include capital goods and reduce truncation error. This paper then uses Monte Carlo analysis to investigate the impact of using different energy mixes (including 100% renewable energy sources) on the overall impacts.

Materials and methods

This section provides description into a novel hybrid LCA method (Sect. 2.1), lists the waste treatment assumptions and system boundaries (Sect. 2.2), outlines the life cycle inventory data sources and compilation methodology of the LCA (Sect. 2.3 and 2.4) and finally describes the methods and materials used for the sensitivity analysis, namely Monte Carlo analysis and energy use scenario analysis (Sect. 2.5 and 2.6).

The hybrid IO-LCA model

The hybrid model was developed in accordance with ISO 14040 (BSI 2006). It consists of four stages: (1) defining the goal and scope of the analysis, (2) conducting an inventory analysis, (3) aggregating inventory results into pre-defined environmental and health impact categories, and (4) interpreting the analysis results. Figure 1 illustrates the main components of the hybrid model and depicts how the IO element was integrated into the basic ISO 14040 LCA schematic process. Additional information on the hybrid IO-LCA model is provided in the online accompanying appendix.

As the hybrid LCA model combines models, two types of data are required: physical data (used to model major processes such as energy input); and monetary data (used to address the gap in inventory data and model processes—usually minor—that could have not been modelled using physical data due to the scarcity of data). These are introduced in Sect. 2.3, with further discussion provided in the online accompanying appendix (Appendix 1).

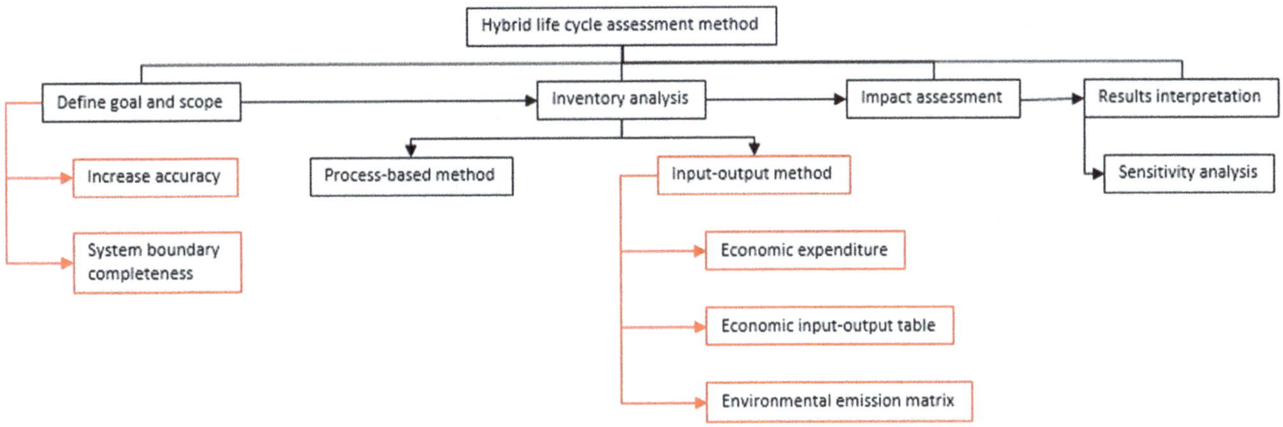

Fig. 1 A schematic illustration of the proposed hybrid LCA method. Boxes in red denote added features and elements related to the IO element of the hybrid model

Waste treatment scenarios

The analysis adopts an expanded system boundary including the construction, operation, maintenance, and decommissioning stages. The time horizon boundaries are defined by the lifespan of the waste management facility: 15 years for composting and 20 years for incineration and AD. The disposal of food waste packaging is excluded due to its insignificant impact (Lebersorger and Schneider 2011; Bernstad Saraiva Schott and Andersson 2015). Diagrams depicting system boundaries of the study options are provided in the online appendices (Figures A.2–A.4). Food waste composition (and moisture content) was defined as the average of the three previous UK compositional analyses (Banks et al. 2011, WRAP 2010, and Zhang et al. 2013) (Table A.7).

Composting

Composting is a long-established technology in the UK which is used comprehensively to treat both food and garden waste, providing a valuable fertiliser as a final output. A standard in-vessel composting process consists of three main stages. Before the food waste is processed, it enters the reception building in order to be shredded. Then, the shredded waste is transferred to stage 1 tunnels where it is aerated for 2–3 weeks. The process should achieve at least 60 °C for 48 continuous hours in order to comply with Animal By-Products Regulations. Sanitised compost is then transferred to stage 2 tunnels where the same programme is repeated. Finally, it is transferred to the maturation pad where the final product is screened and prepared for amenity or agricultural use (Hall et al. 2014). The compost utilisation efficiencies used are: 20% for N, 100% for P, and 100% for K (Andersen et al. 2010) and the compost is considered to be applied on loam soil.

Incineration

A combined heat and electricity (moving grate) incinerator was considered in this study. Gross electricity and heat efficiency considered are 13 and 26%, respectively. Although not a common practice in the UK, combined heat and power technology was considered due to its potential to contribute to the UK's 2020 renewable energy target (ECCC 2016). This assumption is also consistent with recent studies that have investigated the environmental impacts of incinerators in a UK context (Nixon et al. 2013; Jeswani and Azapagic 2016). Additional information on the moving grid incinerator is provided in the appendix.

Anaerobic digestion

Food waste is assumed to be shredded, sieved and then sent to a digestion tank. The output of this biological process is two valuable products: biogas which may be used for either heat or power generation; and digestate that may be utilised as a soil fertiliser (Hall et al. 2014). The anaerobic digestion process involves the breakdown of biodegradable material by micro-organism bacteria in an anaerobic and enclosed environment. As with composting, the anaerobic digestion process must comply with BSI PAS 100 and compost should meet "Quality Control for Compost" requirements (Environment Agency 2010). The digestate substitutes nitrogen, phosphorous and potassium fertilisers with an efficiency of 34.5, 46 and 60%, respectively. Benefits from the contribution made by sulphur, magnesium, and other organic compounds in compost are excluded due to the lack of data (Wallace 2011).

Life cycle inventory data

Monetary data

A thorough literature review, coupled with surveys and personal communications with operators of food waste facilities, was conducted to collect data on expenditure associated with technologies considered. Three leading UK waste management operators were included; these have not been referenced further due to commercial sensitivity. These compiled data were validated and cross-checked with each set to ensure it is representative of those scenarios considered. The data were converted into a basic price in order to match the input output analytical table format. Conversion coefficients used were obtained from Salemdeeb et al. (2016).

The final expenditure of the scenarios, reported in basic price and formatted according to the UK IO format, is available in the online appendix (Table A.2). The breakdown expenditure shows that the majority of the budget is allocated to construction activities; construction expenditure for compost, AD and incineration constitute up to 30, 40 and 60% of the total expenditure, respectively.

Physical data

Physical data, listed in Table 1, were either compiled or calculated based on the information from project documents, literature, or the WRATE database. Upstream and downstream material flows and emissions were collected using existing databases, primarily the Swiss ecoinvent database v2.2 (Econinvent 2014). Capital goods consumed at the construction stage of investigated technologies are also included. Tables that list approximate quantities of capital goods consumed in the construction stage and a list of ecoinvent processes used can be found in the online accompanying appendix.

Life cycle impact assessment

Emissions are aggregated into 14 environmental and health impact categories. Aggregated results were normalised in order to allow for a comparative analysis of the relative importance of each impact category (Table A.6 in the online accompanying appendix). The global warming potential and particulate matter emissions from recycling 1 tonne of food waste are, for example, scaled relative to the per capita greenhouse gas and particulate matter emissions in the year 2010 (and are reported in units of milli-Person equivalents, mPE). Normalisation factors were derived from the

Table 1 Life cycle inventory data of food waste management options

	Materials	Unit	Incineration	Composting[b]	AD[b]
Inputs					
Energy	Food waste	kg	1000	1000	1000
	Process water	kg	218.5	110.8	236
	Woodchip	kg		0.31	
	Electricity (see table)	kWh	3.75	5.78	65
	Diesel	kg	0.929	3.29	0.081
Outputs					
by-products			Electricity efficiency (13%)[c]	Compost (659 kg)	Digestate (550 kg)
			Heat efficiency (26%)		Electricity (260 kWh)[c]
Waste	Wastewater	kg	43		320
	Screening/rejected materials	kg	3	30	56
Process air emissions					
	CO_2	kg	0.003	11	0.26
	CH_4	xE−3 kg	0.05	4.80	34.0
	N_2O	xE−2 kg	0.28	2.70	1.90
	NO_x	xE−2 kg	0.10	0.10	4.40
	CO	xE−3 kg	0.16	59.0	1.50
	MVOC	xE−3 kg	0.003	6.0	24.0

Sources: [a](Kim and Kim 2010) and [b](Hall et al. 2014)

[c]Electricity efficiency is provided for Incineration, and Electricity generation (in kWh) is provided for AD. This is due to Incineration being calculated from calorific values found in the literature, while AD kWh was provided from existing real world sites. We have therefore refrained from providing exact kWh for Incineration

PROSUITE project which was developed specifically for the ILCD method (Blok et al. 2013).

Sensitivity analysis

The study follows a four-step sensitivity analysis approach based on Clavreul et al. (2012). For each food waste disposal technology, stages with the highest environmental burdens were identified using a hotspot analysis. The sensitivity of parameters within key stages was investigated by a perturbation analysis. The sensitivity ratio is calculated using Eq. 1 for all parameters, by varying each parameter by ± 10%. Parameters with the highest sensitivity ratio were selected.

$$\text{Sensitivity Ration (SR)} = \frac{\frac{\Delta\text{Result}}{\text{Initial result}}}{\frac{\Delta\text{Parameter}}{\text{Initial parameter}}} \quad (1)$$

Probability distribution coefficients were then assigned to parameters with $SR > 2$ based on the literature and the results of the "Pedigree quality matrix", an Ecoinvent-recommended method used to check the reliability and robustness of the model created by Weidema and Wesnæs (1996). Table 2 lists sensitivity analysis parameters for all scenarios.

A Monte Carlo analysis was then performed to generate confidence intervals. For each metric, the significance of differences between technologies was also tested using the Monte Carlo method. The model randomly sampled estimates of the mean for each technology and calculated the difference between each, repeating this resampling 1000 times. The analysis tests the possibility of difference between technologies to overlap with zero at a 99% confidence level.

Energy use scenarios

Four electricity-input scenarios were modelled to investigate changes that result due to future policy decisions and energy technology development (Fig. 2). The baseline scenario represented the 2014 UK electricity national grid (DECC 2014). The other three scenarios (scenarios two to four) are based on potential electricity mix pathways (Stamford and Azapagic 2014): scenarios two and three are based on the decarbonisation of electricity by 65% by 2050 (i.e. Ref (65-A) without nuclear energy, and Ref (65-B) with nucelar energy); and scenario four includes 100% of renewable energy with nuclear energy (40%) and offshore wind (23%) (Ref 100). The values used in each electricity mix pathway scenario are based on the average contribution of each energy source to each scenario from 2014 to 2070. As each scenario is the average energy generation mix for this scenario from 2014 to 2070, there is some coal and other non-renewable contribution to the (Ref 100).

Results and discussion

The results of the analysis are summarised in Fig. 3. Overall, composting achieved the best score for 7 out of 14 environmental and health impacts, while AD had the highest environmental burden in 6 impacts and the mean ranking of the three scenarios (1 = best, 3 = worst) were compost: 1.7, AD: 2.1, and incineration: 2.3. A hotspot analysis that shows the contribution of different life cycle stages of each scenario is shown in Fig. 4.

For composting, the operation stage contributed the most to the majority of impacts, in particular depletion of fossil fuels (60%), terrestrial acidification (60%), terrestrial eutrophication (80%), and ozone depletion (90%). The environmental impacts of compost could be offset by its use as a substitute to synthetic fertiliser, the most significant being climate change (55%), freshwater eutrophication (95%), depletion of abiotic resources elements (48%). The treatment of food waste contributes to the majority of environmental burdens associated with AD. These impacts are attributed primarily to two processes, namely the use of auxiliary materials and wastewater treatment. Construction has a slight impact especially regarding eco-toxicity at 17% and carcinogenic human toxicity carcinogenic at 30%. Most burdens are offset by the use of recovered energy to substitute conventional energy sources (i.e. natural gas and coal).

Incineration has a similar environmental hotspot source to those of composting and AD: operation is the stage with the highest environmental burden while energy recovery (both heat and electricity) plays a key role in offsetting these burdens. The significant contribution from operation is the auxiliary materials used in order to control emissions, ammonia injection to control NO_x emissions, lime for control of SO_2 and HCL and activated carbon to capture heavy metals.

The supremacy of composting's results in the majority of the investigated impacts is largely due to the nature of composting itself: it is a technologically simple process that requires considerably low energy and auxiliary material inputs compared to either AD or incineration. In addition, the analysis results show that the majority of the burden in AD and incineration is associated with the treatment of by-products, namely wastewater and methane purification for AD, and Air Pollution Control (APC) for incineration.

The exclusion of the impacts of food collection and transportation in this study also works in favour of composting and AD, as these options require a separate collection system of food waste and, therefore, additional energy input. When treated via incineration, food waste is co-collected and does not require an additional separate food

Table 2 Sensitivity analysis parameters for all scenarios

Scenario	Stage	Process	Unit	Distribution type	Average	Deviation[a]
Incineration	Construction	Tar	kg	Uniform	3.7E−04	1.9E−04
		Steel	kg	Uniform	1.0E−03	6.9E−04
		Cooper	kg	Uniform	1.1E−05	5.9E−06
		Concrete	m^3	Uniform	2.4E−06	2.6E−06
	Operation	Electricity consumption[b]	kWh	Log-normal	3.8E−03	4.0E−03
		Activated carbon	kg	Normal	2.9E−04	1.6E−04
		Diesel consumption	kg	Normal	2.7E−03	4.4E−03
		Urea liquid gas cleaning	kg	Normal	8.6E−04	6.5E−04
		Lime (gas cleaning)	kg	Normal	1.2E−02	2.0E−03
		N_2O	kg	Uniform	2.1E−05	4.1E+00
		SO_2	kg	Uniform	2.9E−05	4.1E+00
		Ammonia	kg	Uniform	6.6E−06	2.1E+00
		Nitrogen oxides NO	kg	Uniform	6.9E−04	2.3E+00
		Carbon monoxide-biogenic	kg	Uniform	3.4E−05	3.5E+00
		Carbon dioxide	kg	Uniform	3.6E−03	4.6E+00
		Hydrogen chloride	kg	Uniform	7.0E−06	1.3E+01
		Hydrogen fluoride	kg	Uniform	5.1E−08	1.4E+01
	Energy recovery	caloric value	MJ	Uniform	6.7E+00	1.9E+00
		Gross electricity efficiency	%	Uniform	1.3E−01	2.0E−02
		Gross heat efficiency	%	Uniform	2.6E−01	5.0E−02
Anaerobic digestion	Construction	Steel	kg	Uniform	3.6E−04	2.7E−04
		Concrete	m[c]	Uniform	2.6E−06	2.7E−06
		Bitumen	kg	Uniform	8.2E−05	1.0E−04
		Polyethylene	kg	Uniform	7.3E−04	7.4E−04
	Operation	Electricity consumption[b]	kWh	Log-Normal	4.6E−02	2.1E+00
		Lime[d]	kg	Normal	1.3E−03	1.3E−04
		Inorganic chemicals[d]	kg	Normal	4.9E−03	4.9E−04
	Digestate	Potassium sulphate, as K_2O, at regional storehouse, RER	kg	Normal	− 1.9E−03	1.9E−04
		Ammonium nitrate phosphate, as N, at regional storehouse, RER	kg	Normal	− 7.1E−04	7.1E−05
	Energy Recovery	Electricity recovery	kWh	Normal	2.6E−01	1.0E−01
Composting	Construction	Steel	kg	Uniform	1.2E−03	4.8E−04
		Concrete	m[c]	Uniform	4.1E−06	3.7E−06
		Aluminum	kg	Uniform	1.0E−04	2.1E−04
		Polyethylene	kg	Uniform	2.2E−05	2.5E−05
	Operation	Electricity consumption[b]	kWh	Log-normal	1.4E−03	2.0E+00
		Diesel	kg	Log-normal	8.7E−03	2.0E+00
	Output utilisation	N fertilizer substitution	kg	Uniform	4.0E−01	2.0E−01

[a]Standard deviation for normal distribution, geometric standard deviation for log-normal distribution, or half width for uniform distribution

[b]GSD was calculated based on average values of IVC and AD

[c]Coefficient of variation is assumed to be 20% of the average value

[d]Coefficient of variation is assumed to be 10% of the average value

waste collection system. Therefore, incineration consumes less energy input compared to composting and AD. Food collection and transportation are significant for the depletion of fossil fuels, elements, and the ozone layer (Jeswani and Azapagic 2016). A separate food collection system would increase the environmental burden across these categories (Burnley et al. 2011). The lack of food collection and transportation in this study is due to this study being a gate-to-grave assessment, meaning food waste collection and transportation were not included in the analysis.

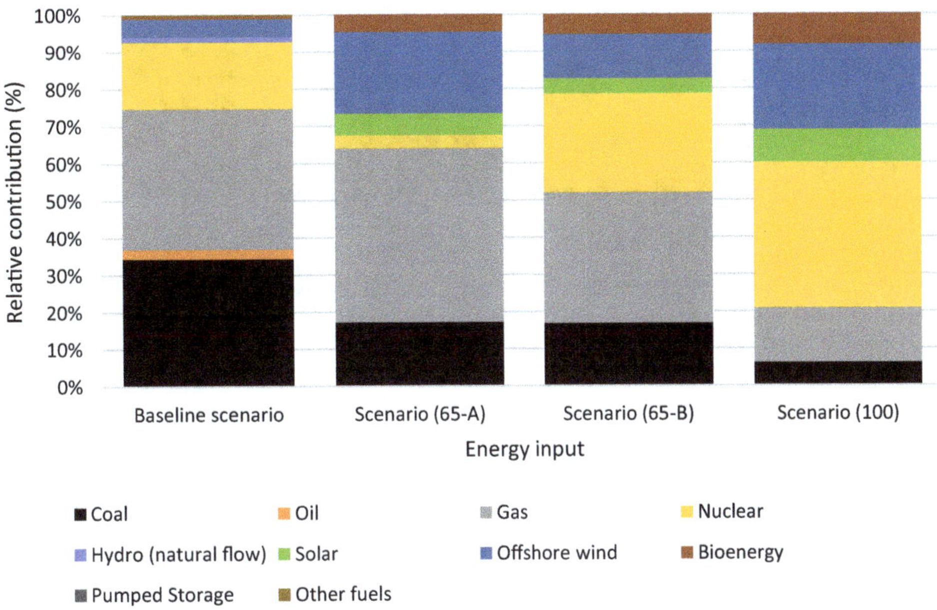

Fig. 2 Electricity-input scenarios considered in this study, adapted from Stamford and Azapagic (2014). The figure above shows the average contribution of each energy source to each scenario from 2014 to 2070

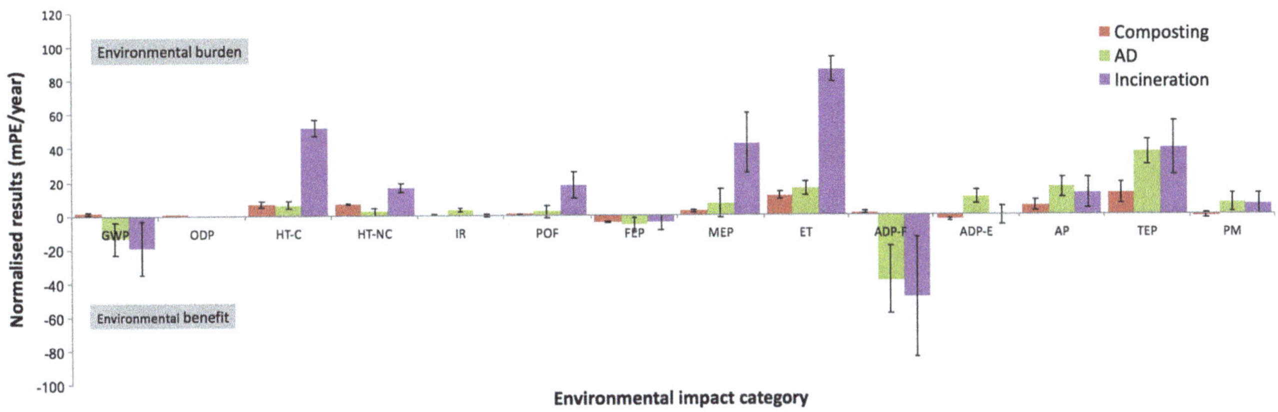

Fig. 3 The normalised environmental and health impacts of three food waste downstream technologies: incineration, composting and AD. Units (mPE) relate a process' emissions in relation to per capita emissions in the EU in 2010. *GWP* global warming potential, *ODP* ozone depletion, *HT-C* emissions of carcinogens, *HT-NC* emissions of non-carcinogenic toxins, *IR* ionising radiation, *POF* photochemical oxidant formation, *FEP* freshwater eutrophication, *MEP* marine eutrophication, *ET* eco-toxicity, *ADP-F* depletion of fossil fuels, *ADP-E* depletion of non-fossil fuel abiotic resources, *AP* terrestrial acidification, *TEP* terrestrial eutrophication, *PM* particulate matter emissions. Error bars show one standard deviation

This was due to the fact that the goal of the assessment was to investigate the environmental impacts of food waste management technologies rather than that of the overall waste management system. Future studies could include this larger system boundary in their analysis.

Although the results of the analysis tend to show that composting is more environmentally friendly than other options, it performs worse than AD in two key environmental impacts: climate change and depletion of fossil fuels. This is attributed primarily to the fact that, unlike AD and incineration options, composting process does not include the generation of energy (in the form of electricity or heat) and, therefore, does not offset the huge quantities of emissions as AD and incineration do. The management of 1 tonne of food waste in a composting facility could emit 74 kg CO_2-eq./FU while the AD and incineration options achieve significant benefits for the same functional unit; estimated average reductions in AD and incineration being − 2400 kg CO_2-eq. and − 3000 kg CO_2-eq., respectively. This unsurprising result is attributed to the fossil fuel energy substituted by the production of energy (in the form of electricity, heat, or both) in both AD and incineration options. If a decarbonised energy mix is substituted, the impacts of composting are reduced.

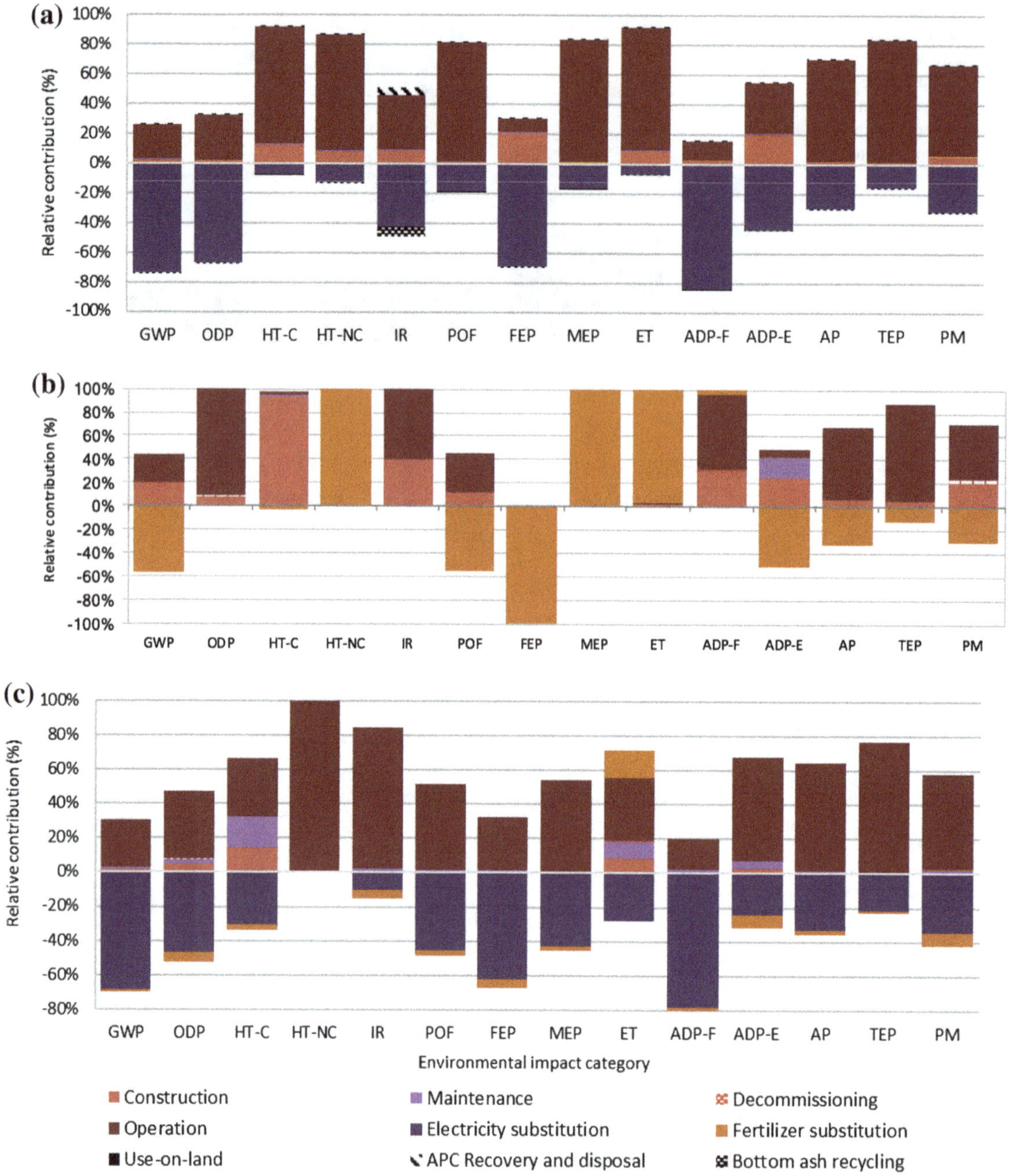

Fig. 4 Hotspot analysis of the analysis results for incineration (**a**), composting (**b**), and AD (**c**).Abbreviations are as follows: *GWP* global warming potential, *ODP* ozone depletion, *HT-C* emissions of carcinogens, *HT-NC* emissions of non-carcinogenic toxins, *IR* ionising radiation, *POF* photochemical oxidant formation, *FEP* freshwater eutrophication, *MEP* marine eutrophication, *ET* eco-toxicity, *ADP-F* depletion of fossil fuels, *ADP-E* depletion of non-fossil fuel abiotic resources, *AP* terrestrial acidification, *TEP* terrestrial eutrophication, *PM* particulate matter emissions

We have assumed the composition and water content of UK food waste via previous studies (Table A.7). As the UK decarbonises, the UK diet may change. This may affect the quality of UK food waste as a feedstock—specifically the water content. A variation in water content will have a large impact on the three treatment methods studied. In case of Incineration, there might not be any positive energy output if the moisture content of food waste goes beyond 50% since more energy is spent in drying which would directly affect the ADP-F (Fig. 3). Similarly the C, H, N, O, S ratio will also have their direct impacts on the environmental benefits and burdens studied here. Future studies could further modify UK food waste and energy generation to match dietary transition.

This study has also not included a comparative cost analysis of the different treatment methods to complement its environmental analysis. Future research could provide a comparative cost analysis of treatment methods under existing and potential decarbonisation scenarios. Furthermore, the study covers three well-established treatment technologies; other food waste treatment technologies are in their "infancy" in the UK waste market and therefore have been excluded. Future research could also broaden scope to compare other waste treatment methods under existing and potential decarbonisation scenarios.

Hybrid LCA vs truncation error

The results of the study demonstrate the ability of the hybrid LCA model to include additional processes and, therefore, reduce truncation error, compared to a conventional process-based LCA method. The adoption of the hybrid model increases estimates of associated GHG emissions with composting, AD and incineration by approximately 26, 10 and 11%, respectively. Thus, the analysis results quantitatively confirm the ability of the hybrid model to reduce truncation error by expanding boundaries of the modelled scenario (Finnveden et al. 2009; Jeswani et al. 2010).

Additional environmental burdens captured are mainly associated with managerial and indirect services such as services of head offices, consulting services, third-party technical services, and testing and maintenance services. These service-based activities are difficult to model using LCA (and, therefore, ignored in the literature), yet contribute substantially to the overall environmental impacts of modelled scenarios. The hybrid model uses the expenditure cost of these activities to estimate their environmental impacts. These expenditure costs are estimated to constitute up to 12% for AD, 10% for compost and 34% for incineration in the construction stage. Expenditure on these services increases for day-to-day operation activities; it accounts for 40% of total expenditure on AD, 67% for compost and 44% for incineration.

The environmental impacts of capital goods

Results show that capital goods constitute up to 20% of the overall environmental impacts in 10 categories for compost and 4 categories for both AD and INC. The study findings support those of previous studies highlighting the significant environmental impacts of capital goods (Finnveden et al. 2005; Brogaard and Christensen 2016). The overall environmental burdens associated with capital goods are presented in Figure A.5.

Capital goods used for the composting scenario contributed mostly to the potential impacts on carcinogenic human toxicity (70%), climate change (45%) and eco-toxicity (40%). These impacts were caused primarily due to the use of cement and steel: the energy input in the production stage of these products contribute the most to climate change while the disposal of steel and cement slag is responsible for the emission of heavy metals and, therefore, contributes to both human toxicity and eco-toxicity (Burchart-Korol 2013; Salas et al. 2016). Significant toxicity impacts are also reported in AD and incineration due to capital goods. The production of steel contributes to both human toxicity and eco-toxicity by 50 and 45% for AD and 13 and 10% for incineration, respectively. In addition, a substantial impact on depletion of abiotic resources (elements) is estimated for both AD and incineration scenarios. This could be explained by the huge quantities of capital goods (in particular steel) required in the construction stage of AD and incineration as these infrastructures are large-scale construction projects compared to composting (Brogaard and Christensen 2016). Significant results were also reported in freshwater eutrophication impacts, in particular in the incineration scenario. This highest impact across all 14 environmental and health categories is caused by excessive use of copper, stainless steel and cement. This finding agrees with those of Brogaard and Christensen (2016) that report a 90% freshwater eutrophication impact due to capital goods.

Comparison with previous literature

Figure 5 provides comparison of ten LCA studies (including this study), reporting greenhouse gas emissions per tonne of food waste processed. The previous studies reported a large variation in the GWP and some broad patterns emerge. Four out of six studies conclude that incineration with energy recovery has the greatest environmental benefits while only two studies conclude in favour of AD. This could be explained due to the high energy input required to operate an AD facility (compared to incineration) and the additional diesel consumption required to set up a separate food waste collection system (Burnley et al. 2011). In addition, variations in the biogas yield and the type of substituted energy adopted in reviewed studies contribute significantly to these discrepancies. For example, Eriksson et al. (2015) reports AD results which are 4 times larger than those reported in this study. This substantial difference could be attributed due to the assumption made by Eriksson and his colleagues that the entire theoretical yield of biogas was produced, while this study is based on actual AD plant figures (see Table 2) Eriksson et al.'s study also assumes that biogas replaces diesel as a fuel for city buses, while this study assumes biogas substitutes UK electricity production by natural gas (61.5%) and coal (38.5%).

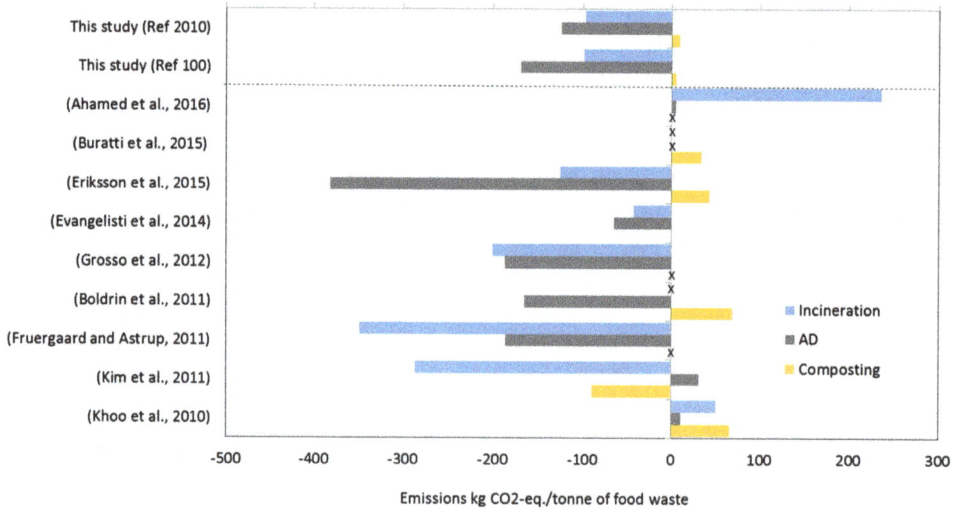

Fig. 5 Results of ten LCA studies (including this study: the energy mix from 2010 (Ref 2010), and 100% of renewable energy with nuclear energy (Ref 100)) reporting greenhouse gas emissions per tonne of food waste. Crosses are marked where a study has not included a technology in their analysis. Where a study reported several scenarios for the same option, the mean value for all scenarios is shown. The averaged Monte Carlo results of this study are shown with bars showing the maximum and minimum results found by Monte Carlo analysis

Table 3 Characterised results of impact categories for the study scenarios

		Disposal technologies					
		Compost		AD		Incineration	
Impact category	Unit	Mean	SD	Mean	SD	Mean	SD
Climate Change	(xE+2 kg CO_2e)	0.1	0.043	− 1.1	0.77	− 1.5	1.3
Stratospheric Ozone Depletion	(xE−7 kg CFC-11e)	6	4.5	− 2.1	13	− 85	86
Human Toxicity, Cancer Effect	(xE−7 CTU)	− 3.5#	0.99	− 3.4#	1.3	28	2.6
Human Toxicity, non-Cancer Effect	(xE−6 CTU)	7.6	0.35	2.5	2.4	18	2.9
Ionizing Radiation, Human Health	(kg U235e)	− 0.13	0.27	0.04	0.02	− 0.37	0.89
Photochemical Ozone Formation	(kg NMVOC)	0.04	0.02	0.12	0.21	1	0.44
Freshwater Eutrophication	(xE−3 kg Pe)	− 2.8#	0.34	− 3.5	2.5	− 2.80#	2.8
Marine Eutrophication	(xE−2 kg Ne)	1.8	0.89	6.5	7.8	40	16
Freshwater Eco-toxicity	(CTU)	7.5	1.4	10	2.8	57	4.8
Depletion of Abiotic Resources-Fossil	(MJ)	74	46	− 2400	1200	− 3000	2200
Depletion of Abiotic Resources-Elements	(xE−4 kg an.-e)	− 1.1	0.22	3.6	1.4	− 0.26	1.9
Acidification	(xE−4 AE)	2.5	1.6	7.9	3	6.3	4.5
Terrestrial Eutrophication	(AE)	1.4	0.72	4.2	0.87	4.5	1.8
Particulate Matter	(xE−2 kgPM2.5e)	− 0.28	0.39	1.9	1.6	1.6	1.7

Both mean and standard deviation (SD) are provided

Abbreviations are listed under Fig. 3

All differences between technologies are statistically significant ($p < 0.01$), except those marked by a #

Robustness of results

The sensitivity of results was investigated by way of a Monte Carlo Analysis. Overall results, listed in Table 3 below and plotted as error bars in Fig. 3, reveal a high level of uncertainty that could lead to a change in the ranking of scenarios in 7 impact categories between AD and incineration, 4 impact categories between AD and composting, 6 impacts between incineration and composting, and 2 impact categories amongst all scenarios studied (i.e., freshwater eutrophication and ozone depletion). The high level of uncertainty makes it difficult to draw a generic conclusion. However, this study has helped to better understand environmental impact patterns. This analysis also highlights the importance of the quality of data used in order to conduct an environmental assessment of a specific technology.

Notwithstanding the large variability in some parameters as shown in Table 3, the indicator values for all metrics are statistically significant from one another ($p < 0.01$), except for the effect of composting and anaerobic digestion on marine eutrophication and non-carcinogenic toxicity.

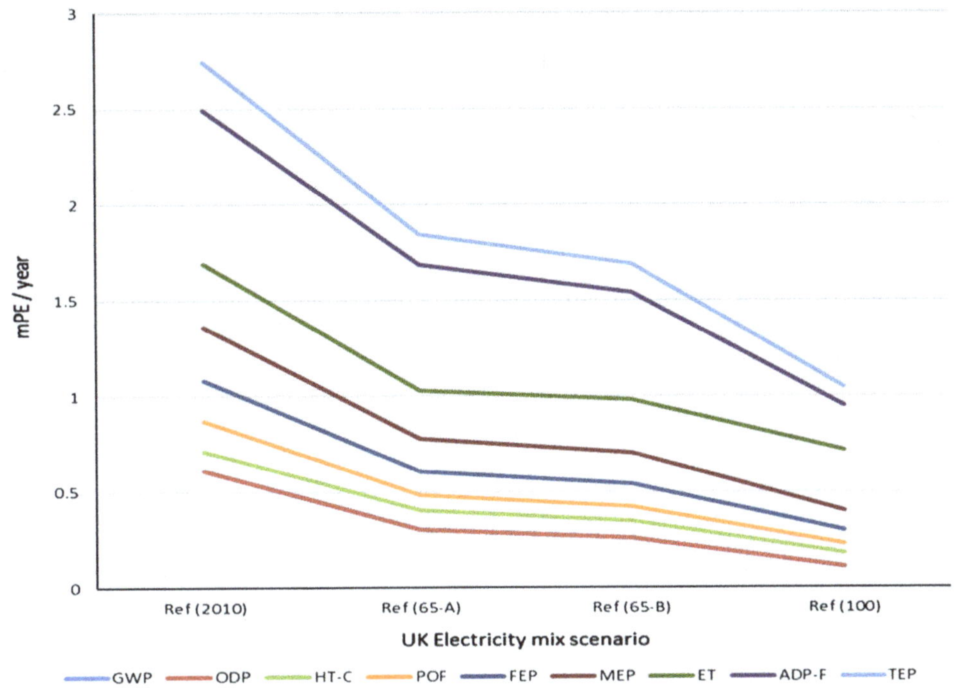

Fig. 6 Normalised environmental burdens of different energy scenarios: 2010 UK electricity mix (Ref 2010), 65% of renewables without nuclear energy (Ref 65-A), and with nuclear energy (Ref 65-B), and 100% of renewable energy with nuclear energy (Ref 100). Abbreviations are listed under Fig. 3

Scenario analysis

Results indicate that the decarbonisation of the UK supply electricity mix leads to reduction in overall environmental burdens across impact categories (Fig. 6). Scenario 4 (Ref100) has reduced the overall impact of global warming potential (GWP) by 82%, while scenario 2 (Ref 65-A) leads to 50% reduction. With regard to Scenario 3 (Ref 65-B), the substitution of natural gas by nuclear energy as a source of low-GHG energy leads to an additional 16% reduction in GWP but does increase the environmental burden in four impact categories: non-carcinogenic human toxicity, ionising radiation (IR), depletion of elements (ADP-E) and particulate matter.

Conclusion

This study introduced a novel hybrid IO-LCA model to evaluate the environmental impacts of downstream waste management infrastructure as they operate currently, or as they could operate under future decarbonised energy use scenarios. Results provided a deeper insight into the overall environmental performance of downstream food waste treatment options. More importantly, the study shows how the decarbonisation of the UK national grid would make composting to look a more attractive treatment option than incineration and AD. The results of this study demonstrate the superiority of hybrid IO LCA as it includes additional processes and reduces truncation error increasing overall captured environmental impacts of composting (26%), AD (10%), and incineration (11%). In addition, the study confirms the significant importance of including capital goods in the list of overall environmental impacts and expanding the system boundaries of the system using hybrid LCA analysis.

Acknowledgements The PhD research project of the RS is funded by the IDB Cambridge International Scholarship. CR is funded from the HEFCE Catalyst-funded N8 AgriFood Resilience Programme and matched funding from the N8 group of Universities.

References

Ahamed A, Yin K, Ng BJH et al (2016) Life cycle assessment of the present and proposed food waste management technologies from environmental and economic impact perspectives. J Clean Prod 131:607–614. https://doi.org/10.1016/j.jclepro.2016.04.127

Andersen JK, Boldrin A, Christensen TH, Scheutz C (2010) Greenhouse gas emissions from home composting of organic household waste. Waste Manag 30:2475–2482. https://doi.org/10.1016/j.wasman.2010.07.004

Banks CJ, Zhang Y, Walker M (2011) Optimising processes for the stable operation of food waste digestion: technical report Defra WR1208. Department for Communities and Local Government

Bernstad A, la Cour Jansen J (2012) Review of comparative LCAs of food waste management systems—Current status and potential improvements. Waste Manag 32:2439–2455. https://doi.org/10.1016/j.wasman.2012.07.023

Bernstad Saraiva Schott A, Andersson T (2015) Food waste minimization from a life-cycle perspective. J Environ Manage 147:219–226. https://doi.org/10.1016/j.jenvman.2014.07.048

Bernstad Saraiva Schott A, Wenzel H, la Cour Jansen J (2016) Identification of decisive factors for greenhouse gas emissions in comparative lifecycle assessments of food waste management—An analytical review. J Clean Prod 119:13–24. https://doi.org/10.1016/j.jclepro.2016.01.079

Blok K, Huijbregts M, Roes L, et al (2013) Development and application of a standardized methodology for the PROspective SUsta İ nability assessment of TEchnologies (PROSU İTE) Report prepared within the EC 7th Framework Project, no: 227078. European Commission, Brussels

Boldrin A, Neidel TL, Damgaard A et al (2011) Modelling of environmental impacts from biological treatment of organic municipal waste in EASEWASTE. Waste Manag 31:619–630. https://doi.org/10.1016/j.wasman.2010.10.025

Brogaard LK, Christensen TH (2016) Life cycle assessment of capital goods in waste management systems. Waste Manag 56:561–574. https://doi.org/10.1016/j.wasman.2016.07.037

BSI (2006) BS EN ISO 14044: environmental management—Life cycle assessment—requirements and guidelines. British Standards Institution, London

Buratti C, Barbanera M, Testarmata F, Fantozzi F (2015) Life cycle assessment of organic waste management strategies: an Italian case study. J Clean Prod 89:125–136. https://doi.org/10.1016/j.jclepro.2014.11.012

Burchart-Korol D (2013) Life cycle assessment of steel production in Poland: a case study. J Clean Prod 54:235–243. https://doi.org/10.1016/j.jclepro.2013.04.031

Burnley S, Phillips R, Coleman T, Rampling T (2011) Energy implications of the thermal recovery of biodegradable municipal waste materials in the United Kingdom. Waste Manag 31:1949–1959. https://doi.org/10.1016/j.wasman.2011.04.015

Clavreul J, Guyonnet D, Christensen TH (2012) Quantifying uncertainty in LCA-modelling of waste management systems. Waste Manag 32:2482–2495. https://doi.org/10.1016/j.wasman.2012.07.008

Cleary J (2009) Life cycle assessments of municipal solid waste management systems: a comparative analysis of selected peer-reviewed literature. Environ Int 35:1256–1266. https://doi.org/10.1016/j.envint.2009.07.009

Curran MA (2012) Life cycle assessment handbook: a guide for environmentally sustainable Products, 1st edn. Scrivener Publishing LLC, Massachusetts

DECC (2014) Digest of United Kingdom energy statistics (DUKES). In: Department of Energy Climate Change

ECCC (2016) 2020 renewable heat and transport targets. House of Commons, London

Econinvent (2014) Swiss Centre for life cycle inventories: Ecoinvent (V2.2). St-Gallen, Switzerland

Eriksson M, Strid I, Hansson P (2015) Carbon footprint of food waste management options in the waste hierarchy—a Swedish case study. J Clean Prod 93:115–125. https://doi.org/10.1016/j.jclepro.2015.01.026

European Commission (2016) Paris agreement—European Commission. http://ec.europa.eu/clima/policies/international/negotiations/paris/index_en.htm. Accessed 8 Mar 2016

Evangelisti S, Lettieri P, Borello D, Clift R (2014) Life cycle assessment of energy from waste via anaerobic digestion: a UK case study. Waste Manag 34:226–237. https://doi.org/10.1016/j.wasman.2013.09.013

FAO (2013) Food Wastage Footprint. Italy, Rome

Finnveden G, Johansson J, Lind P, Moberg Å (2005) Life cycle assessment of energy from solid waste—part 1: general methodology and results. J Clean Prod 13:213–229. https://doi.org/10.1016/j.jclepro.2004.02.023

Finnveden G, Hauschild MZ, Ekvall T et al (2009) Recent developments in life cycle assessment. J Environ Manage 91:1–21. https://doi.org/10.1016/j.jenvman.2009.06.018

Frischknecht R, Althaus H, Bauer C et al (2007) The environmental relevance of capital goods in life cycle assessments of products and services. Int J LCA 2007:7–17

Fruergaard T, Astrup T (2011) Optimal utilization of waste-to-energy in an LCA perspective. Waste Manag 31:572–582. https://doi.org/10.1016/j.wasman.2010.09.009

Grosso M, Nava C, Testori R et al (2012) The implementation of anaerobic digestion of food waste in a highly populated urban area: an LCA evaluation. Waste Manag Res 30:78–87. https://doi.org/10.1177/0734242x12453611

Hall D, Turnbull R, Coleman T et al (2014) WRATE: user manual. Golder Associates Ltd, London

Heijungs R, Guinée JB (2007) Allocation and "what-if" scenarios in life cycle assessment of waste management systems. Waste Manag 27:997–1005. https://doi.org/10.1016/j.wasman.2007.02.013

Jeswani HK, Azapagic A (2016) Assessing the environmental sustainability of energy recovery from municipal solid waste in the UK. Waste Manag 50:346–363. https://doi.org/10.1016/j.wasman.2016.02.010

Jeswani HK, Azapagic A, Schepelmann P, Ritthoff M (2010) Options for broadening and deepening the LCA approaches. J Clean Prod 18:120–127. https://doi.org/10.1016/j.jclepro.2009.09.023

Khoo HH, Lim TZ, Tan RBH (2010) Food waste conversion options in Singapore: environmental impacts based on an LCA perspective. Sci Total Environ 408:1367–1373. https://doi.org/10.1016/j.scitotenv.2009.10.072

Kim M-H, Kim J-W (2010) Comparison through a LCA evaluation analysis of food waste disposal options from the perspective of global warming and resource recovery. Sci Total Environ 408(19):3998–4006

Kim M-HH, Song Y-EE, Song H-BB et al (2011) Evaluation of food waste disposal options by LCC analysis from the perspective of global warming: Jungnang case, South Korea. Waste Manag 31:2112–2120. https://doi.org/10.1016/j.wasman.2011.04.019

Laurent A, Christensen T, Bakas I (2014a) Review of LCA studies of solid waste management systems—part II: methodological guidance for a better practice. Waste Manag 34:589–606. https://doi.org/10.1016/j.wasman.2013.12.004

Laurent A, Clavreul J, Bernstad A et al (2014b) Review of LCA studies of solid waste management systems—part I: lessons learned and perspectives. Waste Manag 34:573–588. https://doi.org/10.1016/j.wasman.2013.10.045

Lebersorger S, Schneider F (2011) Discussion on the methodology for determining food waste in household waste composition studies. Waste Manag 31:1924–1933. https://doi.org/10.1016/j.wasman.2011.05.023

Lenzen M (2001) Errors in conventional and input–output—based life—cycle inventories. J Ind Ecol 4:127–148. https://doi.org/10.1162/10881980052541981

Miller RE, Blair PD (2009) Input-output analysis: foundations and extensions, 2nd edn. Cambridge University Press, Cambridge

Morris J, Scott Matthews H, Morawski C (2013) Review and meta-analysis of 82 studies on end-of-life management methods for source separated organics. Waste Manag 33:545–551. https://doi.org/10.1016/j.wasman.2012.08.004

Nixon JD, Wright DG, Dey PK et al (2013) A comparative assessment of waste incinerators in the UK. Waste Manag 33:2234–2244. https://doi.org/10.1016/j.wasman.2013.08.001

Parry A (2014) UK food waste—historical changes and how amounts might be influenced in the future. Banbury, UK

Quested T, Andrew P (2017) Household food waste in the UK, 2015. Banbury

Salas DA, Ramirez AD, Rodríguez CR et al (2016) Environmental impacts, life cycle assessment and potential improvement measures for cement production: a literature review. J Clean Prod 113:114–122. https://doi.org/10.1016/j.jclepro.2015.11.078

Salemdeeb R, Al-tabbaa A, Reynolds C (2016) The UK waste input—output table: linking waste generation to the UK economy. Waste

Manag Res 34:1089–1094. https://doi.org/10.1177/0734242x16658545

Salemdeeb R, zu Ermgassen EKHJ, Kim MH et al (2017) Environmental and health impacts of using food waste as animal feed: a comparative analysis of food waste management options. J Clean Prod 140:871–880. https://doi.org/10.1016/j.jclepro.2016.05.049

Stamford L, Azapagic A (2014) Life cycle sustainability assessment of UK electricity scenarios to 2070. Energy Sustain Dev 23:194–211. https://doi.org/10.1016/j.esd.2014.09.008

Suh S, Lenzen M, Treloar GJ et al (2006) System boundary selection in life-cycle inventories using hybrid approaches. Environ Sci Technol 38:657–664

United Nations (2015) Adoption of the Paris Agreement. Paris

Wallace P (2011) Digestates: realising the fertiliser benefits for crops and grassland. Waste and Resources Action Programme, Banbury

Weidema BP, Wesnæs MS (1996) Data quality management for life cycle inventories—an example of using data quality indicators. J Clean Prod 4:167–174. https://doi.org/10.1016/s0959-6526(96)00043-1

WRAP (2010) Chemical characterisation of foodwastes collected from welsh local. Waste and Resources Action Programme, Banbury

Zhang Y, Arnold R, Paavola T, Vaz F (2013) Compositional analysis of food waste entering the source segregation stream in four European regions and implications for valorisation via anaerobic digestion. In: Fourteenth international waste management and landfill symposium. CISA Publisher, Cagliari

Permissions

All chapters in this book were first published in IJROWA, by Springer; hereby published with permission under the Creative Commons Attribution License or equivalent. Every chapter published in this book has been scrutinized by our experts. Their significance has been extensively debated. The topics covered herein carry significant findings which will fuel the growth of the discipline. They may even be implemented as practical applications or may be referred to as a beginning point for another development.

The contributors of this book come from diverse backgrounds, making this book a truly international effort. This book will bring forth new frontiers with its revolutionizing research information and detailed analysis of the nascent developments around the world.

We would like to thank all the contributing authors for lending their expertise to make the book truly unique. They have played a crucial role in the development of this book. Without their invaluable contributions this book wouldn't have been possible. They have made vital efforts to compile up to date information on the varied aspects of this subject to make this book a valuable addition to the collection of many professionals and students.

This book was conceptualized with the vision of imparting up-to-date information and advanced data in this field. To ensure the same, a matchless editorial board was set up. Every individual on the board went through rigorous rounds of assessment to prove their worth. After which they invested a large part of their time researching and compiling the most relevant data for our readers.

The editorial board has been involved in producing this book since its inception. They have spent rigorous hours researching and exploring the diverse topics which have resulted in the successful publishing of this book. They have passed on their knowledge of decades through this book. To expedite this challenging task, the publisher supported the team at every step. A small team of assistant editors was also appointed to further simplify the editing procedure and attain best results for the readers.

Apart from the editorial board, the designing team has also invested a significant amount of their time in understanding the subject and creating the most relevant covers. They scrutinized every image to scout for the most suitable representation of the subject and create an appropriate cover for the book.

The publishing team has been an ardent support to the editorial, designing and production team. Their endless efforts to recruit the best for this project, has resulted in the accomplishment of this book. They are a veteran in the field of academics and their pool of knowledge is as vast as their experience in printing. Their expertise and guidance has proved useful at every step. Their uncompromising quality standards have made this book an exceptional effort. Their encouragement from time to time has been an inspiration for everyone.

The publisher and the editorial board hope that this book will prove to be a valuable piece of knowledge for researchers, students, practitioners and scholars across the globe.

List of Contributors

F. I. Achuba
Department of Biochemistry, Delta State University, PMB 1, Abraka, Nigeria

M. O. Ja-anni
Department of Biochemistry, Delta State University, Abraka, Nigeria

Aniek Iriany
The Department of Agrotechnology, Faculty of Agriculture and Animal Husbandry, University of Muhammadiyah Malang, Malang, Indonesia

Mochammad Chanan
The Department of Forestry, Faculty of Agriculture and Animal Husbandry, University of Muhammadiyah Malang, Malang, Indonesia

Gunomo Djoyowasito
The Department of Agricultural Engineering, Faculty of Agricultural Technology, Brawijya University of Malang, Malang, Indonesia

Girish Chander, Suhas P. Wani, S. Gopalakrishnan, Ankita Mahapatra, Swati Chaudhury, C. S. Pawar, Manoj Kaushal and A. V. R. Kesava Rao
Asia, International Crops Research Institute for the Semi-Arid Tropics (ICRISAT), Patancheru, Telangana 502324, India

Hassan El-Zeadani, Mohemed Essalem and Alsanousi Alghali
Department of Crops, Faculty of Agriculture, Sabha University, Sabha, Libya

Jamal Abubaker
Department of Soil and Water, Faculty of Agriculture, Sabha University, Sabha, Libya

T. Gienau and S. Rosenberger
Faculty of Engineering and Computer Science, Osnabrück University of Applied Science, Albrechtstraße 30, 49076 Osnabrück, Germany

U. Brüß
A3 Water Solutions GmbH, Boschstraße 2, 48369 Saerbeck, Germany

M. Kraume
Chair of Chemical and Process Engineering, Technische Universität Berlin, Str. des 17. Juni 135, FH 6-1, 10623 Berlin, Germany

Raya Al-Farsi, Mushtaque Ahmed and Ahmed Al-Busaidi
College of Agricultural and Marine Sciences, Sultan Qaboos University, Muscat, Oman

B. S. Choudri
Center for Environmental Studies and Research, Sultan Qaboos University, Muscat, Oman

J. C. Joardar and M. M. Rahman
Soil, Water and Environment Discipline, Khulna University, Khulna 9208, Bangladesh

Shadia A. Fathy and Magda K. Ezz
Biochemistry Department, Faculty of Science, Ain Shams University, Cairo, Egypt

Abeer E. Mahmoud, Mona M. Rashad and Amira T. Mohammed
Biochemistry Department, Division of Genetic Engineering and Biotechnology Researches, National Research Centre, Dokki, Giza 12622, Egypt

Ramom Rachide Nunes
Academic Unit of Serra Talhada, Federal Rural University of Pernambuco, Serra Talhada, PE, Brazil

Lívia Botacini Favoretto Pigatin, Ticiane Silva Oliveira, Rhaissa Mecca Bontempi and Maria Olímpia Oliveira Rezende
São Carlos Institute of Chemistry, University of São Paulo, São Carlos, Brazil

Najla Dhen, Amin Zouba and Bouthaina AlMohandes Dridi
Laboratory of Biochemistry and Environmental Toxicology, High Institute of Agronomy of Chott-Mariem, University of Sousse, 4042 Chott Mariem, Tunisia

Safa ben Abed
Laboratory of Dry Land Farming and Oasis Cropping, Arid Regions Institute, Medenine, Tunisia

Faouzi Haouala
Department of Biology, College of Sciences, Al Imam Mohammad Ibn Saud Islamic University (IMSIU), Riyadh 11623, Saudi Arabia

C. V. Adubasim
Department of Plant Biology and Biotechnology, University of Benin, Benin City, Nigeria

C. M. Igwenagu, G. O. Josiah, S. E. Obalum, U. M. Okonkwo and I. M. Uzoh
Department of Soil Science, University of Nigeria, Nsukka 410001, Nigeria

S. Sato
Department of Environmental Engineering for Symbiosis, Soka University, 1-236 Tangi-machi, Hachioji-shi, Tokyo 192-8577, Japan

N. Nurhidayati and Indiyah Murwani
Department of Agrotechnology, Faculty of Agriculture, University of Islam Malang, East Java, Indonesia

Masyhuri Machfudz
Department of Agribusiness, Faculty of Agriculture, University of Islam Malang, East Java, Indonesia

Allah Ditta and Juma Muhammad
Department of Environmental Sciences, Shaheed Benazir Bhutto University Sheringal, Dir (Upper), Khyber Pakhtunkhwa 18000, Pakistan

Muhammad Imtiaz
School of Environmental Science and Engineering, Guangzhou University, Guangzhou 510006, China

Sajid Mehmood, Zheng Qian and Shuxin Tu
Microelement Research Center, College of Resources and Environment, Huazhong Agricultural University, Wuhan 430070, China

Raphael Passaglia Azevedo
Agronomy, Universidade Federal de Uberlândia, Av. Para, 1720, Uberlândia, MG, Brazil

Ignacio Hernán Salcedo
Instituto Nacional do Semiárido (INSA), Av. Francisco Lopes de Almeida, S/N, Serrotão, Campina Grande, PB 58434-700, Brazil

Priscila Alves Lima
Instituto Federal do Norte de Minas, Higway BR 367 Almenara/Jequitinhonha, km 111, Rural Zone, Almenara, MG 39900-000, Brazil

Vânia da Silva Fraga
Dpto. de Solos e Engenharia Rural, Universidade Federal da Paraíba, Higway PB 049 km 12, Areia, PB, Brazil

Regina Maria Quintão Lana
Instituto de Ciências Agrárias-ICIAG, Universidade Federal de Uberlândia, Av. Amazonas s/n Blok 2E and 01, Uberlândia, MG, Brazil

Assefa Gonfa
Ethiopia Institute of Agricultural Research, Debre Zeit Agricultural Research Center, Debre Zeit, Ethiopia

Bobe Bedadi and Anteneh Argaw
College of Agriculture and Environmental Sciences, School of Natural Resources Management and Environmental Sciences, Haramaya University, Haramaya, Ethiopia

J. J. Castro-Fernández, A. Serrato-González, M. Ruiz-Montoya and M. J. Díaz
Chemical Engineering Department, Facultad de Ciencias Experimentales, Chemical Process and Product Technology Research Center (Pro2TecS), Universidad de Huelva, Campus El Carmen, 21071 Huelva, Spain

Adebayo Abayomi Olowoake, Oluremi Solomon Osunlola and James Adebayo Ojo
Department of Crop Production, Kwara State University, Malete, Nigeria

Trine A. Sogn, Ivan Dragicevic, Tore Krogstad and Susanne Eich-Greatorex
Faculty of Environmental Sciences and Natural Resource Management, Norwegian University of Life Sciences, Ås, Norway

Vincent G. H. Eijsink
Faculty of Chemistry, Biotechnology and Food Science, Norwegian University of Life Sciences, Ås, Norway

Roar Linjordet
Norwegian Institute of Bioeconomy Research, Ås, Norway

Haifa Rajhi, Mounir Abichou and Ali Rhouma
Institute of Olive Trees, Zarzis 4170, Tunisia

Inès Mnif
Laboratoire d'amélioration des plantes et Valorization des Agroressources, University of Sfax, Road Soukra km 4, 3038 Sfax, Tunisia
National School of Engineers of Sfax, Sfax, Tunisia

Asma Saleem and Bushra Haroon
Department of Environmental Sciences, COMSATS Institute of Information Technology Abbottabad, Abbottabad, Pakistan

Muhammad Irshad
Department of Environmental Sciences, COMSATS Institute of Information Technology Abbottabad, Abbottabad, Pakistan
Arid Land Research Center Tottori University, Hamasaka cho, Tottori 680-0001, Japan

An Ping
Arid Land Research Center Tottori University, Hamasaka cho, Tottori 680-0001, Japan

Zohre Shahryari
Department of Chemical Engineering, Faculty of Engineering, Shahid Bahonar University of Kerman, Kerman, IranSwedish Centre for Resource Recovery, University of Boras, 50190 Borås, Sweden

Ramkumar B. Nair and Mohammad J. Taherzadeh
Swedish Centre for Resource Recovery, University of Boras, 50190 Borås, Sweden

Mohammad H. Fazaelipoor
Department of Chemical and Polymer Engineering, Faculty of Engineering, Yazd University, Yazd, Iran

Payam Setoodeh
School of Chemical and Petroleum Engineering, Shiraz University, Shiraz, Iran

Younes Ghasemi
Department of Pharmaceutical Biotechnology, School of Pharmacy, Shiraz University of Medical Sciences, Shiraz, Iran
Pharmaceutical Sciences Research Center, Shiraz University of Medical Sciences, Shiraz, Iran

Toyin B. Olowoboko, Jamiu O. Azeez and Oluwatoyin A. Babalola
Department of Soil Science and Land Management, Federal University of Agriculture Abeokuta, Abeokuta, Nigeria

Olanrewaju O. Olujimi
Department of Environmental Management and Toxicology, Federal University of Agriculture Abeokuta, Abeokuta, Ogun state, Nigeria

Kassu Tadesse, Asrat Mekonnen, Almaz Admasu, Wubengda Admasu, Dawit Habte, Amare Tadesse and Bahiru Tilahun
Kulumsa Agricultural Research Center, Ethiopian Institute of Agricultural Research, Kulumsa, Ethiopia

Ramy Salemdeeb and Abir Al-Tabbaa
Department of Engineering, University of Cambridge, Trumpington Street, Cambridge CB2 1PZ, UK

Mohammad Bin Daina
Supreme Council for the Environment, Manama, Bahrain

Christian Reynolds
Department of Geography, Faculty of Social Sciences, The University of Sheffield, Sheffield, UK

Abir Al-Tabbaa
Department of Engineering, University of Cambridge, Trumpington Street, Cambridge CB2 1PZ, UK

Aniek Iriany
The Department of Agrotechnology, Faculty of Agriculture and Animal Husbandry, University of Muhammadiyah Malang, Malang, Indonesia

Mochammad Chanan
The Department of Forestry, Faculty of Agriculture and Animal Husbandry, University of Muhammadiyah Malang, Malang, Indonesia

Gunomo Djoyowasito
The Department of Agricultural Engineering, Faculty of Agricultural Technology, Brawijya University of Malang, Malang, Indonesia

Ifeoma Monica Nwajiaku
Faculty of Life and Environmental Science, Shimane University, 1060 Nishikawatsu, Matsue, Shimane 690-8504, Japan
Imo State University, Owerri, Nigeria

John Seye Olanrewaju
Faculty of Life and Environmental Science, Shimanen University, 1060 Nishikawatsu, Matsue, Shimane 690-8504, Japan
New Sawah Project, National Centre for Agricultural Mechanization (NCAM), Ilorin, Nigeria

Mubeen Sarwar
Institute of Agricultural Sciences, University of the Punjab, Lahore 54590, Pakistan

Sumreen Anjum
Department of Botany, University of Agriculture, Faisalabad 38000, Pakistan

Index

www.ingramcontent.com/pod-product-compliance
Lightning Source LLC
LaVergne TN
LVHW060535210926
848485LV00005B/1202
* 9 7 8 1 6 3 9 8 9 4 0 0 0 *